Fiat 127 Owners Workshop Manual

J H Haynes
Member of the Guild of Motoring Writers
and H S H Phelps

Models covered
Fiat 127 C, CL, L Comfort, De Luxe, Special & Standard; 903 cc
Fiat 127 C, CL, L Comfort, Palio, Sport & Super; 1049 cc
Fiat 127 1300 GT; 1301 cc

Covers four- & five-speed manual transmissions
Covers major features of the Fiorino van

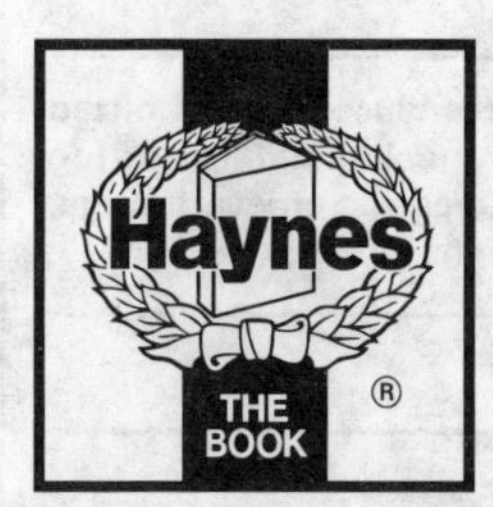

(193-8T6)

ABCDE
FGHIJ
KLMNO
PQR
2

Haynes Publishing Group
Sparkford Nr Yeovil
Somerset BA22 7JJ England

Haynes Publications, Inc
861 Lawrence Drive
Newbury Park
California 91320 USA

Acknowledgements
Thanks are due to the Champion Sparking Plug Company Limited, who supplied the illustrations showing spark plug conditions, to Holt Lloyd Limited who supplied the illustrations showing bodywork repair, and to Duckhams Oils, who provided lubrication data. Thanks are also due to the FIAT Motor Company (UK) Limited and FIAT SpA, Italy, for the provision of technical information and certain illustrations. Sykes-Pickavant provided some of the workshop tools. Lastly, thanks are due to all those people at Sparkford who assisted in the production of this manual.

A book in the **Haynes Owners Workshop Manual Series**

Printed by J. H. Haynes & Co. Ltd., Sparkford, Nr Yeovil, Somerset BA22 7JJ, England

ISBN 0 85696 987 7

British Library Cataloguing in Publication Data
Phelps, H. S. H.
Fiat 127 owners workshop manual. – (Owners Workshop Manual)
1. Fiat automobile
I. Title
629.28'722 TL215.5.F5
ISBN 0-85696-987-7

Contents

About this manual

Its aim

The aim of this manual is to help you get the best value from your vehicle. It can do so in several ways. It can help you decide what work must be done (even should you choose to get it done by a garage), provide information on routine maintenance and servicing, and give a logical course of action and diagnosis when random faults occur. However, it is hoped that you will use the manual by tackling the work yourself. On simpler jobs it may even be quicker than booking the car into a garage and going there twice, to leave and collect it. Perhaps most important, a lot of money can be saved by avoiding the costs a garage must charge to cover its labour and overheads.

The manual has drawings and descriptions to show the function of the various components so that their layout can be understood. Then the tasks are described and photographed in a step-by-step sequence so that even a novice can do the work.

Its arrangement

The manual is divided into thirteen Chapters, each covering a logical sub-division of the vehicle. The Chapters are each divided into Sections, numbered with single figures, eg 5; and the Sections into paragraphs (or sub-sections), with decimal numbers following on from the Section they are in, eg 5.1, 5.2, 5.3 etc.

It is freely illustrated, especially in those parts where there is a detailed sequence of operations to be carried out. There are two forms of illustration: figures and photographs. The figures are numbered in sequence with decimal numbers, according to their position in the Chapter – eg Fig. 6.4 is the fourth drawing/illustration in Chapter 6. Photographs carry the same number (either individually or in related groups) as the Section or sub-section to which they relate.

There is an alphabetical index at the back of the manual as well as a contents list at the front. Each Chapter is also preceded by its own individual contents list.

References to the 'left' or 'right' of the vehicle are in the sense of a person in the driver's seat facing forwards.

Unless otherwise stated, nuts and bolts are removed by turning anti-clockwise, and tightened by turning clockwise.

Vehicle manufacturers continually make changes to specifications and recommendations, and these, when notified, are incorporated into our manuals at the earliest opportunity.

Whilst every care is taken to ensure that the information in this manual is correct, no liability can be accepted by the authors or publishers for loss, damage or injury caused by any errors in, or omissions from, the information given.

Introduction to the FIAT 127

The FIAT 127 is, while being entirely new, the embodiment of several units well-proven on previous FIAT models. The FIAT 127 really does seem to have gone a long way towards solving the eternal designer's problems of space, economy and appeal. To conform to the latest European safety regulations, the 127 body is designed to deform progressively under impact from the front or rear, and the floorpan is specially strengthened to resist outside forces on the body sides. The doors are fitted with burst-proof locks and the interior is designed to eliminate projections which could cause injury in the event of a collision.

Obviously, the antecedents of the 127 range back through the 126 to the FIAT 500, of nostalgic appeal to FIAT afficcionados. But the 127 has greatly improved acceleration, hill-climbing ability and maximum speed. It still, however, retains the same astonishing manoeuvrability and economy. The three-door form with its rear door hinged at the roof and fold-down seat is a useful small 'estate' car offering ample load capacity and ease of access for bulky articles.

The engines of the 127 have been thoroughly tested and tuned, so their natural sporting qualities can be enjoyed without any worries.

The basic design philosophy of the 127 was to enclose the largest possible space within the smallest possible outside dimensions. Fortunately, FIAT had all the answers available, already tested and proven on the FIAT 128: a transversely mounted front engine, front wheel drive, independent four-wheel suspension, long wheelbase and wide track.

All-in-all, one might say an ideal car for present day driving - maximum space and minimum fuel consumption allied to good performance.

FIAT 127 Special

FIAT 1050 CL

FIAT 127 Sport

General dimensions and weights

For modifications, and information applicable to later models, see Supplement at end of manual

Standard, De Luxe and Special

Dimensions

Overall length:	
Standard and De Luxe	11 ft 8 in (3595 mm)
Special	11 ft 10 in (3635 mm)
Overall width	4 ft 11½ in (1527 mm)
Overall height	4 ft 5½ in (1370 mm)
Track:	
Front	4 ft 2 in (1280 mm)
Rear	4 ft 2½ in (1295 mm)
Wheelbase	7 ft 3 in (2225 mm)

Weights

Kerb weight:	
Two-door	1554 lb (705 kg)
Three-door	1565 lb (710 kg)
Fully-laden weight:	
Two-door	2436 lb (1105 kg)
Three-door	2447 lb (1110 kg)
Carrying capacity:	
Two-door	5 persons + 110 lb (50 kg) of luggage
Three-door	5 persons + 110 lb (50 kg) of luggage, or driver + 730 lb (330 kg) of luggage (back seat folded)
Maximum towing capacity	1323 lb (600 kg)

C, L, CL and Sport

Dimensions

Overall length:	
Two-door	11 ft 9 in (3625 mm)
Three-door	11 ft 10 in (3645 mm)
Overall width:	
C, L and CL	4 ft 11½ in (1527 mm)
Sport	5 ft 0 in (1536 mm)
Overall height	4 ft 5 in (1358 mm)
Track:	
C, L and CL:	
Front	4 ft 2 in (1280 mm)
Rear	4 ft 2½ in (1295 mm)
Sport:	
Front	4 ft 2¼ in (1288 mm)
Rear	4 ft 3 in (1303 mm)
Wheelbase	7 ft 3 in (2225 mm)

Weights

Kerb weight:	
900 cc	1565 lb (710 kg)
1050 cc except Sport	1610 lb (730 kg)
1050 cc Sport	1709 lb (775 kg)
Fully laden weight:	
900 cc	2447 lb (1110 kg)
1050 cc except Sport	2492 lb (1130 kg)
1050 cc Sport	2591 lb (1175 kg)
Carrying capacity:	
Two-door	5 persons + 110 lb (50 kg) of luggage
Three-door	5 persons + 110 lb (50 kg) of luggage, or driver + 730 lb (330 kg) of luggage (back seat folded)

Buying spare parts and vehicle identification numbers

Buying spare parts

Spare parts are available from many sources, for example: FIAT garages, other garages and accessory shops, and motor factors. Our advice regarding spare parts is as follows:

Officially appointed FIAT garages - This is the best source of parts which are peculiar to your car and otherwise not generally available (eg; complete cylinder heads, internal gearbox components, badges, interior trim etc). It is also the only place at which you should buy parts if your car is still under warranty: non-FIAT components may invalidate the warranty. To be sure of obtaining the correct parts it will always be necessary to give the partsman your car's engine number, chassis number, and number for spares, and if possible, to take the old part along for positive identification. Remember that many parts are available on a factory exchange scheme - any parts returned should be clean! It obviously makes good sense to go straight to the specialists on your car for this type of part for they are best equipped to supply you. They will also be able to provide their own Fiat service manual for your car should you require one.

Other garages and accessory shops - These are often very good places to buy material and components needed for the maintenance of your car (eg; oil filters, spark plugs, bulbs, fan belts, oils and grease, touch-up paint, filler paste etc). They also sell general accessories, usually have convenient opening hours, charge lower prices and can often be found not far from home.

Motor factors - Good factors will stock all of the more important components which wear out relatively quickly (eg; clutch components, pistons, valves, exhaust systems, brake cylinders/pipes/hoses/seals/shoes and pads etc). Motor factors will often provide new or reconditioned components on a part exchange basis - this can save a considerable amount of money.

Vehicle identification numbers

Modifications are a continuing and unpublicised process in vehicle manufacture quite apart from major model changes. Spare parts manuals and lists are compiled upon a numerical basis, the individual vehicle numbers being essential to correct identification of the component required.

The accompanying illustration shows the orientation of the vehicle identification plates and the caption details the information to be found on each plate.

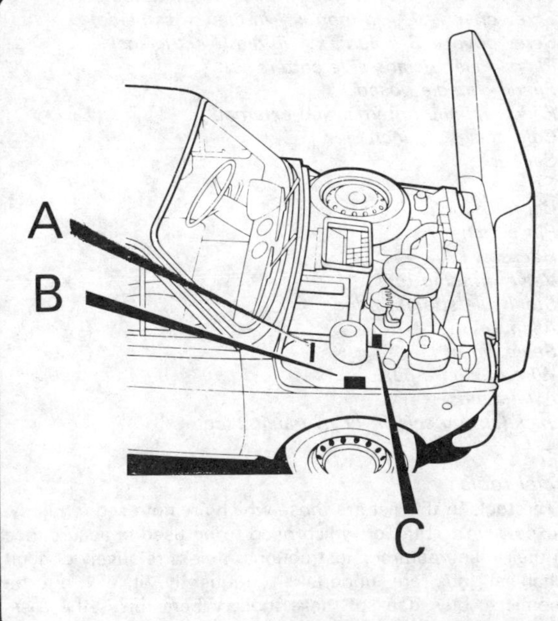

A — Car type (127A) and chassis number
B — Data plate: Type approval reference, car type and number, engine type, number for spares and body colour reference
C — Engine type and number

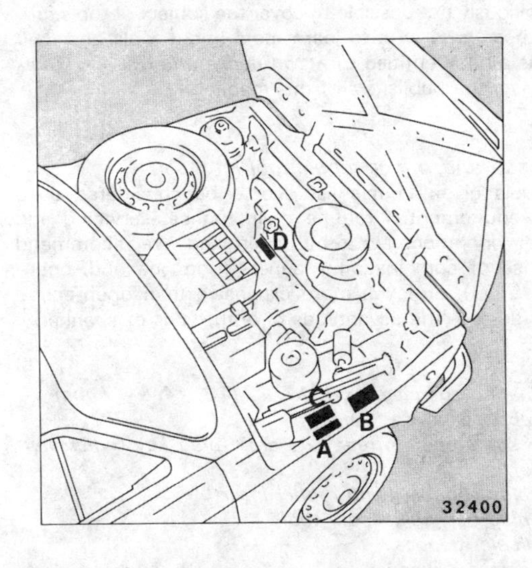

A — Chassis type and number
B — Paintwork colour and make
C — Identification data plate and type approval reference
D — Engine type and number stamped on cylinder block

Tools and working facilities

Introduction

A selection of good tools is a fundamental requirement for anyone contemplating the maintenance and repair of a motor vehicle. For the owner who does not possess any, their purchase will prove a considerable expense, offsetting some of the savings made by doing-it-yourself. However, provided that the tools purchased meet the relevant national safety standards and are of good quality, they will last for many years and prove an extremely worthwhile investment.

To help the average owner to decide which tools are needed to carry out the various tasks detailed in this manual, we have compiled three lists of tools under the following headings: *Maintenance and minor repair, Repair and overhaul,* and *Special.* The newcomer to practical mechanics should start off with the *Maintenance and minor repair* tool kit and confine himself to the simpler jobs around the vehicle. Then, as his confidence and experience grow, he can undertake more difficult tasks, buying extra tools as, and when, they are needed. In this way, a *Maintenance and minor repair* tool kit can be built-up into a *Repair and overhaul* tool kit over a considerable period of time without any major cash outlays. The experienced do-it-yourselfer will have a tool kit good enough for most repair and overhaul procedures and will add tools from the *Special* category when he feels the expense is justified by the amount of use to which these tools will be put.

It is obviously not possible to cover the subject of tools fully here. For those who wish to learn more about tools and their use there is a book entitled *How to Choose and Use Car Tools* available from the publishers of this manual.

Maintenance and minor repair tool kit

The tools given in this list should be considered as a minimum requirement if routine maintenance, servicing and minor repair operations are to be undertaken. We recommend the purchase of combination spanners (ring one end, open-ended the other); although more expensive than open-ended ones, they do give the advantages of both types of spanner.

Combination spanners - 10, 11, 12, 13, 14 & 17 mm
Adjustable spanner - 9 inch
Engine sump/gearbox/rear axle drain plug key (where applicable)
Spark plug spanner (with rubber insert)
Spark plug gap adjustment tool
Set of feeler gauges
Brake adjuster spanner
Brake bleed nipple spanner
Screwdriver - 4 in long x $\frac{1}{4}$ in dia (flat blade)
Screwdriver - 4 in long x $\frac{1}{4}$ in dia (cross blade)
Combination pliers - 6 inch
Hacksaw (junior)
Tyre pump
Tyre pressure gauge
Grease gun
Oil can
Fine emery cloth (1 sheet)
Wire brush (small)
Funnel (medium size)

Repair and overhaul tool kit

These tools are virtually essential for anyone undertaking any major repairs to a motor vehicle, and are additional to those given in the *Maintenance and minor repair* list. Included in this list is a comprehensive set of sockets. Although these are expensive they will be found invaluable as they are so versatile - particularly if various drives are included in the set. We recommend the $\frac{1}{2}$ in square-drive type, as this can be used with most proprietary torque wrenches. If you cannot afford a socket set, even bought piecemeal, then inexpensive tubular box spanners are a useful alternative.

The tools in this list will occasionally need to be supplemented by tools from the *Special* list.

Sockets (or box spanners) to cover range in previous list
Reversible ratchet drive (for use with sockets)
Extension piece, 10 inch (for use with sockets)
Universal joint (for use with sockets)
Torque wrench (for use with sockets)
'Mole' wrench - 8 inch
Ball pein hammer
Soft-faced hammer, plastic or rubber
Screwdriver - 6 in long x $\frac{5}{16}$ in dia (flat blade)
Screwdriver - 2 in long x $\frac{5}{16}$ in square (flat blade)
Screwdriver - $1\frac{1}{2}$ in long x $\frac{1}{4}$ in dia (cross blade)
Screwdriver - 3 in long x $\frac{1}{8}$ in dia (electricians)
Pliers - electricians side cutters
Pliers - needle nosed
Pliers - circlip (internal and external)
Cold chisel - $\frac{1}{2}$ inch
Scriber
Scraper
Centre punch
Pin punch
Hacksaw
Valve grinding tool
Steel rule/straight-edge
Allen keys
Selection of files
Wire brush (large)
Axle-stands
Jack (strong scissor or hydraulic type)

Special tools

The tools in this list are those which are not used regularly, are expensive to buy, or which need to be used in accordance with their manufacturers' instructions. Unless relatively difficult mechanical jobs are undertaken frequently, it will not be economic to buy many of these tools. Where this is the case, you could consider clubbing together with friends (or joining a motorists' club) to make a joint purchase, or borrowing the tools against a deposit from a local garage or tool hire specialist.

The following list contains only those tools and instruments freely available to the public, and not those special tools produced by the vehicle manufacturer specifically for its dealer network. You will find occasional references to these manufacturers' special tools in the text of this manual. Generally, an alternative method of doing the job without the vehicle manufacturers' special tool is given. However, sometimes, there is no alternative to using them. Where this is the case and the relevant tool cannot be bought or borrowed, you will have to entrust the work to a franchised garage.

Valve spring compressor (where applicable)
Piston ring compressor
Balljoint separator
Universal hub/bearing puller
Impact screwdriver
Micrometer and/or vernier gauge
Dial gauge
Stroboscopic timing light
Dwell angle meter/tachometer
Universal electrical multi-meter
Cylinder compression gauge
Lifting tackle
Trolley jack
Light with extension lead

Buying tools

For practically all tools, a tool factor is the best source since he will have a very comprehensive range compared with the average garage or accessory shop. Having said that, accessory shops often offer excellent quality tools at discount prices, so it pays to shop around.

There are plenty of good tools around at reasonable prices, but always aim to purchase items which meet the relevant national safety standards. If in doubt, ask the proprietor or manager of the shop for advice before making a purchase.

Care and maintenance of tools

Having purchased a reasonable tool kit, it is necessary to keep the tools in a clean serviceable condition. After use, always wipe off any dirt, grease and metal particles using a clean, dry cloth, before putting the tools away. Never leave them lying around after they have been used. A simple tool rack on the garage or workshop wall, for items such as screwdrivers and pliers is a good idea. Store all normal wrenches and sockets in a metal box. Any measuring instruments, gauges, meters, etc, must be carefully stored where they cannot be damaged or become rusty.

Take a little care when tools are used. Hammer heads inevitably become marked and screwdrivers lose the keen edge on their blades from time to time. A little timely attention with emery cloth or a file will soon restore items like this to a good serviceable finish.

Working facilities

Not to be forgotten when discussing tools, is the workshop itself. If anything more than routine maintenance is to be carried out, some form of suitable working area becomes essential.

It is appreciated that many an owner mechanic is forced by circumstances to remove an engine or similar item, without the benefit of a garage or workshop. Having done this, any repairs should always be done under the cover of a roof.

Wherever possible, any dismantling should be done on a clean, flat workbench or table at a suitable working height.

Any workbench needs a vice: one with a jaw opening of 4 in (100 mm) is suitable for most jobs. As mentioned previously, some clean dry storage space is also required for tools, as well as for lubricants, cleaning fluids, touch-up paints and so on, which become necessary.

Another item which may be required, and which has a much more general usage, is an electric drill with a chuck capacity of at least $\frac{5}{16}$ in (8 mm). This, together with a good range of twist drills, is virtually essential for fitting accessories such as mirrors and reversing lights.

Last, but not least, always keep a supply of old newspapers and clean, lint-free rags available, and try to keep any working area as clean as possible.

Spanner jaw gap comparison table

Jaw gap (in)	Spanner size
0.250	$\frac{1}{4}$ in AF
0.276	7 mm
0.313	$\frac{5}{16}$ in AF
0.315	8 mm
0.344	$\frac{11}{32}$ in AF; $\frac{1}{8}$ in Whitworth
0.354	9 mm
0.375	$\frac{3}{8}$ in AF
0.394	10 mm
0.433	11 mm
0.438	$\frac{7}{16}$ in AF
0.445	$\frac{3}{16}$ in Whitworth; $\frac{1}{4}$ in BSF
0.472	12 mm
0.500	$\frac{1}{2}$ in AF
0.512	13 mm
0.525	$\frac{1}{4}$ in Whitworth; $\frac{5}{16}$ in BSF
0.551	14 mm
0.563	$\frac{9}{16}$ in AF
0.591	15 mm
0.600	$\frac{5}{16}$ in Whitworth; $\frac{3}{8}$ in BSF
0.625	$\frac{5}{8}$ in AF
0.630	16 mm
0.669	17 mm
0.686	$\frac{11}{16}$ in AF
0.709	18 mm
0.710	$\frac{3}{8}$ in Whitworth; $\frac{7}{16}$ in BSF
0.748	19 mm
0.750	$\frac{3}{4}$ in AF
0.813	$\frac{13}{16}$ in AF
0.820	$\frac{7}{16}$ in Whitworth; $\frac{1}{2}$ in BSF
0.866	22 mm
0.875	$\frac{7}{8}$ in AF
0.920	$\frac{1}{2}$ in Whitworth; $\frac{9}{16}$ in BSF
0.938	$\frac{15}{16}$ in AF
0.945	24 mm
1.000	1 in AF
1.010	$\frac{9}{16}$ in Whitworth; $\frac{5}{8}$ in BSF
1.024	26 mm
1.063	$1\frac{1}{16}$ in AF; 27 mm
1.100	$\frac{5}{8}$ in Whitworth; $\frac{11}{16}$ in BSF
1.125	$1\frac{1}{8}$ in AF
1.181	30 mm
1.200	$\frac{11}{16}$ in Whitworth; $\frac{3}{4}$ in BSF
1.250	$1\frac{1}{4}$ in AF
1.260	32 mm
1.300	$\frac{3}{4}$ in Whitworth; $\frac{7}{8}$ in BSF
1.313	$1\frac{5}{16}$ in AF
1.390	$\frac{13}{16}$ in Whitworth; $\frac{15}{16}$ in BSF
1.417	36 mm
1.438	$1\frac{7}{16}$ in AF
1.480	$\frac{7}{8}$ in Whitworth; 1 in BSF
1.500	$1\frac{1}{2}$ in AF
1.575	40 mm; $\frac{15}{16}$ in Whitworth
1.614	41 mm
1.625	$1\frac{5}{8}$ in AF
1.670	1 in Whitworth; $1\frac{1}{8}$ in BSF
1.688	$1\frac{11}{16}$ in AF
1.811	46 mm
1.813	$1\frac{13}{16}$ in AF
1.860	$1\frac{1}{8}$ in Whitworth; $1\frac{1}{4}$ in BSF
1.875	$1\frac{7}{8}$ in AF
1.969	50 mm
2.000	2 in AF
2.050	$1\frac{1}{4}$ in Whitworth; $1\frac{3}{8}$ in BSF
2.165	55 mm
2.362	60 mm

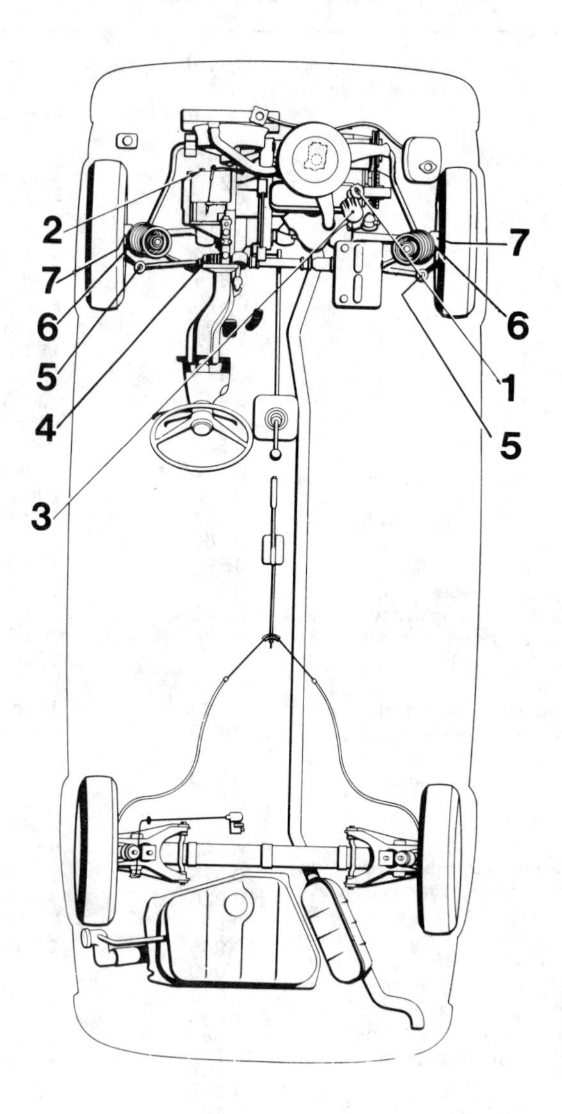

Recommended lubricants and fluids

Component or system	Lubricant type/specification	Duckhams recommendation
1 Engine	Multigrade engine oil, viscosity SAE 15W/40	Duckhams Hypergrade
2 Transmission	Multigrade engine oil, viscosity SAE 15W/40	Duckhams Hypergrade
3 Distributor	Multigrade engine oil, viscosity SAE 15W/40	Duckhams Hypergrade
4 Steering		
Up to Chassis No 2803649	Hypoid gear oil, viscosity SAE 90EP	Duckhams Hypoid 90
From Chassis No 2803650	Fiat grease type K854	Duckhams Adgear 00
5 Trackrod balljoints	Multi-purpose lithium based grease	Duckhams LB 10
6 Constant velocity joints	Lithium based grease with molybdenum disulphide	Duckhams LBM 10
7 Wheel bearings	Multi-purpose lithium based grease	Duckhams LB 10

Routine maintenance

Introduction

1 In the paragraphs that follow is detailed the routine servicing that should be done on the car. This work has the prime aim of making adjustments and lubricating to ensure the least wear and most efficient function. Also by examining the car thoroughly while performing the maintenance tasks, looking the car over, on top and underneath, you have the opportunity to check that all is in order.

2 Every component should be inspected, working systematically over the whole car. Dirt cracking near a nut or a flange can indicate something loose. Fluid leaks will be clearly evident. Electric cables rubbing, or rust appearing through the paint will all be found before they bring on a failure on the road, or if not tackled quickly, a more expensive repair. Thorough inspection on a regular basis prevents the car becoming a danger to yourself or others because of an undetected defect.

3 The tasks to be done are in general those recommended by the vehicle manufacturer, but we have also put in some additional ones which will help to keep your car in good order. For someone getting his service done at a garage it may be more cost-effective to accept component replacement after a somewhat short life in order to avoid labour costs.

4 When you are checking the car, and find something that seems incorrect, look it up in the appropriate Chapter, and repair or renew as necessary.

5 Always drive the car on a road test after a repair, and then check that the repair is holding up all right, and check nuts or hose connections for tightness. Check again after about another 150 miles.

6 Where we call up a routine check, details can be found in the appropriate Chapter.

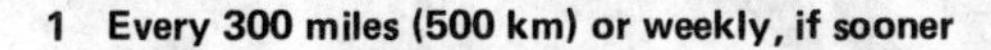

1 Every 300 miles (500 km) or weekly, if sooner

Engine

1 Check the engine oil level. This should be done with the car standing on level ground. Never let the oil level fall below the low level mark, or overfill it. Only replenish with the same type of oil as is in the sump. Use a top quality multi-grade oil.

Cooling system

2 Check the engine coolant level. The normal level can be observed in the expansion tank. The rise as the engine warms up, and contraction on cooling can be seen. The radiator cap should be removed when the engine is cold and the radiator should always be found to be full. Should a suspected leak require the radiator cap to be removed when the engine is hot, it should be allowed to stand first for ten minutes to cool down. Then pressure should be released by gradually turning the filler cap, holding it down, preferably with some rag to protect the hand. Topping-up should be done with a water and antifreeze mix and should not be necessary often. A need to do it frequently

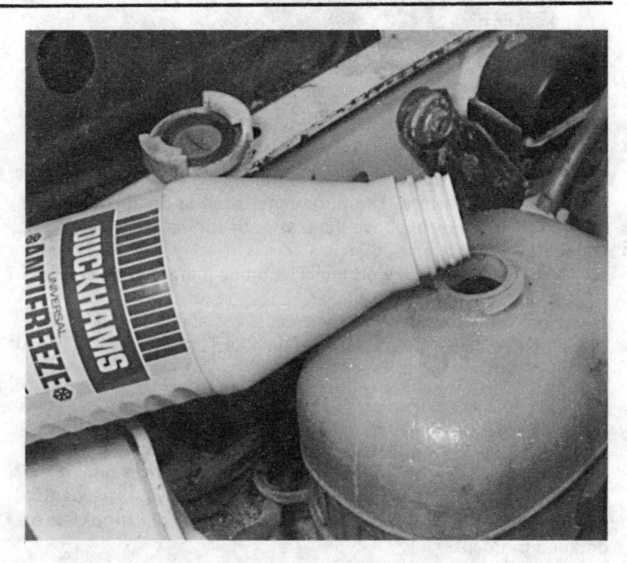

Check the engine coolant level

indicates the expansion tank system is not working, or there is a leak (See Chapter 2 for details of the cooling system).

Braking system

3 Check the level of the brake fluid in the hydraulic reservoir. This can be seen through the plastic. There should be no need for regular topping-up: if it is needed very frequently check the system for leaks.

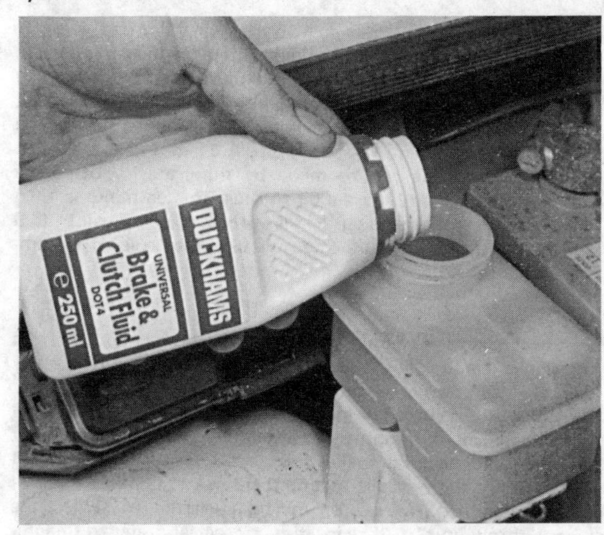

Top-up the brake fluid reservoir

Adjust the generator drive belt

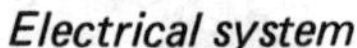

Electrical system

4 Top-up the battery. Use pure water such as distilled. Top-up to 0.25 in (6.35 mm) above the plates, or for 'easy-fill' batteries, the set level.

5 Check all lights are working. The brake lights can be checked by backing up close to a wall.

6 Check the windscreen washer fluid level, and top up if necessary, adding a screen wash additive such as Turtle Wax High Tech Screen Wash.

Suspension

7 Check the tyre pressures. This should be done when they are cold. After running only a few miles the pressure rises due to the heat generated by their flexing. At highway cruising speeds this can be as much as 3 or 4 psi. The tyre pressures cold should be as detailed in Chapter 11.

General

8 Check the engine compartment for leaks of coolant, fuel or oil.

2 Every 3,000 miles (5,000 km) or three monthly, if sooner

In addition to, or instead of, the 300 mile (500 km) maintenance

Engine

1 Change the engine oil if in a cold climate, or the car is used in town for stop-start driving. When the oil is hot from a journey park the car on level ground and put a container for 1 Imp. gallon (4.54 litres) under the engine. Clean and then remove the drain plug in the centre of the sump. A 12 mm hexagon Allen key is needed. One can be made by filing the jaw of a large adjustable spanner. Allow the oil to drain for ten minutes, then clean and replace the plug. Refill with 6 4/5 Imp. pints (3.9 litres) of top quality multi-grade oil (See the table of Recommended Lubricants).

General

2 Check the car underneath. Jack the car up high, or run it up on blocks if you cannot get it on a hoist. Make sure it is really secure. Then crawl underneath. Check the whole car thoroughly for leaks, and damaged or loose parts. If there are seepages of oil from the engine or transmission, watch them at future checks for increase, and check oil level as required. Items that are loose will be shown by the way dirt around them flakes. Look for signs of paint coming off, or rust, on the body, particularly where the wheels throw up dirt. If renovation is necessary refer to Chapter 12.

Top-up the engine oil level

3 Every 6,000 miles (10,000 km) or six months, if sooner

In addition to, or instead of, the 300 mile (500 km) maintenance

Engine

1 Change the engine oil. See the preceding Section.

2 Change the engine oil filter. Whilst the engine oil is draining, put a second drain pan under the filter. Unscrew the filter. If it is tight try a chain wrench. One can be made out of a length of bicycle chain, with a screwdriver pushed through the links. If this cannot be done, hammer a steel bar through the side of the filter element and out the other, to form a 'T' handle. It is quite thin, and easily punctured. Wipe the filter seat clean. Check the new element's sealing washer is in order, smear some engine oil on it. Then screw it in, hand-tight only. When the engine has been refilled with oil, start-up, running the engine gently till the pump has filled the new filter and got up normal pressure. Then check for leaks.

3 Check the engine. Listen to it at idle. Check for leaks of fuel, oil or coolant. Odd noises can be traced either by using some plastic tubing as a listening aid, or a heavy screwdriver with a large handle as sounding rod, put to the ear. Get to know the noises the car makes, so the onset of a new one will be noticed.

Change the engine oil filter

Renew the air cleaner element

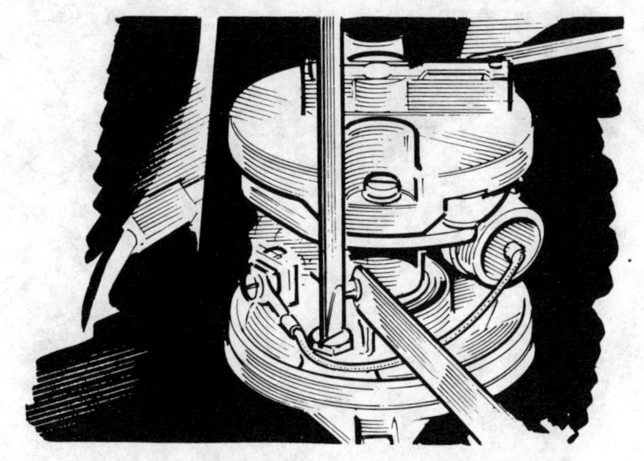

Adjust the contact breaker gap of the distributor

Fuel system

4 Change the air cleaner element, if operating in dusty conditions. Clean the outside of the cleaner. Take off the top by undoing the three wing nuts. Lift out the old element, and throw it away. Wipe out the inside of the cleaner, but make absolutely sure that no trace of dirt or piece of rag goes down the carburettor air intake. Wipe the hands clean: then fit the new element. Check the air cleaner is drawing air from the intake, hot or cold, appropriate to the season, then replace the top.

Ignition system

5 Clean inside of distributor cap and check carbon brush moves freely in recess. Inspect contact breaker points and renew if pitted. Adjust gap to 0.016 in (0.40 mm). Add 5-6 drops of engine oil through the central slot around the spindle. Smear contact breaker cam lightly with grease. Replace the cap.

6 Clean the coil and all leads between the distributor, coil and plugs, and inspect for cracking and deterioration.

7 Remove and inspect the spark plugs. If they appear to be in good condition, check the electrode gaps and reset if necessary, then refit them. If the condition of the plugs is in any way suspect, it would be better to renew them as a set.

8 Check the ignition timing, adjust if necessary (Chapter 4).

Clutch

9 Check the clutch pedal free-travel. It should be 1 in (25 mm). Adjust by undoing the locknut and moving the adjuster nut on the end of the cable on the clutch withdrawal lever on the top of the transmission.

Transmission

10 Check the transmission oil level. Clean around and remove the level plug on the front. Fill to overflowing with the same type of oil as already in it.

Braking system

11 Remove the rear brake drums. Wipe the dust out, but be careful not to inhale it. Check the lining wear and general state of the brakes.

12 Check the front brake pad thickness. If less than 0.08 in (2.0 mm) they must be renewed.

Electrical system

13 Lubricate between the windscreen wiper shafts and bushes with 1 - 2 drops of glycerine. **Do not** use oil. Renew blades if they leave an absolutely clean windscreen streaky.

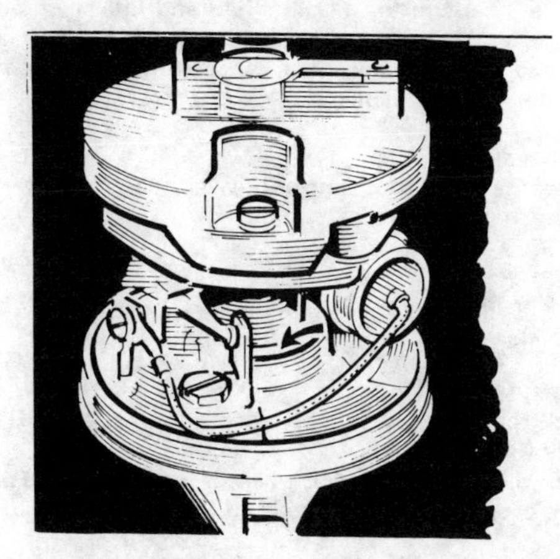

Lubricate the distributor

Suspension

14 Check the suspension for wear and the shock absorbers for proper damping. Clean the areas around springs, and shock absorbers and inspect for damage and fluid leaks. If empty car is not level on level ground, suspect binding shock absorber or weak or broken spring. Bounce each corner: if bouncing is not damped immediately, the shock absorbers need attention. Check for rust, where suspensions are attached. Check rubber bushes for play or deterioration.

General

15 Lubricate all controls, door and bonnet hinges and locks etc; with an oil can filled with engine oil.

4 Every 12,000 miles (20,000 km) or 12 months, if sooner

In addition to, or instead of, the 6,000 mile (10,000 km) maintenance

Engine

1 Check the valve clearances (tappets). Clearance should be checked cold in accordance with Chapters 1 or 13.

2 Check the cylinder compression.

3 Check the condition of the toothed tuning belt (OHC engines). Renew or retension as necessary (Chapter 13).

Adjust the valve clearances

Fuel system

4 Renew the air cleaner element. See the preceding Section.

5 Tune the carburettor. Clean all jets and filter with air blast. Check engine idling, at normal working temperature. If uneven, or if it tends to stop, adjust with the mixture control screw, and throttle stop screw (Chapter 3).

Ignition system

6 Renew the distributor contact breaker points (Chapter 4).

7 Renew the spark plugs.

Driveshafts

8 Check the condition of the inner and outer driveshaft boots (Chapter 7).

Braking system

9 Check the entire system for condition and leaks. Repair as necessary (Chapter 8).

10 Check the handbrake adjustment. If the lever comes up more than four clicks of its ratchet, refer to Chapter 8.

Electrical system

11 Check the battery for security and corrosion. If there is corrosion, deal with it as described in Chapter 9.

12 Check the headlamp aim. Verify that the quick adjustment levers are in the appropriate position, laden or unladen.

13 Check the battery charge rate.

14 Check the tension of the 'V' belt driving the generator. The belt should be tightened so that the longest run of the belt can be pushed down ½ inch (12 mm), pressing hard with one finger. To tighten the belt, slacken the pivot bolts underneath the generator. Slacken the bolt working in the slot on top, and move the generator away from the engine to tension the belt. Hold it whilst the bolt is tightened. Finally retighten the pivot underneath. A slack belt wears due to slip. Overtightening overloads the generator bearings. New belts stretch, so need checking after 100 miles of use.

Steering

15 Check the steering for wear. Hold each side of jacked-up front wheel and rock firmly. If the steering wheel does not immediately respond in both directions, the steering system ball joints are worn and must be replaced. Check boot-type gaiters or bellows of steering system. If damaged, renew them; also renew specified lubricant in the steering rack, as necessary. Where grease nipples are fitted, or can be fitted, to the control arm balljoints, apply three or four strokes of a grease gun containing general purpose grease.

16 Check the wheel alignment (Chapter 10).

Suspension

17 Check the wheel alignment (Chapter 11).

18 Check the hub bearings for condition.

19 Check the tyres for wear and damage. The inner side can be looked at when under the car. Look for cracks or bulges in the sides.

General

20 Check the operation of all doors and windows.

21 Check the whole car over for security and condition of parts.

5 Every 24,000 miles (40,000 km) or two yearly, if sooner

In addition to, or instead of, the 12,000 mile (20,000 km) maintenance

Cooling system

1 Renew the coolant (Chapter 2)

Fuel system

2 Clean the Positive Crankcase Ventilation valve (Chapter 3).

3 Clean the fuel pump filter.

Transmission

4 Change the transmission oil. After a journey, so the oil is warm and thin, park the car on level ground. Put a container that will hold ¾ Imp. gallon (3.4 litres) under the transmission and remove the drain plug on the left flank of the final drive part of the casing. This needs a 12 mm hexagon Allen key, the same as the engine sump. If none is available, the jaw of a large adjustable spanner can be filed to suit. Clean around the plug, remove it, and allow the oil 10 minutes to drain. Clean the plug and replace it. Clean the filter plug on the front of the transmission casing, and remove it. Fill to overflowing with an oil to the specification. This is normally SAE 90, though an oil without the EP quality. Use oil in a plastic dispenser, so the tube can be put straight into the filter hole.

Change the transmission oil

Braking system

5 Renew the brake fluid.

6 Every three years

In addition to, or instead of, the 12,000 mile (20,000 km) maintenance

Engine

1 Renew the toothed timing belt - OHC engines (Chapter 13).

7 Spring and Autumn

1 Wash underneath and check for rust.
2 Shift the air cleaner intake to the seasonal setting.
3 Check exhaust system for leaks, and that hangers and brackets are intact and firmly fixed.

8 Other aspects of routine maintenance

1 **Jacking-up.** Always chock a wheel, on the opposite side to the one being raised, in front and behind. The car's own jack has to be able to work when the car is very low with a flat tyre, so it goes in a socket on the side, taking up both wheels on that side. Using a small jack at one wheel is more secure when work has to be done. There are jacking points reinforced in the centre of the car at front and rear for a trolley jack. Never put a jack under the bodywork or the thin sheet steel will buckle.

2 **Wheel bolts.** These should be cleaned and lightly smeared with grease as necessary during work, to keep them moving easily. If the bolts are stubborn to undo due to dirt and over-tightening, it may be necessary to hold them by lowering the jack till the wheel rubs on the ground. Normally if the wheel brace is used across the hub centre, a foot held against the tyre will prevent the wheel from turning, and so save the wheels and the bolts from the wear when they are slackened with weight on the wheel. After replacing a wheel make a point later of rechecking again for tightness.

3 **Safety.** Whenever working, even partially, under the car, put an additional strong support such as a baulk of timber onto which the car will fall rather than onto you.

4 **Cleanliness.** Keep the mechanical parts of the car clean. It is much more pleasant to work on a clean car. Whenever doing any work, allow time for cleaning. When something is in pieces, components removed improve access to other areas and give an opportunity for a thorough clean. This cleanliness will allow you to cope with a crisis on the road without getting yourself dirty. During bigger jobs when you expect a bit of dirt it is less extreme. When something is taken to pieces there is less risk of ruinous grit getting inside. The act of cleaning focusses your attention on parts, and you are then more likely to spot trouble. Dirt on the ignition parts is a common cause of poor starting. Large areas such as the bulkheads of the engine compartment, should be brushed thoroughly with a water soluble detergent, allowed to soak for about ¼ hour, then carefully hosed with water. Water in the wrong places, particularly the carburettor or electrical components will do more harm than dirt. Detailed cleaning can be done with paraffin (kerosene) and an old paint brush. Petrol cleans better, but remember the hazard of fire, and if used in a confined space, of fumes. Use a barrier cream on the hands.

5 **Waste disposal.** Old oil and cleaning paraffin must be destroyed. It makes a good base for a bonfire, but is dangerous. Never have an open container near a naked flame. Pour the old oil where it cannot run uncontrolled, **before** you light it. Light it by making a "fuse" of newspaper. By buying your oil in one gallon cans you have these for storage of the old oil. The old oil is not household rubbish, so should not be put in the dustbin (trash). Most councils have collection points.

6 **Long journeys.** Before taking the car on a long journey, particularly a long holiday trip, do in advance many of the maintenance tasks that would not normally be due before going. In the first instance do jobs that would be due soon anyway. Then also those that would not come up until well into the trip. Also do the other tasks that are just checks, as a form of insurance against trouble. For emergencies carry on the car some copper wire, plastic insulation tape, plastic petrol pipe, gasket compound such as Hermatite Golden, and repair material of the plastics resin type, such as "Plastic padding". Carry a spare 'V' belt, and some spare bulbs. About 3 ft of electric cable, and some odd metric nuts and bolts should complete your car's emergency repair kit. Also carry a first aid kit: some plasters, germolene ointment, etc; in case you get minor cuts.

7 **On purchase.** If you have bought your 127 brand new you will have the maker's instructions for the early special checks on a new car. If you have just bought a second-hand car, then our advice is to reckon it has not been looked after properly, and so do all the checks, lubrication and other tasks on that basis, assuming all mileage and time tasks are overdue.

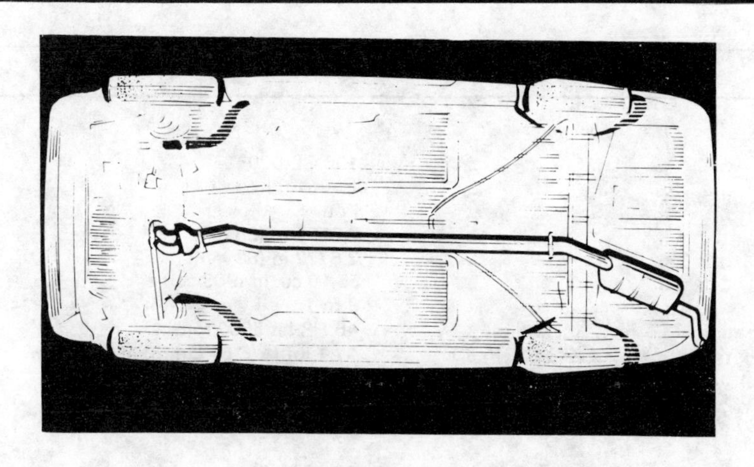

Check the exhaust system

Chapter 1 Engine

For modifications, and information applicable to later models, see Supplement at end of manual

Contents

Specifications

Engine (general)

Engine type	100 GL.000
Cycle	Four-stroke, petrol
No. of cylinders	Four
Bore	2.5591 in (65 mm)
Stroke	2.6772 in (68 mm)
Displacement	55.10 cu. in (903 cc)
Compression ratio	9 to 1
Maximum horsepower, DIN rating	45 HP (at 5600 rpm)
Maximum torque, DIN rating	47.1 lbf ft (6.51 kgf m) at 3000 rpm
Arrangement	Front, transversally mounted

Valve mechanism

Inlet:	
Opens	17° BTDC
Closes	43° ABDC
Exhaust:	
Opens	57° BBDC
Closes	3° ATDC
Valve clearance (cold):	
Inlet	0.006 in (0.15 mm)
Exhaust	0.008 in (0.20 mm)

Cylinder block and connecting rods

	inches	mm
Cylinder bore diameter *	2.5591 to 2.5610	65.000 to 65.050

**Cylinder bores are graded into classes with 0.0004 in. (0.01 mm) progression.*

Tappet bore diameter (standard)	.5516 to .5523	14.010 to 14.028
Diameter of seats for camshaft bushings:		
Timing gear end:		
Class B	1.9882 to 1.9886	50.500 to 50.510
Class C	1.9886 to 1.9890	50.510 to 50.520
Class D	1.9960 to 1.9964	50.700 to 50.710
Class E	1.9964 to 1.9968	50.710 to 50.720
Centre	1.8275 to 1.8287	46.420 to 46.450
Flywheel end	1.4142 to 1.4154	35.921 to 35.951
Crankshaft bearing saddle bore diameter	2.1460 to 2.1465	54.507 to 54.520
Length of rear main bearing bore between thrust ring seats	0.9150 to 0.9173	23.240 to 23.300
Connecting rod bearing seat diameter	1.7187 to 1.7192	43.657 to 43.670
Connecting rod small end bore diameter	0.7853 to 0.7856	19.943 to 19.954
Thickness of standard connecting rod bearings	0.0712 to 0.0714	1.807 to 1.813
Range of undersize connecting rod bearings, for service	0.010; 0.020; 0.030; 0.040	0.254; 0.508; 0.762; 1.016
Piston pin connecting rod fit (Interference)	0.0006 to 0.0015	0.016 to 0.039
Connecting rod bearings fit clearance on crankshaft	0.0010 to 0.0028	0.026 to 0.071
Maximum misalignment between axes of connecting rod small end and big-end (measured at 4.92 in/125 mm from con-rod shank)	± 0.0039	± 0.10

Pistons

	inches	mm
Diameter of standard pistons for service, measured at right angle to piston pin axis (at 1.55 in/39.5 mm from piston head):		
Class A	2.5566 to 2.5570	64.94 to 64.95
Class C	2.5574 to 2.5578	64.96 to 64.97
Class E	2.5582 to 2.5586	64.98 to 64.99
Oversize piston range	0.0079; 0.0157; 0.0236	0.2; 0.4; 0.6
Piston boss bore diameter:		
Grade 1	0.7867 to 0.7868	19.982 to 19.986
Grade 2	0.7868 to 0.7870	19.986 to 19.990
Grade 3	0.7870 to 0.7871	19.990 to 19.994
Piston ring groove width:		
Top groove	0.0703 to 0.0711	1.785 to 1.805
Centre groove	0.0793 to 0.0801	2.015 to 2.035
Bottom groove	0.1558 to 0.1566	3.957 to 3.977
Standard piston pin diameter:		
Grade 1	0.7862 to 0.7863	19.970 to 19.974
Grade 2	0.7863 to 0.7865	19.974 to 19.978
Grade 3	0.7865 to 0.7866	19.978 to 19.982
Oversize piston pin, for service	0.0079	0.2
Piston ring thickness:		
First compression ring	0.0680 to 0.0685	1.728 to 1.740
Second oil ring	0.0779 to 0.0784	1.978 to 1.990
Third oil ring with slots and expander	0.1545 to 0.1550	3.925 to 3.937
Piston fit in cylinder bore, measured at right angle to pin, at 1.55 in/39.5 mm from piston head (fit clearance of new parts)	0.0020 to 0.0028	0.050 to 0.070
Fit clearance of piston pin in piston	0.0003 to 0.0006	0.008 to 0.016
Piston ring side clearance in piston groove:		
First compression ring: clearance of new parts	0.0018 to 0.0030	0.045 to 0.077
Second oil ring: clearance of new parts	0.0010 to 0.0022	0.025 to 0.057
Third oil ring: clearance of new parts	0.0008 to 0.0020	0.020 to 0.052
Piston ring end-gap:		
First compression ring	0.0079 to 0.0138	0.20 to 0.35
Second oil ring	0.0079 to 0.0138	0.20 to 0.35
Third oil ring	0.0079 to 0.0138	0.20 to 0.35
Oversize piston ring range, for service	0.0079; 0.0157; 0.0236	0.2; 0.4; 0.6

Camshaft, tappets and rocker mechanism

	inches	mm
Diameter of bores for camshaft bushings in crankcase:		
Timing gear end:		
Class B	1.9882 to 1.9886	50.500 to 50.510
Class C	1.9886 to 1.9890	50.510 to 50.520
Class D	1.9961 to 1.9965	50.700 to 50.710
Class E	1.9965 to 1.9969	50.710 to 50.720
Centre	1.8275 to 1.8287	46.420 to 46.450
Flywheel end	1.4142 to 1.4152	35.921 to 35.951

	inches	mm
Outside diameter of bushings:		
Timing gear end:		
Class B	1.9876 to 1.9882	50.485 to 50.500
Class C	1.9880 to 1.9886	50.495 to 50.510
Class D	1.9955 to 1.9961	50.685 to 50.700
Class E	1.9959 to 1.9965	50.695 to 50.710
Centre	1.8320 to 1.8335	46.533 to 46.571
Flywheel end	1.4185 to 1.4200	36.030 to 36.068
Inside diameter of bushings, finished in bores:		
Timing gear end	1.4971 to 1.4981 *	38.025 to 38.050 *
Centre	1.7088 to 1.7096	43.404 to 43.424
Flywheel end	1.2215 to 1.2223	31.026 to 31.046

** This bushing is supplied for service finish reamed in the inside diameter. It is held in place by means of a screw.*

	inches	mm
Fit between bushings and bores in crankcase:		
Timing gear end (fit clearance)	0 to 0.001	0 to 0.025
Centre (interference)	0.0033 to 0.0059	0.083 to 0.151
Flywheel end (interference)	0.0031 to 0.0058	0.079 to 0.147
Diameter of camshaft journals:		
Timing gear end	1.4951 to 1.4961	37.975 to 38.000
Centre	1.7060 to 1.7070	43.333 to 43.358
Flywheel end	1.2194 to 1.2205	30.975 to 31.000
Fit between bushings and camshaft journals:		
Timing gear end	0.0010 to 0.0030	0.025 to 0.075
Centre	0.0018 to 0.0036	0.046 to 0.091
Flywheel end	0.0010 to 0.0028	0.026 to 0.071
Diameter of standard tappet bore in crankcase	0.5516 to 0.5523	14.010 to 14.028
Outside diameter of standard tappet	0.5505 to 0.5512	13.982 to 14.000
Range of oversize tappets, for service	0.002 to 0.004	0.05 to 0.10
Fit between tappets and bores: clearance of new parts	0.0004 to 0.0018	0.010 to 0.046
Diameter of bore in rocker arm shaft supports	0.5910 to 0.5916	15.010 to 15.028
Diameter of rocker arm shaft	0.5897 to 0.5902	14.978 to 14.990
Fit between supports and rocker arm shaft (clearance of new parts)	0.0004 to 0.0016	0.010 to 0.040
Diameter of rocker arm bore	0.5910 to 0.5917	15.010 to 15.030
Fit between rocker arms and shaft (clearance of new parts)	0.0008 to 0.0020	0.020 to 0.052

Cylinder head

	inches	mm
Valve guide bore in cylinder head	0.5099 to 0.5109	12.950 to 12.977
Valve guide outside diameter	0.5122 to 0.5130	13.010 to 13.030
Valve guide oversize on O.D. for service	0.0079	0.2
Inside diameter of valve guides, fitted in cylinder head	0.2765 to 0.2772	7.022 to 7.040
Valve guide fit in head (interference)	0.0013 to 0.0032	0.033 to 0.080
Valve stem diameter	0.2748 to 0.2756	6.982 to 7.000
Valve stem fit in valve guide (clearance)	0009 to .0023	0.022 to 0.058
Valve seat angle in cylinder head	$45^\circ \pm 5'$	
Valve face angle	$45^\circ 30' \pm 5'$	
Valve head diameter:		
Intake	1.1417	29
Exhaust	1.0236	26
Max. valve run-out on a full turn, guided on stem, with dial indicator set at centre of contact face	0.0012	0.03
Width of valve seats in cylinder head (contact surface) - inlet and exhaust	0.0512 to 0.0591	1.3 to 1.5

Valve springs

	Inner spring	Outer spring
Spring height under a load of 54 lbs (24.5 kg)	–	1.437 in (36.5 mm)
Spring height under a load of 12.1 lbs (5.5 kg)	1.279 in (32.5 mm)	–
Minimum allowable load referred to the above heights	10.8 lbs (4.9 kg)	48.5 lbs (22 kg)

Crankshaft and main bearings

	inches	mm
Main bearing journal standard diameter	1.9994 to 2.0002	50.785 to 50.805
Main bearing saddle bore	2.1460 to 2.1465	54.507 to 54.520
Standard main bearing shell thickness	0.0721 to 0.0723	1.831 to 1.837
Undersize main bearing shells, for service	0.010; 0.020; 0.030; 0.040	0.254; 0.508; 0.762; 1.016
Crankpin standard diameter	1.5741 to 1.5750	39.985 to 40.005
Main bearing to journal fit (clearance of new parts)	0.0011 to 0.0029	0.028 to 0.073
Length of centre main journal (shoulder to shoulder)	1.1056 to 1.1071	28.080 to 28.120
Width of centre main bearing saddle bore, between thrust ring seats	0.9150 to 0.9173	23.240 to 23.300
Thickness of thrust rings for centre saddle bore	0.0909 to 0.0929	2.310 to 2.360
Thickness of oversize thrust rings	0.0959 to 0.0979	2.437 to 2.487
Crankshaft end-play, with thrust rings installed: clearance of new parts	0.0024 to 0.0102	0.06 to 0.26

	inches	mm
Maximum allowable misalignment of crankshaft journals	0.0024 *	0.06 *
* *Total reading on dial gauge.*		
Maximum allowable misalignment of crankpins in relation to crankshaft journals	± 0.02	± 0.5
Maximum out of round of crankshaft journals and crankpins, after grinding	0.0002	0.005
Maximum taper of crankpins and journals, after grinding	0.0002	0.005
Squareness of flywheel face to crankshaft centreline (max. allowable tolerance with dial indicator set laterally at a distance of about 1.22 in/31 mm from crankshaft rotation axis)	0.001	0.025
Flywheel:		
Parallel relationship of clutch disc resting face to crankshaft mounting flange: max. allowable tolerance	0.004	0.1
Squareness of above faces to rotation axis: max. allowable tolerance	0.004	0.1

Lubrication system

Oil type/specification	Multigrade engine oil, viscosity SAE 15W/40 (Duckhams Hypergrade)	
Oil filter type (up to March 1981)	Champion C117	
Oil pump	Gear type	
Pump drive	From camshaft	
Oil pressure relief valve	Built in oil pump	
Clearance between upper side of gears and housing surface ...	0.0008 to 0.0041 in.	0.020 to 0.105 mm
Clearance between gears and pump housing	0.0020 to 0.0055 in.	0.05 to 0.14 mm
Fit between driveshaft guide bushing and bore in crankcase (interference at all times)	0.0010 to 0.0027 in.	0.025 to 0.070 mm
Clearance of driveshaft in bushing press fitted in crankcase ...	0.0010 to 0.0024 in.	0.025 to 0.062 mm
Clearance between drive gear shaft and bore in pump housing ...	0.0005 to 0.0020 in.	0.013 to 0.050 mm
Clearance of pin in driven gear	0.0004 to 0.0020 in.	0.010 to 0.050 mm
Lash of meshed drive and driven gears	0.0031 in.	0.08 mm
Clearance between driveshaft gear and camshaft gear	0.0023 in.	0.06 mm
Full-flow oil filter with bypass safety valve	Cartridge type	
Low oil pressure indicator sending unit	Electrical, closes at 2.8 to 8.5 psi (0.2 to 0.6 kg/cm^2)	
Oil pressure at 212°F (100°C)	43 to 57 psi	3 to 4 kg/cm^2
Sump capacity including filter	6.5 pints (3.7 litres)	

Pressure relief valve spring

	Before engine no. 2 635 011	From engine no. 2 635 011
Length of seated spring under a load of 10.16 ± 0.33 lbs (4.61 ± 0.15 kg)	0.7874 in (22.5 mm)	1.41 in (36 mm)
Minimum allowable load referred to length of seated spring ...	9.5 lbs (4.3 kg)	5.07 lbs (2.3 kg)

Torque wrench settings

	Thread size (metric)	lb f ft	kg fm
Bolt, flywheel to crankshaft	M8	29	4
Bolt, connecting rod bearing cap	M8 x 1	29	4
Bolt, self-locking, driven gear and fuel pump drive lobe to camshaft	M10 x 1.25	36	5
Cylinder head bolt:			
Before engine no. 2 835 584	M9	36	5
From engine no. 2 835 584	M9	43	6
Bolt, crankshaft main bearing cap	M10 x 1.25	51	7
Nut, water pump and generator drive pulley to crankshaft ...	M18 x 1.5	72	10
Bolt, securing sump to crankcase, to timing gear cover and flywheel end main bearing oil seal cover	M6	7	1
Nut, self-locking, rocker arm shaft support to cylinder head stud ...	M10 x 1.25	29	4
Thermal switch	M16 (x 1.5) (Taper)	36	5
Spark plug	M14 x 1.25	25	3.5
Nut, power engine mounting insulator to body bolt, engine side ...	M10 x 1.25	25	3.5
Bolt, power engine mounting insulator to rail, transmission side ...	M8	11	1.5
Bolt, engine mounting rail to body, transmission side	M8	18	2.5
Nut, engine mounting insulator to transmission case	M8	18	2.5
Bolt, power plant anchor rod	M8	18	2.5

1 General description

The FIAT 127 engine is a four cylinder overhead valve type and is mounted transversely together with the transmission, on three mountings at the front of the car.

Two valves per cylinder are mounted at a slight angle in the cast cylinder head and run in pressed-in valve guides. They are operated by rocker arms and pushrods from the camshaft which is located at the base of the cylinder bores.

The cylinder head has the exhaust ports on one side whilst the carburettor is bolted to the upper face on the opposite side.

The cylinder block and crankcase are cast together and a separate oil sump is secured to the underside of the crankcase, this containing the engine oil not in circulation.

The pistons are made from anodised aluminium and have solid skirts. One compression and two oil control rings are fitted. The gudgeon pin is a press fit in the connecting rod little end. Renewable steel backed bearing shells are fitted to the big-ends.

At the front of the engine is a double row chain driving the camshaft via chain wheels fitted to the crankshaft and camshaft. The camshaft is supported by three steel backed white metal bushes. If these are replaced it is necessary to ream the bushes in position.

The overhead valves are operated by means of rocker arms mounted on the rocker shaft running along the top of the cylinder head. The rocker arms are activated by pushrods and tappets which, in turn, rise and fall in accordance with the cams on the camshaft. The valves are held closed by double valve springs.

The statically and dynamically balanced crankshaft is supported by three renewable main bearings. Crankshaft end-float is controlled by four semi-circular thrust washers, two of which are located on each side of the centre bearing.

A pressure relief valve fitted in the lubrication system controls the maximum oil pressure when the oil is cold. Any surplus oil is returned to the sump. To warn the driver should loss of oil pressure occur, a pressure switch is fitted to the main oil gallery to indicate low oil pressure.

The OHC engine is dealt with in Chapter 13.

2 Main operations possible with engine in place

The following major operations can be carried out on the engine with it in place but, because of the ease of engine removal, any work of a major nature should be carried out with the engine on the bench:

1. *Removal and replacement of the cylinder head assembly.*
2. *Removal and replacement of the sump.*
3. *Removal and replacement of the big-end bearings.*
4. *Removal and replacement of the pistons and connecting rods.*
5. *Removal and replacement of the timing chain and gears.*
6. *Removal and replacement of the flywheel (with the transmission removed).*
7. *Removal and replacement of the engine mounting.*
8. *Removal and replacement of the oil pump.*
9. *Removal and replacement of the crankshaft (flywheel end) oil seal.*

3 Major operations requiring engine removal

The following major operations can be carried out only with the engine out of the car and on the bench:

1. *Removal and replacement of the main bearings.*
2. *Removal and replacement of the crankshaft.*
3. *Removal and replacement of the camshaft.*

4 Engine removal - preliminary work

1 Before starting work, decide how the engine is to be

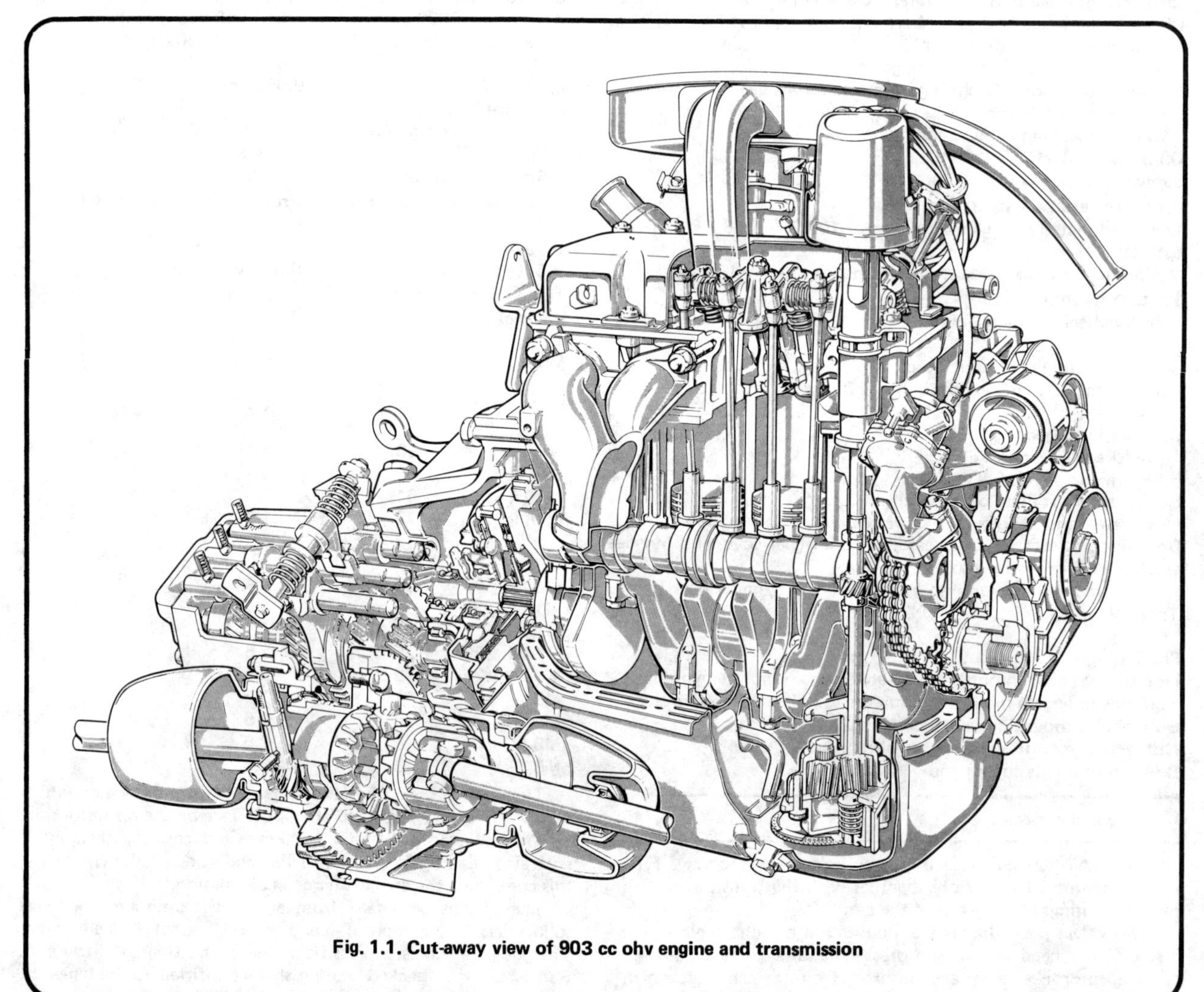

Fig. 1.1. Cut-away view of 903 cc ohv engine and transmission

removed. If the car is an early model, secure lifting tackle to the engine, and disconnect the cylinder head tie bar and the right-hand engine mounting. Remove the lower support member from under the car, disconnect the driveshafts and lift the engine and transmission out of the car.

If the car has a chassis number later than 1104290 the engine is normally disconnected, lowered and removed from the car. If decent tackle is available to lift the engine, the car can be jacked-up, and then the engine lowered and pulled from underneath. Without such tackle, the car must be jacked or blocked up before doing the preliminary work, then lowered as far as possible. Blocks are then put under the engine, the mountings disconnected, and the car jacked-up again to allow the engine to be pulled clear. This latter method is not recommended, because it will be difficult to reach the supporting bracket under the left end of the engine, and it is probably less safe. A tackle can be slung from a strong beam, or rigged over the front of the car using stout timber as shearlegs. If using ropes, get enough pulley blocks to make a tackle with a purchase of 8 to 1.

2 Drain the engine oil, and the transmission oil too, if the transmission is going to be dismantled.

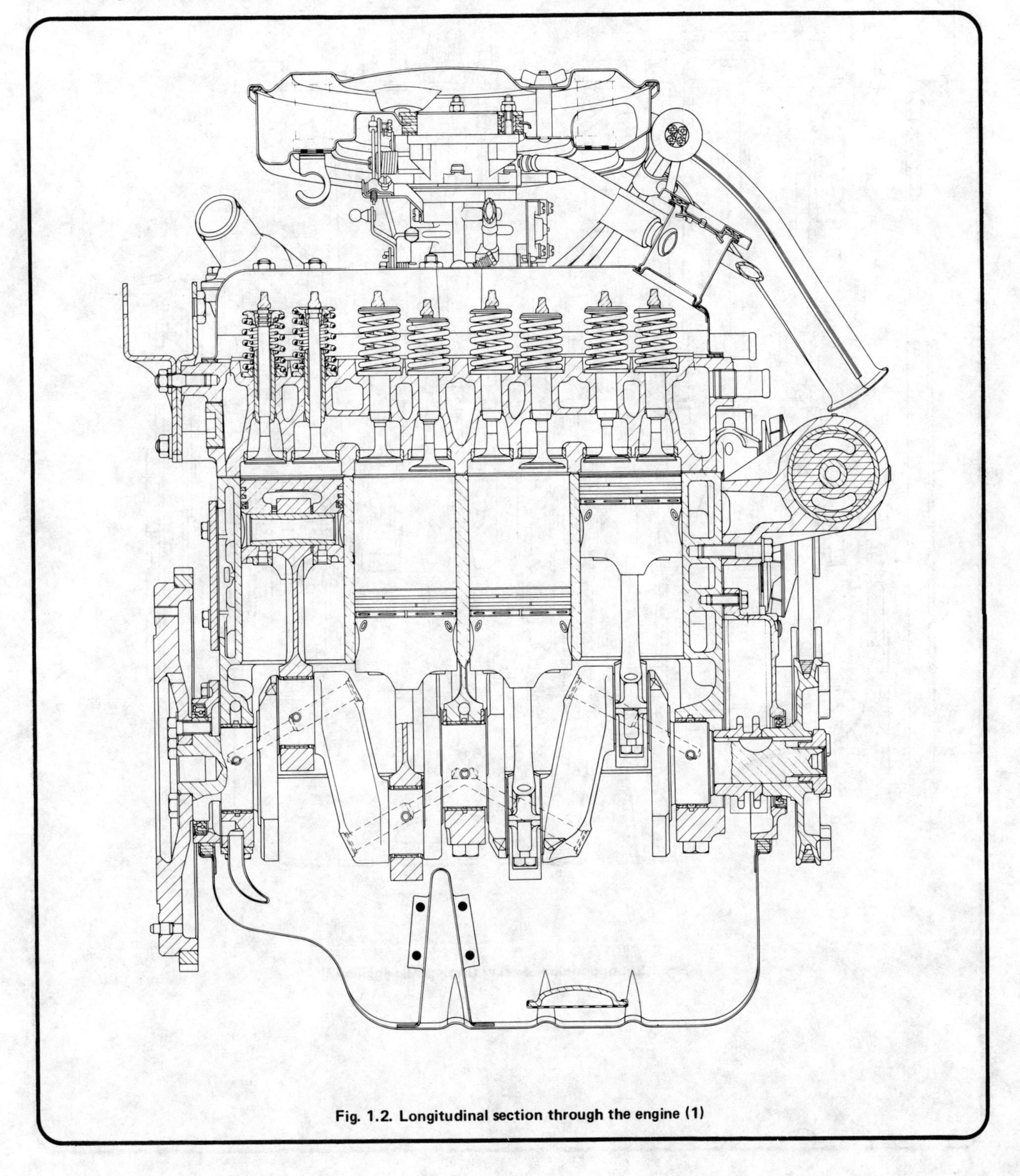

Fig. 1.2. Longitudinal section through the engine (1)

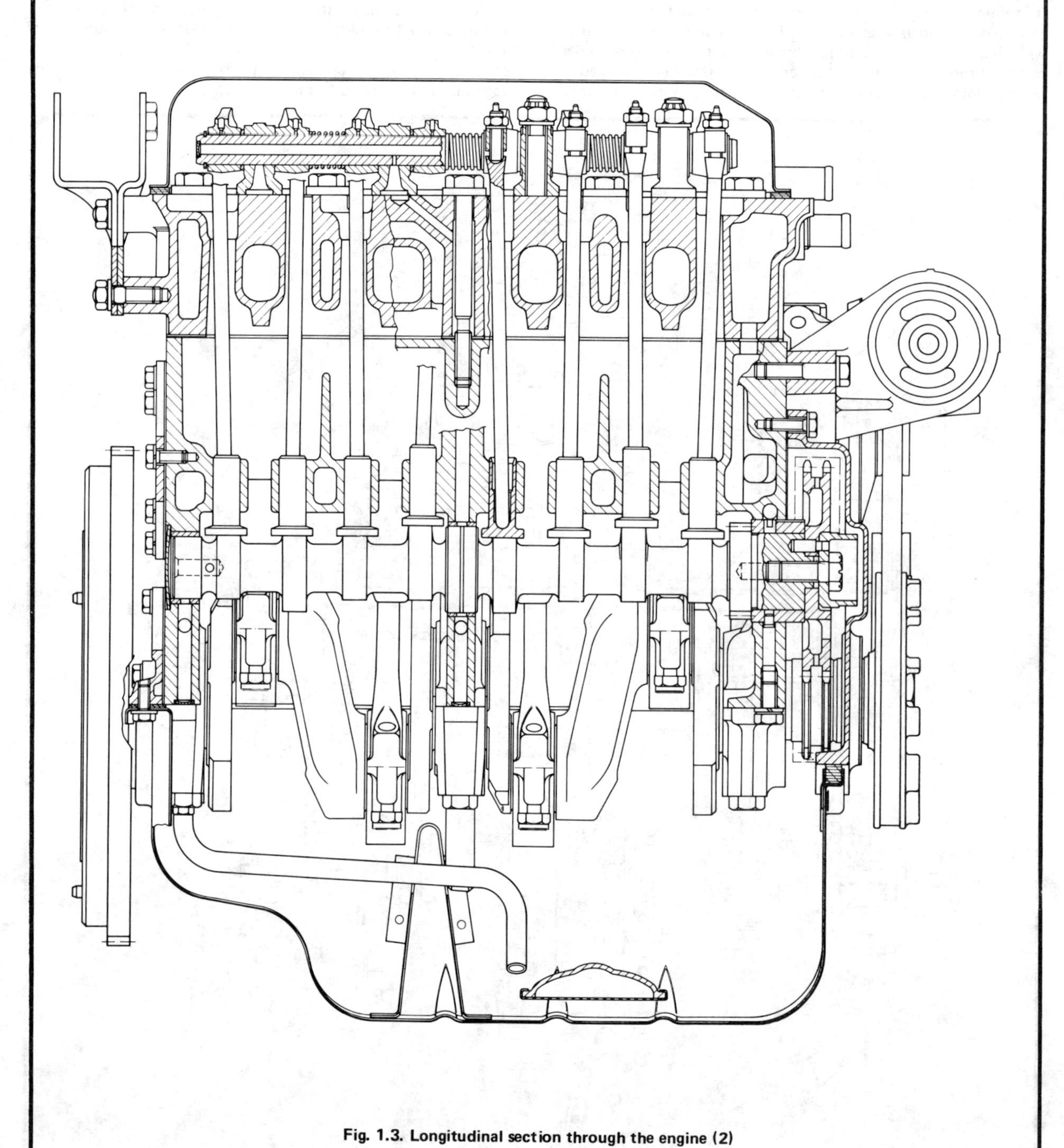

Fig. 1.3. Longitudinal section through the engine (2)

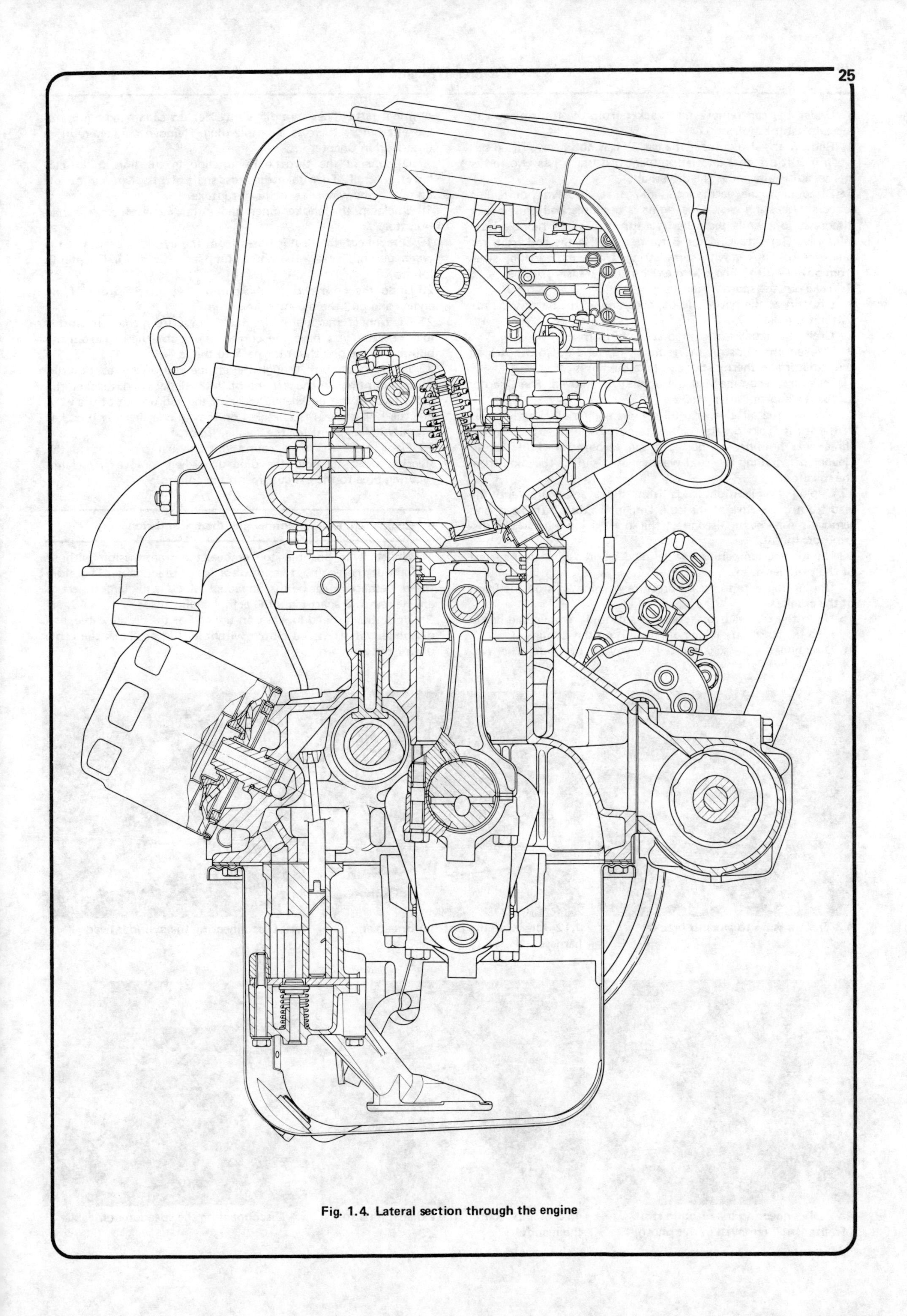

Fig. 1.4. Lateral section through the engine

3 Under the car, remove the bracket from the transmission to the exhaust (photo).
4 Remove the shield under the car. This is not always fitted but when it is it presents a small problem because it has two rather inaccessible bolts above the driveshaft.
5 Disconnect the gearchange linkage at the two points just above the steering gear. One point is a simple ball and socket disconnection, while the other is a spring clip and flexible bush assembly. Refer to Chapter 6 for details. Tie the rod from the gearlever to the car with some string. Undo the earthing strap from power unit to body, at the rear of the transmission.
6 Take out the spare wheel.
7 Disconnect the battery leads, and remove the battery clamp. Lift out the battery.
8 Drain the radiator at the tap at the bottom.
9 Slacken the clips on the radiator top and bottom hoses, at the radiator and thermostat, and take the hoses off.
10 It is now recommended the bonnet is removed. See Chapter 12 for the appropriate method.
11 Remove the air cleaner. Pull off the engine oil breather pipe at the front. Take off the lid, and remove the element. Undo the three nuts holding the body to the carburettor. Pull the air cleaner off, pulling the cold weather inlet out of the pipe from the manifold.
12 Unplug the ignition leads from the spark plugs, and the lead from the centre of the coil. Unclip the distributor cap, and remove the harness. Disconnect the low tension wire at the distributor. (photo)
13 Unplug the temperature gauge sender unit's wire on the front of the cylinder head.
14 Unplug the wire to the oil pressure warning light on the front of the crankcase.
15 Undo the red lead to the starter solenoid, remove the heavy cable to the starter by taking off the nut on the terminal.
16 Disconnect the generator. If a dynamo, undo the two terminal nuts. These are different sizes, so cannot be confused On alternators there is a simple plug. Remove the generator as described in Chapter 9.
17 Disconnect the throttle by unclipping the link at the carburettor itself. Then take the cross shaft off its pivot and tuck it up out of the way near the heater intake.
18 Unclamp the choke inner and outer cables from the carburettor.
19 Disconnect the heater hoses from the cylinder head and the water pump. Remove the water pump, as described in Chapter 2. (photo)
20 Undo the hose clip on the pump inlet on the front of the engine and pull the pipe off the pump.
21 Disconnect the fuel recirculatory pipe from the carburettor inlet. Ensure you have the correct pipe; not the one from the pump, but the one that runs back to the tank.
22 Take out the bolt though the engine end of the reaction rod at its mounting on the left end of the cylinder head. Slacken the other end on the bulkhead, and pull the rod up out of the way. Put the bolt back in the bracket on the cylinder head as it makes a handy lifting point for the engine. (photo)
23 Tie a piece of string round the exhaust pipe, and to the steering rack, to hold the pipe up. Then remove the clamp securing pipe to the exhaust manifold. (photo)

5 Engine removal - disconnecting the transmission

1 Whilst it is possible to remove the transmission without taking out the engine, the converse is not practicable. Therefore the driveshafts must be disconnected which is the largest part of engine removal apart from the actual engine lift.
2 Undo the nut and locknut on the end of the clutch cable, and disconnect it from the clutch withdrawal lever. Tuck the cable out of the way.

4.3 Transmission to exhaust bracket

4.12 Removing the distributor cap and harness

4.19 Disconnecting the cylinder head hoses

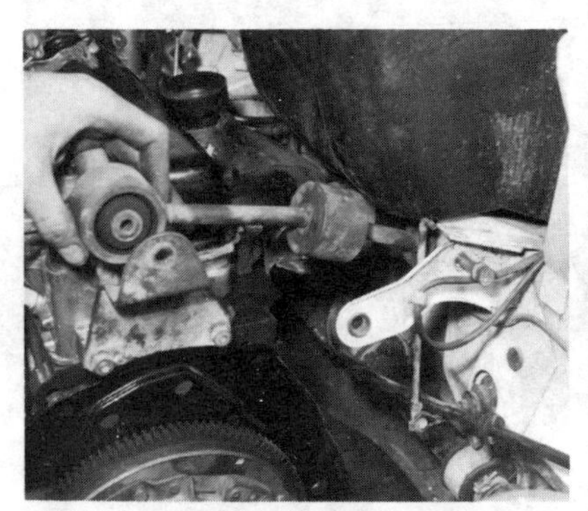

4.22 Disconnecting the reaction strut (Transmission removed in this photo)

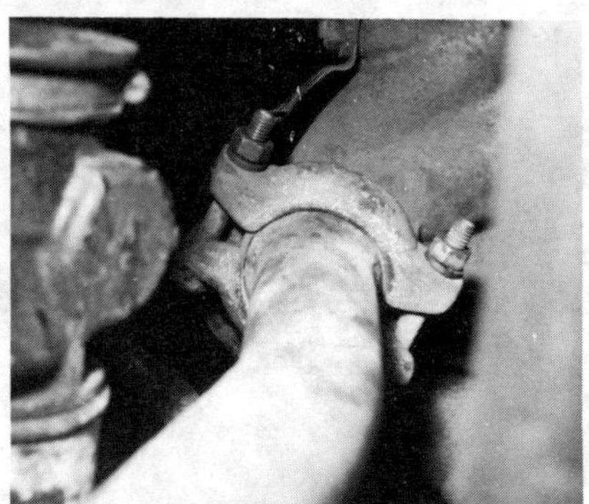

4.23 Clamp securing the exhaust pipe to the manifold

5.4 Disconnecting the speedometer cable

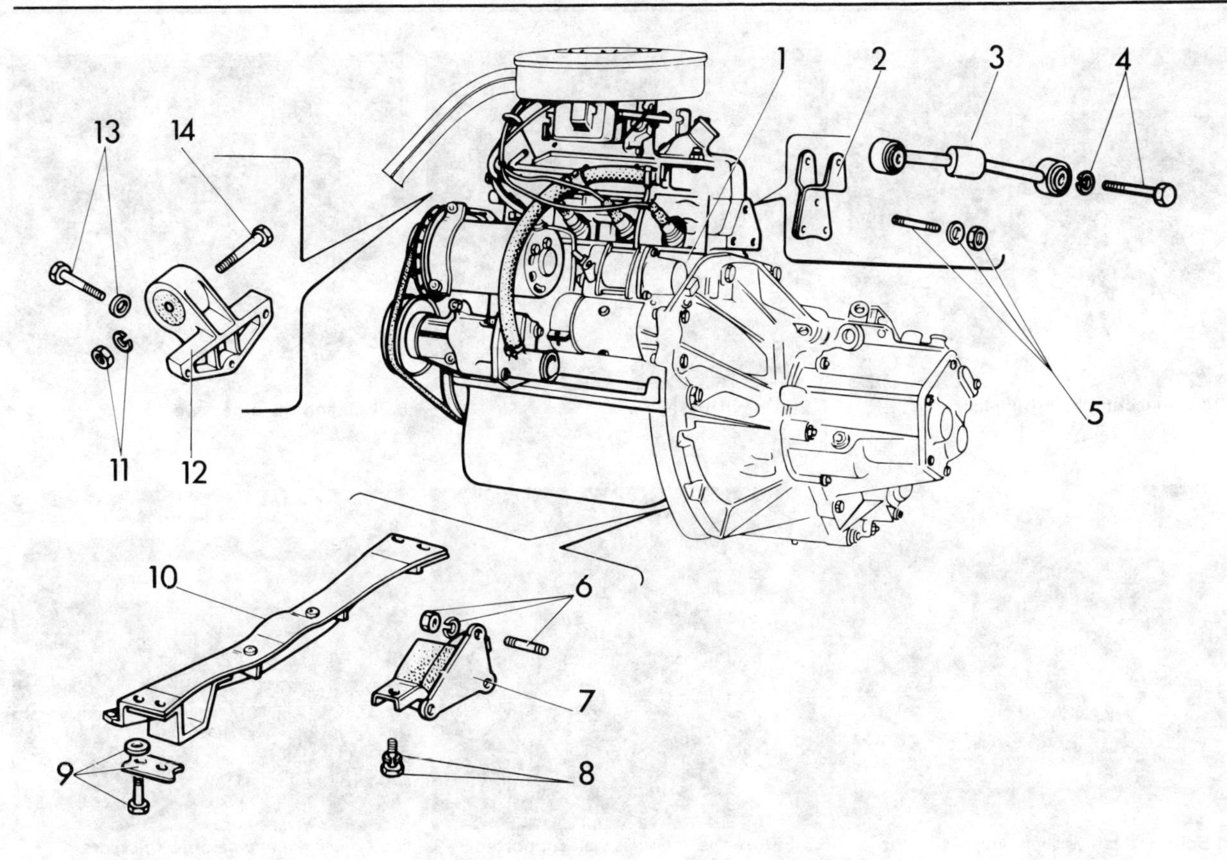

Fig. 1.5. Engine and transmission mountings

1 Power plant
2 Reaction strut anchor
3 Reaction strut
4 Bolt and spring washer, reaction strut attachment to body
5 Nut, flat washer and stud, reaction strut anchor attachment to engine
6 Nut, spring washer and stud, mount attachment to transmission
7 Mount
8 Bolt and spring washer, mount attachment to support rail
9 Bolt, lockwasher and flat washer, support rail attachment to body
10 Power plant support rail
11 Engine mount, right side
12-14 Nut, spring washer and bolt, engine mount attachment to body
13 Bolt and washer, mount attachment to engine

3 Remove the shield on top of the transmission. (This one is not fitted to some cars).

4 Disconnect the speedometer from the transmission casing. (photo)

5 Now the driveshafts must be disconnected. The various methods are given in Chapter 7. If deciding to do it by taking out the complete shaft as an assembly, by withdrawing the constant velocity joint from the hub, first remove the hub cap from the wheel, and whilst the car is still on the ground, undo the nuts on the outer end of the driveshafts..

6 Jack-up the car until the wheels, hanging free on the suspension, are 1 foot (30 cm) clear of the ground. Put the firm supports under the strong points of the body just behind the wheel arches. Put extra blocks in case the car should shift. There will not be room to work underneath, and pull the power unit clear after it has been lowered to the ground.

7 Clean the area around the inner end of the driveshafts, so when the rubber boots are removed, no dirt will get in.

8 Remove the driveshafts, as described in Chapter 7.

9 It is then convenient, as it gives more room, to take the suspension completely out of the way. Undo the pivot bolts at the inner end of the suspension arms. Take out the two bolts through the two mountings for the anti-roll bar at the front of the car. The whole assembly is now free, and can be put out of the way whilst the engine is being lowered.

6 Engine removal - lowering out the power unit

1 The engine and transmission should by now be disconnected, only the actual mountings being left in place. Check that nothing has been forgotten.

2 Connect the lifting tackle to the engine. One eye is adjacent to the generator, while the other is just in front of the thermostat, where the reaction strut was disconnected. Arrange the sling so that the power unit will hang horizontally. If using ordinary rope, put shackles on the lifting eyes as they are sharp, and could cut the rope. (photos)

3 Take the weight of the power unit on the tackle.

4 Bend back the tab washers and take out the two bolts at each end of the bracket under the transmission. Remove the bracket. (photo)

5 Take out the bolt through the mounting bracket at the right end of the engine. Adjust the tackle to ease the load on the bolt so it is free to be pulled out. (photo)

6 Recheck nothing is now still connecting the power unit to the car.

7 Place the spare wheel under the car, onto which the power unit can be lowered, to make it easier to slide it away from under the car.

8 Lower the power unit down on the spare wheel. (photo)

6.2A Connecting the lifting tackle

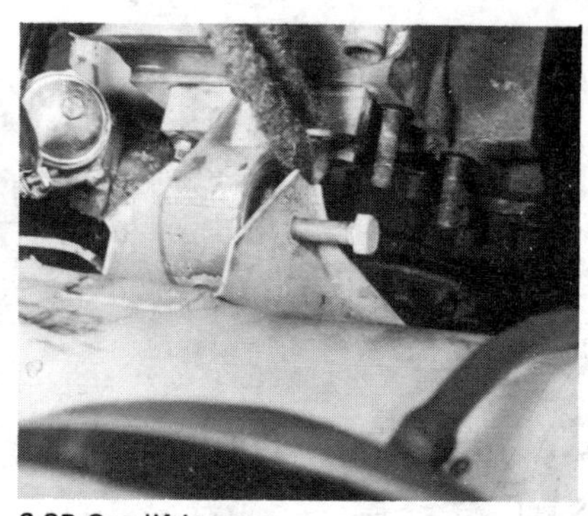

6.2B One lifting eye ...

6.2C ... and the other two

6.4 Bolts and tab washers supporting the bracket under the transmission

6.5 Ensure the engine is well supported before removing this bolt

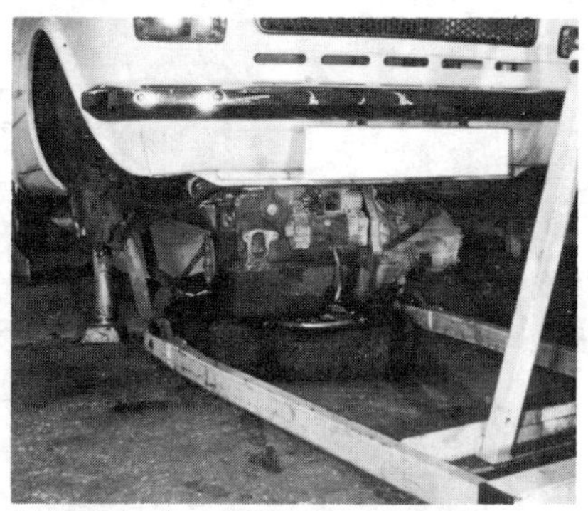

6.8 Lower the engine onto the spare wheel and drag it clear

9 It is then convenient, as there is sufficient room, to slide the engine out through the front, or it might come out from under the wheel arch. However, if the car could not originally be jacked up to the recommended height, transfer the tackle to the front, and lift up the car to allow the engine to be pulled clear.

7 Separating the engine and transmission

1 Remove the starter motor (3 x 13 mm bolts).
2 Remove the three bolts that pass through the engine endplate and secure the mounting bracket to the transmission.
3 Support the transmission so its weight does not hang on the shaft through the clutch.
4 Remove the remaining nuts from the studs mating the transmission to the engine.
5 Pull the transmission off the engine, keeping it straight so the shaft comes out of the clutch without putting any weight on it.
6 Remove the clutch assembly from the flywheel, as described in Chapter 5.

8 Engine replacements - exchange and 'Short' engines

1 Having removed the engine it must be further dismantled, either in preparation to exchange for an official replacement unit, or stripped so overhaul work can be done on it.
2 Exchange units come without all the external components, such as generator, water pump, manifolds, etc. 'Short' engines are also available without other things, such as cylinder head, flywheel and oil pump.
3 In the Sections that follow the removal of these items is covered as for a major overhaul, though with comment as to how a particular component can be removed in isolation.

9 Engine - dismantling (general)

1 Stand the engine on the floor and thoroughly clean down the outside of the engine. Remove all traces of oil and congealed dirt. A good grease solvent will make the job much easier, as, after the solvent has been applied and allowed to stand for a time, a vigorous jet of water will wash off all the solvent together with the grease and filth. If the dirt is thick and deeply embedded, work the solvent into it with a stiff paintbrush.
2 Finally wipe down the exterior of the engine with a rag and only then, when it is quite clean, should the dismantling process begin.
3 Stand the engine on a strong bench so as to be at a comfortable working height. Failing this it can be stripped down on the floor.
4 During the dismantling process, the greatest care should be taken to keep the exposed parts free from dirt. As the engine is stripped, clean each part in a bath of paraffin or petrol.
5 Never immerse parts with oilways in paraffin, e.g. the crankshaft, but to clean, wipe down carefully with a petrol dampened rag. Oilways can be cleaned out with nylon pipe cleaners. If an air line is available, all parts can be blown dry and the oilways blown through as an added precaution.
6 Re-use of old gaskets is false economy and can give rise to oil and water leaks, if nothing worse. To avoid the possibility of trouble after the engine has been reassembled always use new gaskets throughout.
7 Do not throw the old gaskets away as it sometimes happens that an immediate replacement cannot be found and the old gasket is then very useful as a template. Hang up the gaskets on a suitable nail or hook as they are removed.
8 To strip the engine, it is best to work from the top downwards. The engine oil sump provides a firm base on which the engine can be supported in an upright position. When the stage is

reached where the crankshaft must be removed, the engine can be turned on its side and all other work carried out with it in this position.
9 Wherever possible, replace nuts, bolts and washers finger-tight from wherever they were removed. This helps avoid later loss and muddle. If they cannot be replaced then lay them out in such a fashion that it is clear from where they came.

10 Engine - removing ancillary components

Before basic engine dismantling begins it is necessary to strip it of ancillary components as follows:

Distributor
Thermostat
Carburettor
Exhaust manifold
Fuel pump

It is possible to strip all these items with the engine in the car if it is merely the individual items that require attention. Presuming the engine to be out of the car and on the bench, follow the procedure described below.
1 Undo and remove the nuts and washers securing the exhaust manifold and hot air ducting to the cylinder head.
2 Scribe a mark on the distributor pedestal and cylinder to act as a datum for refitting. Undo and remove the distributor pedestal clamping nut, washer and bracket. Pull the distributor up out of the engine. Lift out the distributor drive gear. With long enough fingers, one of these can be wedged inside it. Otherwise, sharpen a wooden stick to a suitable taper, and jam it in the gear. Then pull it up out of the block.
3 Remove the oil pressure warning sender. Take off the oil filter element, and throw it away.
4 Detach the fuel inlet pipe to the carburettor float chamber. Undo and remove the carburettor mounting nuts or bolts and washers, and lift away the carburettor and drip tray. Recover the gasket between the drip tray and spacer block.
5 Disconnect the drip line from the side of the spacer block, then remove the four nuts and washers securing the spacer block. Remove the spacer and its gasket.
6 To remove the thermostat, remove the bolt, two nuts and spring washers securing the upper half of the thermostat housing. Part the two halves of the housing and lift away the thermostat and gasket. The lower half of the housing may now be lifted from two studs in the cylinder head.
7 Undo and remove the two nuts and spring washers securing the fuel pump to the side of the timing cover. Recover the two gaskets, spacer and pump actuating rod and keep in a safe place.
8 Remove the three nuts and washers that secure the engine mounting bracket to the cylinder block.
9 The engine is now stripped of all ancillary components and is ready for major dismantling to begin.

11 Cylinder head - removal (engine on bench)

1 With the engine out of the car and standing upright on the bench or on the floor remove the cylinder head as follows:
2 Unscrew the four nuts securing the rocker cover to the top of the cylinder head and lift away the spring washers and metal packing pieces. Remove the rocker cover and cork gasket.
3 Unscrew the four rocker pedestal securing nuts in a progressive manner. Lift away the four nuts and spring washers and ease the valve rocker assembly from the cylinder head studs.
4 Remove the pushrods, keeping them in the relative order in which they were removed. The easiest way to do this is to push them through a sheet of thick paper or thin card in the correct sequence.
5 Unscrew the cylinder head securing bolts half a turn at a time in the reverse order to that shown in Fig.1.19; don't forget the one within the inlet manifold. When all the bolts are no longer under tension they may be screwed off the cylinder head one at a time. This will also release a section of the cooling system pipe secured by two of the bolts. (photo)
6 The cylinder head may now be lifted off. If the head is jammed, try to rock it to break the seal. Under no circumstances try to prise it apart from the cylinder block with a screwdriver or cold chisel as damage may be done to the faces of the head or block. If the head will not readily free, turn the engine over by the flywheel as the compression in the cylinders will often break the cylinder head joint. If this fails to work, strike the head sharply with a plastic headed hammer, or with a wooden hammer, or with a metal hammer with an interposed piece of wood to cushion the blows. Under no circumstances hit the head directly with a metal hammer as this may cause the casting to fracture. Several sharp taps with the hammer, at the same time pulling upwards, should free the head. Lift the head off and place on one side.

12 Cylinder head - removal (engine in car)

To remove the cylinder head with the engine still in the car the following additional procedures should be carried out before those listed in the previous Section:
1 For safety reasons disconnect the battery positive and negative terminals.
2 Refer to Chapter 2 and drain the cooling system.
3 Refer to Chapter 3 and remove the carburettor, air cleaner and spacer block.
4 Undo and remove the five nuts and washers securing the exhaust manifold and hot air ducting to the cylinder head.
5 Detach the cable from the temperature indicator sender unit.
6 Refer to Chapter 4 and remove the distributor and spark plug leads.
7 Refer to Chapter 2 and remove the thermostat and housing from the cylinder head.
8 Remove the clips connecting the water hoses to the cylinder head, just above the support bracket.
9 Disconnect the single bolt that secures the reaction strut to the mounting bracket by the thermostat.
10 Note the electrical cable connections to the rear of the dynamo or alternator and disconnect the cables. Undo the three mountings and lift away the generator.
11 The procedure is now the same as for removing the cylinder head when the engine is on the bench. One tip worth nothing is that should the cylinder head refuse to free easily, the battery can be reconnected and the engine turned over on the solenoid switch. Under no circumstances turn the ignition on with the carburettor and distributor fitted as otherwise the engine might fire.

11.5 Showing the cooling system pipe secured by two of the cylinder head bolts

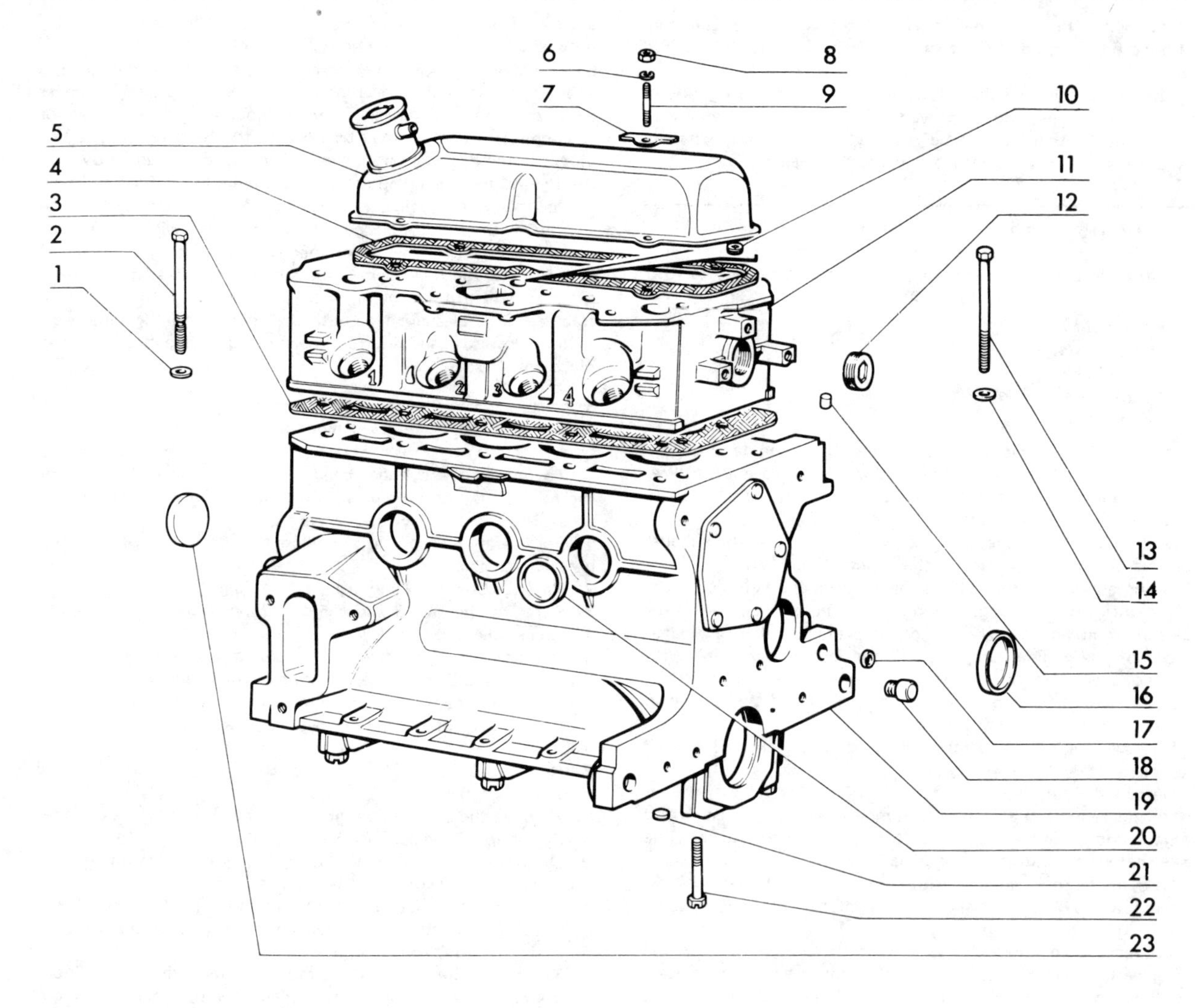

Fig. 1.6. Cylinder head and crankcase components

1 *Washer*	6 *Washer*	12 *Plug*	18 *Dowel*
2 *Cylinder head bolt (waisted)*	7 *Plate*	13 *Cylinder head bolt*	19 *Cylinder block*
3 *Head gasket*	8 *Nut*	14 *Washer*	20 *Plug*
4 *Rocker cover gasket*	9 *Stud*	15 *Dowel*	21 *Plug*
5 *Rocker cover*	10 *Plug*	16 *Plug*	22 *Bolt*
	11 *Cylinder head*	17 *Plug*	23 *Plug*

13 Valves - removal

1 The valves can be removed from the cylinder head by the following method: With a valve spring compressor, compress each spring in turn until the two halves of the collets can be removed. Release the compressor and remove the valve spring cap, inner and outer springs, spring lower seating and valve.
2 If, when the valve spring compressor is screwed down, the valve spring cap refuses to free and expose the split collets, do not continue to screw down on the compressor as there is a likelihood of damaging it.
3 Gently tap the top of the tool directly over the cap with a light hammer. This will free the cap. To avoid the compressor jumping off the spring cap when it is tapped, hold the compressor firmly in position with one hand.
4 It is essential that the valves are kept in their correct sequence unless they are so badly worn that they are to be renewed. If they are going to be kept and used again, place them in a sheet of card having eight holes numbered 1 to 8 corresponding with the relative positions the valves were in when fitted. Also keep the valve springs, caps etc, in correct order.

14 Valve guides - removal

If it is wished to remove the valve guides they can be removed from the cylinder head in the following manner. Place the cylinder head upside down on the bench and with a suitable diameter drift carefully drive the guides from the cylinder head. It should be noted that the inlet and exhaust valve guides are not interchangeable.

15 Rocker assembly - dismantling

1 To dismantle the rocker assembly, release the rocker shaft securing circlip from each end of the shaft and slide off the rocker arms and pedestals. Keep in the order in which the parts

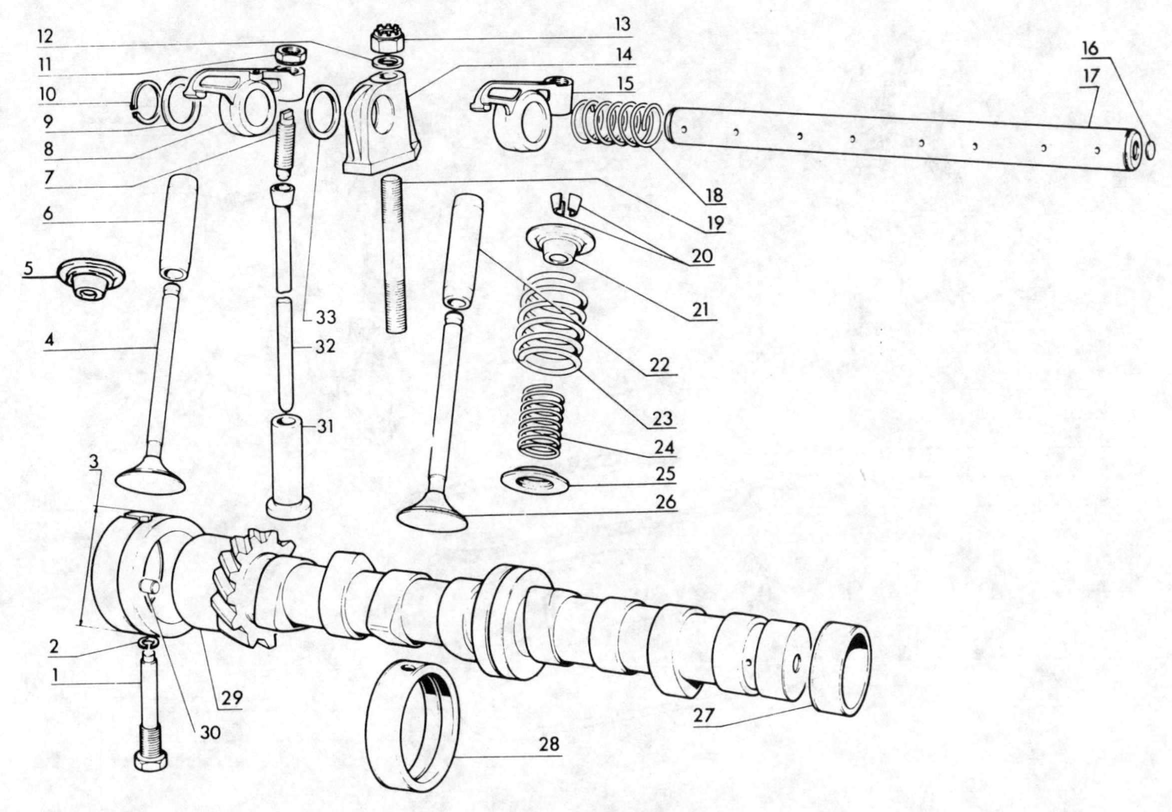

Fig. 1.7. Valve gear

1 Bush locating bolt
2 Washer
3 Bush
4 Valve - exhaust
5 Spring
6 Valve guide
7 Adjuster - tappet
8 Rocker
9 Thrust washer
10 Circlip
11 Locknut
12 Washer
13 Locknut
14 Pedestal
15 Rocker
16 Plug
17 Rocker shaft
18 Spring
19 Stud
20 Collets
21 Spring retainer
22 Valve guide
23 Outer valve spring
24 Inner valve spring
25 Spring seat
26 Inlet valve
27 Bush
28 Bush
29 Camshaft
30 Locating dowel
31 Tappet
32 Pushrod
33 Washer

were removed so that they can be reassembled in their original positions.
2 The interior drillings of the rocker shaft should be cleaned to ensure there is no sludge build up causing restricted oil flow.
3 Check that the small lubrication hole in the rocker is free, using a piece of thin wire as a probe.

16 Sump - removal

To remove the engine oil sump, undo and remove the securing bolts/nuts and lift away the sump. If it has stuck on the gasket, carefully tap the side of the mating flange to break the seal. Recover the joint washer and remove any signs of old jointing compound. Note the cork inserts in the recesses at either end of the sump.

17 Oil pump - removal

1 With the engine oil sump removed, undo and remove the two bolts and spring washers that secure the return pipe from the front main bearing cap.
2 Undo and remove the two bolts and spring washers securing the oil pump housing to the underside of the crankcase.
3 Carefully ease the oil pump assembly from the crankcase.
4 Recover the gasket between the oil pump housing and underside of the crankcase.

18 Timing cover, sprockets and chain - removal

1 First, remove the engine support bracket, secured by three

18.1 Disconnecting the engine support bracket

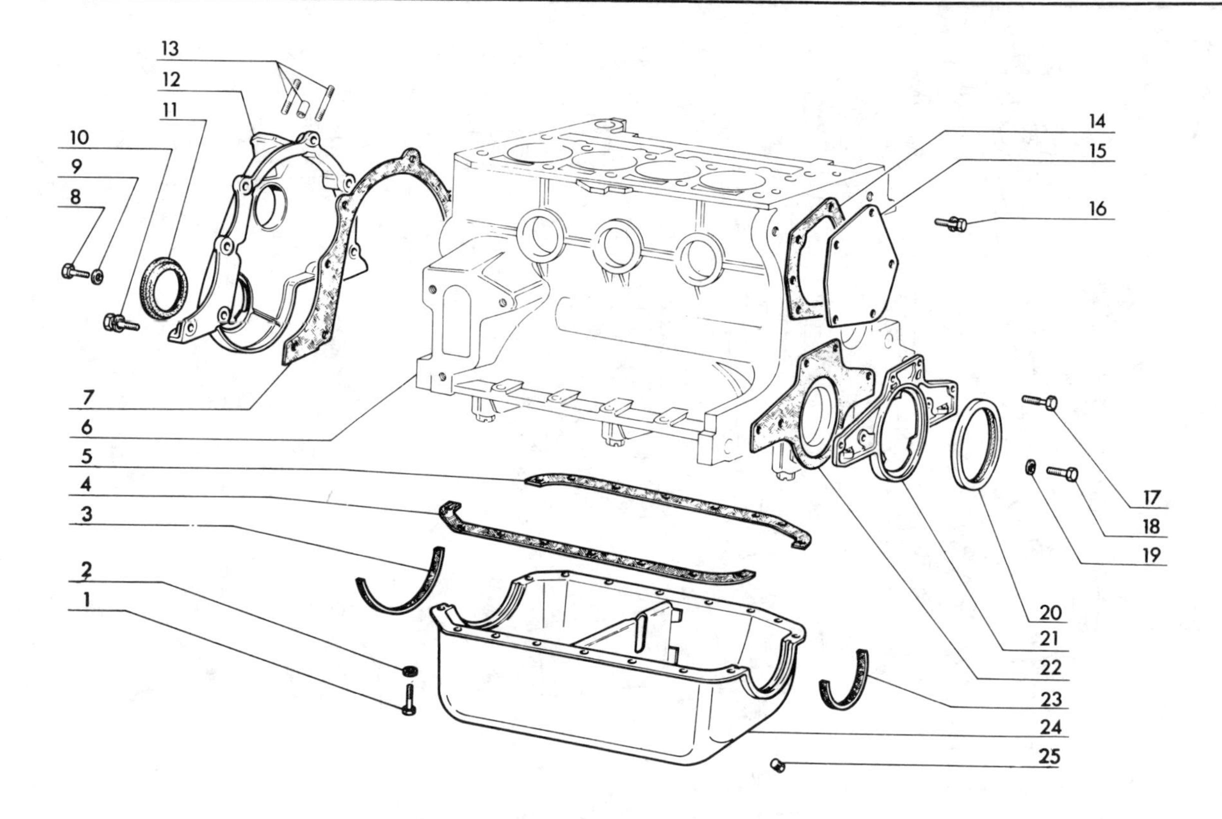

Fig. 1.8. Crankcase and sump gaskets and seals

1 *Sump bolt*
2 *Washer*
3 *Gasket*
4 *Gasket*
5 *Gasket*
6 *Cylinder block*
7 *Timing cover gasket*
8 *Bolt*
9 *Washer*
10 *Bolt and washer*
11 *Oil seal*
12 *Timing cover*
13 *Dowels and bush for fuel pump*
14 *Gasket*
15 *Cover*
16 *Bolt and washer*
17 *Bolt (waisted)*
18 *Bolt*
19 *Washer*
20 *Oil seal*
21 *Seal carrier*
22 *Carrier gasket*
23 *Gasket*
24 *Sump*
25 *Drain plug*

bolts, above the timing cover. Then undo and remove the seven timing cover securing bolts and spring washers. If the sump is still in position note that the two front bolts locate in the end of the timing cover so these must also be removed. (photo)

2 Undo and remove the camshaft sprocket securing bolt; this will also release the fuel pump drive cam from the end of the camshaft. Note the two timing marks on the camshaft and crankshaft sprockets.

3 Using two tyre levers, carefully ease the two sprockets forwards away from the crankcase. Lift away the two sprockets and timing chain.

4 Remove the Woodruff key from the crankshaft nose with a pair of pliers and note how the channel in the pulley is designed to fit over it. Place the Woodruff key in a container, eg; a jam jar, as it is a very small part and can easily become lost. The camshaft sprocket is located on the camshaft by a dowel peg.

19 Camshaft and tappets (cam followers) - removal

1 It is impossible to remove the camshaft with the engine in position in the car. However, it is unusual for the camshaft on its own to require attention. Usually it will be necessary to carry out other major work as well and in this case the engine will be out of the car.

2 Undo and remove the camshaft front bush securing screw. The camshaft may now be withdrawn. Take great care to remove the camshaft gently and, in particular, ensure that the cam peaks do not damage the camshaft bearings as the shaft is pulled forwards.

3 Lift out the tappets and place in order so that they may be refitted in their original positions.

20 Pistons, connecting rods and big-end bearings - removal

The piston and connecting rod assemblies may be removed with the engine in the car. For this it will be necessary to remove the cylinder head as described in Section 12 and also the engine oil sump. Removing the sump is not easy because the engine support bracket passes right underneath it. Therefore, before the sump can be removed, the transmission must be well supported by blocks or jacks, and then the support bracket removed.

1 Undo and remove the big-end cap retaining bolts and keep them in their respective order for correct refitting.

2 Check that the connecting rod and big-end bearing cap assemblies are correctly marked. Normally the numbers 1–4 are stamped on adjacent sides of the big-end caps and connecting rods, indicating which cap fits on which rod and which way round the cap fits. If no numbers or lines can be found then, with a sharp screwdriver, scratch mating marks across the joint from the rod to the cap. One line for connecting rod No. 1 (timing chain end), two for connecting rod No. 2 and so on. This will ensure that there is no confusion later as it is most important that the caps go back in the correct position on the connecting rods from which they were removed.

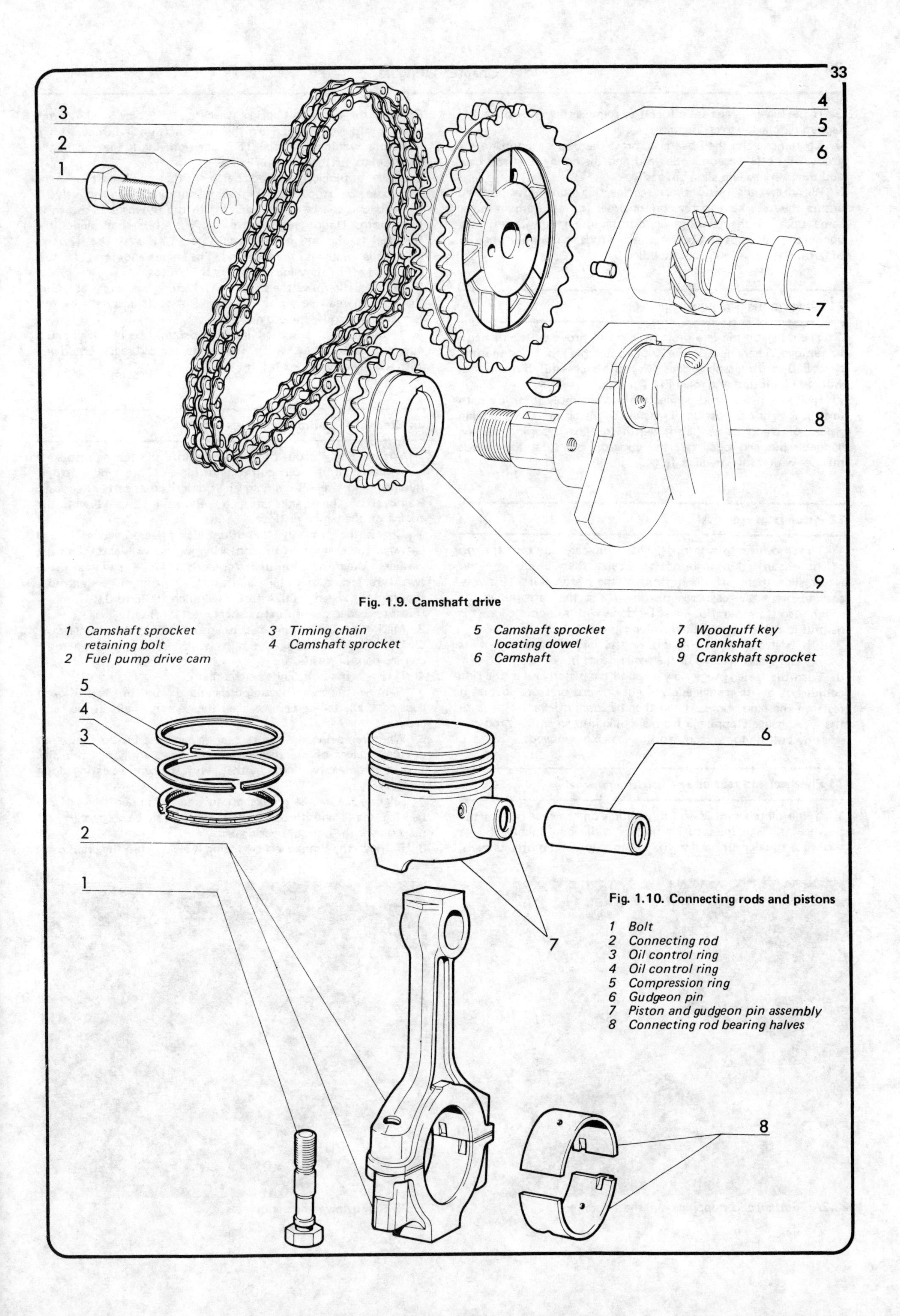

Fig. 1.9. Camshaft drive

1 *Camshaft sprocket retaining bolt*
2 *Fuel pump drive cam*
3 *Timing chain*
4 *Camshaft sprocket*
5 *Camshaft sprocket locating dowel*
6 *Camshaft*
7 *Woodruff key*
8 *Crankshaft*
9 *Crankshaft sprocket*

Fig. 1.10. Connecting rods and pistons

1 *Bolt*
2 *Connecting rod*
3 *Oil control ring*
4 *Oil control ring*
5 *Compression ring*
6 *Gudgeon pin*
7 *Piston and gudgeon pin assembly*
8 *Connecting rod bearing halves*

3 If the big-end caps are difficult to remove they may be gently tapped with a soft faced hammer.
4 To remove the shell bearings, press the bearing opposite the groove in both the connecting rod and the connecting rod caps and the bearings will slide out easily.
5 Withdraw the pistons and connecting rods upwards and ensure they are kept in the correct order for replacement in the same bore. If the bearings do not require renewal, refit the connecting rod caps and bearings to the rods to minimise the risk of getting the caps and rods muddled.

21 Gudgeon pins - removal

1 The gudgeon pin is a press fit and can prove a little difficult to remove. There is a special tool to do this job if the method described in paragraph 2 does not prove successful, but this tool will only be found at a main FIAT agent.
2 Heat the piston and connecting rod assembly in an oil bath until it is as hot as possible. Transfer the piston onto two parallel pieces of wood of sufficient depth to allow clearance of the gudgeon pin and push out the gudgeon pin in a downwards manner with a soft metal drift.

22 Piston rings - removal

1 To remove the piston rings, slide them carefully over the top of the piston, taking care not to scratch the aluminium alloy. Never slide them off the bottom of the piston skirt. It is very easy to break the cast iron piston rings if they are pulled off roughly so this operation must be done with extreme care. It is helpful to make use of an old feeler gauge.
2 Lift one end of the piston ring to be removed out of its groove and insert the end of the feeler gauge under it.
3 Turn the feeler gauge slowly round the piston and as the ring comes out of its groove apply slight upward pressure so that it rests on the land above. It can then be eased off the piston with the feeler gauge stopping it from slipping into an empty groove if it is any but the top piston ring that is being removed.

23 Flywheel and rear oil seal carrier - removal

1 If the clutch unit is still in position on the rear face of the flywheel, slacken the securing bolts in a diagonal manner half a turn at a time ensuring that the cover is not held by the dowels. Remove the bolts and spring washers and lift away the clutch cover and driven plate. Turn the crankshaft until numbers 1 and 4 pistons would be in the TDC position. Mark the top of the flywheel to assist correct refitting.
2 Place a block of wood between the crankshaft and the crankcase to stop the crankshaft rotating. Alternatively, a toothed clip can be fitted adjacent to the flywheel as shown in the photo. Undo and remove the six bolts that secure the flywheel to the end of the crankshaft. Lift away the flywheel bolt thrust plate and the flywheel. The engine endplates can now be pulled off of their locating dowels. (photos)
3 Undo and remove the six bolts and spring washers that secure the rear oil seal carrier to the crankcase. Lift away the oil seal carrier and recover the old gasket.
4 If the oil seal has been leaking it should be removed and a new one fitted. Note which way round the old seal is fitted and with a screwdriver lever it out.

24 Crankshaft and main bearings - removal

With the engine out of the car, remove the timing sprockets and chain, sump, oil pump and big-end bearings, pistons, flywheel and rear oil seal carrier as described in earlier Sections. Removal of the crankshaft can only be attempted with the engine on the bench or floor.
1 Check the crankshaft endfloat using feeler gauges placed between the crankshaft main bearing journal wall and the thrust washers. Move the crankshaft forwards as far as it will go using two tyre levers to obtain a maximum reading. The endfloat should not exceed 0.0024 to 0.0102 inch (0.06 to 0.26 mm). If this is exceeded new thrust washers must be fitted.
2 Mark the main bearing caps to ensure correct replacement.
3 Undo by one turn the bolts which hold the three main bearing caps in position.
4 Unscrew the bolts and remove them.
5 Remove the main bearing caps and the bottom half of each bearing shell, taking care to keep the bearing shells in the right caps.
6 When removing the centre bearing cap note the bottom semi-circular halves of the thrust washers, one half lying on either side of the main bearing. Lay them with the centre bearing along the correct side.
7 Slightly rotate the crankshaft to free the upper halves of the bearing shells and thrust washers which can be extracted and placed over the correct bearing cap.
8 Remove the crankshaft by lifting it away from the crankcase.

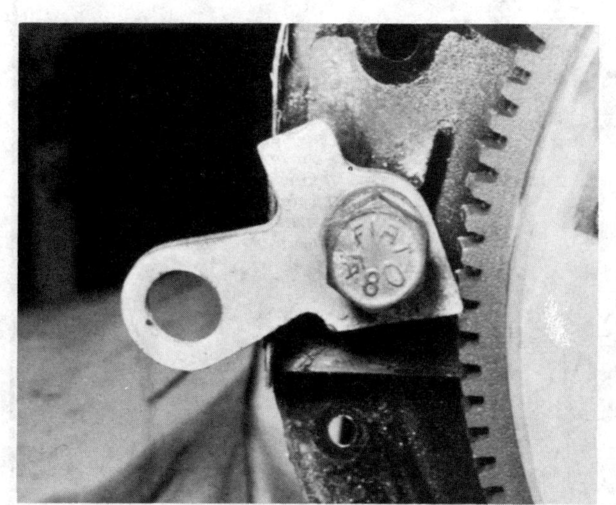

23.2A Toothed clip stopping crankshaft rotating

23.2B Removing engine endplates

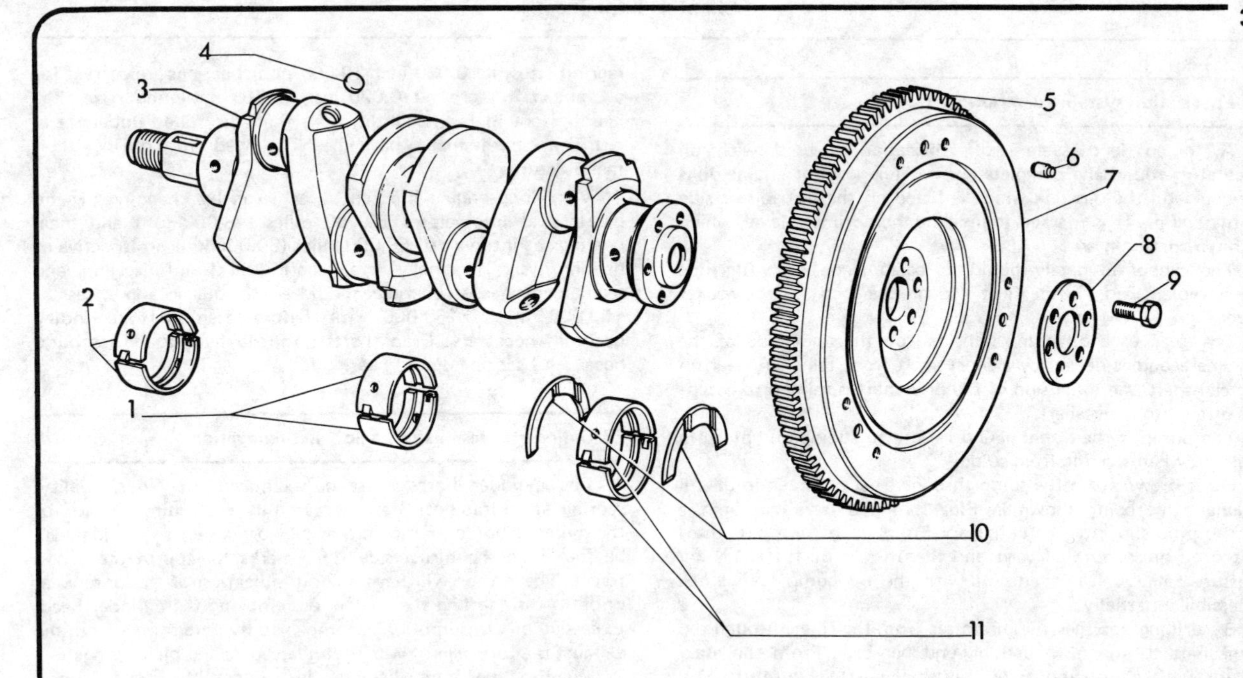

Fig. 1.11. Crankshaft and flywheel

1 Centre main bearing halves
2 Main bearing halves - timing case end
3 Crankshaft
4 Plug
5 Starter ring
6 Dowel
7 Flywheel
8 Thrust plate
9 Bolt
10 Thrust washers
11 Main bearing halves - flywheel end

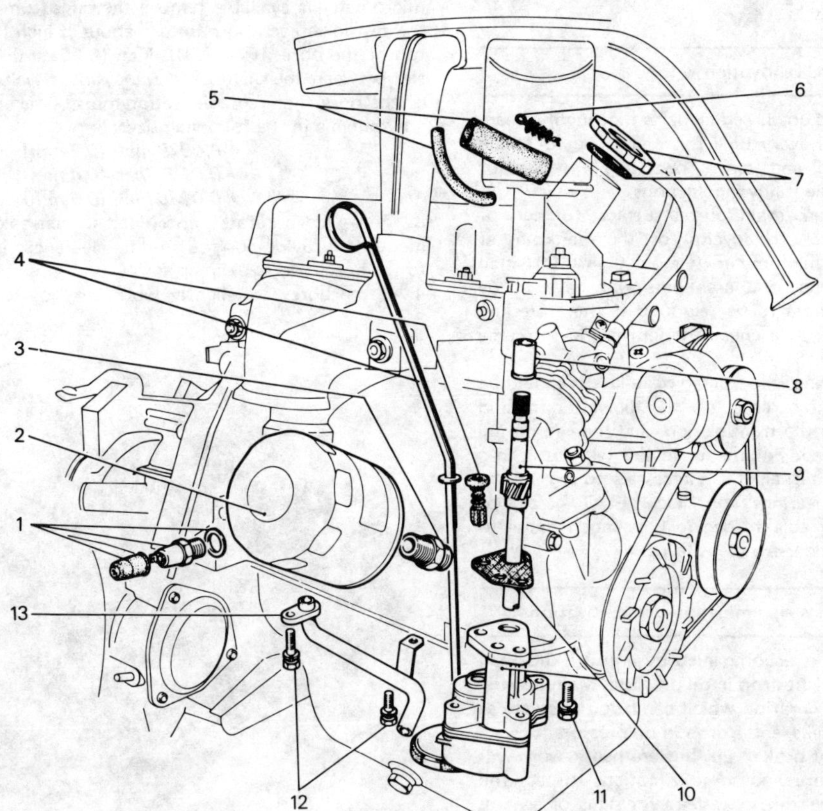

Fig. 1.12. Engine lubrication system components (Sec. 25)

1 Low oil pressure sending unit
2 Cartridge-type oil filter
3 Filter connector
4 Seal and dipstick
5 Breather hoses
6 Backfire suppressor
7 Oil filler cap and gasket
8 Bushing
9 Oil pump shaft
10 Oil pump body and gasket
11 Oil pump body to crankcase screw
12 Bolts and spring washers for oil return pipe
13 Oil return pipe

25 Lubrication system - description

1 A forced feed system of lubrication is used with oil circulating round the complete engine. The level of engine oil is indicated on the dipstick which is fitted on the distributor side of the engine. It is marked to indicate the optimum level, which is the maximum mark.
2 The level of oil ideally should not be above or below this line. Oil is replenished via the filler cap on the front of the rocker cover.
3 The gear type oil pump is bolted on the underside of the crankcase and is driven by a short shaft from the skew gear on the camshaft. An extension of the driveshaft is splined to accept the distributor driveshaft.
4 The pump is the non-draining variety to allow rapid pressure build up when starting from cold.
5 Oil is drawn from the sump through a gauze screen in the oil strainer, this being shown in Fig. 1.14 and is drawn up the pickup pipe and into the oil pump. From the oil pump it passes to the oil pressure relief valve and then to the oil filter. The oil pressure relief valve is integral with the oil pump and is not accessible externally.
6 By drilling passages the oil passes from the filter through the crankshaft to the main and big-end bearings. From the main gallery, which also supplies the camshaft bearings, a return pipe takes the oil back to the sump.
7 From the centre camshaft bearing oil is fed to the overhead rocker gear, returning via two oilways to the camshaft to lubricate the tappets and thereafter to the sump.
8 An oil pressure sender unit is fitted into the main oil gallery to indicate low oil pressure.

26 Engine - examination and renovation

1 With the engine stripped down and all parts thoroughly clean, it is now time to examine everything for wear. The following items should be checked and where necessary renewed or renovated as described in the following Sections.
2 Examine the crankpin and main journal surfaces for signs of scoring or scratches. Check the ovality of the crankpins at different positions with a micrometer. If more than 0.001 inch (0.025 mm) out of round, the crankpins will have to be reground. They will also have to be reground if there are any scores or scratches present. Also check the journals in the same fashion.
3 If the centre main bearing is suspected of failure it should be immediately investigated by removing the power unit and separating the engine from the transmission, and then removing the centre main bearing cap. Failure to do this will result in a badly scored centre main journal. If it is necessary to regrind the crankshaft and fit new bearings your local FIAT garage or engineering works will be able to decide how much metal to grind off and the correct undersize shells to fit.

27 Big-end and main bearings - examination and renovation

1 Big-end bearing failure is accompanied by a noisy knocking from the crankcase and a slight drop in oil pressure. Main bearing failure is accompanied by vibration which can be quite severe as the engine speed rises and falls, and a drop in oil pressure.
2 Bearings which have not broken up, but are badly worn will give rise to low oil pressure and some vibration. Inspect the big-ends, main bearings and thrust washers for signs of general wear, scoring, pitting and scratches. The bearings should be matt grey in colour. With lead-indium bearings, should a trace of copper colour be noticed the bearings are badly worn as the lead bearing material has worn away to expose the indium underlay. Whatever the condition of the old shells it is advisable to fit new shells whenever the engine is dismantled.
3 The undersizes available are designed to correspond with the regrind sizes, ie; 0.020 inch (0.508 mm) bearings are correct for a crankshaft reground 0.020 inch (0.508 mm) undersize. The bearings are in fact, slightly more than the stated undersize as running clearances have been allowed for during their manufacture.
4 Very long engine life can be achieved by changing big-end bearings at intervals of 30,000 miles (48,000 km) and main bearings at intervals of 50,000 miles (80,000 km), irrespective of bearing wear. Normally, crankshaft wear is infinitesimal and regular changes of bearings should ensure mileages in excess of 100,000 miles (160,000 km) before crankshaft regrinding becomes necessary. Crankshafts normally have to be reground because of scoring due to bearing failure.

28 Cylinder bores - examination and renovation

1 The cylinder bores must be examined for taper, ovality, scoring and scratches. Start by carefully examining the top of the cylinder bores. If they are at all worn a very slight ridge will be found on the thrust side. This marks the top of the piston travel. The owner will have a good indication of the bore wear prior to dismantling the engine, or removing the cylinder head. Excessive oil consumption accompanied by blue smoke from the exhaust is a sure sign of worn cylinder bores and piston rings.
2 Measure the bore diameter just under the ridge with a micrometer and compare it with the diameter at the bottom of the bore, which is not subject to wear. If the difference between the two measurements is more than .006 inch (0.15 mm) then it will be necessary to fit special piston rings or to have the cylinders rebored and fit oversize pistons and rings. If no micrometer is available remove the rings from a piston and place the piston in each bore in turn about ¾ inch (20 mm) below the top of the bore. If an 0.010 inch (0.25 mm) feeler gauge can be slid between the piston and the cylinder wall on the thrust side of the bore then remedial action must be taken. Oversize pistons are available in the following sizes:

+ 0.0079 inch (0.2 mm)
+ 0.0157 inch (0.4 mm)
+ 0.0236 inch (0.6 mm)

3 These are accurately machined to just below these measurements to provide correct running clearances in bores bored out to the exact oversize dimensions.
4 If the bores are slightly worn but not so badly worn as to

Fig. 1.13. Arrows show identification letters of cylinder bore grades, according to their actual diameter (Sec. 29)

justify reboring them, special oil control rings can be fitted to the existing pistons which will restore compression and stop the engine burning oil. Several different types are available and their manufacturer's instructions concerning their fitting must be followed closely.

29 Pistons and piston rings - examination and renovation

1 If the old pistons are to be refitted, carefully remove the piston rings and then thoroughly clean them. Take particular care to clean out the piston ring grooves. At the same time do not scratch the aluminium in any way. If new rings are to be fitted to the old pistons, then the top ring should be stepped so as to clear the ridge left above the previous top ring. If a normal but oversize new ring is fitted it will hit the ridge and break, because the new ring will not have worn in the same way as the old, which will have worn in unison with the ridge.
2 Before fitting the rings on the pistons each should be inserted approximately 3 inches (76 mm) down the cylinder bore and the gap measured with a feeler gauge. This should be as detailed in the 'Specifications' at the beginning of this Chapter. It is essential that the gap is measured at the bottom of the ring travel, as if it is measured at the top of a worn bore and gives a perfect fit, it could easily seize at the bottom. If the ring gap is too small rub down the ends of the ring with a very fine file until the gap, when fitted, is correct. To keep the rings square in the bore for measurement, line each up in turn with an old piston in the bore upside down, and use the piston to push the ring down about 3 inches (76 mm). Remove the piston and measure the piston ring gap.
3 When fitting new pistons and rings to a rebored engine the ring gap can be measured at the top of the bore as the bore will now not taper. It is not necessary to measure the side clearance in the piston ring grooves with rings fitted, as the groove dimensions are accurately machined during manufacture. When fitting new rings it may be necessary to have the grooves widened by machining to accept the new wider rings. In this instance the manufacturer's representative will make this quite clear and will supply the address to which the pistons must be sent for machining.
4 When new pistons are fitted, take great care to fit the exact size best suited to the particular bore of your engine. FIAT go one stage further than merely specifying one size of piston for all standard bores. Because of very slight differences in cylinder machining during production it is necessary to select just the right piston for the bore. A range of different sizes are available either from the piston manufacturer's or from the dealer for the particular model of car being repaired.
5 Examination of the cylinder block face will show, adjacent to each bore, a letter stamped in the metal. Careful examination of the piston underside will reveal a matching letter. These are the standard piston sizes and will be the same for all four bores. If standard pistons are to be refitted or standard low compression pistons changed to standard high compression pistons, then it is essential that only pistons with the same letter are used. (Fig. 1.13).

30 Camshaft and camshaft bearings - examination and renovation

1 Carefully examine the camshaft bearings for wear. If the bearings are obviously worn or pitted or the metal underlay just showing through, then they must be renewed. This is an operation for your local FIAT agent or automobile engineering works as it demands the use of specialized equipment. The bearings are removed using a special drift after which the new bearings are pressed in, care being taken that the oil holes in the bearings line up with those in the block. With another special tool the bearings are then reamed in position.
2 The camshaft itself should show no sign of wear, but, if very slight scoring marks on the cams are noticed, the score marks can be removed by very gentle rubbing down with very fine emery cloth or an oil stone. The greatest care should be taken to keep the cam profiles smooth.

31 Valves and seats - examination and renovation

1 Examine the heads of the valves for pitting and burning, especially the heads of the exhaust valves. The valve seatings should be examined at the same time. If the pitting on the valves and seats is very slight the marks can be removed by grinding the seats and valves together with coarse, and then fine, valve grinding paste. Where bad pitting has occurred to the valve seats it will be necessary to recut them and fit new valves. If the valve seats are so worn that they cannot be recut, then it will be necessary to fit new valve seat inserts. These latter two jobs should be entrusted to the local FIAT agent or automobile engineering works. In practice it is very seldom that the seats are so badly worn that they require renewal. Normally, it is the valve that is too badly worn for replacement, and the owner can easily purchase a new set of valves and match them to the seats by valve grinding.
2 Valve grinding is carried out as follows: Place the cylinder head upside down on a bench, with a block of wood at each end to give clearance for the valve stems. Alternatively place the head at 45° to a wall with the combustion chambers facing away from the wall.
3 Smear a trace of coarse carborundum paste on the seat face and apply a suction grinder tool to the valve heads. With a semi-rotary action, grind the valve head to its seat, lifting the valve occasionally to redistribute the grinding paste. When a dull matt even surface finish is produced on both the valve seat and the valve, then wipe off the paste and repeat the process with fine carborundum paste, lifting and turning the valve to redistribute the paste as before. A light spring placed under the valve head will greatly ease this operation. When a smooth unbroken ring of light grey matt finish is produced, on both valve and valve seat faces, the grinding operation is complete.
4 Scrape away all carbon from the valve head and the valve stem. Carefully clean away every trace of grinding compound, taking great care to leave none in the ports or in the valve guides. Clean the valves and valve seats with a paraffin soaked rag, then with a clean rag, and finally, if an air line is available blow the valves, valve guides and valve ports clean.

32 Timing sprockets and chain - examination and renovation

1 Examine the teeth on both the crankshaft sprocket and the camshaft sprocket for wear. Each tooth forms an inverted 'V' with the gear wheel periphery and if worn, the side of each tooth under tension will be slightly concave in shape when compared with the other side of the tooth, ie; one side of the inverted 'V' will be concave when compared with the other. If any sign of wear is present the sprockets must be renewed.
2 Examine the links of the chain for side slackness and particularly check the self-tensioning links for freedom of movement. Renew the chain if any slackness is noticeable when compared with a new chain. It is a sensible precaution to renew the chain at about 30,000 miles (48,000 km) and at a lesser mileage if the engine is stripped down for a major overhaul. The actual rollers on a very badly worn chain may be slightly grooved.

33 Rockers and rocker shaft - examination and renovation

1 Thoroughly clean out the rocker shaft. As it acts as the oil passages for the valve gear, clean out the oil holes and make sure they are quite clear. Check the shaft for straightness by rolling it on a flat surface. It is most unlikely that it will deviate from normal, but, if it does, then a judicious attempt must be made to straighten it. If this is not successful purchase a new shaft. The surface of the shaft should be free from any worn ridges caused

by the rocker arms. If any wear is present, renew the rocker shaft. Wear is likely to have occurred only if the rocker shaft oil holes have become blocked.

2 Check the rocker arms for wear of the rocker bushes, for wear at the rocker arm face which bears on the valve stem, and for wear of the adjusting ball ended screws. Wear in the rocker arm bush can be checked by gripping the rocker arm tip and holding the rocker arm in place on the shaft, noting if there is any lateral rocker arm shake. If any shake is present, and the arm is very loose on the shaft, remedial action must be taken. It is recommended that any worn rocker arm be taken to your local FIAT agent or automobile engineering works to have the old bush drawn out and a new bush fitted.

3 Check the tip of the rocker arm where it bears on the valve head, for cracking or serious wear on the case hardening. If none is present the rocker arm may be refitted. Check the pushrods for straightness by rolling them on a flat surface.

34 Tappets (cam followers) - examination and renovation

Examine the bearing surface of the tappets which lie on the camshaft. Any indentation in this surface or any cracks indicate serious wear and the tappets should be renewed. Thoroughly clean them out, removing all traces of sludge. It is most unlikely that the sides of the tappets will be worn, but, if they are a very loose fit in their bores and can be readily rocked, they should be discarded and new tappets fitted. It is unusual to find worn tappets and any wear present is likely to occur only at very high mileages.

35 Flywheel starter ring - examination and renovation

1 If the teeth on the flywheel starter ring gear are badly worn, or if some are missing, then it will be necessary to remove the ring. This is achieved by splitting the old ring using a cold chisel, after partially cutting through with a hacksaw. The greatest care must be taken not to damage the flywheel during this process.

2 To fit a new ring gear, heat it gently and evenly with an oxyacetylene flame until a temperature of approximately 350°C (662°F) is reached. This is indicated by a light metallic blue surface colour. With the ring gear at this temperature, fit it to the flywheel. The ring gear should be either pressed or lightly tapped gently onto its register and left to cool naturally, when the contraction of the metal on cooling will ensure that it is a secure and permanent fit. Great care must be taken not to overheat the ring gear, as if this happens its temper will be lost.

3 Alternatively, your local FIAT agent or local automobile engineering works may have a suitable oven in which the ring gear can be heated. The normal domestic oven will give a temperature of about 250°C (482°F) only, at the very most, except for the latest self-cleaning type which will give a higher temperature. With the former it may just be possible to fit the ring gear with it at this temperature, but it is unlikely and no great force should have to be used.

36 Oil pump - dismantling, examination and renovation

1 With the oil pump away from the engine, undo and remove the four bolts and spring washers securing the suction horn to the pump housing. This will also release the plate that retains the gears in the top half of the housing.

2 Thoroughly clean all the component parts in petrol and then check the gear endfloat and tooth clearances in the following manner:

3 With the two gears in position in the pump, place the straight edge of a steel rule across the joint face of the housing, measure the gap between the bottom of the straight edge and the top of the gears with a feeler gauge. The clearance should be 0.0008 to 0.0041 in (0.020 to 0.105 mm).

4 Check the side clearance between the gears and oil pump housing. This should be 0.0019 to 0.0055 in (0.05 to 0.14 mm).

5 Check the oil pressure relief valve spring for cracks or evidence of weakening. The spring height, non-loaded, should be 40.2 mm (1.583 in). Under a load of 5 kg (11.02 lb) the spring height should be 21 mm (0.827 in). Later pumps, fitted to engine number 2635011 onwards, have a modified spring, the height of which should be, non-loaded, 44.5 mm (1.752 in). Under a load of 4.5 kg (9.92 lbs) the spring height should be 29 mm (1.142 in). Renew the spring if these readings are not achieved.

6 If the pump is worn generally, it is sound policy to obtain a replacement pump since the degree of overhaul that can be done at home is limited.

37 Cylinder head - decarbonisation

1 This operation can be carried out with the engine either in or out of the car. With the cylinder head off, carefully remove with a wire brush and blunt scraper all traces of carbon deposits from the combustion spaces and the ports. The valve stems and valve guides should also be freed from any carbon deposits. Wash the combustion spaces and ports down with petrol and scrape the cylinder head surface free of any foreign matter with the side of a steel rule or a similar article. Take care not to scratch the surfaces.

2 Clean the pistons and top of the cylinder bores. If the pistons are still in the cylinder bores then it is essential that great care is taken to ensure that no carbon gets into the cylinder bores as this could scratch the cylinder walls or cause damage to the piston and rings. To ensure that this does not happen first turn the crankshaft so that two of the pistons are at the top of the bores. Place clean non-fluffy rag into the other two bores or seal them off with paper and masking tape. The waterways and pushrod holes should also be covered with a small piece of masking tape to prevent particles of carbon entering the cooling system and damaging the water pump, or entering the lubrication system and causing damage to a bearing surface.

3 There are two schools of thought as to how much carbon ought to be removed from the piston crown. One is that a ring of carbon should be left around the edge of the piston and on the cylinder bore wall as an aid to keep oil consumption low. Although this is probably true for early engines with worn bores, on later engines the tendency is to remove all traces of carbon during decarbonisation.

4 If all traces of carbon are to be removed, press a little grease into the gap between the cylinder walls and the two pistons which are to be worked on. With a blunt scraper carefully scrape away the carbon from the piston crown, taking care not to scratch the aluminium. Also scrape away the carbon from the surrounding lip of the cylinder wall. When all carbon has been removed, scrape away the grease which will now be contaminated with carbon particles, taking care not to press any into the bores. To assist prevention of carbon build-up the piston crown can be polished with a metal polish such as 'Brasso'. Remove the rags or masking tape from the other two cylinders and turn the crankshaft so that the two pistons which were at the bottom are now at the top. Place non-fluffy rag into the other two bores or seal them off with paper and masking tape. Do not forget the waterways and oilways as well. Proceed as previously described.

5 If a ring of carbon is going to be left round the piston then this can be helped by inserting an old piston ring into the top of the bore to rest on the piston and ensure that carbon is not accidentally removed. Check that there are no particles of carbon in the cylinder bores. Decarbonising is now complete.

38 Valve guides - examination and renovation

1 Examine the valve guides internally for wear. If the valves are a very loose fit in the guides and there is the slightest suspicion of lateral rocking, then new guides will have to be fitted. If the

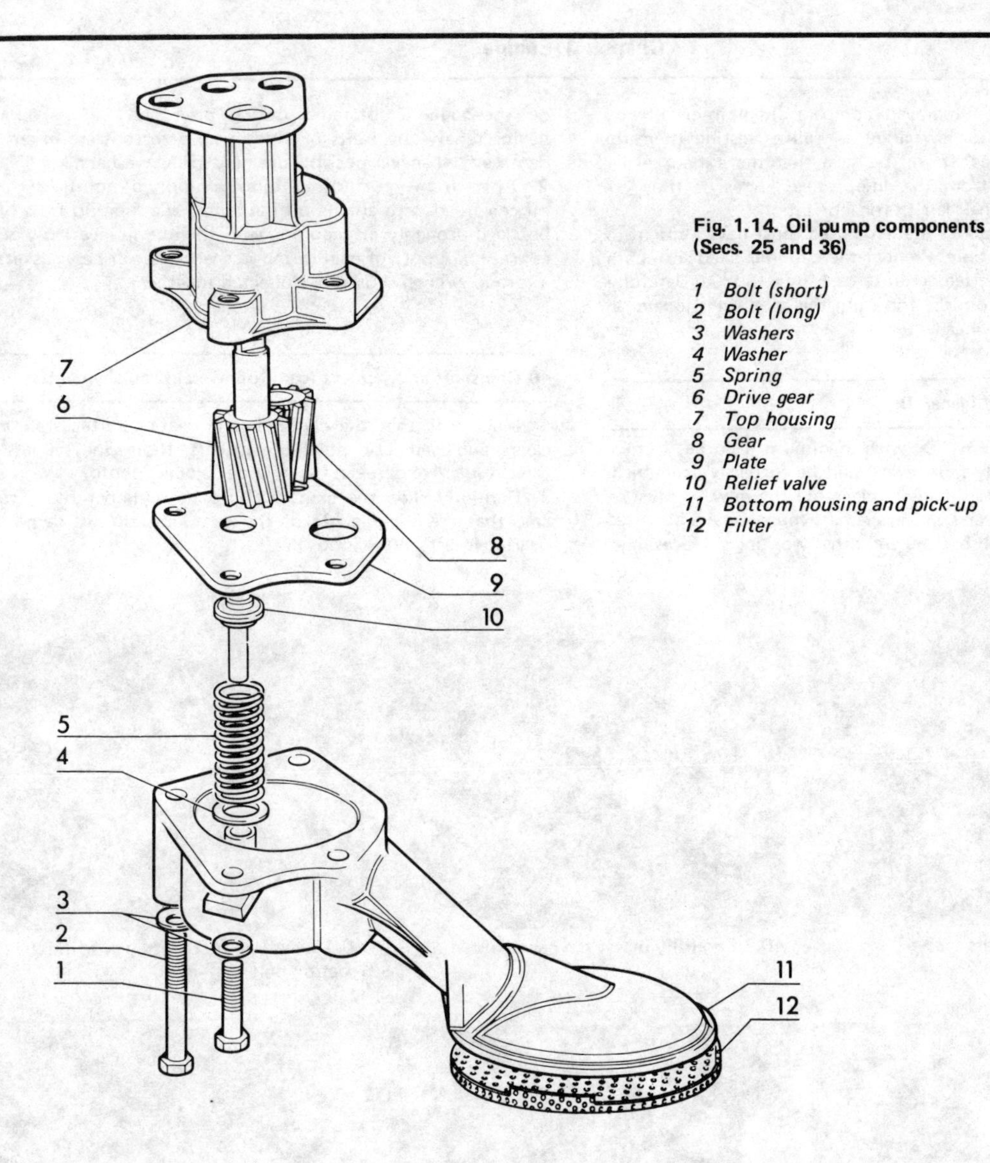

Fig. 1.14. Oil pump components (Secs. 25 and 36)

1 Bolt (short)
2 Bolt (long)
3 Washers
4 Washer
5 Spring
6 Drive gear
7 Top housing
8 Gear
9 Plate
10 Relief valve
11 Bottom housing and pick-up
12 Filter

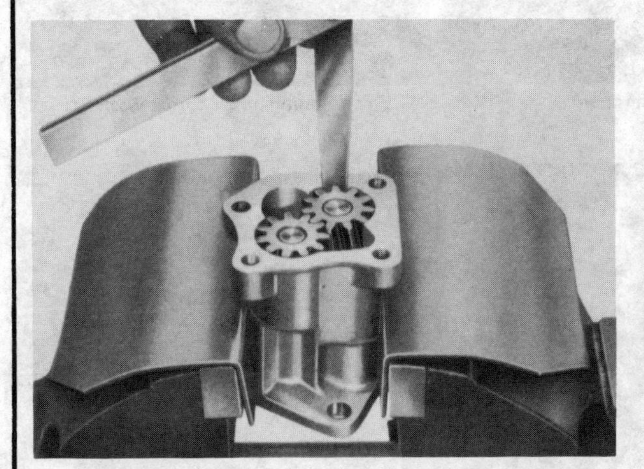

Fig. 1.15. Checking oil pump gears side clearance (Sec. 36)
Clearance should be 0.0019 to 0.0055 in (0.05 to 0.14 mm)

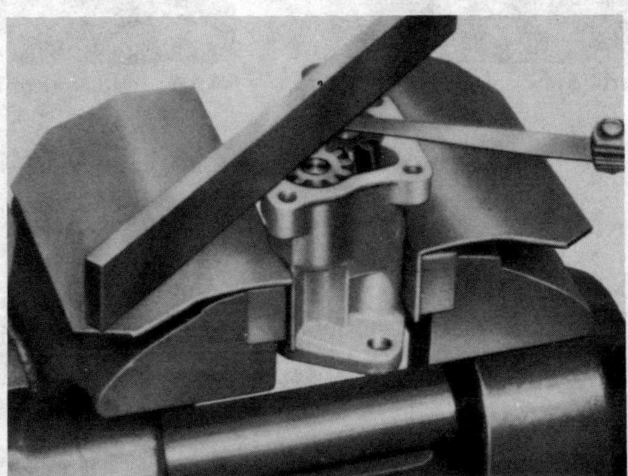

Fig. 1.16. Checking clearance between oil pump gears and cover (Sec. 36)
Clearance should be 0.0008 to 0.0041 in (0.020 to 0.105 mm)

valve guides have been removed compare them internally by visual inspection with a new guide as well as testing them for rocking with the valves. It will be seen that the exhaust valve guides are different from the inlet valve guides as they are threaded throughout their length for lubrication.

When fitting new guides the drift used must have a spigot to locate the new guide bore. Preheat the cylinder head in a bath or oven to a maximum temperature of 176°F (80°C). Carefully drive new guides in from the top until the snap rings locate on the face of the cylinder head.

39 Engine - reassembly (general)

1 To ensure maximum life with minimum trouble from a rebuilt engine, not only must every part be correctly assembled, but everything must be spotlessly clean, all the oilways must be clear, locking washers and spring washers must always be fitted where indicated and all bearing and other working surfaces must be thoroughly lubricated during assembly. Before assembly begins renew any bolts or studs whose threads are in any way damaged; whenever possible use new spring washers.

2 Apart from your normal tools, a supply of non-fluffy rag, an oil can filled with engine oil (an empty washing up fluid plastic bottle thoroughly cleaned and washed out will invariably do just as well), a supply of new spring washers, a set of new gaskets and a torque wrench should be collected together.

40 Camshaft and tappets (cam followers) - replacement

1 Make sure that the camshaft and bearing surfaces are really clean, and well lubricate the camshaft. Insert the tappets into their respective bores in the cylinder block. (photo)

2 Carefully slide the camshaft into the cylinder block taking care that the sharp edges of the cam lobes do not damage the bearing insert finish. (photo)

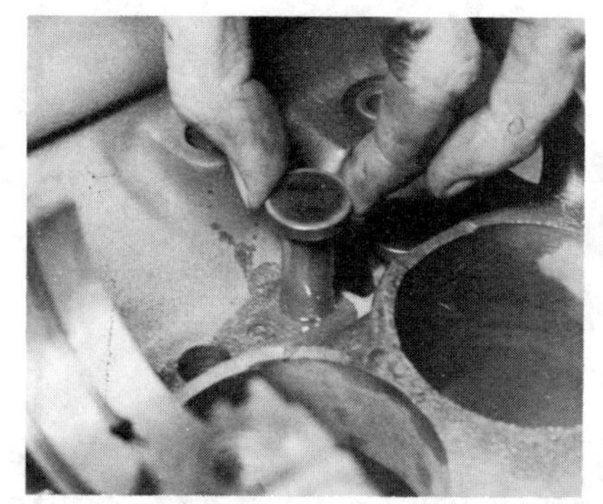

40.1 Inserting the tappets

40.2 Carefully insert the camshaft

40.4 Locating the front bush with its locating bolt

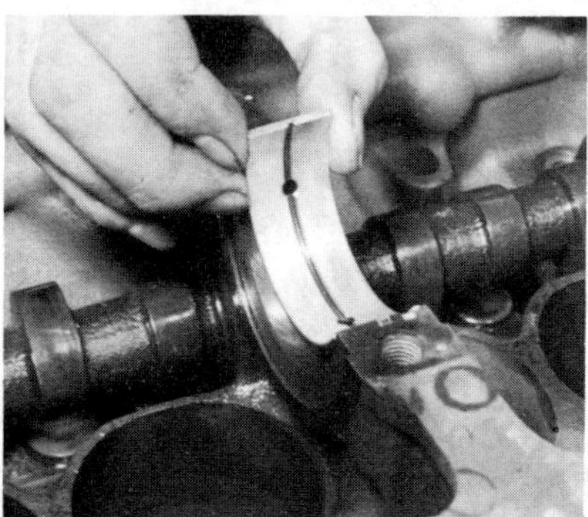

41.1A Fitting main bearing shells ...

41.1B ... and lubricating them

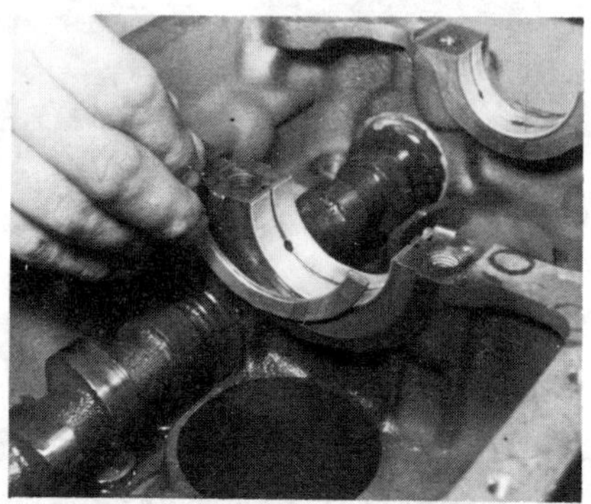

41.6 Positioning upper thrust washers

41.7 Lowering crankshaft into position

41.9 Fitting main bearing caps

41.13 Tightening the main bearing bolts

3 Lubricate the timing gear end bushing and then line up the bushing hole with the hole in the crankcase. Carefully push the bushing fully home.
4 Fit the front bushing dowel bolt and spring washer and tighten securely. This will lock the bushing in position. (photo)
5 Turn the camshaft several times to make sure that it rotates freely in its bushings.

41 Crankshaft - replacement

Ensure that the crankcase is thoroughly clean and that all oilways are clear. A thin twist drill is useful for cleaning them out. If possible blow them out with compressed air. Treat the crankshaft in the same fashion and then inject engine oil into the crankshaft oilways.

Commence work on rebuilding the engine by replacing the crankshaft and main bearings.
1 If the old main bearing shells are to be re-used (a false economy to do so, unless they are virtually new), fit the upper halves of the main bearing shells to their location in the crankcase, after wiping the location clean. (photos)
2 Note that on the back of each bearing is a tab which engages in locating grooves in either the crankcase or the main bearing cap housings.
3 If new bearings are being fitted, carefully wipe away all traces of protective grease with which they are coated.
4 With the three upper bearing shells securely in place, wipe the lower bearing cap housings and fit the three lower shell bearings to their caps ensuring that the right shell goes into the right cap if the old bearings are being refitted.
5 Wipe the recesses either side of the centre main bearing which locate the upper halves of the thrust washers.
6 Smear some grease onto the plain sides of the upper halves of the thrust washers and carefully place them in their recesses (photo).
7 Generously lubricate the crankshaft journals and the upper and lower main bearing shells and carefully lower the crankshaft into position. Make sure that it is the right way round (photo).
8 Lubricate the crankshaft journals, injecting oil into the oilways to ensure adequate lubrication upon the initial start of the engine.
9 Fit the main bearing caps into position ensuring that they locate properly. The mating surfaces must be spotlessly clean or the caps will not seat correctly (photo).
10 When replacing the centre main bearing cap ensure that the thrust washers, generously lubricated, are fitted with their oil grooves facing outwards and the locating tab of each washer in the slot in the bearing cap.
11 Replace the main bearing cap bolts and screw them up finger-tight.
12 Test the crankshaft for freedom of rotation. Should it be very stiff to turn, or possess high spots, a most careful inspection must be made, preferably by a skilled mechanic with a micrometer to trace the cause of the trouble. It is very seldom that any trouble of this nature will be experienced when fitting the crankshaft.
13 Tighten the main bearing bolts to a torque wrench setting of 51 ft lb (7 kg m) and recheck the crankshaft for freedom of rotation (photo).
14 Using feeler gauges check the crankshaft endfloat which should not exceed 0.0024 to 0.0102 inch (0.06 to 0.26 mm). Oversize thrust washers are available.

42 Piston rings - replacement

1 Check that the piston ring grooves and oilways are thoroughly clean and unblocked. Piston rings must always be fitted over the head of the piston and never from the bottom.
2 The easiest method to use when fitting rings is to wrap a 0.020 inch (0.50 mm) feeler gauge round the top of the piston and place the rings one at a time, starting from the bottom oil control ring, over the feeler gauge.
3 The feeler gauge, complete with ring can then be slid down the piston over the other piston ring grooves until the groove is reached. The piston ring is then slid gently off the feeler gauge into the groove. Set all ring gaps 120° to each other.
4 An alternative method to fit the rings is by holding them slightly open with the thumb and both index fingers. This method requires a steady hand and great care as it is easy to open the ring too much and break it.
5 The two top rings are suitably marked to ensure that they are not fitted the wrong way round. The lettering should be to the top of the piston.

43 Pistons and connecting rods - reassembly

If the same pistons are being used, then they must be mated to the same connecting rod with the same gudgeon pin. If new pistons are being fitted it does not matter with which connecting rod they are used, but all pistons should be of the same weight within 0.9 ozs (2.5 g). Weight can be reduced by removing up to 0.2 in (5 mm) of metal from the lower surfaces of the gudgeon pin bosses.

Because the gudgeon pin is a tight fit in the connecting rod it will be necessary to heat the rod to 240°C before inserting the pin, using a special tool to draw it into position. Alternatively, it is possible to heat the rod and carefully drive the gudgeon pin into position, provided that it is not too tight a fit.
1 Using the tool, first remove the knurled nut and slide off the sleeve. Fit the gudgeon pin over the tool.
2 Next slide the spacer onto the tool and secure with the knurled nut.
3 Secure the connecting rod in the vice. Place the piston the correct way round on the connecting rod.
4 Carefully drift the gudgeon pin into the little end until it is central in the connecting rod.
5 The second method of fitting the gudgeon pin is to heat the piston in an oil bath and then place it on its side on soft wood blocks. Fit the connecting rod into the piston making sure it is the correct way round. The number stamped on the connecting rod must face towards the side away from the offset. Refer to Fig. 1.18 for more details.
6 Using a soft metal drift, carefully drive the gudgeon pin in until it is central in the connecting rod.
7 Whichever method is used make sure that the piston is free to move axially on the connecting rod. A little stiffness is permissible when new parts have been fitted.

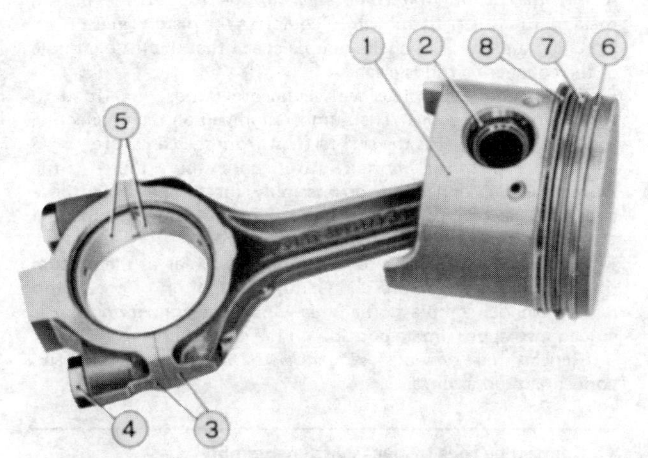

Fig. 1.17. Piston and connecting rod assembly

1 Piston
2 Piston pin
3 Connecting rod shank and cap
4 Connecting rod cap screw
5 Bearing
6 Compression ring
7 Oil ring
8 Slotted oil ring with expander

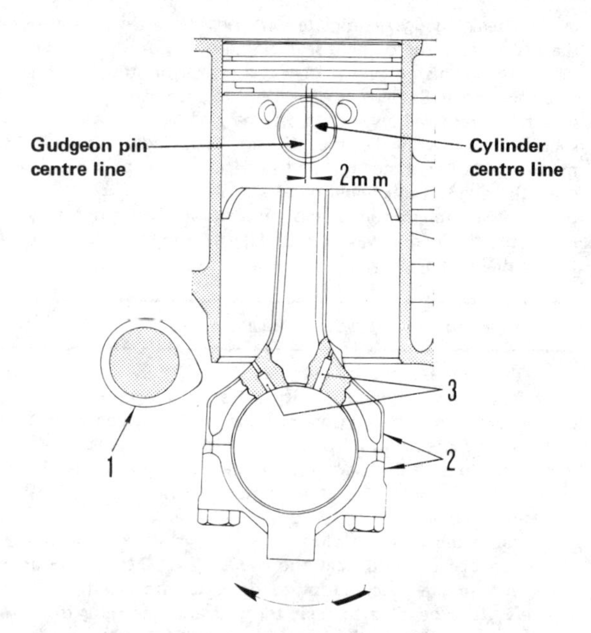

Fig. 1.18. Correctly fitted piston and connecting rod assembly (Sec. 43)

1 Camshaft
2 Location of connecting rod matching number to cylinder
3 Oil squirt holes
Arrow shows direction of rotation of front viewed engine

Note: The gudgeon pin is 0.08 in (2 mm) offset towards camshaft. When assembling piston to connecting rod small-end, make sure that the number stamped on connecting rod (2) faces toward side away from piston bore offset.

44 Pistons - replacement

1 Lay the piston and connecting rod assemblies in the correct order ready for refitting into their respective bores.
2 With a wad of clean non-fluffy rag wipe the cylinder bores clean.
3 Position the piston rings so that their gaps are 120° apart and then lubricate the rings.
4 Fit the piston ring compressor to the top of the piston, making sure it is tight enough to compress the piston rings.
5 Using a piece of fine wire double check that the little jet hole in the connecting rod is clean.
6 The pistons, complete with connecting rods, are fitted to their bores from above. The number stamped on the connecting rod must face towards the side farthest from the camshaft.
7 As each piston is inserted into its bore ensure that it is the correct piston/connecting rod assembly for the particular bore and that the connecting rod is the right way round. Lubricate the piston well with clean engine oil. (photo)
8 The piston will slide into the bore only as far as the bottom of the piston ring compressor. Gently tap the top of the piston with a wooden or plastic hammer whilst the connecting rod is guided into approximate position on the crankshaft.
9 Repeat the previous sequence for all four piston and connecting rod assemblies.

45 Connecting rods to crankshaft - reassembly

1 Wipe clean the connecting rod half of the big-end bearing and the underside of the shell bearing. Fit the shell bearing in position with its locating tongue engaged with the corresponding groove in the connecting rod.
2 If the old bearings are nearly new and are being refitted then ensure they are replaced in their correct locations in the correct rods.
3 Generously lubricate the crankpin journals with engine oil, and turn the crankshaft so that the crankpin is in the most advantageous position for the connecting rod to be drawn into it.
4 Wipe clean the connecting rod bearing cap and back of the shell bearing and fit the shell bearing in position ensuring that the locating tongue at the back of the bearing engages with the locating groove in the connecting rod cap.
5 Generously lubricate the shell bearing and offer up the connecting rod bearing cap to the connecting rod. (photo)
6 Fit the connecting rod bolts and tighten in a progressive manner to a final torque wrench setting of 32.5 lb f ft (4.5 kg fm). (photo)

46 Timing sprockets, chain and cover - replacement

1 Wipe the nose of the crankshaft and refit the Woodruff key. Make sure it is seating fully, and parallel with the crankshaft nose.
2 Note the alignment mark on the crankshaft sprocket. This mark will be required in subsequent operations.
3 Using a tubular drift carefully tap the crankshaft sprocket into position on the nose of the crankshaft.
4 Turn the crankshaft until the mark is pointing towards the camshaft.
5 Turn the camshaft until, when the camshaft sprocket is fitted on the dowel of the camshaft, the alignment mark in the form of a dot lines up with the crankshaft sprocket mark.
6 Fit the timing chain to the crankshaft sprocket and insert the camshaft into the timing chain, selecting the right position to ensure the alignment marks still are in alignment (photo). Ensure the self-tensioning links are inside, adjacent to the cylinder block. (photos)
7 Place the camshaft sprocket onto the camshaft.
8 Secure the camshaft sprocket by fitting the special cam, that drives the fuel pump, on its locating dowel. Fit the camshaft retaining bolt. (photo)
9 Tighten the camshaft securing bolt as firmly as possible. The bolt should be tightened to a torque wrench setting of 36 lb f ft (5 kg fm).
10 If the timing cover oil seal showed signs of leaking before engine overhaul the old seal should be removed and a new one fitted.
11 Using a screwdriver, carefully remove the old oil seal, working from the rear of the cover.
12 Fit the new seal making sure it is inserted squarely, and tap home with a hammer.
13 Lubricate the oil seal with engine oil.
14 With all traces of old gasket and jointing compound removed from the timing cover and cylinder block mating faces, smear a little grease onto the timing cover mating face and fit a new gasket in position.
15 Fit the timing cover to the cylinder block and lightly tighten the seven securing bolts, and spring washer. Ensure that the fuel pump pushrod bush is in place in the cover. (photo)
16 Wipe the hub of the pulley and carefully place into position on the crankshaft nose. It should locate on remaining portion of the Woodruff key. It may be necessary to adjust the position of the timing cover slightly in order to centralise the oil seal relative to the pulley hub. (photo)
17 Tighten the timing cover securing bolts in a diagonal and progressive manner.
18 Refit the hub securing nut and tighten fully. (photo)
19 Replace the engine support and secure with two bolts and washers. Note the lifting eye that fits under one of the bolts. (photo)

47 Oil pump - refitting

1 Refit the oil return pipe to the main bearing at the flywheel

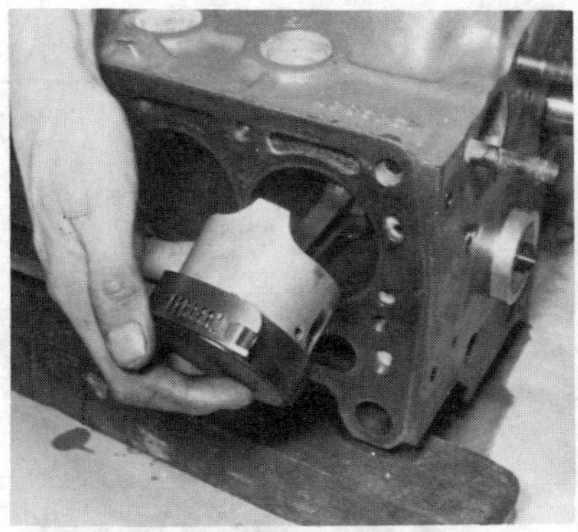

44.7 Easing the piston in with the rings compressed

45.5 Fitting the connecting rod bearing cap ...

45.6 ... and torquing up

46.6A Note the self-tensioning links on the inside of the chain

46.6B Fitting the sprockets and timing chain, so that when in position ...

46.6C ... the timing marks line up like this

46.8 Fitting the fuel pump drive cam and the securing bolt

46.15 Positioning the timing cover

46.16 Locating the crankshaft pulley on the Woodruff key

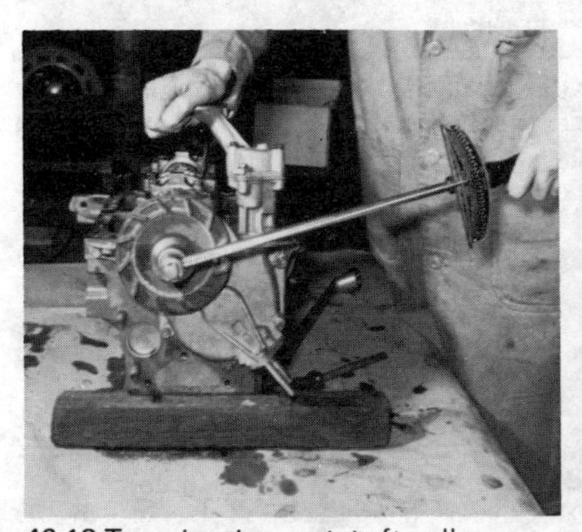

46.18 Torquing the crankshaft pulley securing nut

46.19 Replacing the engine support bracket and lifting eye

47.1 Fitting the oil return pipe

47.2 Inserting the oil pump driveshaft

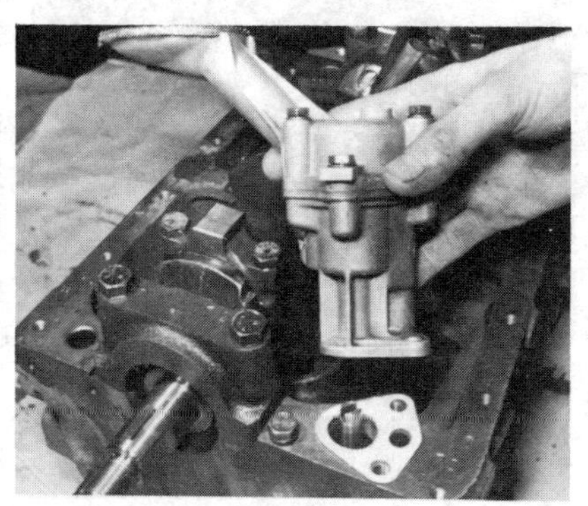
47.4 Engaging the pump with the driveshaft

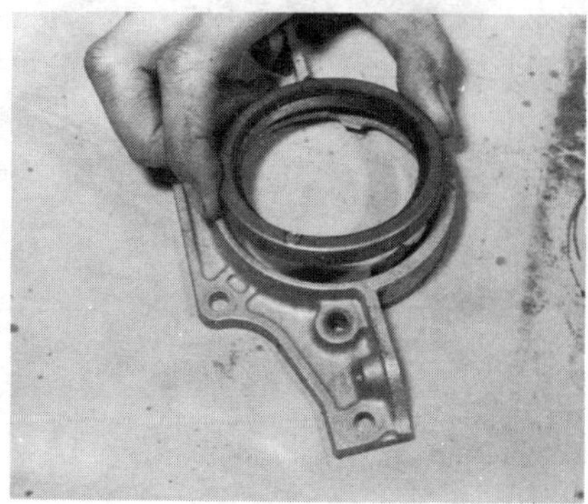
48.3 Fitting new oil seal in the carrier

end: the flange locates over the oil hole in the top of the bearing and is secured by a bolt and washer. It is further secured, by a bolt and bracket, to the centre bearing cap. (photo)

2 Insert the pump driveshaft, splined end first, so that it mates with the camshaft drive. (photo)

3 Fit a new oil pump to crankcase gasket.

4 Mate the oil pump to the driveshaft, ensuring that the slot in the driveshaft engages with the dog on the pump shaft. (photo)

5 Secure the pump with its two bolts and washers.

48 Crankshaft rear oil seal and carrier - refitting

1 If the crankshaft rear oil seal showed signs of leaking before engine overhaul the oil seal should be renewed.

2 Using a screwdriver carefully remove the old oil seal.

3 Position the new seal squarely on the carrier and with a hammer tap it fully home (photo).

4 Wipe the end of the crankshaft and lubricate the oil seal. Smear the carrier mating face with a little grease and place a new gasket in position.

5 Very carefully ease the oil seal over the end of the crankshaft and secure with the bolts and spring washers. (photo)

49 Sump - refitting

1 Place a new cork gasket between the jaws of a vice and compress the gasket

2 Carefully ease the cork gasket into the locating flange at one end of the sump pressing

3 Using a pair of side cutters or a sharp knife, trim excess cork gasket from each end until there is about 1/8 inch (3.2 mm) protruding at each end. Whilst this is being done check that the cork gasket is still seating fully, otherwise it will be cut too short

4 Make sure that the mating faces of the crankcase and sump are free of old gasket and jointing compound.

5 Smear a little grease onto the crankcase mating face and place the two halves of the sump gasket in position (photo).

6 Carefully place the sump on the new gasket (photo).

7 Secure the sump in position with the bolts/nuts and plain washers. These should be tightened in a progressive and diagonal manner (photo).

50 Flywheel - refitting

1 Refit the top half of the metal endplate assembly. Position it over the locating dowel/bushes on the end of the crankcase. Then position the lower half of the endplate assembly. (photo)

2 Turn the crankshaft until it is in the TDC position for number 1 and 4 cylinders. Place the flywheel in position on the end of the crankshaft. The bolt holes are offset, so it can only go on one way.

3 Fit the securing bolt thrust plate to the flywheel and then the six securing bolts. (photo)

4 Place a block of wood between the crankshaft and the crankcase, to stop the crankshaft rotating. Tighten the six bolts in a progressive and diagonal manner to a torque wrench setting of 36 kb f ft (5 kg fm).

5 Replace the clutch assembly to the flywheel, as described in Chapter 5. (photo)

51 Valves and valve springs - reassembly

To refit the valves and valve springs to the cylinder head, proceed as follows:

1 Rest the cylinder head on its side and fit each valve, spring lower seating, inner spring and outer spring, in turn, wiping down and lubricating each valve stem as it is inserted into the same valve guide from which it was removed. (photos)

2 Fit the valve spring cup to each valve spring. (photo)

3 With the base of the valve spring compressor on the valve head, compress the valve spring until the two cotters can be slipped into place in the cotter grooves. (photos)

4 Gently release the compressor. Repeat this procedure until all eight valves and valve springs are fitted.

52 Rocker shaft - reassembly

1 Fit the circlip to one end of the rocker shaft and slide on the rockers, pedestals and spacer springs in the order shown in Fig. 1.7.

2 When all are in place secure the components on the rocker shaft with the second circlip.

3 Lubricate the rockers to ensure no damage upon initial starting of the engine.

53 Cylinder head - replacement

After checking that both the cylinder block and cylinder head mating surfaces are perfectly clean, generously lubricate each cylinder with engine oil.

1 Always use a new cylinder head gasket as the old gasket will be compressed and not capable of giving a good seal. (photo)

2 Never smear grease on the gasket as, when the engine heats up, the grease will melt and may allow compression leaks to develop.

3 The cylinder head gasket cannot be fitted incorrectly due to its asymmetrical shape, but the word 'ALTO' should be uppermost in any event.

4 The locating dowels should be refitted to the front right and left-hand side cylinder head securing bolt holes. (photo)

5 Carefully fit the cylinder head gasket to the top of the cylinder block.

6 Lower the cylinder head onto the gasket, taking care not to move the position of the gasket. (photo)

48.5 Positioning the oil seal and carrier over the crankshaft

49.5 One of the gasket halves in position with a small blob of sealant in the corner to ensure a good tight fit

49.6 Lowering sump into position: note the cork gasket slightly protruding from its housing

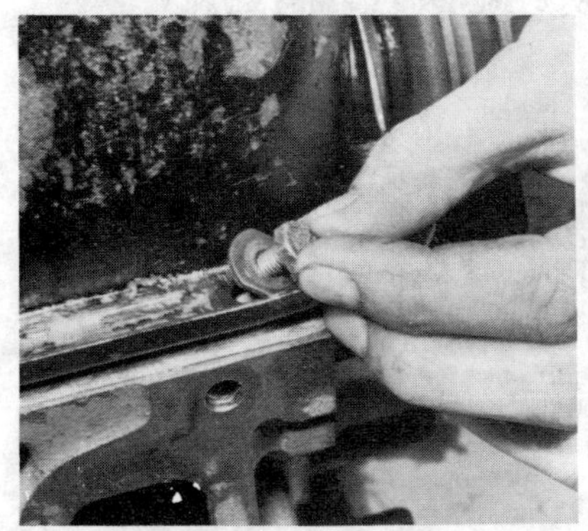
49.7 Securing the sump

50.1 Locating the endplate assembly on the dowels

50.3 Securing the flywheel with thrust plate and bolts

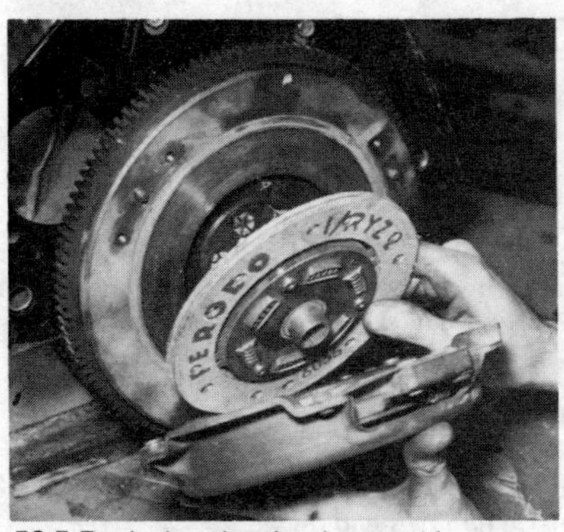
50.5 Replacing the clutch assembly to the flywheel

51.1A Inserting the valve ...

51.1B ... lower spring seating ...

51.1C ... inner and outer springs ...

51.2 ... and then the top spring cup

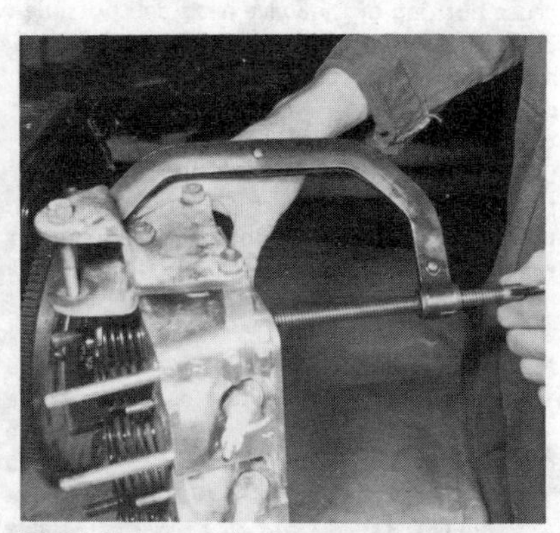
51.3A Using a spring compressor ...

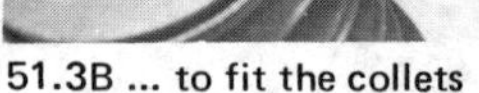
51.3B ... to fit the collets

53.1 Positioning new head gasket

53.4 One of the locating dowels

7 Fit the cylinder head securing bolts and tighten all finger-tight. One bolt is located within the inlet manifold. Refit the water temperature sending unit and note that part of the water pipe cooling system, the copper pipe portion, is secured to the cylinder head bolts adjacent to the sending unit. (photos)
8 Tighten the cylinder head securing bolts in a progressive manner in the order shown in Fig. 1.19 to the specified final torque wrench setting.

54 Pushrods and rocker shaft - refitting

1 With the cylinder head in position, fit the pushrods in the same order in which they were removed. Ensure that they locate properly in the stems of the tappets and lubricate the pushrod ends before fitment. (photo)
2 Fit the rocker shaft over the four studs in the cylinder head and lower onto the cylinder head. Make sure the ball ends of the rockers locate in the cups of the pushrods. (photo)
3 Fit the four nuts and washers to the rocker shaft pedestal studs and tighten in a progressive manner to the torque wrench setting given in the Specifications.

55 Rocker arm/valve clearance - adjustment

1 The valve adjustments should be made with the engine cold. The importance of correct rocker arm/valve stem clearances cannot be overstressed as they vitally affect the performances of the engine.
2 If the clearances are set too open, the efficiency of the engine is reduced as the valves open later and close earlier than was intended. If, on the other hand the clearances are set too close there is a danger that the stem and pushrods will expand upon heating and not allow the valves to close properly which will cause burning of the valve head and possible warping.
3 If the engine is in the car, to gain access to the rockers undo and remove the four nuts, spring washers and metal packing pieces (photo).
4 Carefully lift away the rocker cover.
5 It is important that the clearance is set when the tappet of the valve being adjusted is on the heel of the cam (ie; opposite the peak). This can be done by carrying out the adjustments in the following order, which also avoids turning the crankshaft more than necessary:

Valve fully open	Check and adjust
Valve No. 8	Valve No. 1*
Valve No. 6	Valve No. 3
Valve No. 4	Valve No. 5
Valve No. 7	Valve No. 2
Valve No. 1*	Valve No. 8
Valve No. 3	Valve No. 6
Valve No. 5	Valve No. 4
Valve No. 2	Valve No. 7

** Timing chain end*

6 The correct valve clearance is given in technical data at the beginning of this Chapter. It is obtained by slackening the hexagonal locknut with a spanner while holding the ball pin against rotation with a spanner as shown in this photograph. Then still pressing down with the ring spanner, insert a feeler gauge of the required thickness between the valve stem and the rocker arm and adjust the ball pin until the feeler gauge will just move in and out without nipping. Then, still holding the ball pin in the correct position, tighten the locknut. Re-check the clearance after tightening the locknut.
7 An alternative method is to set the gaps with the engine running at idle speed. Although this method may be quicker more practice is needed and it is no more reliable.

56 Engine - final assembly

1 Make sure the mating faces of the cylinder head and thermostat lower housing are clean of old gasket and jointing compound and slide a new gasket over the two studs.
2 Place the thermostat lower housing in position on the cylinder head.
3 Fit the thermostat to the lower housing so that the frame is situated between the two studs.
4 Make sure the mating faces of the thermostat upper and lower housing are free of old gasket and jointing compound and fit a new gasket.
5 Fit the thermostat upper housing, and secure in position with the two nuts and one bolt, all with spring washers. (photo)
6 Fit a new oil filter: lightly rub the rubber sealing ring, on the face of the filter, with engine oil. Ensure the mating face on the cylinder block is clean. Tighten by hand only.
7 Make sure the mating faces of the cylinder head and carburettor spacer are free of old gasket and jointing compound. Fit a new gasket and place the carburettor spacer on the cylinder head. (photo)
8 Secure the carburettor spacer with the four nuts and spring washers. Refit the drip pipe to the spacer. spacer.
9 Fit a new gasket to the top of the carburettor mounting.
10 Next place the drip tray on the carburettor mounting. (photo)
11 Place a second new gasket on the drip tray and position the carburettor on the mounting. (photo)
12 Secure the carburettor with the two nuts and spring washers and tighten fully.
13 Slide the petrol feed pipe onto the carburettor union pipe marked with an arrow.
14 Fit a new gasket to the petrol pump mounting boss on the side of the timing cover and slide in the spacer.
15 Insert the pump actuating pushrod into the centre hole of the spacer.
16 Fit a second new gasket to the petrol pump spacer and place the pump over the gasket. It is most important that the size of these gaskets is correct and the correct method of selecting them is detailed in Chapter 3.

53.6 Lowering the head

53.7A Fitting the head bolts. Note the temperature sending unit and the cooling pipe locating brackets under the heads of two bolts either side of the sender unit

53.7B Note the one bolt in the inlet manifold

53.8 Torquing up the head bolts

Fig. 1.19. Correct sequence for tightening-up cylinder head bolts (Secs. 11 and 53)

54.1 Inserting the pushrods ...

54.2 ... and positioning the rocker shaft

55.3 Rocker cover retaining washer and packing piece

55.6 Adjusting a valve clearance

56.5 New gasket and thermostat upper housing being fitted

56.7 Fit a new gasket, then the carburettor spacer

56.10 Assembling the drip tray and another gasket ...

56.11 ... and then the carburettor

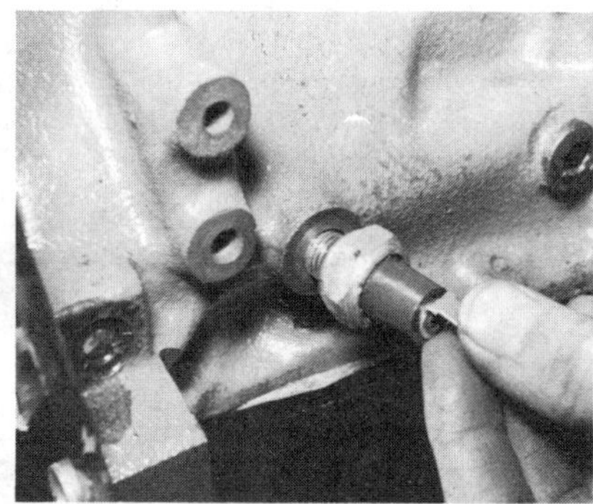
56.18 Fitting the oil pressure sender

56.21A Replacing the exhaust manifold ...

17 Secure the pump with the two nuts and spring washers.
18 Fit the oil pressure warning light switch into its location in the oil gallery. (photo)
19 Fit the petrol feed pipe from the pump to the carburettor at the pump union pipe.
20 Fit a new exhaust manifold gasket to the side of the cylinder head.
21 Refit the exhaust manifold and hot air ducting and secure in position with the five nuts and spring washers. (photos)
22 Refit the distributor but do not tighten the clamp until the timing has been reset as described in Chapter 4. (photo)
23 Fit the water temperature indicator thermal transmitter and union, using new sealing washers on either side of the union. Tighten to a torque wrench setting of 28.9 lb f ft (4 kg fm).
24 Refit the engine to the transmission if the transmission was removed in unit with the engine. Full details will be found in Chapter 6.

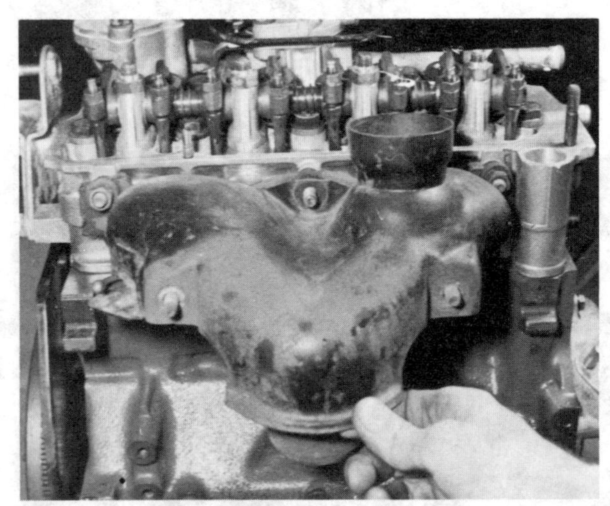
56.21B ... and the hot air ducting

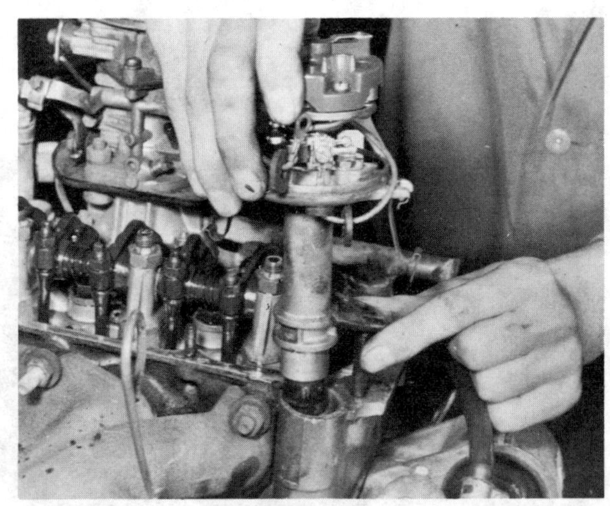
56.22 Inserting the distributor

57 Preparation for refitting the engine

1 With the engine reassembled on the bench, it must have the transmission fitted so that it is ready as a power-unit to go back in the car.
2 The clutch must have been centralised as described in Chapter 5. Wipe clean and dry the splines on the end of the transmission input shaft, and in the clutch friction plate. Lubricant would, in time, get sticky, and give clutch drag.
3 Offer up the transmission to the engine, holding it square. Slide the stud on the rear into the tubular dowel, and the shaft through the clutch. An assistant can help by turning the engine over slightly by a spanner on the crankshaft pulley nut, to allow the splines to line up and engage.
4 As the transmission goes into place, keep the weight supported, so it does not hang on the clutch.
5 Fit the bolts and the nut holding the engine and transmission together. Tighten the top one first so that it takes the weight of the transmission.
6 Fit the shield to the bottom of the flywheel housing, and then the bearer bracket that will mate with the power plant support bracket when the engine is back in the car.
7 Fit the starter motor.

58 Refitting the engine to the car

1 With the engine and transmission joined, move them into position under the car. The support bracket should already be on, so that the power unit can be bolted into place as soon as it has been lifted. Connect up the lifting tackle. If this is not very manoeuverable, make sure the power unit is precisely below its proper position, and the tackle lifting vertically.
2 Lift the engine straight up until the hole in the right bearer is lined up. As the engine comes up check nothing is fouling.
3 Fit the right bearer bolt. To get the hole lined up accurately put a screwdriver through, and use this to pull it straight. The bolt can then push out the screwdriver.
4 Get the support bracket up into place. A jack under the transmission supplementing the lifting tackle will help to control the movement accurately.
5 Fit the bolts to secure the support bracket to the car. Put all four in loosely first, then tighten them all, then lock them with the tab washer.
6 Reconnect the steady rod from the bulkhead to the cylinder head.
7 Remove the lifting tackle.
8 Connect the exhaust pipe to the manifold. Use some hard setting exhaust compound.
9 Under the car, refit the exhaust bracket to the transmission, reconnect the gear linkage, and the earthing strip.
10 On top reconnect the throttle linkage and choke cable. Adjust the latter so it pushes the choke fully open.
11 Refit the water pump and connect up the cooling system and heater hoses.
12 Reconnect the clutch cable, and adjust it to give 1 inch (12 mm) free-play at the pedal, ensuring the grommet at the end of the outer cable is fitted properly into its seat.
13 Reconnect the speedometer cable.
14 Connect the heavy cable to the starter, and its red solenoid wire.
15 Reconnect the wires to the water temperature and oil pressure senders.
16 Refit the generator and connect the cable leads to it.
17 Refit the fuel pipe to the mechanical pump, and the main feed line and re-circulatory pipe to the carburettor.
18 Refit and reconnect the front driveshafts, as described in Chapter 7.
19 Refit the wheels, and lower the car to the ground.
20 Refit the water hoses, and the bonnet and radiator.
21 Fill the cooling system with water.
22 Fill the engine sump and the transmission with oil.

59 Engine - initial start-up after overhaul or major repair

1 Make sure that the battery is fully charged and that all lubricants, coolant and fuel are replenished.
2 If the fuel system has been dismantled it will require several revolutions of the engine on the starter motor to pump the petrol up to the carburettor. An initial 'prime' of about 1/3 of a cupful of petrol poured down the air intake of the carburettor will help the engine to fire quickly, thus relieving the load on the battery. Do not overdo this, however, as flooding may result.
3 As soon as the engine fires and runs, keep it going at a fast tickover only (no faster) and bring it up to normal working temperature.
4 As the engine warms up there will be odd smells and some smoke from parts getting hot and burning off oil deposits. The signs to look for are leaks of water or oil which will be obvious, if serious. Check also the exhaust pipe and manifold connections as these do not always 'find' their exact gas tight position until the warmth and vibration have acted on them and it is almost certain that they will need tightening further. This should be done, of course, with the engine stopped.
5 When normal running temperature has been reached, adjust the engine idle speed as described in Chapter 3.
6 Stop the engine and wait a few minutes to see if any lubricant or coolant is dripping out when the engine is stationary.
7 Road test the car to check that the timing is correct and that the engine is giving the necessary smoothness and power. Do not race the engine – if new bearings and/or pistons have been fitted it should be treated as a new engine and run in at a reduced speed for the first 300 miles (500 km).

For 'Fault diagnosis - engine' see next page.

60 Fault diagnosis - engine

When investigating starting and uneven running faults do not be tempted into a snap diagnosis. Start from the beginning of the check procedure and follow it through. It will take less time in the long run. Poor performance from an engine in terms of power and economy is not normally diagnosed quickly. In any event the ignition and fuel systems must be checked first before assuming any further investigation needs to be made.

Symptom	Reason/s	Remedy
Engine will not turn over when starter switch is operated	Flat battery	Check that battery is fully charged and that all connections are clean and tight.
	Bad connections at solenoid switch and/or starter motor	
	Defective solenoid	Remove starter to repair solenoid.
	Starter motor defective	Remove and overhaul starter motor.
Engine turns over normally but fails to fire and run	No spark at plugs	Check ignition system according to procedures given in Chapter 4.
	No fuel reaching engine	Check fuel system according to procedures given in Chapter 3.
	Too much fuel reaching the engine (flooding)	Check the fuel system.
Engine starts but runs unevenly and misfires	Ignition and/or fuel system faults	Check the ignition and fuel systems as though the engine had failed to start.
	Incorrect valve clearances	Check and reset clearances.
	Burnt out valves Blown cylinder head gasket	Remove cylinder head and examine and overhaul as necessary.
	Worn out piston rings Worn cylinder bores	Remove cylinder head and examine pistons and cylinder bores. Overhaul as necessary.
Lack of power	Ignition and/or fuel system faults	Check the ignition and fuel systems for correct ignition timing and carburettor settings.
	Incorrect valve clearances	Check and reset the clearances.
	Burnt out valves Blown cylinder head gasket	Remove cylinder head and examine and overhaul as necessary.
	Worn out piston rings Worn cylinder bores	Remove cylinder head and examine pistons and cylinder bores. Overhaul as necessary.
Excessive oil consumption	Oil leaks from crankshaft rear oil seal, cover gasket and oil seal, rocker cover gasket, oil filter gasket, sump gasket, sump plug washer	Identify source of leak and renew seal as appropriate.
	Worn piston rings or cylinder bores resulting in oil being burnt by engine (smoky exhaust is an indication)	Fit new rings or rebore cylinders and fit new pistons, depending on degree of wear.
	Worn valve guides and/or defective valve stem seals	Remove cylinder heads and recondition valve stem bores and valves and seals as necessary.
Excessive mechanical noise from engine	Wrong valve to rocker clearances	Adjust valve clearances.
	Worn crankshaft bearings	Inspect and overhaul where necessary.
	Worn cylinders (piston slap)	Inspect and overhaul where necessary.
	Slack or worn timing chain and sprockets	Renew all timing mechanism.
Unusual vibration	Broken engine/gearbox mounting	Renew mounting.
	Misfiring on one or more cylinders	Check ignition system.
	Main bearing wear	Check and renovate as necessary.

Chapter 2 Cooling, heating and exhaust systems

For modifications, and information applicable to later models, see Supplement at end of manual

Contents

Specifications

Cooling system capacity	8¾ Imp. pints (5 litres)
Operating pressure	7 psi (0.5 kg/cm^2)
Antifreeze proportion	Follow makers instructions
Fan	
Fan power	55 watts
Fan cut in temperature	194°F to 201°F (90°C to 94°C)
Fan switch off temperature	185°F to 192°F (85°C to 89°C)
Thermostat	
Thermostat starts to open	185°F to 192°F (85°C to 89°C)
Travel fully open (212°F, 100°C)	0.29 in. (7.5 mm)
Water pump	
Maximum bearing end play	0.005 in. (0.12 mm)
Clearance impeller/housing	0.031 to 0.047 in. (0.8 to 1.2 mm)
Heater fan power	20 watts

1 General description

1 The engine is cooled by water, and in most respects the cooling system is conventional. The radiator is at the front, and is normally cooled by the car's forward motion through the air. However, since the engine is transverse, the fan cannot be driven in the usual way. Instead, an electric fan is mounted on the radiator. It cuts in automatically when the temperature is sufficiently high, such as in traffic, or climbing mountains. A high temperature warning light gives indication of when the coolant temperature is excessive.

2 The rubber 'V' belt, usually called a fan belt, drives the water pump, and is tensioned at the generator in the usual way. This belt is dealt with under Routine Maintenance and in the generator Section of the electrical Chapter.

3 The fan is controlled by a temperature sensor in the bottom of the radiator. This controls a relay, mounted just below the battery, so that the sensor contacts do not have to carry the full fan current load.

4 The heater is mounted behind the engine on the right. The heater uses the engine coolant, and the temperature is controlled by regulating the flow of the water. The heater cannot be properly drained, so when flushing the system, this must be allowed for.

5 The exhaust is a simple system, running down the centre of the floor, and then across at the rear. There is one rigid bracket to the transmission at the front; at the rear flexible brackets allow the pipe to move in accord with the vibration. A single silencer is fitted at the end of the pipe.

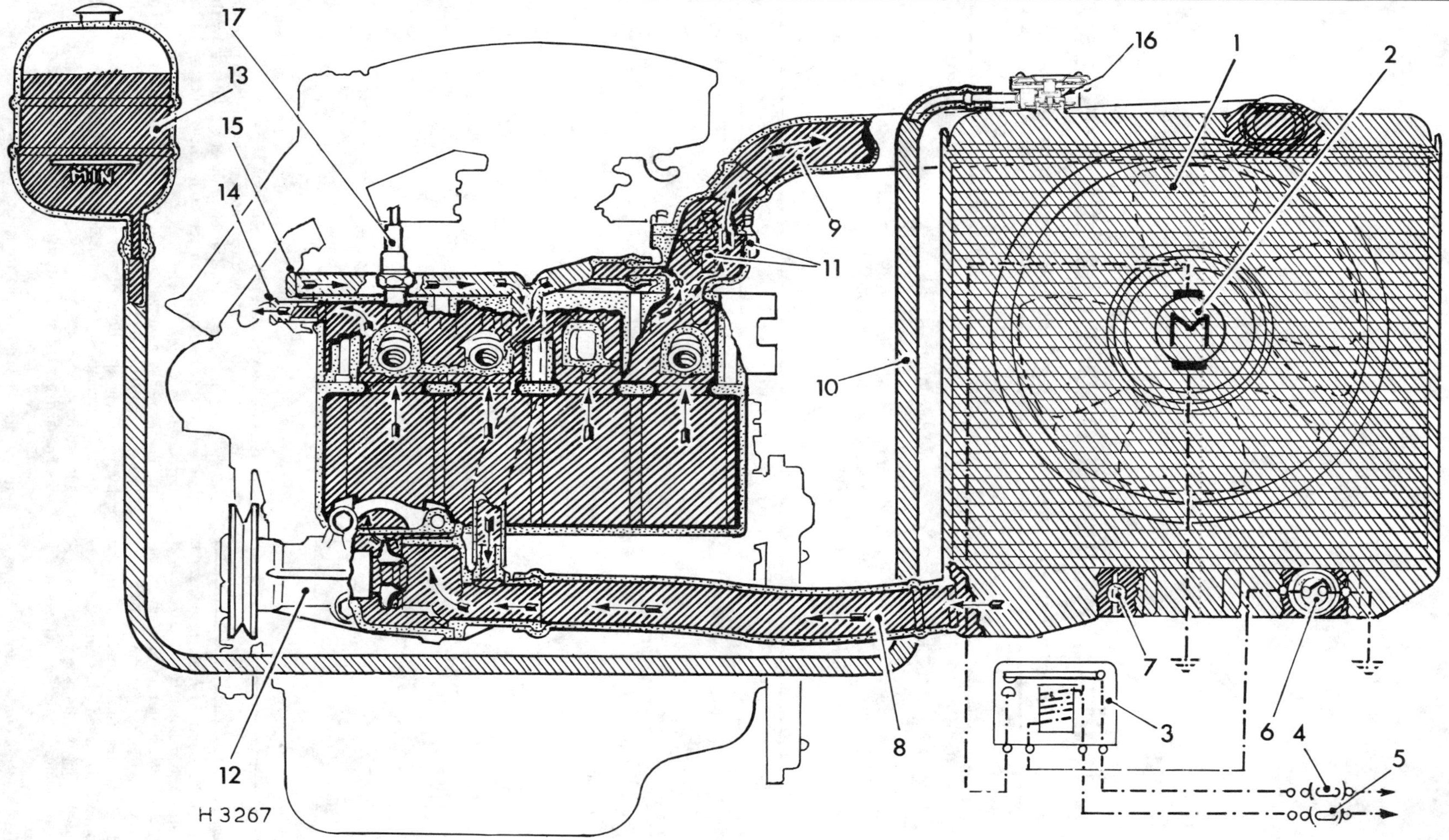

H 3267

Fig. 2.1. Cooling system operational diagram

1 *Radiator*
2 *Electrofan*
3 *Electrofan motor relay*
4 *16-amp fuse for electrofan motor*
5 *8-amp fuse for electrofan motor excitation winding*
6 *Thermal switch*
7 *Drain cock*
8 *Water hose, radiator to pump*
9 *Water hose, engine to radiator*
10 *Line connecting radiator to expansion tank*
11 *Outlet with thermostat*
12 *Water pump*
13 *Expansion tank*
14 *Water hose to heater*
15 *Water hose, heater to engine*
16 *Radiator cap with double acting (pressure and vacuum) vent valve*
17 *Water temperature thermal sending unit*

2 Coolant level check

1 Under normal circumstances the coolant level can be seen at a glance in the expansion tank on the right of the engine compartment. When cold the level will be lower than when hot. The cold level should be maintained about 2.5 ins. (6 cm) above the minimum mark on the expansion tank.
2 If the level does not rise and fall in the expansion tank as the engine heats up and cools, then there is an air leak between the tank and the radiator, so the coolant cannot be drawn back in. This leak can either be in the fixing of the expansion pipe, or the seal of the cap on the radiator.
3 If there is suspicion that the expansion system is not keeping the radiator full at all times, the radiator itself should be checked. Anyway, this is a good idea at intervals. If possible this should be done when the engine is cold. If the engine is hot, and there is a crisis, switch off, and allow the engine at least ten minutes to cool slightly. The radiator cap must not be taken off when the temperature of the coolant is above boiling point. The cap is pressurised, so the coolant will not actually boil though above the normal boiling point. But when the cap is opened the pressure is released, and violent expansion with boiling will give grave risk of scalding. Once it has cooled, open the cap to the partway position, with some rag covering the hand, and allow the pressure to escape gradually. Whenever the cap is taken off, hot or cold, the radiator header tank should be full, and an air space indicates the expansion system is not working.
4 The system should need topping-up very seldom. A need for regular topping-up indicates a leak. Topping-up should be done with the antifreeze/water mix.
5 If the system has been drained, recheck the level after the engine has been warmed up. There is likely to have been a pocket of air, so the level will fall. Also check the heater works, and has not got an air-lock.

3 Antifreeze and inhibitors

1 In cold climates antifreeze is needed for two reasons. In extreme cases if the coolant in the engine freezes solid it could crack the cylinder block or head. But also in cold weather, with the circulation restricted by the thermostat, and what warm water is getting to the radiator being at the top, the bottom of the radiator could freeze, so blocking circulation completely, and thus causing the coolant trapped in the engine to boil.
2 The antifreeze should be mixed in the proportions advocated by the makers according to the climate. The normal proportion in temperate climates is 25% antifreeze by volume, with 33.1/3% for colder. This mix should be used for topping-up too, otherwise the mixture will gradually get weaker.
3 Antifreeze should be left in through the summer because it has an important secondary function; to act as an inhibitor against corrosion. In the cooling system are many different metals that are vulnerable; in particular the aluminium cylinder head. In contact with the coolant this sets up electrolytic corrosion, accentuated by any dirt in the system. Reputable antifreeze of a suitable formula must be used and should obviously include an anti-corrosion inhibitor. Whilst FIAT approve other oils as well as their own, their instructions specify only their own antifreeze.
4 After about two years the effectiveness of the antifreeze's inhibitor is used up. It must then be discarded, and the system flushed, and refilled with new. Mix the new with water, half and half. Pour this in, then finally fill to over-flowing with water.
5 In warm climates free from frost, an inhibitor should be used. Again a reputable make giving full protection must be chosen with renewal every two years. Inhibitors with dyes are useful for finding leaks, and on some makes the dye shows when the inhibiting ability is exhausted.

4 Flushing the cooling system

1 Despite the use of inhibitors, the cooling system collects dirt and sludge. Whenever circumstances, such as renewing the antifreeze, permit, or when it has been drained for such a job as removing the cylinder head, the system should be flushed. Even though the drain taps on radiator and engine are opened, the whole system will not drain, and there is no way of draining the heater.
2 After a journey, so all sludge is well stirred up, open the tap at the bottom of the radiator, and on the rear side of the cylinder block (directly below the carburettor). Also, move the heater control lever (top) fully to the left. Remove the filler cap from the radiator. Put a hose in the filler cap, and turn it on when the hot water is still flowing out so that the change to cold water is gradual.
3 Disconnect the heater pipe at the cylinder head. Turn the heater full on. Transfer the hose to the heater connection. If possible use a short length of metal pipe to connect the hose direct to the heater hose. If none is available, once the hose is on the heater union, and the radiator has had a good flush, turn off the radiator drain so that more flow is available to go through the heater circuit, and ensure water comes out of the disconnected heater hose.
4 Leave the heater hose disconnected whilst refilling the system, till the new coolant begins to come out of the connection. This will avoid the possibility of air-locks.
5 On old cars, or ones using very hard water, or without inhibitors, the water passages may become coated with hard deposits that will not come out with normal flushing. In this case, use a good proprietary cleaning agent, such as Holts Radflush or Holts Speedflush. It is important that the manufacturer's instructions are followed carefully. The regular renewal of antifreeze should prevent further scaling and contamination of the system.
6 When the engine has been stripped for a job such as 'decarbonising' there will be dirt that has fallen into the water passage. So when refilling after such work put normal water in first, run the engine, and then flush out, before putting in the antifreeze or inhibitor.
7 After refilling, start-up and allow the engine to warm up. Check the heater works, proving there is no air-lock. Switch off. Check the level: it is likely to have fallen, as some air pockets will now have been displaced by coolant.

5 Cooling system leaks

1 The fitting of hoses, and their freedom from leaks, will be better if whenever any of the simple clips originally fitted by FIAT are removed, they are replaced with screw worm ones.
2 New hose connections settle in, and the clips will need retightening after 500 miles and again at 2,000 miles.
3 Clean antifreeze is 'searching', and will leak at weak spots. The system should be very carefully inspected before refilling with antifreeze.
4 Hoses should ideally be replaced before failure - so don't leave it to the last minute! This particularly applies to the top hose to the radiator, and the two to the heater, as these have the hottest water, and also flex a lot. Unless nearly new, such hoses should be renewed when some task demands their removal.
5 Coolant leaks from the cylinder head gasket can give some strange symptoms. Water can get in the oil. This can give a rising oil level. Very quickly the oil becomes creamy coloured with the emulsified water. Oil can get into the water, though this is less common due to the pressurised cooling system. If the water is leaking into the combustion chambers, the water will show as excess vapour in the exhaust. Water vapour is always visible in the exhaust when cold, but a leaking head gasket can make it continue when hot. Unlike oil smoke it is white, and very whispy, blown more quickly by the wind.
6 Minor leaks from parts of the engine or the radiator can be successfully cured with proprietary sealants that are put in the

coolant. Leaking hoses must be replaced. Bad leaks in metal parts must be properly mended.

6 Thermostat

1 The thermostat is needed in all climates to give quick warm-up, and in cold ones to prevent the engine running too cold all the time. It also ensures hot water for the heater.
2 The thermostat has a double acting valve to close off the bypass once the main passage is open. If there was no bypass system the coolant would be static, so the top of the engine would heat too much compared with the bottom.
3 The thermostat is of the wax capsule type. If it fails it is most likely to do so in the closed position, and give rise to immediate and serious overheating. Should it stick in the open position the engine will run cold, and take a long time to warm up. On normal cars without temperature gauges this will only show by poor output from the heater.
4 If there is sudden severe overheating on a journey then the thermostat is suspect.
5 To remove the thermostat, first allow the engine to cool well below boiling point. Then open the radiator drain tap, and let out just over half the coolant, to get the level below the bottom of the thermostat.
6 Disconnect the top radiator hose at the thermostat. With a suitable spanner now remove the two nuts, one bolt and associated washers that secure the cover on the thermostat housing.
7 Separate the two halves of the housing and recover the joint washer. Lift out the thermostat.
8 Test the thermostat for correct functioning by suspending it on a string in a saucepan of cold water together with a thermometer. Heat the water and note the temperature at which the thermostat begins to open. Check with Specifications.
9 With the thermostat fully open measure the distance between the valve and thermostat body. Check with Specifications.
10 Discard the thermostat if it opens too early or too late and also if the valve does not open fully.
11 Allow the thermostat to cool down and check if the valve seats fully.
12 Refitting the thermostat is the reverse sequence to removal.
13 Always ensure that the mating faces of the two halves of the thermostat housing are clean and flat. If excessive corrosion is evident a new housing should be obtained. A new paper joint must always be used.
14 If the thermostat jams shut on the road, it must be removed then and there. Trying to continue with the engine overheating will damage it. If no replacement is available, then the engine must be run without the thermostat. There will not then be any valve to shut the bypass. There will be a tendency for the circulation to take the shortest route, and not go through the radiator, resulting in overheating. If driving on after removing the thermostat, the temperature warning light may come back on. The cure for this will be to seal off the bypass passage in the thermostat housing. A wooden plug should be cut, and driven into the hole dividing the upper and lower parts of it. Alternatively, a more secure block could be made by clamping two washers together from either side.

7 Electric fan

1 When the car is halted, it is possible to hear the fan cut in. An enthusiastic owner may like to rig up a warning light to show when it is working: such a person will probably fit a temperature gauge too, which will enable the correct functioning of the whole system to be monitored properly.
2 It is worthwhile checking that the fan is working, so that it does not remain with a defect undetected for some time, and result in a crisis. To check the fan operation the temperature should be raised by running the engine at a fast idle with the car stationary, preferably after climbing a steep hill.
3 If the fan does not cut in, then refer to Chapter 9 for electrical fault finding in general. One test is to short-circuit the sensor contacts; this should result in the relay operating and thus running the fan (with ignition on). Apart from this one can use a test lamp on terminal 30/51 of the relay, which should be permanently live; and terminal 87 which should be live when the sensor is operated or short-circuited manually. If not, then the relay is defective and must be changed.
4 Note that only the sensor circuit to the relay is controlled by the ignition switch. The main feed through the relay is direct. If the relay system fails the leads can be temporarily joined together, but then will need disconnecting when the engine is switched off.

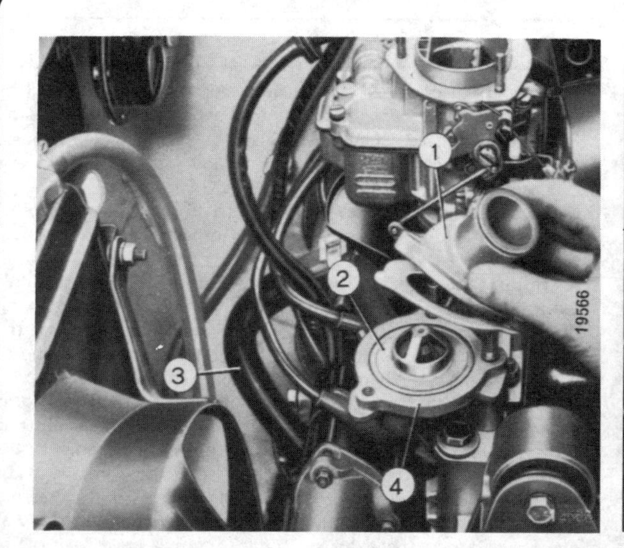

Fig. 2.2. Removing the thermostat (Sec. 6)

1 Water outlet cover
2 Thermostat
3 Hose from outlet to pump
4 Outlet housing thermostat

Fig. 2.3. Thermostat housing details (Sec. 6)

1 Outlet
2 Water return hole from outlet to pump
3 Water passage from cylinder head to outlet

Fig. 2.4. Radiator and electrofan in the vehicle (Secs. 7 and 8)

1 Water hose from engine to radiator
2 Water hose from radiator ro expansion tank
3 Drain cock
4 Electrofan control thermal switch
5 Electrofan

8 Radiator - removal and repair

Note: *If the reason for removing the radiator is concern over coolant loss, note that minor leaks may be repaired by using a radiator sealant, such as Holts Radweld, with the radiator* in situ.

1 To remove the radiator first drain the coolant. Then disconnect the leads to the fan sensor at the bottom of the radiator and the fan leads to the relay and earth terminals. Take off the two hoses from the top and the bottom of the radiator. Undo the small pipe from the filler to the expansion tank. (photo)

2 Undo the two mounting bolts at the top of the radiator.

3 Lift the radiator up out of the rubber padded rest at the bottom. (photos)

4 If the radiator core is leaking, then it is best to get this repaired by professional radiator repairers. There are specialist firms who do this. The core is made of thin metal, and easily damaged by an unskilled repairer. If the leak occurs en-route a temporary repair can be made with a proprietary product.

5 When the radiator is removed, the opportunity should be taken to clean it inside and out. The bottom tank should be swilled out, the debris being tipped out of the bottom hose union, and the radiator not turned upside down, lest the thin passages get blocked. If the radiator is old and dirty, and has to remain off the car for a day or so, it must not be allowed to dry out. The unions must be blocked and the filler cap kept on. Otherwise the deposits will harden, and prove impossible to remove. Hose dead insects etc, out of the air passages. Carefully straighten any of the cooling fins that might have been bent.

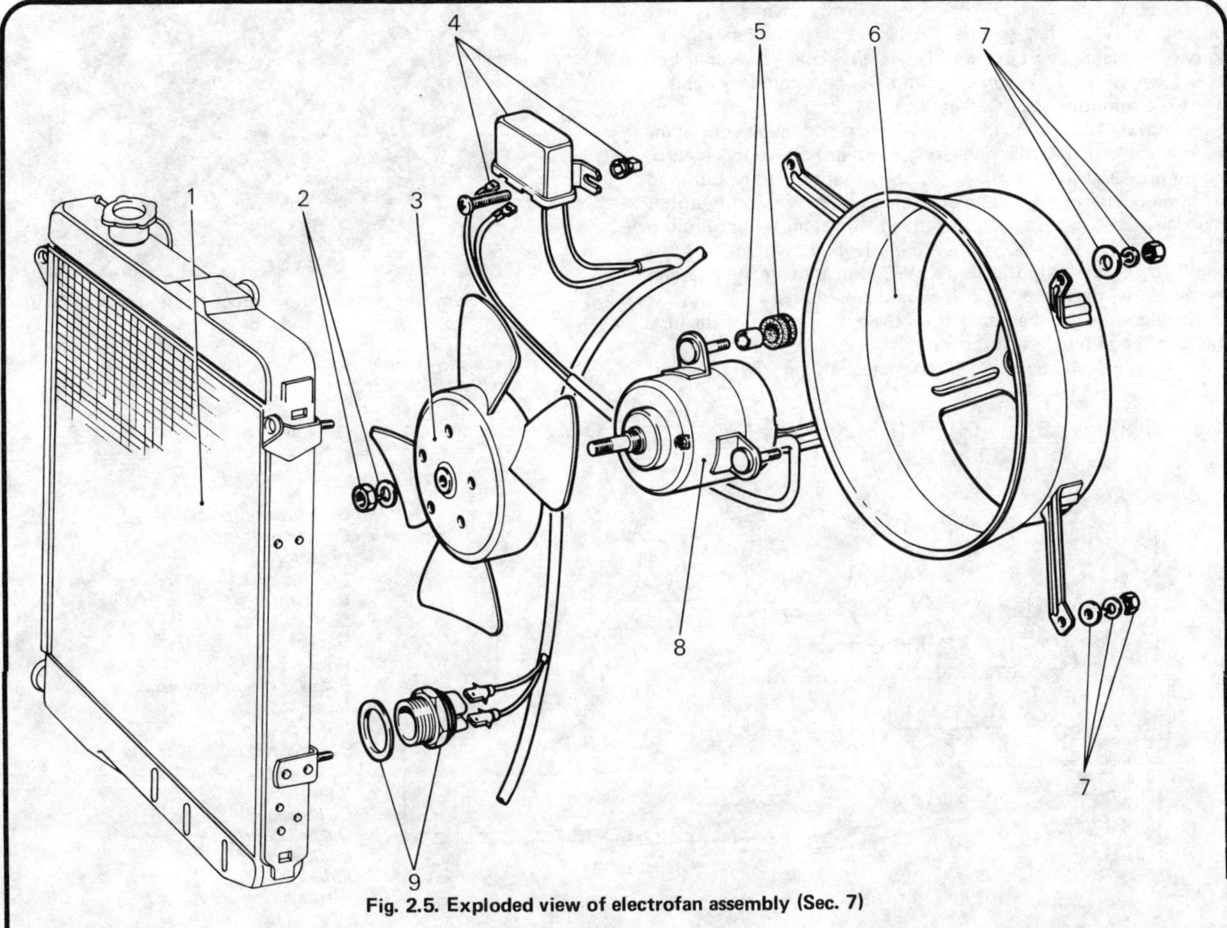

Fig. 2.5. Exploded view of electrofan assembly (Sec. 7)

1 Radiator
2 Nut and spring washer, fan to motor
3 Fan
4 Relay with screw and block for attachment to body
5 Spacer and grommet
6 Air conveyor
7 Nuts, spring washers and flat washers for air conveyor attachment to radiator
8 Motor
9 Thermal switch with seal for relay

8.1 Showing the wires to the temperature sensor at the bottom of the radiator

8.3A Lifting the radiator out of the vehicle

8.3B When the radiator is replaced ensure the base fits in the padded support

9 Overhaul of the water pump

1 The water pump is likely to need overhaul for worn or noisy bearings, or if a seal is leaking. There is a drain hole between the seal and the bearings so any leak can get out without contaminating the bearing grease. Seal leaks are usually worse when the engine is not running. Once started a leak is likely to get worse quickly, so should be dealt with as soon as possible. Worn bearings are likely to be noted first, due to noise. To check them the pulley should be rocked firmly, when any free movement can be felt despite the belt. But if the bearings are noisy yet there is not apparently any free-play, then the belt should be removed so that the pump can be rotated by hand to check the smoothness of the bearings.

2 Removal of the pump is straightforward and involves draining the coolant and then disconnecting the two hoses connected to the pump. Next, release the three bolts that secure the pump to the engine when it should be possible to lift it away, although it may also be necessary to remove the engine mounting. Sometimes it is easier to slacken the drivebelt first.

3 Whenever the pump is dismantled, even though the bearings are the cause, the gland (or seal) should be replaced.

4 Having removed the pump from the engine, take off the nuts and split the two halves of the pump.

5 The pump shaft is an interference fit in the impeller,

Fig. 2.6. Removing water pump impeller using special tool (Sec. 9)

1 Water pump body
2 Impeller
3 Puller

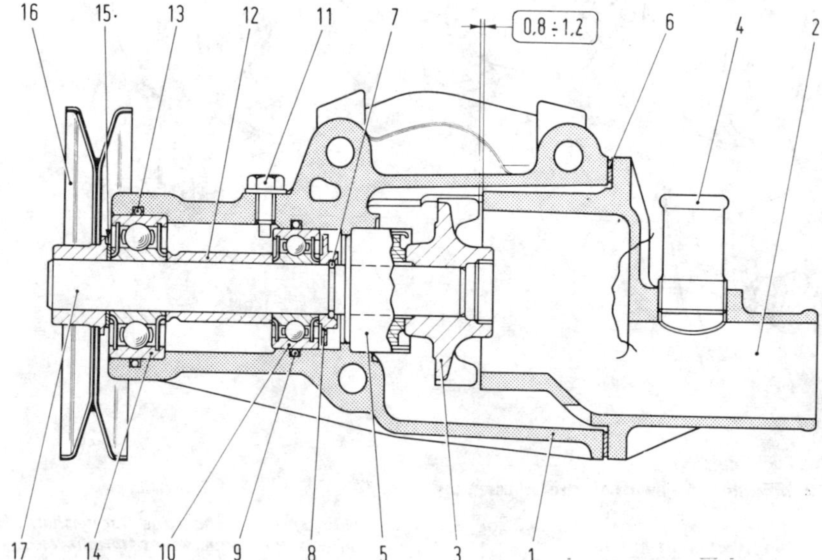

1 Water pump body
2 Pump cover
3 Impeller
4 Connector for water hose from outlet to pump
5 Seal
6 Gasket
7 Circlip
8 Bearing shoulder washer
9 Inner seal
10 Inner bearing
11 Bearing retainment screw and lock washer
12 Spacer
13 Outer seal
14 Outer bearing
15 Lock washer
16 Pulley
17 Water pump shaft

Fig. 2.7. Sectional view of water pump (Sec. 9)

bearings, and pulley boss. How the pump is dismantled depends on whether only the seal needs replacement, or the bearings as well, and what puller or press is available to get everything apart.
6 Assuming complete dismantling is required, proceed as follows. Using a suitable puller (Fig. 2.6), remove the impeller from the driveshaft.
7 Take out the bearing stop screw.
8 From the impeller end, press the shaft with the bearings out of the cover half of the housing.
9 Press the shaft out of the bearings and pulley, taking off the spacer, the circlip, the shouldered ring, and the lockwasher.
10 Do not immerse the bearings in cleaning fluid. They are 'sealed'. Liquid will get in, but a thorough clean will be impracticable, and it will be impossible to get new grease in.
11 Check all the parts. Get a new main seal, two bearing seals and a new gasket. Scrape all water deposits out of the housing and off the impeller.
12 To reassemble start by inserting the new bearing seals in their grooves by each bearing. Fit the circlip to the shaft, then the shouldered ring, bearings, spacer and lockwasher. Fit the shaft and bearing assembly into the cover. Fit the stop screw. Press on the pulley.
13 Fit the new gland (seal), seating it in its location in the cover. Press the impeller onto the shaft. FIAT have special press tools to achieve the correct clearances. Without these the impeller must be put on part way, and then the housing held in place to see how far the impeller must go down the shaft to give the correct clearance of 0.031 to 0.047 in. (0.78 to 1.19 mm) between the impeller vanes and pump cover.
14 The impeller clearance can be checked through the water passage in the side of the pump.

10 Exhaust system

1 If the engine is to be removed the exhaust must be disconnected at the manifold. Even if the exhaust system is being replaced as a separate entity it is advisable to start at the manifold. First, disconnect the support bracket to the transmission; then the front end of the pipe must be supported by string to save over-stressing the pipe and mountings at the rear.
2 When a leak occurs in the exhaust system, it may be possible to use a good proprietary repair kit to seal it. Holts Flexiwrap and Holts Gun Gum exhaust repair systems can be used for effective repairs to exhaust pipes and silancer boxes, including ends and bends. Holts Flexiwrap is an MOT-approved permanent exhaust repair. However, if the leak is large, or if serious damage is evident, it may be better to renew the exhaust complete. Stainless steel replacements, although twice the cost of mild steel ones, are well worth the extra outlay, as service life is virtually unlimited.
3 When fitting a new exhaust, use new clips and flexible mountings.

11 Heater

1 The heater draws water from the engine and returns it to the water pump.
2 The heater hoses will often need to be disconnected for work on the engine. But there is no need to remove the heater itself. If working in the area below it, access can usually be gained by leaning across the car from the left. However, the air intake can be removed, without the heater itself, quite readily.
3 The item most likely to need attention is the water valve. This may fail internally, not stopping all flow, so in hot weather cold ventilation is impossible, or it may develop a water leak to the outside.
4 When working on other parts of the cooling system the water valve to the heater should be open; that is the top lever fully to the left to allow flow for draining, flushing or filling.
5 To remove the valve it is best to remove the heater body. Drain the coolant. Disconnect the two heater hoses at the heater.
6 Disconnect the link to the air admission shutter in the intake at the bolt and nut, halfway along. Undo the clips holding the air intake to the heater radiator. Take out the screw in the centre of the air intake, and lift it out.
7 Undo the cable from the valve. Take out the heater radiator.
8 Take the valve off the radiator.
9 The foregoing dismantling is also necessary to reach the heater fan.

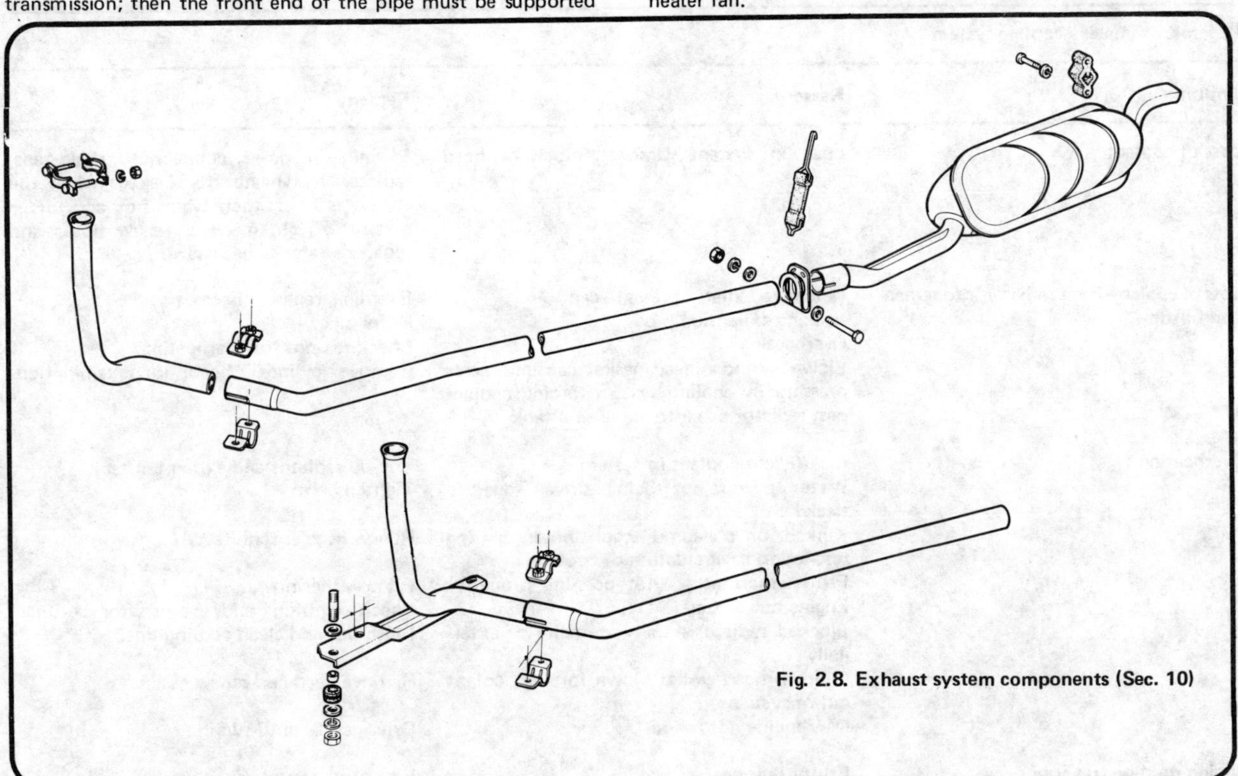

Fig. 2.8. Exhaust system components (Sec. 10)

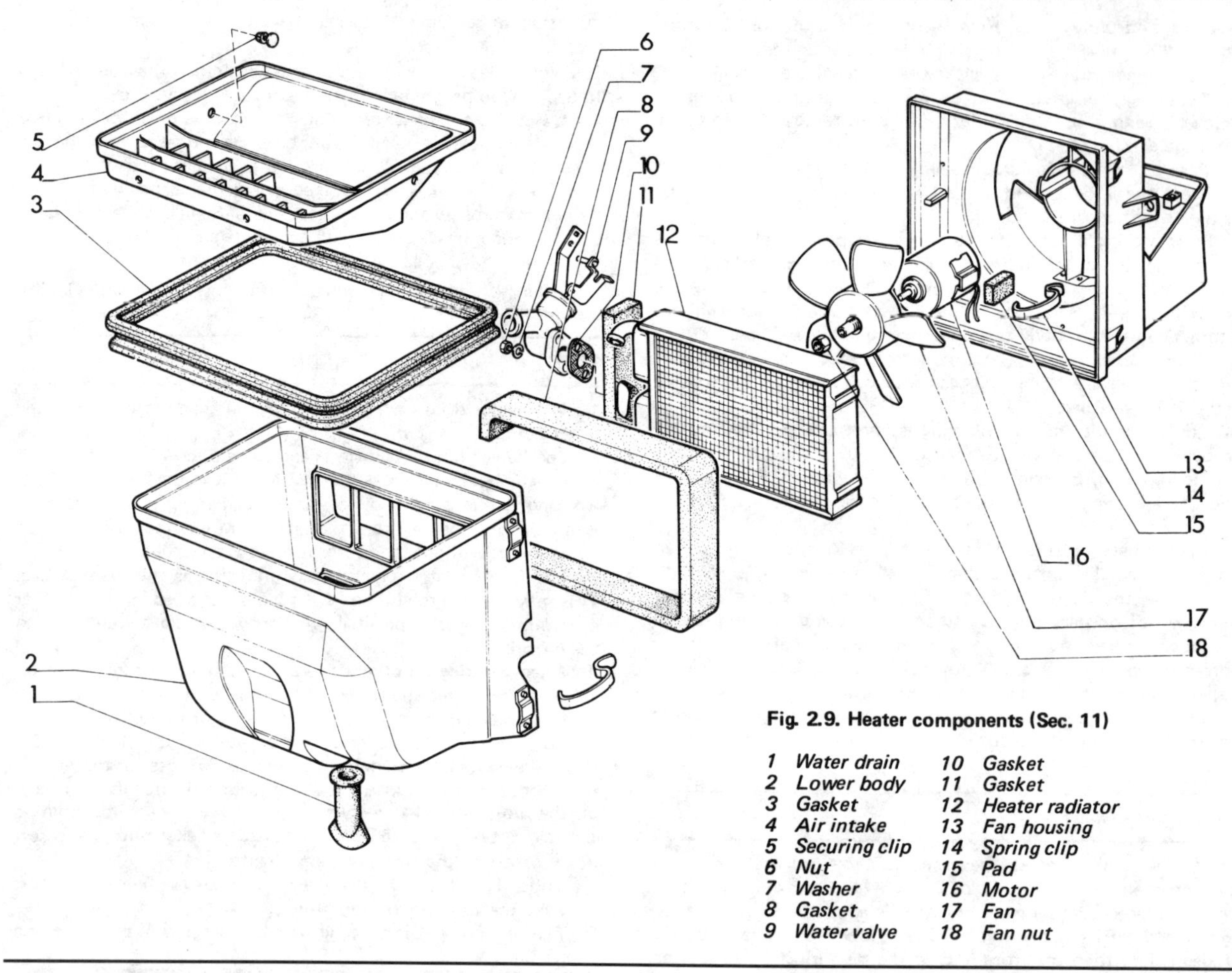

Fig. 2.9. Heater components (Sec. 11)

1	*Water drain*	10	*Gasket*
2	*Lower body*	11	*Gasket*
3	*Gasket*	12	*Heater radiator*
4	*Air intake*	13	*Fan housing*
5	*Securing clip*	14	*Spring clip*
6	*Nut*	15	*Pad*
7	*Washer*	16	*Motor*
8	*Gasket*	17	*Fan*
9	*Water valve*	18	*Fan nut*

12 Fault diagnosis - cooling system

Symptom	Reason/s	Remedy
Loss of coolant	Leak in system. Cracked block or head	Examine all hoses, connections, drain taps, radiator and heater for leakage when the engine is cold, then when hot and under pressure. Tighten clips, renew hoses and repair radiator as necessary.
Loss of coolant from radiator into expansion bottle	Defective radiator pressure cap	Examine, renew if necessary.
	Leaking expansion pipe	Refix pipe.
	Overheating	Check reasons for overheating.
	Blown cylinder head gasket causing excess pressure in cooling system forcing coolant past radiator cap into expansion tank	Remove cylinder head for examination.
Overheating	Insufficient coolant in system	Top-up radiator and expansion tank.
	Water pump not being driven due to slack belt	Tighten belt.
	Kinked or collapsed water hoses causing restriction to circulation of coolant	Renew hose as required.
	Faulty thermostat (not opening properly)	Fit new thermostat.
	Engine out of tune	Check ignition and carburettor settings.
	Blocked radiator either internally or externally	Flush out and clean cooling fins.
	Cylinder head gasket blown forcing coolant out of system	Remove head and renew gasket.
	New engine not run-in	Drive gently until run-in.
Engine running too cool	Faulty thermostat	Fit new thermostat.

Chapter 3 Fuel system and carburation

For modifications, and information applicable to later models, see Supplement at end of manual

Contents

Specifications

Fuel pump

Control plunger stroke	0.0945 in (2.4 mm)
Feed pressure at 4000 rpm of crankshaft	2.8 to 4.2 psi (0.2 to 0.3 kg/cm^2)

Air cleaner

Disposable paper catridge (Champion W107), warm and cold air intakes

Fuel filter

Champion L101

Fuel tank

Location	Under rear structure
Capacity	6.7 gallons (30.5 litres)
Petrol grade	Premium (4 star)

Carburettor

Type	Weber or Holley 321BA20 (early models), 301BA22/35 or Solex C30 DI 40 (later models)
Barrel	1.260 in (32 mm)
Choke	Throttle type
Primary venturi	0.945 in (24 mm)
Auxiliary venturi	0.1378 in (3.5 mm)
Main jet diameter	0.0531 in (1.35 mm)
Idle jet diameter	0.0177 in (0.45 mm)
Main air metering jet	0.0591 in (1.50 mm)
Idle air metering jet	0.0669 in (1.70 mm)
Pump jet diameter	0.0157 in (0.40 mm)
Pump discharge orifice	0.0236 in (0.60 mm)
Well	F 52
Needle valve seat diameter	0.0590 in (1.50 mm)
Power fuel device:	
Fuel jet diameter	0.0295 in (0.75 mm)
Mixture orifice diameter	0.0787 in (2.00 mm)
Throttle opening with choke in	0.0256 to 0.0275 in (0.65 to 0.70 mm)
Pump capacity (for ten strokes)	0.165 to 0.195 cu. in (2.7 to 3.2 cu. cm)
Float weight	0.39 oz (11 g)
Float level:	
Distance between float and cover with gasket, in vertical position	0.236 in (6 mm)
Travel	0.275 in (7 mm)

1 General description

1 The fuel system is conventional in most respects, although it has several refinements that allow it to comply with the latest emission control regulations.
2 A 6.7 gallon (30.5 litre) tank is positioned at the rear of the vehicle under the floor; a mechanical pump draws fuel from this tank and delivers it to a down-draught carburettor mounted on a spacer on the integral inlet manifold.
3 A return pipe from the carburettor returns to the tank, petrol that is not used by the carburettor. This prevents excess petrol evaporating from the carburettor in hot weather, particularly at high altitudes, which would cause fuel vapour locks.
4 The carburettor is fitted with a valve that is used in a positive crankcase ventilation system and assists in recirculating oil vapours and gases from the crankcase back into the cylinders, where they are burnt, rather than being emitted to the atmosphere.

2 Fuel pump - description

1 The mechanically-operated fuel pump is bolted to the timing cover and is actuated through a rocker arm and pushrod. One end of the pushrod is in contact with an eccentrically mounted cam that drives the pushrod through the timing cover; the other end is in contact with the rocker arm. The rocker arm is attached to the diaphragm pullrod.
2 As the engine camshaft rotates, the eccentric on the camshaft moves the pushrod outwards which, by means of the rocker arm, causes the diaphragm pullrod to move downwards against the action of the return spring.
3 This creates sufficient vacuum in the pump chamber to draw in fuel from the fuel tank through the filter gauze and non-return inlet valve.
4 The rocker arm is held in constant contact with the pushrod and eccentric on the camshaft by an anti-rattle spring and, as the engine camshaft continues to rotate, the eccentric allows the rocker arm inner end to return upwards. The diaphragm spring is thus free to push the diaphragm upwards, forcing the fuel in the pump chamber out to the carburettor float chamber through the non-return outlet valve.
5 When the float chamber in the carburettor is full, the float chamber needle valve will close so preventing further flow from the fuel pump.
6 The pressure in the delivery line will hold the diaphragm downwards against the pressure of the diaphragm spring and it will remain in this position until the needle valve in the float chamber opens to admit more petrol.

3 Fuel pump filter - removal and refitting

1 Unscrew the cover mounting screw and lift away the screw, fibre washer and cover. (photo)
2 Lift out the filter gauze from the pump upper body.
3 Inspect the filter gauze for sediment and, if dirty, clean it with petrol and a soft brush.
4 Check the condition of the gasket and renew it if it has hardened or distorted.
5 Replacement is the reverse sequence to removal. Tighten the centre screw just sufficiently to ensure a fuel-tight joint.

4 Fuel pump - removal and refitting

1 Remove the inlet and outlet fuel lines from the pump. Plug the ends of the pipes with a piece of tapered wood, such as a pencil, to stop dirt ingress or loss of fuel.
2 Undo and remove the two pump securing nuts and spring washers. Carefully lift away the fuel pump. Recover the two paper gaskets, spacer, pump actuating pushrod and bush from the pushrod bore. (photos)
3 Refitting the fuel pump is the reverse sequence to removal, but care is necessary to use the right thickness of gasket. Protrusion of the pushrod beyond the surface of the outer gasket at the start of the stroke should be 0.039 to 0.059 inch (1 to 1.5 mm) with a 0.027 inch (0.7 mm) thick gasket fitted between the timing cover and the insulating block. If the protrusion is too small use a 0.012 inch (0.3 mm) thick gasket, and if it is too great fit a 0.047 inch (1.2 mm) thick gasket. This will ensure that the pump diaphragm operates correctly throughout its range.

5 Fuel pump - testing

Presuming that the fuel lines and unions are in good condition and that there are no leaks anywhere, check the performance of the fuel pump in the following manner: Disconnect the fuel pipe at the carburettor inlet union, and the high tension lead to the coil and, with a suitable container or a large rag in position to catch the ejected fuel, turn the engine over on the starter motor solenoid. A good spurt of petrol should emerge from the end of the pipe every second revolution.

6 Fuel pump - dismantling

1 Refer to Fig. 3.1 and unscrew the securing bolt from the centre of the cover and lift the cover away. Note the fibre washer under the head of the bolt.
2 Remove the fine mesh filter gauze.
3 If the condition of the diaphragm or valves is suspect or for

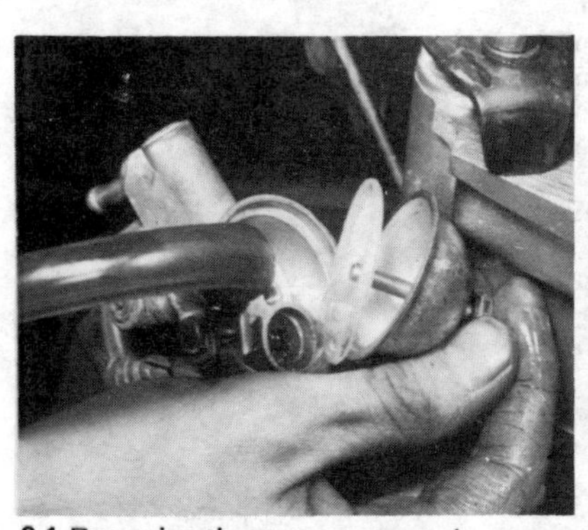
3.1 Removing the pump cover and filter

4.2A Lifting the pump off of the studs

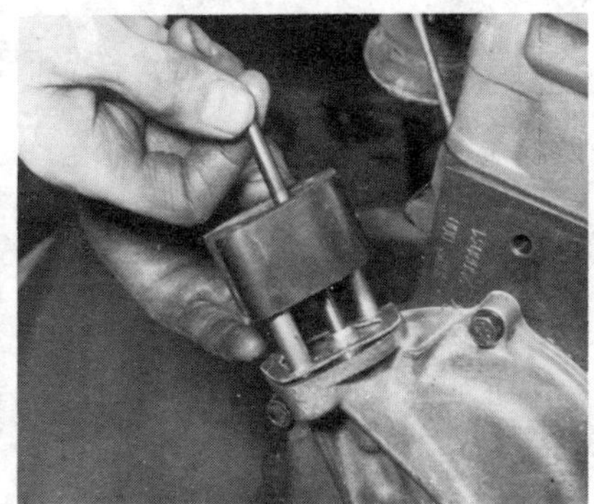
4.2B Pulling out the pushrod and spacer. The top gasket has already been removed but the lower one is in position

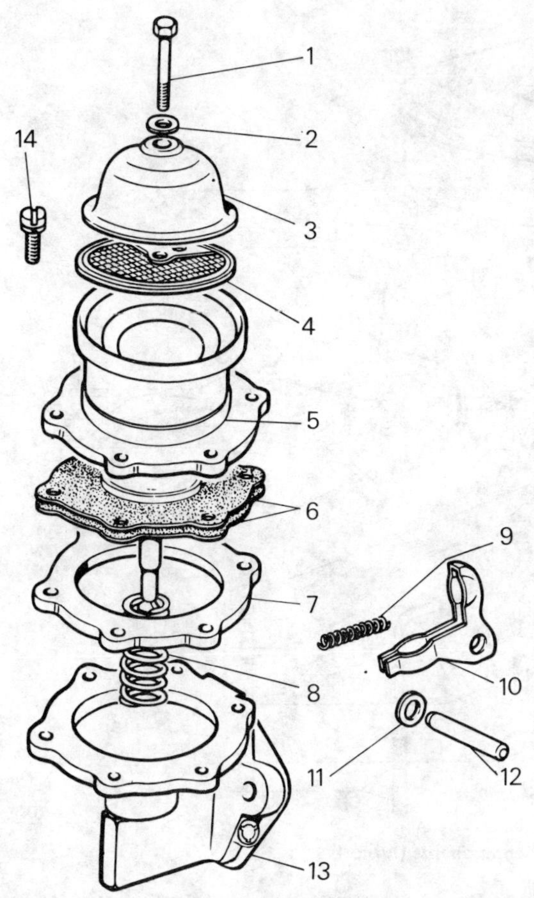

Fig. 3.1. Exploded view of fuel pump (Sec. 6)

1 Cover screw
2 Flat washer
3 Pump cover
4 Filter
5 Upper body
6 Diaphragm
7 Spacer
8 Diaphragm spring
9 Rocker arm spring
10 Rocker arm
11 Shoulder washer
12 Rocker arm shaft
13 Lower body
14 Upper and lower bodies attachment screw

Fig. 3.2. Fuel pump and gaskets

1 Fuel pump
2 Pushrod
3 Gasket 0.012 in (0.3 mm) thick
4 Insulator
5 Variable thickness gasket 0.012,0.027,0.047 in (0.3, 0.7,1.2 mm) for pump drive adjustment

any other reason it is wished to dismantle the pump fully, proceed as follows: Mark the upper and lower flanges of the pump that are adjacent to each other so that they may be refitted in their original positions.

4 Undo and remove the six screws and spring washers which hold the two halves of the pump body together. Separate the two halves with great care, ensuring that the diaphragm does not stick to either of the two flanges. This also releases the spacer between the two halves of the pump.

5 To release the diaphragm, depress the centre and turn it 90°. This will release the diaphragm pull-rod from the stirrup in the operating link, and its associated spring.

6 Do not remove the rocker arm and pivot from the body base unless there are signs of excessive wear - in which case it would probably be more economical to obtain an exchange pump.

7 To remove the valve assemblies from the body centre section they must be prised out carefully past the stakes which locate them. Remove the sealing ring fitted behind each valve. On some later models the valves are not available as seperate spares. Instead, the complete upper half of the pump must be replaced if a valve is faulty.

7 Fuel pump - examination and reassembly

1 Check the condition of the cover sealing washer, and if it has hardened, distorted or broken, it must be renewed. The diaphragm should be checked similarly and replaced if faulty. Clean the pump and agitate the individual parts of the valve assemblies (keep them in their respective sets) to clean them. It is unlikely that the pump body will be damaged, but check for fractures and cracks. Renew the cover if distorted by over-tightening.

2 Reassembly of the pump is the reverse sequence to removal. Take care to ensure that the valves are correctly assembled to the upper housing.

8 Carburettor operation (typical)

1 The carburettor is of the single throat downdraught type with a barrel bore of 32 mm (1.26 inch) at the throttle valve. It is fitted with an accelerator pump and a manually operated choke.

2 For normal running, the fuel (see Fig. 3.4) passes through the needle valve to the bowl. Here the float, hinged on its pin, regulates the rise and fall of the needle so that the fuel level in the bowl remains constant; the needle is attached to the float tongue by means of the recall hook.

The fuel then passes from the bowl into the well, via the main jet. It is then mixed with air drawn in through the air bleed jet and the air jet and passed through the holes in the emulsion tube. The air/fuel emulsion is then ejected through the spray outlet to reach the carburation area, consisting of the auxiliary and primary venturi.

The carburettor is provided with an **enrichment circuit.**

Fuel passes from the bowl through the passage and the jet and mixes with air drawn in via the calibrated orifice.

This mixture passes through the passage and the calibrated orifice into the carburettor throat during fast running.

3 Fig. 3.4 also shows the crankcase ventilation device in the idle running (A) and normal running (B) positions. This device consists of a rotating valve, entrained by the throttle shaft and the lever. This valve channels the blow-by gas from the duct, via the slot, to below the throttle valve. During idle running, gas is drawn in through the calibrated orifice.

4 When idling, or progressing from idle to normal running, fuel passes from the well to the idle jet, via the passage. It is then emulsified with air drawn in through the calibrated bush, the passage and the idling feed orifice, which can be adjusted by means of the screw. The air/fuel mixture enters the carburettor throat downstream from the throttle.

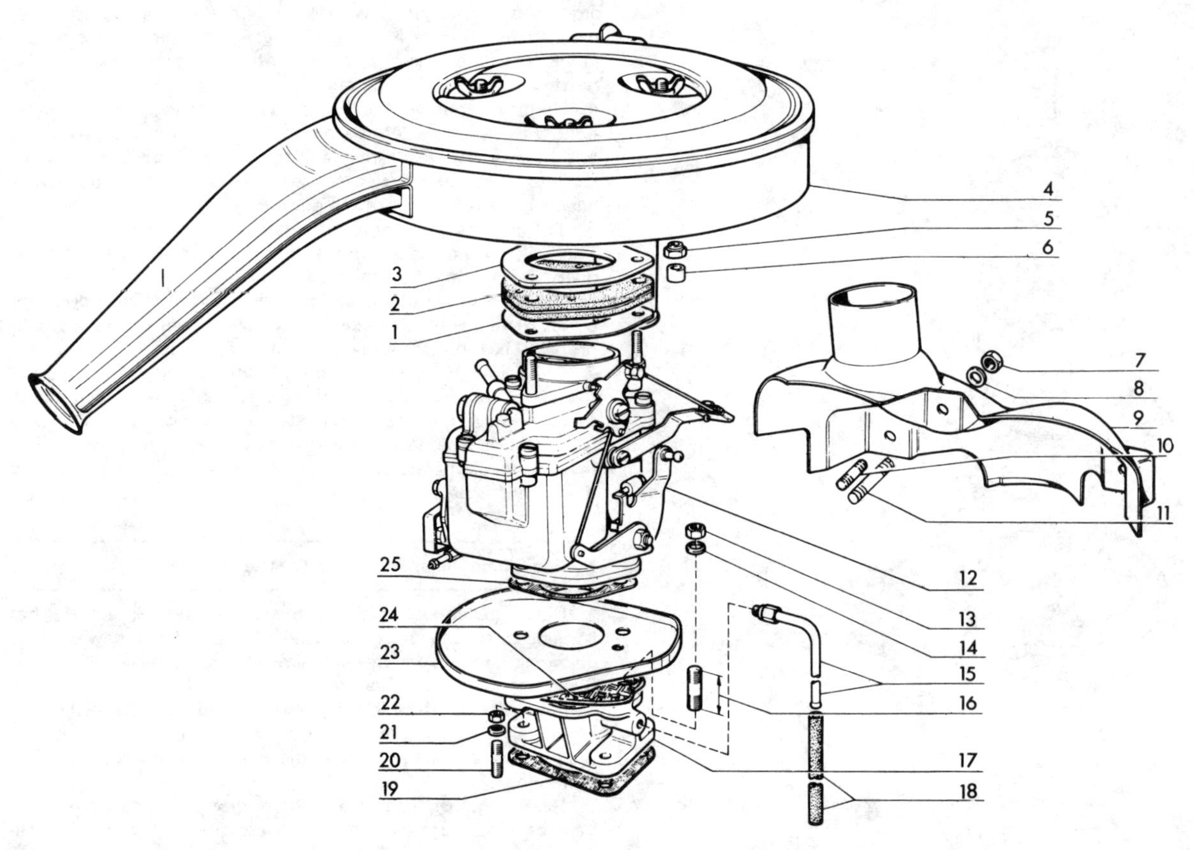

Fig. 3.3. Carburettor and associated components (typical)

1 Plate
2 Gasket
3 Plate
4 Filter body
5 Locknut
6 Bush/dowel
7 Nut
8 Washer
9 Hot air ducting
10 Stud
11 Stud
12 Carburettor
13 Nut
14 Washer
15 Drain pipe
16 Stud
17 Carburettor spacer block
18 Drain pipe - flexible tube
19 Gasket
20 Stud
21 Washer
22 Nut
23 Driptray
24 Gasket
25 Gasket

As the throttle is gradually opened from the idling position, the mixture is also drawn in via the progression orifices, so that a steady increase in power is given.

5 Quick response is needed for acceleration. When the throttle is opened by means of the linkage and the spring, the diaphragm injects fuel into the carburettor throat via the passage, the delivery valve and the nozzle.

Abrupt throttle movements are stored by the spring and the delivery of fuel is prolonged.

Excess fuel delivered by the accelerator pump is fed back into the bowl through the calibrated bushing, together with fumes from the pump chamber.

When the throttle is closed, the lever releases the diaphragm, allowing the suction spring to draw fuel from the bowl via the ball valve.

6 To start from cold, a rich mixture is needed, as the fuel is not evaporating properly.

When the choke control is operated, the lever is in position A (Fig. 3.4) the choke throttle blocks the carburettor air inlet, while the throttle is opened a little way by means of lever (45) via lever (44) and rod (46).

The spray outlet therefore provides a suitable mixture for rapid starting.

When the engine has started, the induction vacuum in the carburettor causes the throttle butterfly to partially open against the action of the spring and the mixture is rich enough to ensure easy running. As the engine begins to warm up the throttle

Fig. 3.4. Schematic diagrams of carburettor operation (Sec. 8) 

1 Calibrated bushing
2 Air bleed jet
3 High speed orifice passage
4 High speed orifice
5 Spray nozzle
6 Auxiliary venturi
7 Venturi
8 Throttle shaft
9 Throttle valve
10 Emulsion tube
11 Emulsion tube well
12 Main jet
13 Fuel bowl
14 Float
15 Hook, needle return to float tang
16 Float hinge pin
17 Valve needle
18 Needle valve
19 Blow-by gas duct
20 Rotary valve
21 Calibrated orifice, for blow-by gas suction at idle
22 Slot, for blow-by gas flow when cruising
23 Idle air metering bushing
24 Idle jet
25 Fuel passage
26 Idle passage
27 Transfer orifice
28 Idle adjusting screw
29 Idle feed bushing
30 Idle air calibrated orifice
31 Pump jet nozzle
32 Delivery valve
33 Ball valve
34 Intake control spring
35 Diaphragm
36 Accelerating pump control lever
37 Calibrated bushing, excess fuel discharge
38 Throttle return spring
39 Throttle control lever
40 Fuel delivery passage
41 Choke throttle
42 Choke control lever extension
43 Calibrated spring
44 Choke control lever
45 Fast idle control lever
46 Link

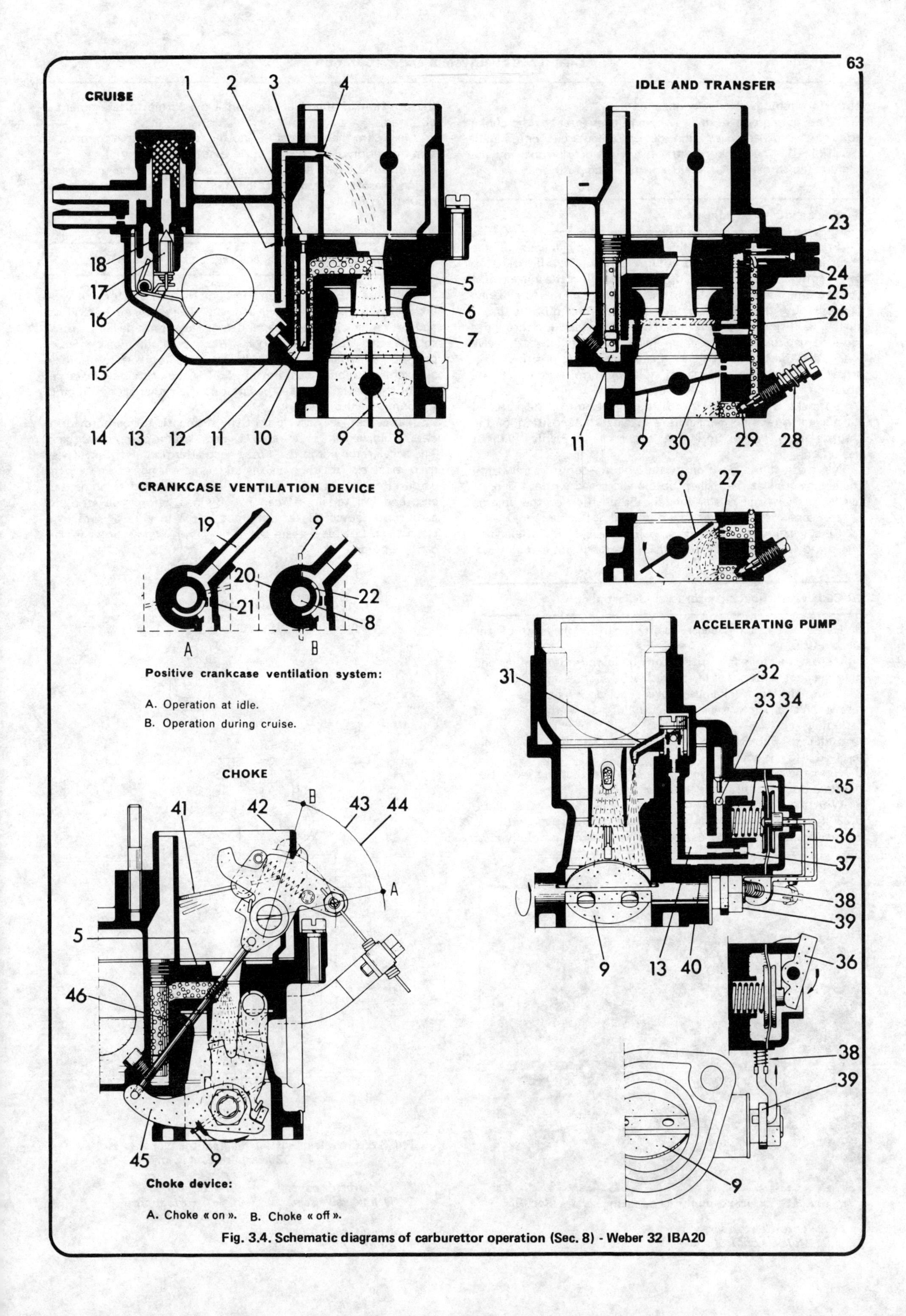

Positive crankcase ventilation system:

A. Operation at idle.
B. Operation during cruise.

Choke device:

A. Choke « on ». B. Choke « off ».

Fig. 3.4. Schematic diagrams of carburettor operation (Sec. 8) - Weber 32 IBA20

butterfly should be gradually opened.

When the correct running temperature is reached, the choke control is pushed back, the device is cut out completely (position B). The choke throttle is now held fully open by the rod and the throttle takes up its normal idling position.

9 Engine idle speed adjustment

Engine idle speed adjustment is made by means of the throttle stop screw and the volume screw. The throttle stop screw controls the amount of opening of the throttle valve whilst the volume screw regulates the volume of mixture being delivered by the idling mixture passage. The mixture has a further addition of air drawn in by engine vacuum through a reduced gap between the main barrel wall and throttle valve when the throttle valve is in the engine idling position. This enables the correct degree of idle mixture richness to be obtained.

To make the adjustment, run the engine until it has reached its normal operating temperature. Adjust the position of the throttle stop screw until the engine runs evenly without hesitation.

Next adjust the position of the volume control screw until the engine runs at the highest obtainable speed without signs of hesitation. On many later models the position of the volume control screw is factory-sealed and should not be tampered with.

Finally adjust the throttle stop screw again until the engine idle speed is correct with the ignition warning light just glowing.

10 Carburettor float chamber level - adjustment

1 Remove the air cleaner assembly from the top of the carburettor.
2 Detach the two fuel pipes from the carburettor unions and plug the ends with pieces of tapered wood such as a pencil.
3 Undo the float chamber cover and upper body securing screws and spring washers. Carefully lift away the cover and gasket.
4 Refer to Fig. 3.6 and check that the needle valve is fully screwed into the cover.
5 Hold the cover in the vertical position so that the weight of the float does not influence the valve needle.
6 With the float arm just touching the ball, the float should be 0.23 in. (6 mm) from the mating face of the cover with the gasket in position. If any adjustment is necessary carefully alter the position of the float arms until the correct measurement is obtained.
7 Make sure the arm is at 90° to the needle and free to move so that operation of the needle valve is not impaired.
8 Refit the cover and gasket and secure with the screws and spring washers. Reconnect the fuel lines to the carburettor unions and finally replace the air cleaner assembly.

11 Carburettor - dismantling and reassembly (general)

1 With time the component parts of the carburettor will wear and petrol consumption will increase. The diameter of drillings and jets may alter, and air and fuel leaks may develop round spindles and other moving parts. Because of the high degree of precision involved, in the authors opinion it is best to purchase an exchange rebuilt carburettor. This is one of the few instances where it is better to take the latter course rather than to rebuild the component oneself.
2 It may be necessary to partially dismantle the carburettor to clear a blocked jet or to renew the accelerator pump diaphragm. The accelerator pump itself may need attention and gaskets may need renewal. Providing care is taken there is no reason why the carburettor may not be completely reconditioned at home, but ensure a full repair kit can be obtained before you strip the carburettor down. **Never** poke out jets with wire or similar to clean them but blow them out with compressed air or air from a car tyre pump.

Fig. 3.5. Typical carburettor adjusting screws (Sec. 9)

1 Throttle stop screw
2 Volume control screw

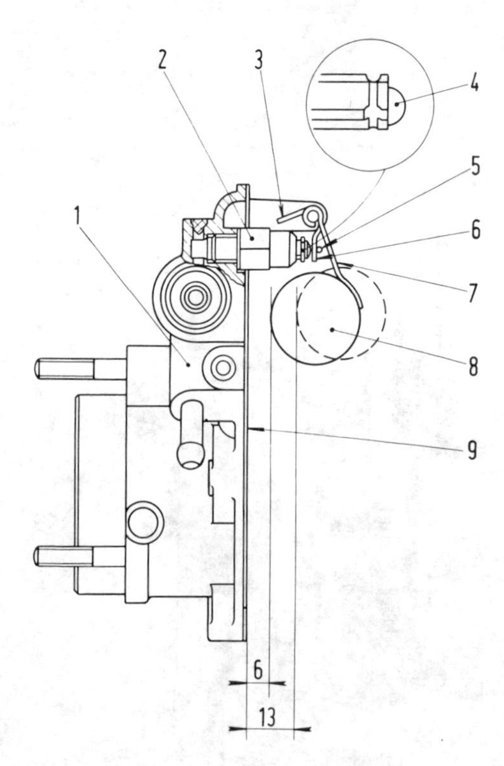

Fig. 3.6. Float level setting diagram for Weber or Holley 32 IBA 20 carburettors (Sec. 10)

1 Carburettor cover
2 Needle valve seat
3 Lug
4 Needle with movable ball
5 Return hook
6 Tongue
7 Float arm
8 Float
9 Gasket

12 Carburettor - partial dismantling and cleaning

1 This information is aimed, basically, at those owners who feel that tampering with a carburettor can lead to troubles, rather than cure them, and so would prefer to only check the more obvious things.
2 To clean the carburettor, first remove the air cleaner, then disconnect the throttle and choke connections. Note the exact spot at which the choke cable is clamped. This is important when reassembling. (photo)
3 Remove the filter plug. Undo the fuel pipe to the carburettor, and return pipe to the fuel tank. Both of these pipe ends should now be plugged to prevent the ingress of dirt. Unclip the choke to throttle interconnection (photo).
4 Undo the screws holding the top to the carburettor body.
5 Lift the top up, carefully, so as to bring the float out of the float chamber without bending it. Put it down where it cannot get damaged, or the float pivot come out.
6 Mop out the fuel, and any sediment, from the bottom of the float chamber. Be careful the dirt is not swilled into any jet ducts.
7 If the carburettor is found to be very dirty, then the jets must be removed. These should be blown through to clean. In extreme cases, it will be necessary to remove the carburettor, so that it can be cleaned under good conditions, completely dismantled on a bench. Wash out the filter in petrol and dry out; do not wipe with a fluffy rag. Replace the filter and all jets, then reassemble the carburettor, but check the correct float level (Section 10), before replacing the top. Reconnect the choke cable exactly in the position it was disconnected.
8 The maintenance of the crankcase fume extraction system is covered in Section 18.

13 Air cleaner

1 For most markets a paper element filter is used. This will need replacement at the periods specified in Routine Maintenance.
2 The paper element is changed by undoing the top of the cleaner, and lifting out the element. When handling the new one ensure no dirt is put on the inside surface. Note carefully the position of the rubber sealing ring. (photo)
3 If the air cleaner body is to be removed for dismantling purposes, first pull off the oil breather pipes. Then lift out the element to reveal the nuts that secure the body to the top of the carburettor. Remove the nuts and lift off the body, then remove the gaskets and two plates on the carburettor studs. Clean the inside of the body. (photo)

12.2 Throttle and choke connections are arrowed

12.3 The inlet pipe is shown by the arrow: the other pipe is the fuel return to tank

13.2 Removing the air cleaner lid

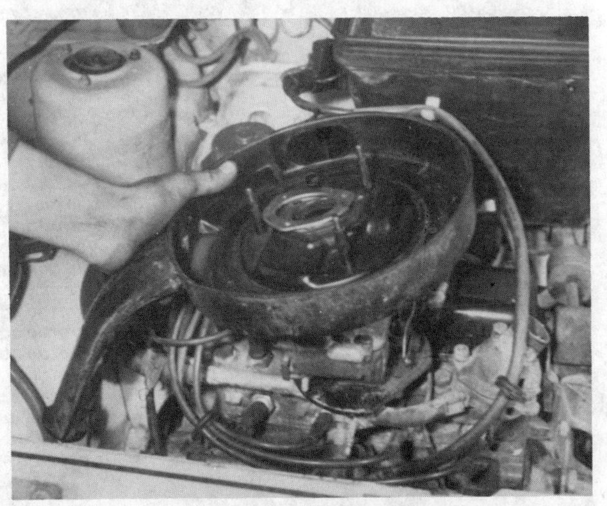
13.3 Lifting the air filter body off of the carburettor

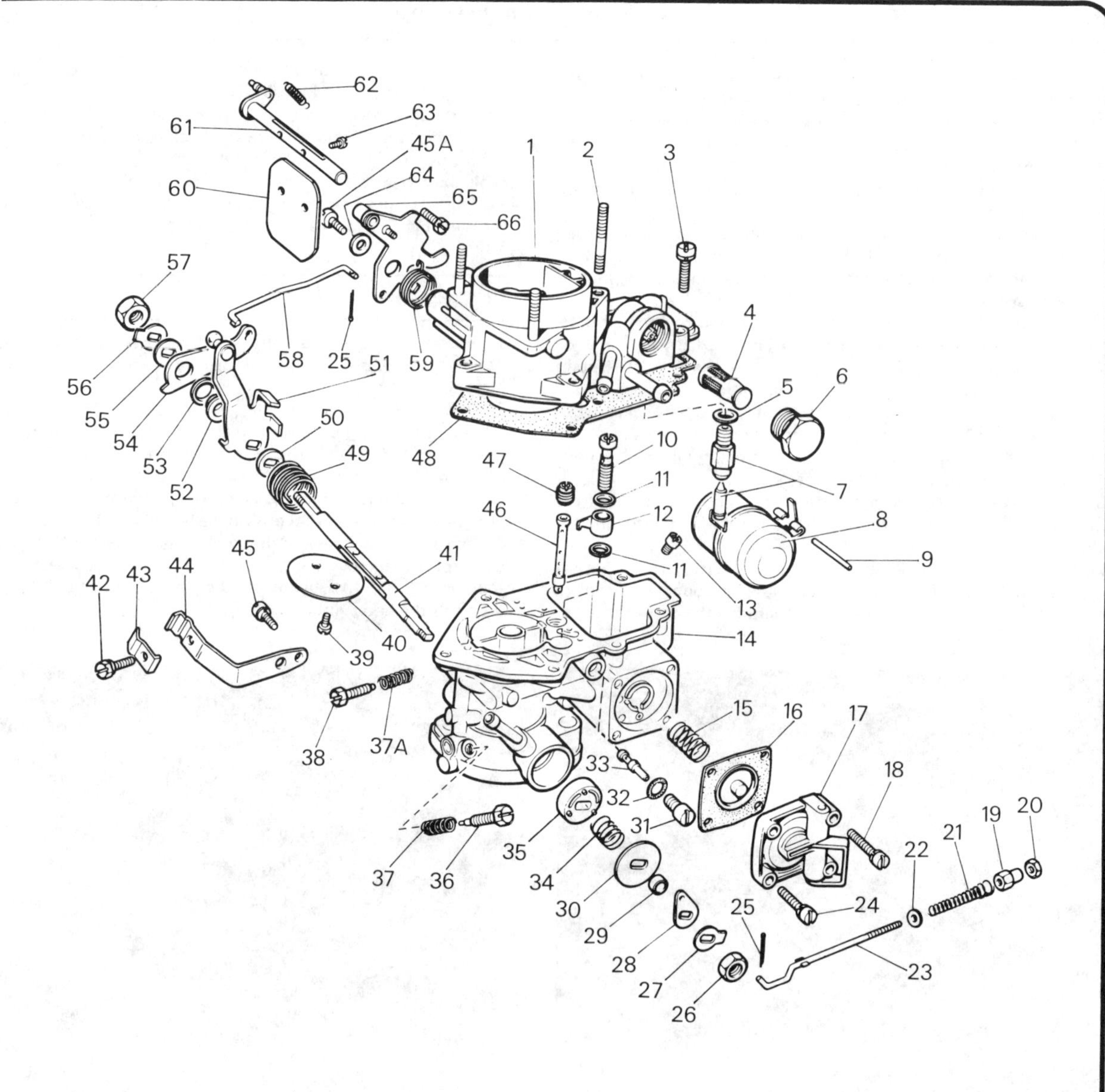

Fig. 3.7. Exploded view of Weber 32 IBA 20 carburettor

1 Carburettor cover
2 Stud
3 Cover screw
4 Filtering element
5 Needle valve seal
6 Filter inspection plug
7 Needle valve
8 Float
9 Float hinge pin
10 Pump delivery valve
11 Pump jet seal
12 Pump jet
13 Main jet
14 Carburettor body
15 Pump spring
16 Pump diaphragm
17 Pump cover
18 Pump cover screw
19 Pump rod bushing
20 Bushing nut
21 Pump delivery extension spring
22 Pump rod washer
23 Pump rod
24 Pump cover screw
25 Pump rod cotter pin
26 Pump lever nut
27 Lock washer
28 Pump lever
29 Spacer
30 Shoulder washer
31 Idle jet holder
32 Idle jet holder seal
33 Idle jet
34 Valve spring
35 Valve
36 Idle mixture adjusting screw
37 Idle mixture adjusting screw spring
37a Throttle adjusting screw spring
38 Throttle adjusting screw
39 Throttle attachment screw
40 Throttle
41 Mainshaft
42 Clip screw
43 Sheath retaining clip
44 Choke bowden support
45 Choke cable support screw
45a Choke lever pivot screw
46 Emulsion tube
47 Air corrector jet
48 Carburettor cover gasket
49 Throttle lever spring
50 Shoulder washer
51 Throttle lever
52 Fast idle lever bushing
53 Friction washer
54 Fast idle lever
55 Shoulder washer
56 Lockwasher
57 Throttle lever nut
58 Fast idle rod
59 Choke lever spring
60 Choke throttle
61 Choke throttle shaft
62 Choke throttle adjusting spring
63 Choke throttle screw
64 Bushing
65 Choke control lever
66 Choke control cable screw

4 The air cleaner has two air inlets. One is the normal one for cold air, by a pipe drawing from the front of the car. In cold weather to reduce the possibility of carburettor icing, the air can be drawn from the hot area around the exhaust manifold. The top must be removed to make the change. In warm climates the reference letter 'E' should line up with arrow 'D'. In cold climates the letter 'I' should line up with arrow 'D'.

14 Carburettor - removal and dismantling

1 Do not undertake this job until you have bought a manifold gasket, and set of gaskets, washers, and diaphragms for the carburettor. Very often the carburettor will be removed as part of a more general overhaul, such as a cylinder head removal. It should only be necessary to remove the carburettor for its own sake on cars that have large mileages on the clock.
2 To commence dismantling, remove the spare wheel and the air cleaner. Then unclip the throttle linkage at the throttle, and unclip the cross rod from its pivot on the engine. Unclamp the inner and outer choke cable (photos).
3 Take off the fuel inlet pipe at the float chamber, and the recirculatory pipe from its adjacent connector.
4 Take off the two nuts holding the carburettor down to the drip tray. Hold the drip tray so its gasket is not torn, and lift the carburettor off. If required, the drip tray and its gasket can be lifted off the spacer. The spacer can be removed by releasing the four nuts and washers from the studs and lifting it off. Inspect the drain hole in the spacer to ensure it is not blocked. (photo)
5 Cover the inlet manifold to prevent anything falling into the engine.
6 Before dismantling the carburettor, refer to Fig. 3.7, the exploded diagram of the carburettor.
7 Take off the external fittings, and then the screws holding the two halves together.
8 Clean out the float chamber, and the filter. Then take out all the jets, laying them out in systematic order so they will not get confused.
9 Inspect the jets, needle valve, actuating arm and float top for excessive wear or damage, also the diaphragms for any signs of fatigue, or tears. Also shake the float to ensure it does not contain any petrol. **Note:** It is suggested that all jets, diaphragms, float needle valve actuating arm and needle valve are renewed. This may seem a needless expense, but it is obvious that a certain degree of wear will have taken place in all these components and this will lower performance and economy. Therefore the small outlay for these new parts will soon be recouped in lower running costs. Even if the other parts are not renewed the needle valve assembly certainly should be, as a worn needle valve will allow excessive fuel into the carburettor, making the mixture over-rich. This will not do the engine or your pocket any good!
10 The carburettor body and all components should be washed in clean petrol, and then blown dry. If an air jet is not available in any form, the parts can be wiped dry with a soft non-fluffy rag. Jets can be blown through by mouth to clear obstructions, or poked through with a nylon bristle. **Do not** scour the carburettor body or components with a wire brush of any variety, or poke through the jets with wire. The reason for this is that the carburettor is a precision instrument made of relatively soft metal (ie; body - aluminium, jets - brass) and wire will scratch this material, possibly altering the performance of the carburettor.
11 Reassembly is the reverse of dismantling, but before replacing the air filter, check that the throttle flap opens fully when the pedal is depressed: if not, adjust the linkage. Also check that the choke flap is fully open when the choke control is depressed: adjust if necessary.

15 Petrol tank - removal and refitting

1 The petrol tank is positioned at the rear of the car under the floor. Access is made a lot easier if the rear of the car is raised on ramps, but this should preferably be done when the fuel has been drained.
2 For safety reasons, disconnect the battery positive (+) terminal.
3 Place a container under the petrol tank drain plug and unscrew and remove the drain plug. When all the petrol has drained out replace the drain plug.
4 Slacken the hose clip and disconnect the filler neck hose from the fuel tank.
5 Detach the two feed lines from the fuel tank and also the tank breather tube from the tank. This is not always easy to do at this stage, so it can be left until the tank is partially lowered; it all hinges, really, on what sort of access you have.
6 Scrape off the mud and underseal that will probably be thickly encrusted around the tank retaining nuts; loosen the nuts and lower the tank slightly. If the lines and hoses have not previously been disconnected, now is the time to do it. You can also disconnect the electrical connections after first recording their terminal positions and colours. Light-blue/yellow to terminal 'T', and Grey-red to terminal 'W'.
7 Lower the tank and ease away from under the car.

16 Petrol tank - cleaning

1 With time it is likely that sediment will collect in the bottom of the tank. Condensation, resulting in rust and other impurities will usually be found in the fuel tank of any car more than three or four years old.
2 When the tank is removed it should be vigorously flushed out and turned upside down and, if facilities are available, steam cleaned.
3 Never weld or bring a naked light close to an empty fuel tank unless it has been steamed out for at least two hours or washed internally with boiling water and detergent several times. If using

14.2A Using screwdriver to prise the clip off one end of the cross-rod ...

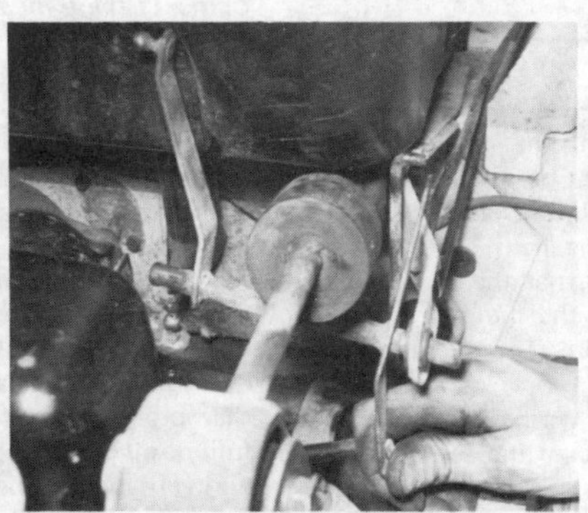
14.2B ... then unclip the other end and the cross-rod can be pulled out

14.4 Inspect the spacer drain hole to ensure its not blocked

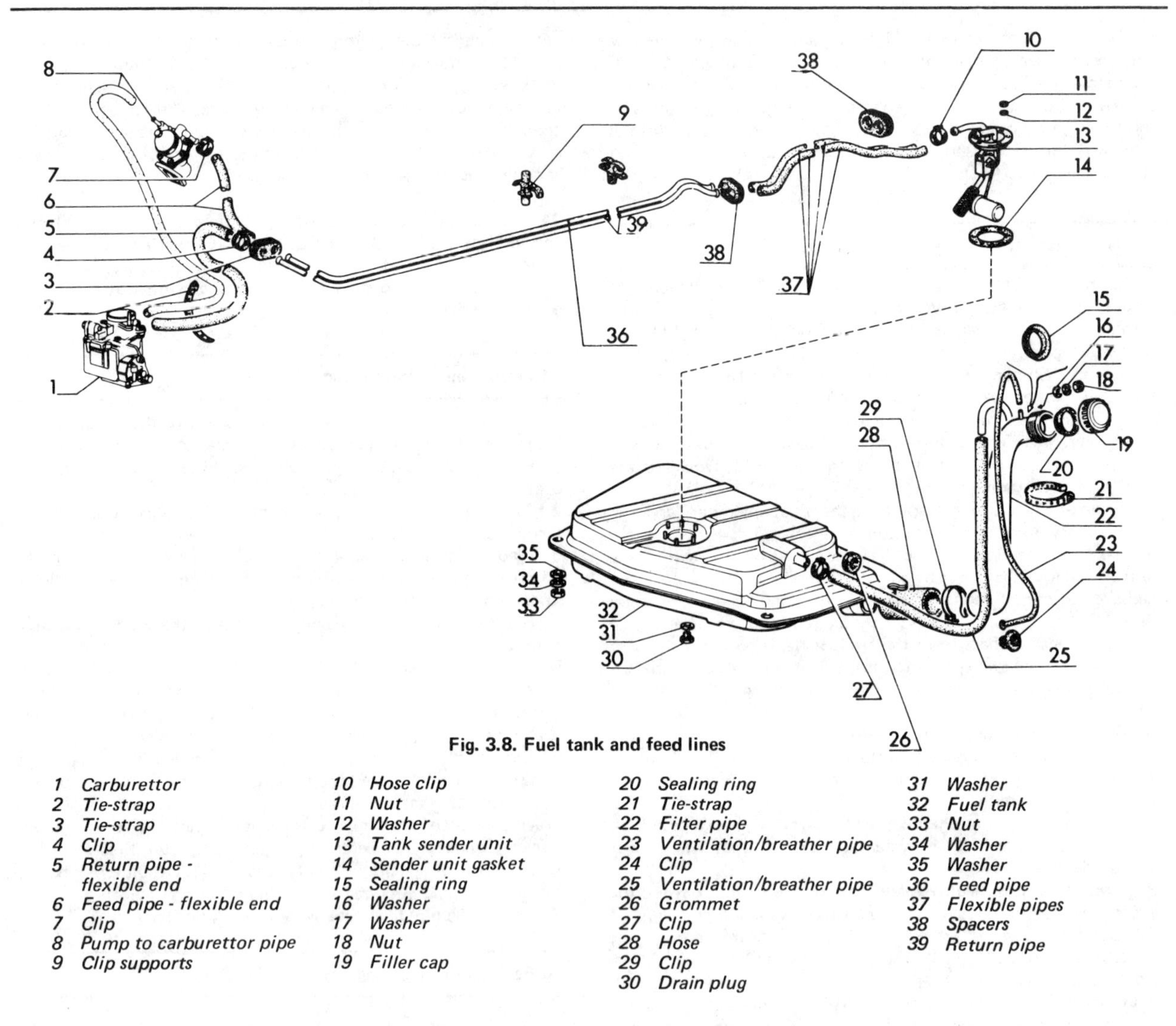

Fig. 3.8. Fuel tank and feed lines

1 *Carburettor*
2 *Tie-strap*
3 *Tie-strap*
4 *Clip*
5 *Return pipe - flexible end*
6 *Feed pipe - flexible end*
7 *Clip*
8 *Pump to carburettor pipe*
9 *Clip supports*
10 *Hose clip*
11 *Nut*
12 *Washer*
13 *Tank sender unit*
14 *Sender unit gasket*
15 *Sealing ring*
16 *Washer*
17 *Washer*
18 *Nut*
19 *Filler cap*
20 *Sealing ring*
21 *Tie-strap*
22 *Filter pipe*
23 *Ventilation/breather pipe*
24 *Clip*
25 *Ventilation/breather pipe*
26 *Grommet*
27 *Clip*
28 *Hose*
29 *Clip*
30 *Drain plug*
31 *Washer*
32 *Fuel tank*
33 *Nut*
34 *Washer*
35 *Washer*
36 *Feed pipe*
37 *Flexible pipes*
38 *Spacers*
39 *Return pipe*

the latter method, finally fill the tank with boiling water and detergent and allow to stand for at least three hours.

4 Temporary repairs to the fuel tank to stop leaks are best carried out using resin adhesives and hardeners as supplied in most accessory shops. In cases of repairs being done to large holes, fibre glass mats or perforated zinc sheet may be required, to give area support. If any soldering, welding or brazing is contemplated, the tank must be steamed out, as described in paragraph 3.

17 Fuel gauge sender unit

1 If the fuel gauge fails to give a reading with the ignition on or reads 'FULL' all the time, then a check must be made to see if the fault is in the gauge, sender unit, or wire in between.

2 Turn the ignition on and disconnect the light blue/yellow wire from the fuel tank sender unit. Check that the fuel gauge needle is on the empty mark. To check if the fuel gauge is in order, now earth the fuel tank sender unit wire. This should send the needle to the full mark.

3 If the fuel gauge is in order check the wiring for leaks or loose connections. If none can be found, then the sender unit will be at fault and must be replaced.

4 The sender unit is readily accessible when the tank has been removed, so we recommend that this is done before attempting to remove the sender unit.

5 Remove the six nuts and washers that secure the sender unit. Before the unit can be lifted out it will probably be necessary to prize up the unit flange from its gasket. Ease out the unit carefully to avoid bending the float arm.

6 If the accuracy of the sender unit is in doubt measure the resistance between the terminal 'T' (Light-blue/yellow wire) on the sender unit, and the body of the unit (earth). Pivot the float arm to simulate a full tank: the resistance should be high - about 90 ohms. Let the float fall to its lowest point, ie; simulating an empty tank: the resistance should now be low, about 3 to 8 ohms. If these measurements are not achieved the unit should be renewed.

7 The fuel reserve supply indicator light on the instrument panel should light when the level in the tank has fallen to less than ½ to 1 gallon (2.25 to 4.55 litres). If it is not functioning, check the circuit between terminal 'W' on the sender unit and the lamp on the instrument panel. When the capacity of the tank has fallen to the level given above, if one test lead of an ohmmeter is applied to terminal 'W' and the other lead connected to earth, a full scale deflection of the meter should be observed. If not, then the sender unit is faulty.

8 If the gauge reading or reserve light operation are faulty, but the preceding tests have proved that the sender unit is functioning satisfactorily, then it only remains to check the wiring between the sender unit and instrument panel in more detail.

9 When replacing the unit, fit a new gasket between the unit and the tank.

18 Positive crankcase ventilation system

1 The PCV system is essentially a device for recirculating oil vapours and gases from the crankcase back into the cylinders where they are burnt, rather than allowing them to escape to the atmosphere. The positive crankcase ventilation system comprises a valve on the carburettor, two short lengths of piping and the air cleaner. Basically, it functions as follows: With the engine running at cruise power, the crankcase fumes and blow-by gases are drawn up into the rocker cover from which they pass to the metering valve on the carburettor, via the air cleaner.
2 When the engine is at cruise power it is sufficient to send the fumes direct in the main stream of air. But the flow needs better control when the engine is idling.
3 To give this control there is a valve on the carburettor, by the throttle, that has a calibrated orifice when in the idling position.
4 It is a FIAT recommended procedure that this valve is cleaned as a part of Routine Maintenance.
5 Release the valve by undoing the nut on the end of the throttle spindle.
6 The valve must be cleaned in a carbon solvent, and washed off rather than scraped. FIAT recommend 70% paraffin (kerosene) and 30% butylcellosolva.
7 Clean the lengths of breather pipe and wash out the anti-backfire device, in the thicker pipe, in petrol.

19 Fuel system - fault diagnosis

There are three main types of fault the fuel system is prone to, and they may be summarised as follows:-

a) Lack of fuel at engine
b) Weak mixture
c) Rich mixture

A Lack of Fuel at Engine

1 If it is not possible to start the engine, first positively check that there is fuel in the fuel tank, and then check the ignition system as detailed in Chapter 4. If the fault is not in the ignition system then disconnect the fuel inlet pipe to the carburettor and turn the engine over. The inlet pipe is the one marked with an inward pointing arrow; the other pipe is the recirculatory pipe back to the tank.
2 If petrol squirts from the end of the inlet pipe, reconnect the pipe and check that the fuel is getting to the float chamber. This is done by unscrewing the bolts from the top of the float chamber, and lifting the cover just enough to see inside.
3 If fuel is there then it is likely that there is a blockage in the starting jet, which should be removed and cleaned.
4 No fuel in the float chamber is caused either by a blockage in the pipe between the pump and float chamber or a sticking float chamber valve. Alternatively the gauze filter at the top of the float chamber may be blocked. Remove the securing nut and check that the filter is clean. Washing in petrol will clean it.
5 If it is decided that it is the float chamber valve that is sticking, remove the fuel inlet pipe, and lift the cover, complete with valve and floats, away.
6 Remove the valve spindle and valve and thoroughly wash them in petrol. Petrol gum may be present on the valve or valve spindle and this is usually the cause of a sticking valve. Replace the valve in the needle valve assembly, ensure that it is moving freely, and then reassemble the float chamber. It is important that the same washer be placed under the needle valve assembly as this determines the height of the floats and therefore the level of petrol in the chamber. Check the float level as indicated in Section 10.
7 Reconnect the fuel pipe and refit the air cleaner.
8 If no petrol squirts from the end of the pipe leading to the carburettor then disconnect the pipe leading to the inlet side of the fuel pump. If fuel runs out of the pipe then there is a fault in the fuel pump, and the pump should be checked as has already been detailed.
9 No fuel flowing from the tank when it is known that there is fuel in the tank indicates a blocked pipe line. The line to the tank should be blown out. Remember that this model also has a return line back to the tank. It is unlikely that the fuel tank vent would become blocked, but this could be a reason for the reluctance of the fuel to flow. To test for this, blow into the tank down the fill orifice. There should be no build up of pressure in the fuel tank, as the excess pressure should be carried away down the vent pipe.

B Weak Mixture

1 If the fuel/air mixture is weak there are six main clues to this condition:

a) The engine will be difficult to start and will need much use of the choke, stalling easily if the choke is pushed in.
b) The engine will overheat easily.
c) If the spark plugs are examined (as detailed in Section 8, Chapter 4), they will have a light grey/white deposit on the insulator nose.
d) The fuel consumption may be light.
e) There will be a noticeable lack of power.
f) During acceleration and on the over-run there will be a certain amount of spitting back through the carburettor.

2 As the carburettors are of the fixed jet type, these faults are invariably due to circumstances outside the carburettor. The only usual fault likely in the carburettor is that one or more of the jets may be partially blocked. If the car will not start easily but runs well at speed, then it is likely that the starting jet is blocked, whereas if the engine starts easily but will not rev, then it is likely that the main jets are blocked.

3 If the level of petrol in the float chamber is low this is usually due to a sticking valve or incorrectly set floats.
4 Air leaks either in the fuel lines, or in the induction system should also be checked for.
5 The fuel pump may be at fault as has already been detailed.

C Rich Mixture

1 If the fuel/air mixture is rich there are also six main clues to this condition:

a) If the spark plugs are examined they will be found to have a black sooty deposit on the insulator nose.
b) The fuel consumption will be heavy.
c) The exhaust will give off a heavy black smoke, especially when accelerating.
d) The interior deposits on the exhaust pipe will be dry, black and sooty (if they are wet, black and sooty this indicates worn bores, and much oil being burnt).
e) There will be a noticeable lack of power.
f) There will be a certain amount of back-firing through the exhaust system.

2 The faults in this case are usually in the carburettor and the most usual is that the level of petrol in the float chamber is too high. This is due either to dirt behind the needle valve, or a leaking float which will not close the valve properly, or a sticking needle.
3 With a very high mileage (or because someone has tried to clean the jets out with wire), it may be that the jets have become enlarged.
4 If the air correction jets are restricted in any way the mixture will tend to become very rich.
5 Occasionally it is found that the choke control is sticking or has been maladjusted.
6 Again, occasionally the fuel pump pressure may be excessive so forcing the needle valve open slightly until a higher level of petrol is reached in the float chamber.

Chapter 4 Ignition system

For modifications, and information applicable to later models, see Supplement at end of manual

Contents

Specifications

Distributor

Type	Marelli or Ducellier
Timing marks	On bellhousing at 10°, 5° and 0° BTDC
Timing (static)	10° BTDC
Timing (dynamic)	10° BTDC at 850 rpm (vacuum hose disconnected)
Centrifugal advance	Starts at 900 rpm $\pm$ 200 rpm. Complete at 4,700 rpm = $28^o \pm 2^o$
Firing order	1 - 3 - 4 - 2
Contact breaker gap	0.015 - 0.017 in (0.37 - 0.43 mm)
Contact dwell angle	$55^o \pm 3^o$
Contact breaker spring pressure	19.4 $\pm$ 1.76 ozs (550 $\pm$ 50 g)
Distributor rotation	Clockwise
Lubricant type/specification	Multigrade engine oil, viscosity SAE 15W/40 (Duckhams Hypergrade)

Spark plugs

Make and type	Champion RN9YCC or RN9YC
Electrode gap:	
Champion RN9YCC	0.032 in (0.8 mm)
Champion RN9YC	0.028 in (0.7 mm)

HT leads

903 cc (right-angle connectors)	Champion CLS5 (boxed set)
903 cc (straight connectors)	Champion CLS8 (boxed set)

Coil

Types and resistances at 68^oF (20^oC)	**Primary**	**Secondary**
Marelli BE 200B	3.1 – 3.4 ohm	6,750 – 8,250 ohm
Martinetti G 52S	3.0 – 3.3 ohm	6,500 – 8,000 ohm
Bosch 0221.102.049	3.0 – 3.4 ohm	7,000 – 9,300 ohm

Condenser	0.20 - 0.25 microfarads at 50 - 1000 Hz
Resistance terminals to earth	About 50 megohms at 500 V DC

1 General description

1 For the engine to run correctly it is necessary for an electrical spark to ignite the fuel/air mixture in the combustion chamber at exactly the right moment in relation to engine speed and load. The ignition system is based on feeding low tension current to the coil where it is converted to high tension current. This high voltage is powerful enough to jump the spark plug gap in the cylinders under high pressures, providing that the system is in good condition and that all adjustments are correct.

2 The ignition system is divided into two circuits, the low tension circuit and the high tension circuit.

The low tension (sometimes known as the primary) circuit consists of the battery; lead to the control box; lead to the ignition switch; lead from the ignition switch to the low tension or primary coil windings, and the lead from the low tension coil windings to the contact breaker points and condenser in the distributor.

The high tension circuit consists of the high tension or secondary coil windings, the heavy ignition lead from the centre of the coil to the centre of the distributor cap, the rotor arm, and the spark plug leads and spark plugs.

The system functions in the following manner: High tension voltage is generated in the coil by the interruption of the low tension circuit. The interruption is effected by the opening of the contact breaker points in this low tension circuit. High tension voltage is fed from the centre of the coil, via the carbon brush in the centre of the distributor cap, to the rotor arm of the distributor.

The rotor arm revolves at half engine speed inside the distributor cap, and each time it comes in line with one of the four metal segments in the cap, which are connected to the spark plug leads, the opening of the contact breaker points causes the high tension voltage to build up, jump the gap from the rotor arm to the appropriate metal segment, and so via the spark plug lead to the spark plug, where it finally jumps the spark plug gap before going to earth.

3 The ignition is advanced and retarded automatically, to ensure the spark occurs at just the right instant for the particular load at the prevailing engine speed.

The ignition advance is controlled mechanically. The mechanical governor mechanism comprises two weights, which move out from the distributor shaft as the engine speed rises, due to centrifugal force. As they move outwards they rotate the cam relative to the distributor shaft, and so advance the spark. The weights are held in position by two springs and it is the tension of the springs which is largely responsible for correct spark advancement.

4 Lack of maintenance of the ignition system is the prime cause of difficult starting. In particular, cleanliness of the HT leads and distributor cap is vital. See Routine Maintenance.

2 Contact breaker - adjustment

1 To adjust the contact breaker points so that the correct gap is obtained, first release the two clips securing the distributor body, and lift away the cap. Clean the inside and outside of the cap with a dry cloth. It is unlikely that the four segments will be badly burned or scored, but if they are, the cap must be renewed. If only a small deposit is on the segments it may be scraped away using a small screwdriver. (photo)

2 Push in the carbon brush located in the top of the cap several times to ensure that it moves freely. The brush should protrude by at least a quarter of an inch.

3 Gently prise open the contact breaker points to examine the condition of their faces. If they are rough, pitted or dirty, it will be necessary to remove them for resurfacing, or for replacement points to be fitted.

4 To make adjustment easier, note the position of the rotor arm relative to the automatic timing control table. Undo the two securing screws and lift away the rotor.

5 It will be noticed that it is not possible to fit the rotor the wrong way round due to the round and square pegs.

6 Presuming the points are satisfactory, or they have been cleaned or replaced, measure the gap between the points by turning the engine over until the contact breaker arm is on the peak of one of the four cam lobes. An 0.014 to 0.017 inch (0.37 to 0.43 mm) feeler gauge should now just fit between the points. (photo).

7 If the gap varies from this amount, slacken the contact plate securing screw and adjust the contact gap by inserting a screwdriver in the notched hole of the stationary point and table and turn in the required direction, to increase or decrease the gap.

8 Tighten the securing screw and check the gap again.

9 Replace the rotor and secure with the two screws. Refit the distributor cap and clip the spring blade retainers into position.

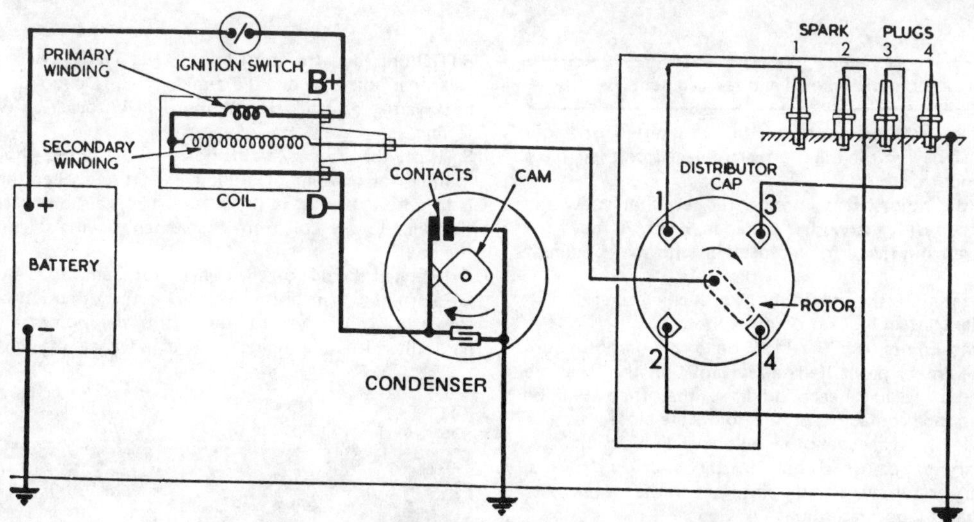

Fig. 4.1. Schematic diagram of typical ignition system (Sec. 1)

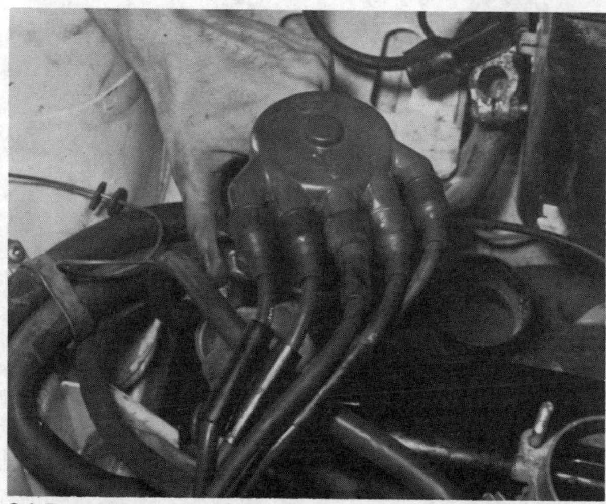

2.1 Removing the distributor cap

2.6 Measuring the points gap

3.2 Disconnecting the LT and condenser cables

3.4 Lifting away the contact breaker points

3 Contact breaker points - removal and replacement

1 If the contact breaker points are burned, pitted or badly worn, they must be removed and either replaced or their faces must be filed smooth.
2 To remove the points, first unscrew the terminal screw and remove it together with the washer under its head. Remove the low tension cable from the terminal, together with the condenser cable. (photo)
3 Unscrew and remove the contact breaker locking screw.
4 Lift away the contact breaker points. (photo)
5 To reface the points, rub the faces on a fine carborundum stone or on fine emery paper. It is important that the faces are rubbed flat and parallel to each other so that there will be complete face to face contact when the points are closed. One of the points will be pitted and the other will have deposits on it.
6 It is necessary to remove completely the built up deposits, but not necessary to rub the pitted point right to the stage where all the pitting has disappeared, though obviously if this is done it will prolong the time before the operation of refacing the points has to be repeated.
7 Refitting the contact breaker points is the reverse sequence to removal. It will be necessary to adjust the points gap as described in Section 2.

4 Ignition timing resetting

1 As the contact breaker heel wears, and with the variations inherent in different sets of contacts, the timing must be reset whenever the contact breaker points are cleaned or replaced. If this is not done engine power and efficiency will be lowered, and idling may be uneven or unreliable. If the timing is retarded the idle will be smooth, but slow. If advanced, it will tend to be fast, but uneven. Variations of at least 3° are needed to make the idling speed obviously incorrect. But such variations will be showing more important loss of efficiency in fuel consumption, power, and exhaust emissions if measured with suitable equipment. If the timing has been completely lost, as opposed to needing resetting, refer first to the next Section.
2 For static timing, first take out the spare wheel and remove the rubber shroud from the timing hole in the bellhousing. Then turn the engine over by putting the car in gear and moving it forwards, till the first, the 10° mark on the bellhousing is opposite the spot on the flywheel. Note that the first mark on the bellhousing is the 10° BTDC one, the second mark, the 5° BTDC one, and the third one 0° (ie; TDC). These positions are clearly marked below the timing hole. Do not turn the engine backwards, as backlash in the distributor drive will upset the timing.
3 With the engine at this position the points should be at the moment of opening. The ideal way to see when the points open is to wire a 12 volt bulb across the contact breaker points, using the wire to the coil from the switch, having taken the wire off the coil.
4 Slacken the nut on the distributor clamping plate.
5 Turn the distributor slightly in the direction of rotation of the cam (clockwise) to make sure the points are shut. Then carefully turn it anticlockwise to advance the ignition, against

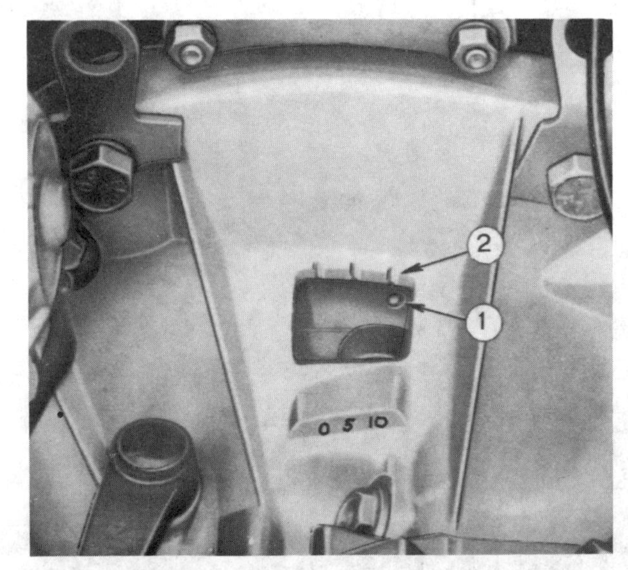

Fig. 4.2. Ignition timing marks (Sec. 4)

1 Spot on flywheel
2 10° BTDC timing mark above the aperture

Note: By carefully looking at the marks **below** the aperture it can be seen that the first mark is the 10° BTDC mark, the second the 5° BTDC, and the third the TDC mark

the direction of rotation, until the points open, as shown by the light going out. If not using a light, then with the ignition on, a spark can sometimes be seen at the points, or if an ammeter is fitted, an assistant can watch this for a flicker.
6 Reclamp the distributor.
7 Now recheck, by turning the engine over, forwards again, till the timing spot on the flywheel is coming up to the mark again. Watch the timing light and the flywheel. Turn the engine smoothly and slowly, and see where the spot was when the timing light went out. It should of course be by the 10° BTDC mark.
8 The timing can be set with the engine running, using a stroboscope. This method is more accurate, and such accuracy is needed to meet the American emissions regulations.
9 Connect up the timing light to No 1 cylinder plug lead in accordance with the makers instructions. Start up the engine. Check the idle speed is correct with a tachometer. Shine the light on the timing marks. Slacken the distributor, and move it as required to get the correct relationship of the spot on the flywheel with the mark on the bellhousing as 'frozen' by the light. Reclamp the distributor. Speed up the engine, and check the automatic advance is working.
10 Unless an accurate tachometer is in use, setting the timing by stroboscope will be inaccurate, as the automatic advance is varying the timing as the engine speeds up.

5 Distributor - removal and replacement

1 If the distributor is removed, the timing will be lost. It is quite straightforward to set it up again, but do not take it out unless you realise what is involved.
2 If there is no need to rotate the engine whilst the distributor is out, then the problem can be eased. To do this, turn the engine over until the timing spot on the flywheel is opposite the timing mark on the bellhousing (the first, 10° BTDC mark). Note to which segment the rotor arm is pointing. It should be cylinder No 1.
3 Undo the nut on the clamp holding the distributor to the engine. Take off the clamp.
4 Pull the distributor out of the engine. Take care as the spindle comes out of the cylinder block in case the drive gear has come as well.
5 If the gear did not come out and it is needed, then a suitable extractor is a piece of wood a little fatter than a pencil, sharpened to a gradual taper.
6 When refitting the distributor, first oil the gear, and put it back in place with the piece of wood/finger.
7 Reset the contact breaker points, as this is much easier done with the distributor on the bench. (See Section 2).
8 If the engine has not been turned over since the distributor was removed, put the rotor arm at the appropriate position for cylinder No 1.
9 If the engine has been rotated, turn it again until No 1 cylinder is on the compression stroke, ie; with the plug removed and a finger over the plug hole, pressure can be felt as the piston rises (and the inlet and outlet valves should be closed). Next, check that the flywheel timing mark is lined up with the 10° mark on the bellhousing. Set the distributor so that the rotor arm will point to the segment in the cap for No 1 cylinder.
10 Slide the distributor down into the engine, turning the spindle slightly to engage the splines in the gear.
11 Fit the clamping plate, and put on the nut finger tight.
12 Wire up the timing light and set the static timing as described in Section 4. If it is intended to use a stroboscope, still set the timing now, so it is accurate enough to start the engine.
13 Refit the distributor cap, checking that all the HT leads go to the correct plugs, in the firing order 1—3—4—2.

6 Distributor - dismantling and inspection

1 Apart from the contact points the other parts of a distributor which deteriorate with age and use, are the cap, the rotor, the shaft bushes, and the bob weight springs.
2 The cap must have no flaws or cracks and the 4 HT terminal contacts should not be severely corroded. The centre spring-loaded carbon contact is replaceable. If in any doubt about the cap buy a new one.
3 The rotor deteriorates minimally but with age the metal conductor may corrode. It should not be cracked or chipped and the metal conductor must not be loose. If in doubt renew it. Always fit a new rotor if fitting a new cap. Assuming the cover and rotor have been removed, proceed with dismantling as follows:
4 Release the low tension and condenser cables from the terminal on the contact set, then remove the contact set by releasing the retaining screws.
5 If it is necessary to remove the centre shaft from the distributor body, use a small parallel pin punch to drive out the pin securing the collar to the bottom of the centre shaft. Recover the washer from under the collar and slide out the centre shaft.
6 Check the contact breaker points as described in Section 2.
7 Examine the balance weights and pivot pins for wear, and renew the weights or centre shaft if a degree of wear is found.
8 Examine the centre shaft and the fit of the cam assembly on the shaft. If clearance is excessive compare the items with new units, and renew either, or both, if they show excessive wear.
9 If the shaft is a loose fit in the distributor bushes and can be seen to be worn it will be necessary to fit a new shaft and bushes.
10 Examine the length of the balance weight springs and compare them with new springs. If they have stretched, they should be renewed.
11 Reassembly is a straightforward reversal of the dismantling process, but there are several points which should be noted.
12 Lubricate the balance weights and other parts of the mechanical advance mechanism, and the distributor centre shaft, with engine oil during assembly. Do not oil excessively but ensure these parts are adequately lubricated.
13 Check the action of the weights in the fully advanced and fully retarded positions and ensure they are not binding.
14 Finally, set the contact breaker gap to the correct clearance (refer to Section 2).

Fig. 4.3. Removing the distributor (Sec. 5)

1 Ignition distributor
2 Bracket attachment nut
3 Distributor bracket

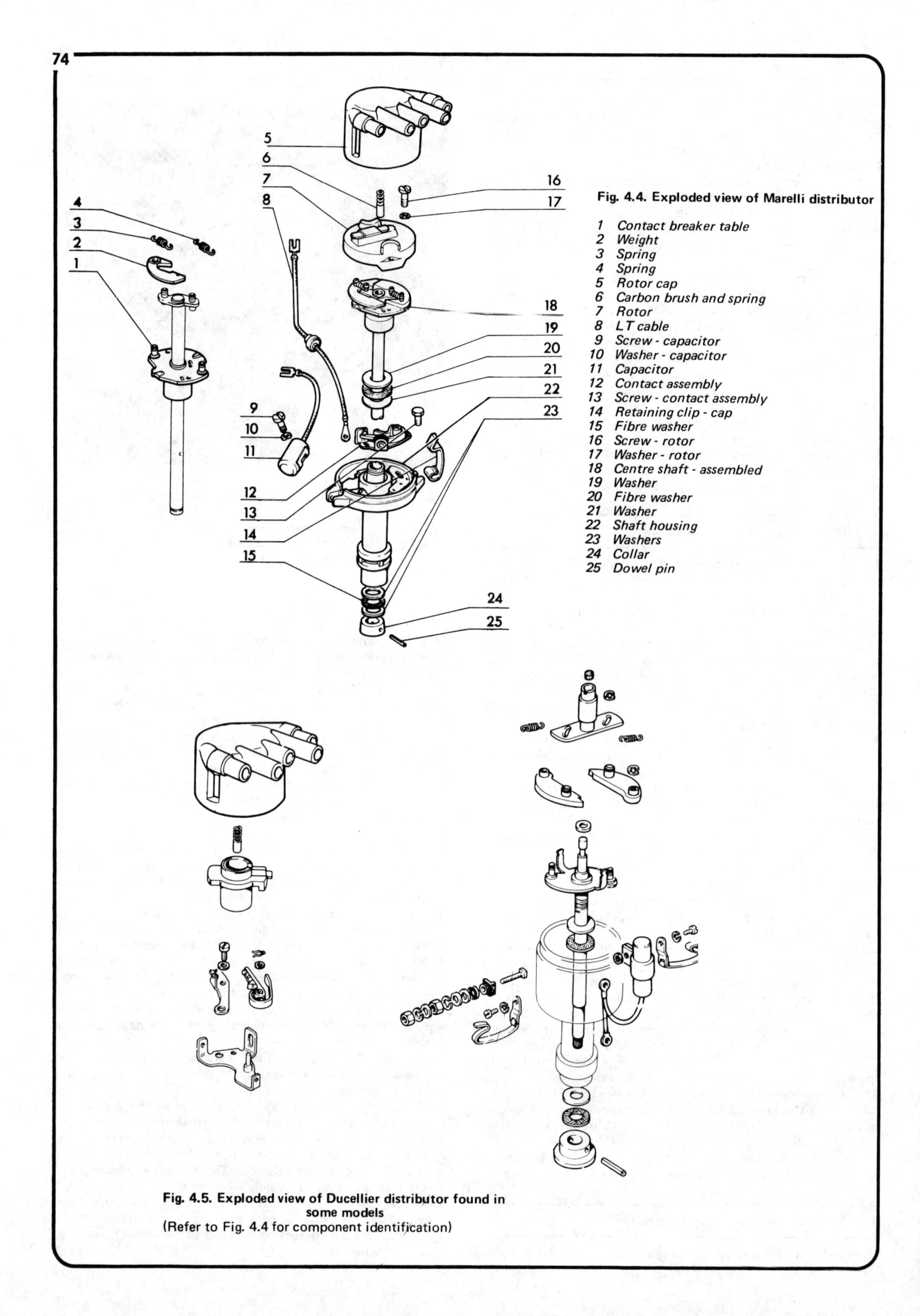

Fig. 4.4. Exploded view of Marelli distributor

1 *Contact breaker table*
2 *Weight*
3 *Spring*
4 *Spring*
5 *Rotor cap*
6 *Carbon brush and spring*
7 *Rotor*
8 *LT cable*
9 *Screw - capacitor*
10 *Washer - capacitor*
11 *Capacitor*
12 *Contact assembly*
13 *Screw - contact assembly*
14 *Retaining clip - cap*
15 *Fibre washer*
16 *Screw - rotor*
17 *Washer - rotor*
18 *Centre shaft - assembled*
19 *Washer*
20 *Fibre washer*
21 *Washer*
22 *Shaft housing*
23 *Washers*
24 *Collar*
25 *Dowel pin*

Fig. 4.5. Exploded view of Ducellier distributor found in some models
(Refer to Fig. 4.4 for component identification)

Are your plugs trying to tell you something?

Normal.
Grey-brown deposits, lightly coated core nose. Plugs ideally suited to engine, and engine in good condition.

Heavy Deposits.
A build up of crusty deposits, light-grey sandy colour in appearance.
Fault: Often caused by worn valve guides, excessive use of upper cylinder lubricant, or idling for long periods.

Lead Glazing.
Plug insulator firing tip appears yellow or green/yellow and shiny in appearance.
Fault: Often caused by incorrect carburation, excessive idling followed by sharp acceleration. Also check ignition timing.

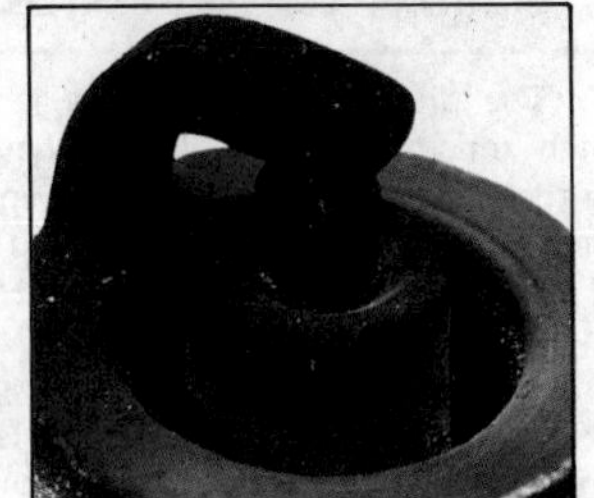

Carbon fouling.
Dry, black, sooty deposits.
Fault: over-rich fuel mixture.
Check: carburettor mixture settings, float level, choke operation, air filter.

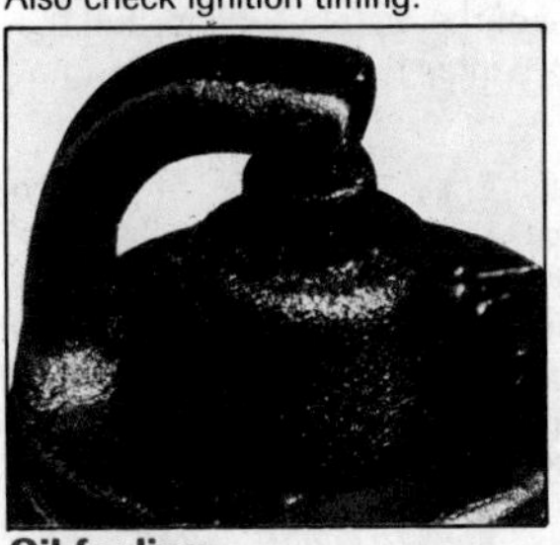

Oil fouling.
Wet, oily deposits. Fault: worn bores/piston rings or valve guides; sometimes occurs (temporarily) during running-in period.

Overheating.
Electrodes have glazed appearance, core nose very white – few deposits. Fault: plug overheating. Check: plug value, ignition timing, fuel octane rating (too low) and fuel mixture (too weak).

Electrode damage.
Electrodes burned away; core nose has burned, glazed appearance. Fault: pre-ignition. Check: for correct heat range and as for 'overheating'.

Split core nose.
(May appear initially as a crack). Fault: detonation or wrong gap-setting technique. Check: ignition timing, cooling system, fuel mixture (too weak).

WHY DOUBLE COPPER IS BETTER FOR YOUR ENGINE.

Unique Trapezoidal Copper Cored Earth Electrode
50% Larger Spark Area
Copper Cored Centre Electrode

Champion Double Copper plugs are the first in the world to have copper core in both centre and earth electrode. This innovative design means that they run cooler by up to 100°C – giving greater efficiency and longer life. These double copper cores transfer heat away from the tip of the plug faster and more efficiently. Therefore, Double Copper runs at cooler temperatures than conventional plugs giving improved acceleration response and high speed performance with no fear of pre-ignition.

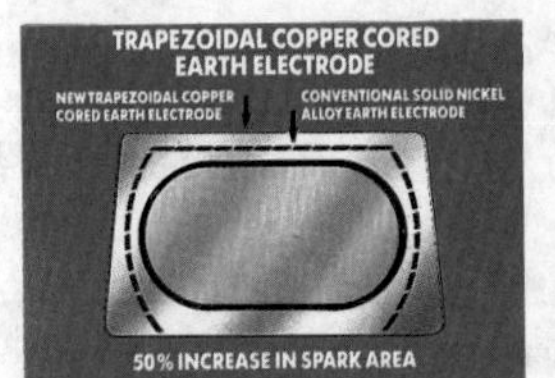

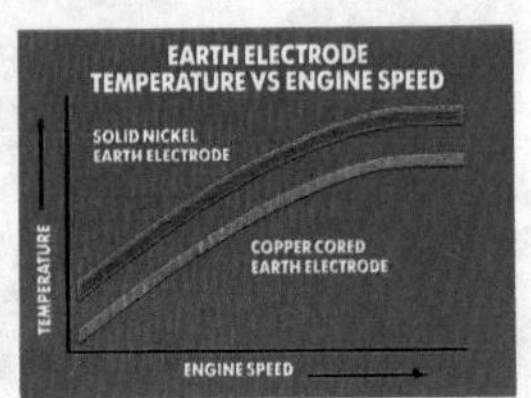

Champion Double Copper plugs also feature a unique trapezoidal earth electrode giving a 50% increase in spark area. This, together with the double copper cores, offers greatly reduced electrode wear, so the spark stays stronger for longer.

 **FASTER COLD STARTING**

 **FOR UNLEADED OR LEADED FUEL**

 **ELECTRODES UP TO 100°C COOLER**

 **BETTER ACCELERATION RESPONSE**

 **LOWER EMISSIONS**

 **50% BIGGER SPARK AREA**

 **THE LONGER LIFE PLUG**

Plug Tips/Hot and Cold.
Spark plugs must operate within well-defined temperature limits to avoid cold fouling at one extreme and overheating at the other.
Champion and the car manufacturers work out the best plugs for an engine to give optimum performance under all conditions, from freezing cold starts to sustained high speed motorway cruising.
Plugs are often referred to as hot or cold. With Champion, the higher the number on its body, the hotter the plug, and the lower the number the cooler the plug. For the correct plug for your car refer to the specifications at the beginning of this chapter.

Plug Cleaning
Modern plug design and materials mean that Champion no longer recommends periodic plug cleaning. Certainly don't clean your plugs with a wire brush as this can cause metal conductive paths across the nose of the insulator so impairing its performance and resulting in loss of acceleration and reduced m.p.g.
However, if plugs are removed, always carefully clean the area where the plug seats in the cylinder head as grit and dirt can sometimes cause gas leakage.
Also wipe any traces of oil or grease from plug leads as this may lead to arcing.

7 Condenser - removal, testing and replacement

1 The purpose of the condenser (sometimes known as capacitor) is to ensure that when the contact breaker points open there is no sparking across them which would weaken the spark and cause rapid deterioration of the points.
2 The condenser is fitted in parallel with the contact breaker points. If it develops a short circuit, it will cause ignition failure as the points will be prevented from interrupting the low tension circuit.
3 If the engine becomes very difficult to start or begins to misfire whilst running and the breaker points show signs of excessive burning, then suspect the condenser has failed. A further test can be made by separating the points by hand with the ignition switched on. If this is accompanied by a bright spark at the contact points it is indicative that the condenser has failed.
4 Without special test equipment the only sure way to diagnose condenser trouble is to replace a suspected unit with a new one and note if there is any improvement.
5 To remove the condenser from the distributor, release the screw which secures it to the breaker plate; and slacken the contact breaker terminal screw enough to remove the wire connection tag.
6 When fitting the condenser it is vital to ensure that the fixing screw is secure and the condenser tightly held. The lead must be secure on the terminal with no chance of short circuiting.

8 Spark plugs and high tension leads

1 The correct functioning of the spark plugs is vital for the correct running and efficiency of the engine. It is essential that the plugs fitted are appropriate for the engine (the correct type is specified at the beginning of this Chapter). If this type is used, and the engine is in good condition, the spark plugs should not need attention between scheduled service renewal intervals. Spark plug cleaning is rarely necessary, and should not be attempted unless specialised equipment is available, as damage can easily be caused to the firing ends.

2 At intervals of 6,000 miles/6 months the plugs should be removed and inspected. If they appear to be in good condition, the electrode gaps should be checked and adjusted if necessary, then the plugs refitted. If their condition is in any way suspect, it would be better to renew them as a set. The condition of the spark plugs will also tell them much about the overall condition of the engine.

3 If the insulator nose of the spark plug is clean and white, with no deposits, this is indicative of a weak mixture, or too hot a plug. (A hot plug transfers heat away from the electrode slowly - a cold plug transfers it away quickly).

4 If the tip of the insulator nose is covered with sooty black deposits, then this is indicative that the mixture is too rich. Should the plug be black and oily, then it is likely that the engine is fairly worn, as well as the mixture being rich.

5 If the insulator nose is covered with light tan to greyish brown deposits, then the mixture is correct and it is likely that the engine is in good condition and correctly tuned.

6 Clean around the plug seats in the cylinder head before removing the plugs to prevent dirt getting into the cylinder.

7 The spark plug gap is of considerable importance, as, if it is too large or too small the size of the spark and its efficiency will be seriously impaired. The spark plug gap should be set to the gap shown in the Specifications for the best results.
8 To set it, measure the gap with a feeler gauge, and then bend open, or close, the outer plug electrode until the correct gap is achieved. The centre electrode should never be bent as this may crack the insulation and cause plug failure, if nothing worse.
9 When replacing the plug see that the washer is intact and carbon free, also the shoulder of the plug under the washer. Make sure also that the plug seat in the cylinder head is quite clean.

10 Note that, as the cylinder head is of aluminium alloy, it is recommended that a little anti-seize compound (such as Copaslip) is applied to the plug threads before they are fitted. Screw in the plugs by hand until the sealing washer just seats, then tighten by no more than a quarter-turn using a spark plug spanner.

11 Replace the leads from the distributor in the correct firing order, which is 1–3–4–2; No 1 cylinder being the one nearest the drivebelt for the generator and water pump.

12 The plug leads require no routine attention other than being kept clean and wiped over regularly. At intervals, say twice yearly, pull each lead off the plug in turn and also from the distributor cap. Water can seep down into these joints giving rise to a white corrosive deposit which must be carefully removed from the brass connectors at the end of each cable, using a product such as Duckhams DPP, to prevent corrosion and dispel damp.

9 Distributor cap, rotor arm, and HT leads

1 The distributor cap, leads, and the rotor arm distribute the high tension to the plugs. They should all last long mileages without replacement. But in a very short mileage their state can be responsible for frustration and annoyance.
2 If the components of the HT circuit are dirty, they will attract damp. This damp will allow the HT current to leak away and will give difficulty in starting the engine when cold. The oil will be thick when the engine is cold, making the starter take more voltage and current, yet only cranking the engine slowly. The voltage drop due to the starter's demands means reduced voltage in the primary circuit, and a corresponding fall in that in the secondary, the HT, just at the moment a strong spark is needed to fire the cold cylinder.

3 Spraying the ignition leads with a water dispersant aerosol such as Holts Wet Start can be very effective. The leads, the spark plug insulators, the outside and the inside of the distributor cap, and the rotor arm must be clean and dry. In winter this will need doing at least at 3,000 mile intervals.

4 Some plug leads are 'resistive'. They are made of cotton impregnated with carbon, to give the necessary radio interference suppression without separate suppressors. After a time flexing makes the leads break down inside. If difficult starting persists, then new leads, with normal metal core, and suppressor end fittings, could effect a cure. End fittings should not be taken off the resistive wire, as it is difficult to fit them so that a good contact is achieved.

10 Coil

1 Coils normally last the life of the car. The most usual reason for a coil to fail is after being left with the ignition switched on but the engine not running. There is then constant current flowing, instead of the intermittent flow when the contact breaker is opening. The coil then overheats, and the insulation is damaged.
2 The contact breaker points should preferably not be flicked without a lead from the coil centre to some earth, otherwise the opening of the points will give a HT spark which, finding no proper circuit, could break down the insulation in the coil. When connecting a timing light for setting the ignition, this should come from the switch side of the coil, the coil itself being disconnected.
3 If the coil seems suspect after fault finding, the measurement of the resistance of the primary and secondary windings can establish the matter definitely. But if an ohmmeter is not available, then it will be necessary to try a new one.

11 Ignition system - fault diagnosis

There are two main symptoms indicating ignition faults. Either the engine will not start or fire, or the engine is difficult to start and misfires. If it is a regular misfire, ie; the engine is only running on two or three cylinders, the fault is almost sure to be in the secondary, or high tension, circuit. If the misfiring is intermittent, the fault could be in either the high or low tension circuits. If the engine stops suddenly, or will not start at all it is likely that the fault is in the low tension circuit. Loss of power and overheating, apart from faulty carburation settings, are normally due to faults in the distributor or incorrect ignition timing.

Engine fails to start

1 If the engine fails to start and it was running normally when it was last used, first check there is fuel in the petrol tank. If the engine turns over normally on the starter motor and the battery is evidently well charged, then the fault may be in either the high or low tension circuits. First check the HT circuit. **Note:** If the battery is known to be fully charged, the ignition comes on, and the starter motor fails to turn the engine, **check the tightness of the leads on the battery terminals** and also the secureness of the earth lead **connection to the body**. It is quite common for the leads to have worked loose, even if they look and feel secure. If one of the battery terminal posts gets very hot when trying to work the starter motor this is a sure indication of a faulty connection to that terminal.

2 One of the most common reasons for bad starting is wet or damp HT leads or distributor – FIAT distributor caps are particularly prone to internal condensation. Remove the distributor cap. If condensation is visible internally, dry the cap with a rag, and wipe over the leads. Refit the cap. Alternatively, using a moisture dispersant such as Holts Wet Start can be very effective in starting the engine. To prevent the problem recurring, Holts Damp Start can be used to provide a sealing coat, so excluding any further moisture from the ignition system. In extreme difficulty, Holts Cold Start will help to start a car when only a very poor spark occurs.

3 If the engine still fails to start, check that current is reaching the plugs, by disconnecting each plug lead in turn at the spark plug end, and holding the end of the cable about 3/16 inch (5 mm) away from the cylinder block. Have an assistant run the engine in short bursts.

4 Sparking between the end of the cable and the block should be fairly strong with a regular blue spark. (Hold the lead with rubber to avoid electric shocks). If current is reaching the plugs, then remove them and clean and regap them. The engine should now start.

5 If there is no spark at the plug leads take off the HT lead from the centre of the distributor cap and hold it to the block as before. Spin the engine on the starter once more. A rapid succession of blue sparks between the end of the lead and the block indicate that the coil is in order and that the distributor cap is cracked, the rotor arm faulty or the carbon brush in the top of the distributor cap is not making good contact with the spring on the rotor arm. Possibly the points are in bad condition. Clean and reset them as described in this Chapter.

6 If there are no sparks from the end of the lead from the coil, then check the connections of the lead to the coil and distributor head, and if they are in order, check out the low tension circuit.

7 Detach the thin cable from terminal 'D' on the coil. This wire connects the distributor to the coil. Next remove the other end of this thin cable from the distributor.

8 Connect a 0–20 voltmeter or a test lamp between these two terminals. Switch on the ignition and turn the engine over slowly in the normal direction of rotation, the voltmeter or test lamp shows a reading/lights up, when the contact breaker points close, and then shows no reading/light goes out, when they are open, it is a sure sign that the low tension circuit is satisfactory. Should the meter fail to show a reading or the light does not come on, inspect and clean the contact breaker points and check all connections for tightness.

9 If there is still a fault in the low tension circuit, switch on the ignition and systematically check the low tension circuit using the relevant wiring diagram, to be found in Chapter 10, as a guide. For these tests a 0–20 voltmeter is best but a test lamp may be used. The meter should show a reading of approximately 12 volts.

10 Connect the voltmeter between ignition switch terminal '30' and earth. No reading indicates a fault in the battery to ignition switch cable.

11 Next connect the voltmeter between ignition switch terminal '54' and earth. Switch on the ignition. No reading indicates an internal fault in the ignition switch. A new switch should be fitted.

12 Connect the voltmeter between ignition switch terminal '50' and earth. Move the ignition switch to the start position. No reading indicates an internal fault in the ignition switch. A new switch should be fitted.

13 Connect the voltmeter between ignition coil terminal 'B' and earth. No reading indicates a fault in the cable between the ignition switch and ignition coil.

14 Next disconnect the cable from ignition coil terminal 'D' and connect the voltmeter to this terminal and to earth. No reading indicates that there is a fault in the primary winding of coil. A new coil should be fitted.

15 Disconnect the thin cable from the terminal in the distributor and connect the voltmeter between the end of this cable and earth. No reading indicates a loose connection or damaged cable.

16 Finally connect the voltmeter across the contact breaker points. If no reading is obtained the condenser should be suspect.

Engine misfires

1 If the engine misfires regularly, run it at a fast idling speed and short out each of the plugs in turn by placing a short screwdriver across from the plug terminal to the cylinder. Ensure that the screwdriver has a **wooden** or **plastic insulated handle.**

2 No difference in engine running will be noticed when the plug in the defective cylinder is short circuited. Short circuiting the working plugs will accentuate the misfire.

3 Remove the plug lead from the end of the defective plug and hold it about 3/16 inch (5 mm) away from the block. Restart the engine. If the sparking is fairly strong and regular the fault must lie in the spark plug.

4 The plug may be loose, the insulation may be cracked, or the points may have burnt away giving too wide a gap for the spark to jump. Worse still, one of the points may have broken off. Either renew the plug, or reset the gap, and then test it.

5 If there is no spark at the end of the plug lead, or if it is weak and intermittent, check the ignition lead from the distributor to the plug. If the insulation is cracked or perished, renew the lead. Check the connections at the distributor cap.

6 If there is still no spark, examine the distributor cap carefully for tracking. This can be recognised by a very thin black line running between two or more electrodes, or between an electrode and some other part of the distributor. These lines are paths which now conduct electricity across the cap, thus letting it run to earth. The only answer is a new distributor cap.

7 Apart from the ignition timing being incorrect, other causes of misfiring have already been dealt with under the Section dealing with the failure of the engine to start.

8 If the ignition timing is too far retarded, it should be noted that the engine will tend to overheat, and there will be a quite noticeable drop in power. If the engine is overheating and the power is down, and the ignition timing is correct, then the carburettor should be checked, as it is likely that this is where the fault lies. See Chapter 3 for details on this.

Chapter 5 Clutch

For modifications, and information applicable to later models, see Supplement at end of manual

Contents

Specifications

Clutch

Type	Single, dry plate, diaphragm spring
Lining outer diameter	6.30 in. (160 mm), early models; 6.69 in. (170 mm), later models
Lining inner diameter	4.33 in. (110 mm), early models; 4.72 in. (120 mm), later models
Maximum runout of driven plate friction linings	0.01 in. (0.25 mm)

Clutch pedal free-travel, corresponding to a clearance of 0.079 in (2 mm) between friction ring and throwout sleeve ... ... 1 in. (25 mm)

Travel of release flange, corresponding to a pressure plate displacement not less than 0.055 in. (1.4 mm) ... ... 0.315 in. (8 mm)

Withdrawal mechanism ... ... Cable

Torque wrench settings

	lb f ft	kg f m
Clutch to flywheel	11	1.5
Clutch withdrawal fork to pivot	18	2.5
Clutch pedal pivot nut	11	1.5

1 General description

1 The vehicle is fitted with a single dry plate diaphragm clutch. The unit comprises a steel cover which is dowelled and bolted to the face of the flywheel and contains the pressure plate, pressure plate diaphragm spring and fulcrum rings.
2 The clutch disc is free to slide along the splined gearbox input shaft and is held in position between the flywheel and the pressure plate by the pressure of the pressure plate spring. Friction lining material is riveted to the clutch disc and it has a spring cushioned hub to absorb transmission shocks.
3 The circular diaphragm spring is mounted on shouldered pins and held in place in the cover by two fulcrum rings and rivets.
4 The clutch is actuated by cable.
5 Unlike in-line engines and gearboxes, the shaft through the clutch into the gearbox does not have a spigot bearing in the end of the crankshaft.

2 Clutch - removal and inspection

1 Remove the transmission (see Chapter 6).
2 Mark the position of the clutch cover relative to the flywheel.
3 Slacken off the bolts holding the cover to the flywheel in a diagonal sequence, undoing each bolt a little at a time. This keeps the pressure even all round the diaphragm spring and prevents distortion. When all the pressure is released on the bolts, remove them, lift the cover off the dowel pegs and take it off together with the friction plate which is between it and the flywheel.
4 Examine the diaphragm spring for signs of distortion or fracture.
5 Examine the pressure plate for signs of scoring or abnormal wear.
6 If either the spring or the plate is defective it will be necessary to replace the complete assembly with an exchange unit. The assembly can only be taken to pieces with special equipment and, in any case, individual parts of the assembly are not obtainable as regular spares.
7 Examine the friction plate for indications of uneven wear and scoring of the friction surfaces. Contamination by oil will also show as hard and blackened areas which can cause defective operation. If there has been a leak from engine or transmission this must be cured before reassembling the clutch. If the clearance between the heads of the securing rivets and the face of the friction lining material is less than 0.020 in. (0.5 mm), it is worthwhile to fit a new plate. Around the hub of the friction disc are six springs acting as shock absorbers between the hub and the friction area. These should be intact and tightly in position.
8 The face of the flywheel should be examined for signs of scoring or uneven wear and, if necessary, it will have to be renewed or reconditioned. See Chapter 1 for details of flywheel removal.
9 Also check the clutch withdrawal bearing (Section 4) before reassembly.
10 Clutch parts are relatively cheap compared with the labour costs, so it is best to replace them if in any doubt.

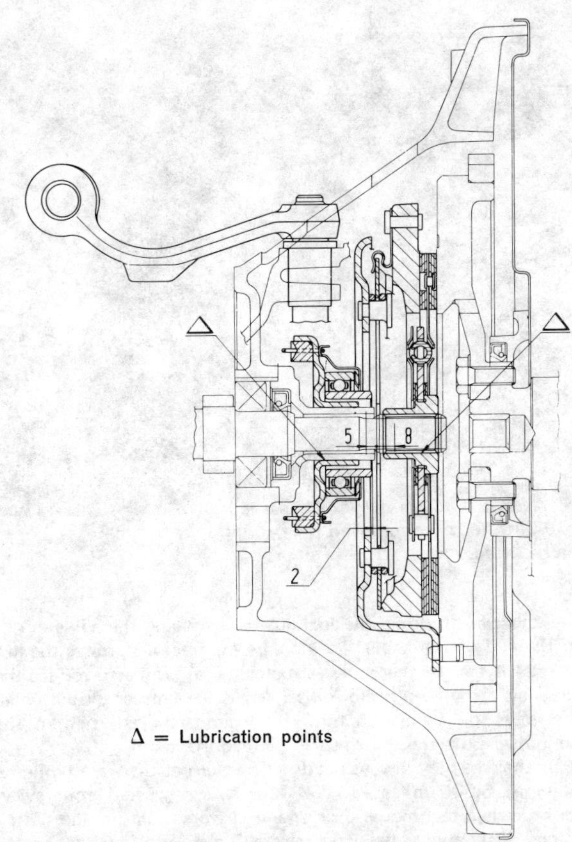

Fig. 5.1. Section through clutch and release mechanism

0.079 in (2 mm)	=	*Clearance to be obtained by adjusting release control cable*
0.197 in (5 mm)	=	*Maximum acceptable movement of diaphragm spring with worn clutch disc linings*
0.315 in (8 mm)	=	*Declutching travel of diaphragm spring*

Fig. 5.2. Centralising the clutch plate using FIAT tool (Sec. 3)

1 Clutch assembly
2 Tool
3 Flywheel
4 Retaining bolts

3 Clutch - refitting

1 Hold the clutch friction, or driven plate up against the flywheel. The disc holding the cushion springs and the longer boss must be away from the flywheel.
2 Put the clutch cover assembly over the plate, lining up the dowel holes, and fit the bolts. Tighten the bolts finger-tight, just enough to take the weight of the friction plate between the pressure plate and the flywheel.
3 Now centralise the clutch plate. The splines in the plate must be accurately lined up with the end of the crankshaft, otherwise when the clutch is clamped up tight, and the friction plate cannot be moved, the shaft will not fit through, and reassembly of transmission to engine will be impossible. FIAT use a special mandrel that fits into the end of the crankshaft. The alignment can be done satisfactorily using a metal or wooden bar, and checking carefully by eye.
4 Tighten the bolts round the edge of the clutch cover carefully and diagonally, gradually compressing the clutch spring.
5 Refit the transmission as described in Chapter 6.

4 Clutch withdrawal or release mechanism

1 To release the clutch, so that the drive is disengaged, the pressure plate is 'withdrawn' from the flywheel. The driver's pedal action, through the cable, pulls on the withdrawal lever on the outside of the transmission case, and the fork inside pushes the withdrawal bearing against the clutch.
2 The pivot for the lever fork should outlast the car.
3 The withdrawal or release bearing should not give trouble either. But should it fail the transmission must be removed from the car to reach it. So when the transmission is out, even though for some other work, the bearing should be looked at very critically, and if there is any doubt it should be renewed.
4 The bearing is clipped to the withdrawal fork; when the clip is released the bearing can be removed. It should not be immersed in cleaning fluid, as it is sealed. Liquid will get in, but cannot be dried out; nor can new lubricant be inserted. The

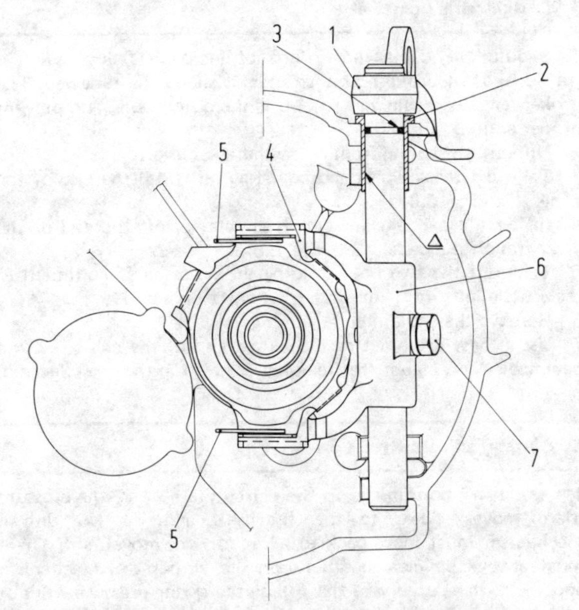

Fig. 5.3. Clutch release mechanism components (Sec. 4)

1 Clutch release lever and shaft
2 Bushing
3 Seal
4 Release sleeve
5 Sleeve retaining springs to release fork
6 Fork
7 Bolt and washer, fork to release shaft

4.4A The bearing is shown clipped to the release fork

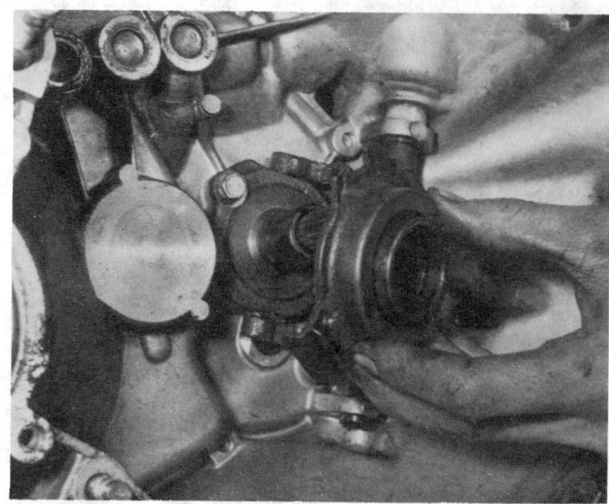

4.4B With the clips removed the bearing can be lifted off

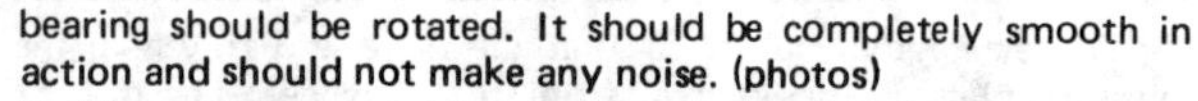

bearing should be rotated. It should be completely smooth in action and should not make any noise. (photos)

5 When fitting a new bearing the clips and the bearing surfaces of the fork should be lubricated with grease.

6 If the bearing is suspect, but the transmission is still in the car, the bearing failure can only be assessed by noise. If the actual bearings are breaking up, this will be shown by a squawking or groaning noise when the clutch pedal is pressed down. If the bearing is loose, it may rattle at speed. This may be quietened if the pedal is lightly pressed to hold the bearing. The car should not be driven in this condition, for if the bearing breaks up, repair is likely to be expensive, as other parts will be damaged.

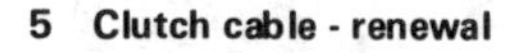

5 Clutch cable - renewal

1 Should the cable show signs of incipient failure such as frayed ends of broken cable strands, it should be renewed. The broken ends of cable may jam up in the outer cable, and prevent proper action.

2 Unhook the spring from the withdrawal lever.

3 Take off the locknut and adjusting nut from the end of the cable. (See Fig. 5.4).

4 Inside the car take the circlip or retainer off the end of the cable pin on the pedal. (See. Fig. 5.5).

5 Take out the two bolts holding the outer cable abutment to the scuttle in front of the driver, from the engine side.

6 Remove the cable.

7 Use a new grommet at the clutch end of the cable. Fit the new cable, and adjust the free-play as given in the next Section.

6 Clutch - adjustment

1 As the friction linings on the clutch plate wear, the pressure plate moves closer to the flywheel, and the withdrawal mechanism must move back to allow for this movement. There must always be clearance between the withdrawal mechanism and the clutch, otherwise the full clutch spring pressure will not be available, and the clutch will slip, and quickly ruin itself. Also the life of the withdrawal bearing will be shortened.

2 The small clearance of 0.07 in. (2 mm) at the clutch itself, appears as 1 in. (25 mm) at the pedal.

3 This free-movement at the pedal can be felt as movement against the pull-off spring on the clutch withdrawal lever, before the firmer pressure needed to free the clutch itself.

4 Adjust by undoing the locknut, and moving the adjuster nut on the end of the cable. To increase the free-play, move the nut nearer the end of the cable (anticlockwise), while to reduce the free-play turn the nut clockwise. When the correct clutch pedal free-travel of 1 in. (25 mm) has been achieved, tighten the locknut. Ensure the return spring is reconnected.

5 If the free-play always needs adjustment at the stated mileage intervals, 6,000 miles (10,000 km) it is suggested you review your driving technique. For many drivers wear in the clutch linings just about balances stretch and wear in the clutch withdrawal mechanism, and no adjustment is ever needed.

Fig. 5.4. Clutch cable adjustment details (Sec. 6)

1 *Cable*
2 *Locknut*
3 *Adjustment head*
4 *Release fork return spring*

Fig. 5.5. Clutch pedal, cable and release mechanism details

1 *Return spring*
2 *Release sleeve to fork retaining spring*
3 *Clutch*
4 *Release sleeve*
5 *Bolt and lockwasher*
6 *Clutch release lever and shaft*
7 *Seal*
8 *Bushing*
9 *Fork*
10 *Adjustment head*
11 *Locknut*
12 *Flat washer*
13 *Cable*
14 *Pedal support*
15 *Nut*
16 *Washers*
17 *Retainer*
18 *Cable attachment bolt*
19 *Pedal rubber pad*
20 *Pedal*

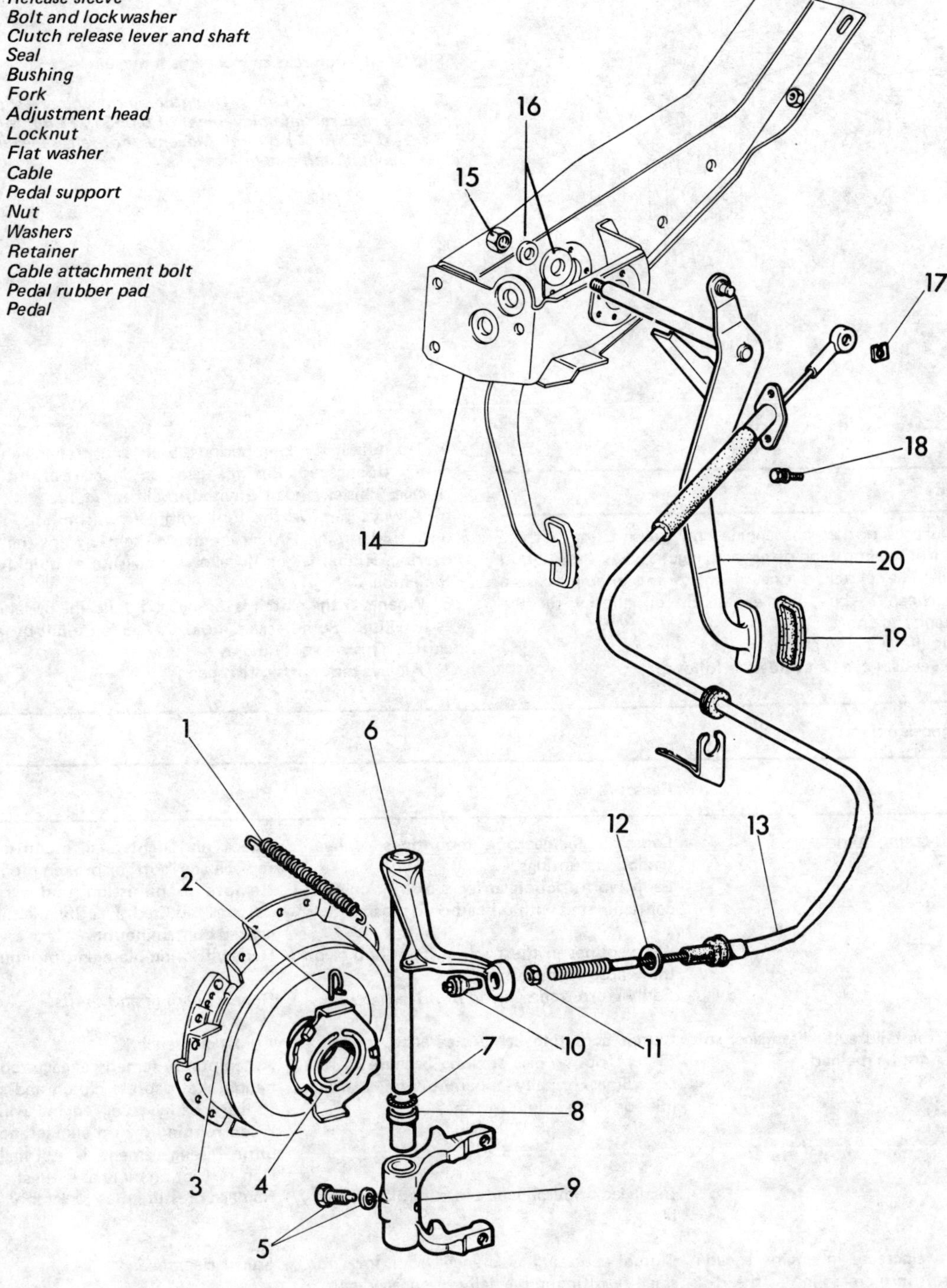

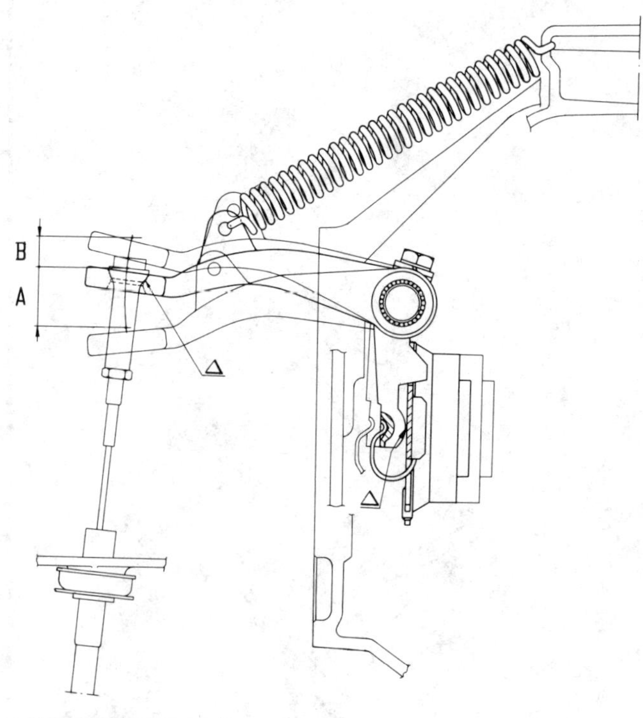

Fig. 5.6. Clutch release mechanism movement

A = 0.905 in (23 mm). Declutching travel, corresponding to a pressure plate movement of 0.055 in (1.4 mm)
B = 0.453 in (11.5 mm). Movement of release lever with worn out clutch disc linings

7 Clutch life

1 Some clutches outlast the engine and transmission. Others require attention almost as often and regularly as the 12,000 mile service. The clutch is one of the prime examples of a component cheap to buy, but expensive (or lengthy for the home mechanic) to fit.
2 The short life is due to abuse.
3 Common abuses can be avoided as follows:
4 Do not sit for long periods, such as at traffic lights, with the clutch disengaged. Put the gearbox in neutral until the lights change. This extends the withdrawal bearing life.
5 Always remove the foot completely from the clutch pedal once under way. Riding the clutch lightly ruins the withdrawal bearing. Riding it heavily allows it to slip, so quickly wears out the lining.
6 Whenever the clutch is disengaged, hold the pedal down as far as it will go. Some drivers hold the car on a hill by slipping the clutch. This wears it quickly.
7 Always move off in first gear.

8 Fault diagnosis - clutch

Symptom	Reason/s	Remedy
Judder when taking up drive	Loose engine/gearbox mountings or over-flexible mountings	Check and tighten all mounting bolts and replace any 'soft' or broken mountings.
	Badly worn friction surfaces or friction plate contaminated with oil carbon deposit	Remove transmission and replace clutch parts as required. Rectify the oil leak which caused contamination.
	Worn splines in the friction plate hub or on the gearbox input shaft.	Renew friction plate and/or input shaft.
	Badly worn transmission or driveshafts	Renew bearings and shafts
Clutch drag (or failure to disengage) so that gears cannot be meshed	Clutch actuating cable clearance too great	Adjust clearance.
	Clutch friction disc sticking because of rust on splines (usually apparent after standing idle for some length of time)	As temporary remedy engage top gear, apply handbrake, depress clutch and start engine. (If very badly stuck engine will not turn). When running rev up engine and slip clutch until disengagement is normally possible. Renew friction plate at earliest opportunity.
	Damaged or misaligned pressure plate assembly	Replace pressure plate assembly.
Clutch slip - (increase in engine speed does not result in increase in car speed - especially on hills)	Clutch actuating clearance from fork too small resulting in partially disengaged clutch at all times	Adjust clearance.
	Clutch friction surfaces worn out (beyond further adjustment of operating cable) or clutch surfaces oil soaked	Replace friction plate and remedy source of oil leakage.
	Damaged clutch spring	Fit reconditioned assembly.

Chapter 6 Transmission

For modifications, and information applicable to later models, see supplement at end of manual

Contents

Specifications

Speeds ... Four forward and one reverse

Synchromesh (spring ring type) ... 1st, 2nd, 3rd and 4th gears

Gear type

Forward	Constant mesh, helical toothed
Reverse	Straight toothed, with sliding idler gear

Gear ratios

First	Early models 3.636 : 1, later models 3.910 : 1
Second	2.055 : 1
Third	1.348 : 1
Fourth	Early models 0.977 : 1, later models 0.963 : 1
Reverse	3.615 : 1

Final drive gears ... Helical-toothed

Final drive ratio ... Early models 4.692 : 1, later models 4.071 : 1

Overall ratios

Gears	1st	2nd	3rd	4th	Reverse
Reduction ratio	17.06	9.64	6.32	4.58	16.96

Differential case bearings

Number	Two
Bearing type	Taper roller
Bearing preload setting	By shims

Side to planet gear backlash adjustment ... By thrust washers; selected from range 0.7 mm to 1.3 mm

Tolerances

Gear backlash	0.004 - 0.008 in (0.1 - 0.2 mm)
Clearance reverse idler to shaft	0.003 - 0.006 in (0.08 - 0.15 mm)
Clearance other gears to second shaft	0.001 - 0.003 in (0.04 - 0.08 mm)

Ball bearings

Side play; maximum	0.002 in (0.05 mm)
End play; maximum	0.02 in (0.50 mm)

Differential bearings

Differential preload: Interference of	0.003 in (0.08 mm)
Preload shims: Seven thicknesses	0.04 mm to 1.0 mm

Lubricant

Type/specification	Multigrade engine oil, viscosity SAE 15W/40 (Duckhams Hypergrade)
Capacity	4.25 Imp. pints (2.4 litres)

Torque wrench settings

	lb f ft	kg f m
Differential halves/crownwheel	51	7
Bolts securing selector forks and levers to rods (Size M6) (10 mm AF)	14½	2
Other nuts/bolts (Size M6) (10 mm AF)	7	1
Nuts, Gearbox casing to clutch housing	18	2.5
Nuts, differential flange to transmission	18	2.5
Bolt and nuts, Transmission to engine	58	8

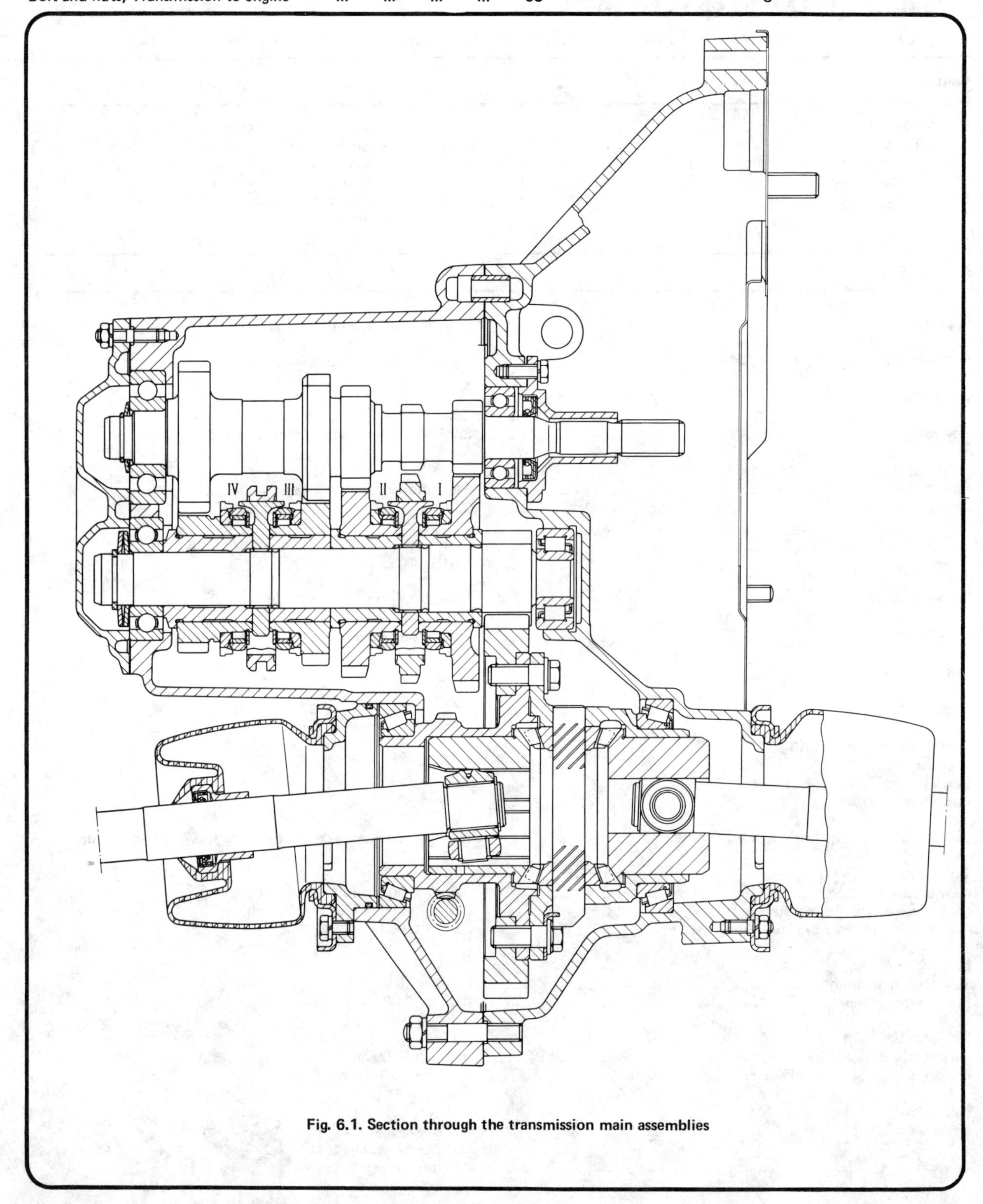

Fig. 6.1. Section through the transmission main assemblies

1 General description

1 Two items of the transmission, the clutch and the driveshafts, are dealt with in separate Chapters. In this one the combined gearbox, final drive and differential are described.

2 Fig. 6.1 shows the three-shaft layout of the transmission. Because the components are arranged so differently from

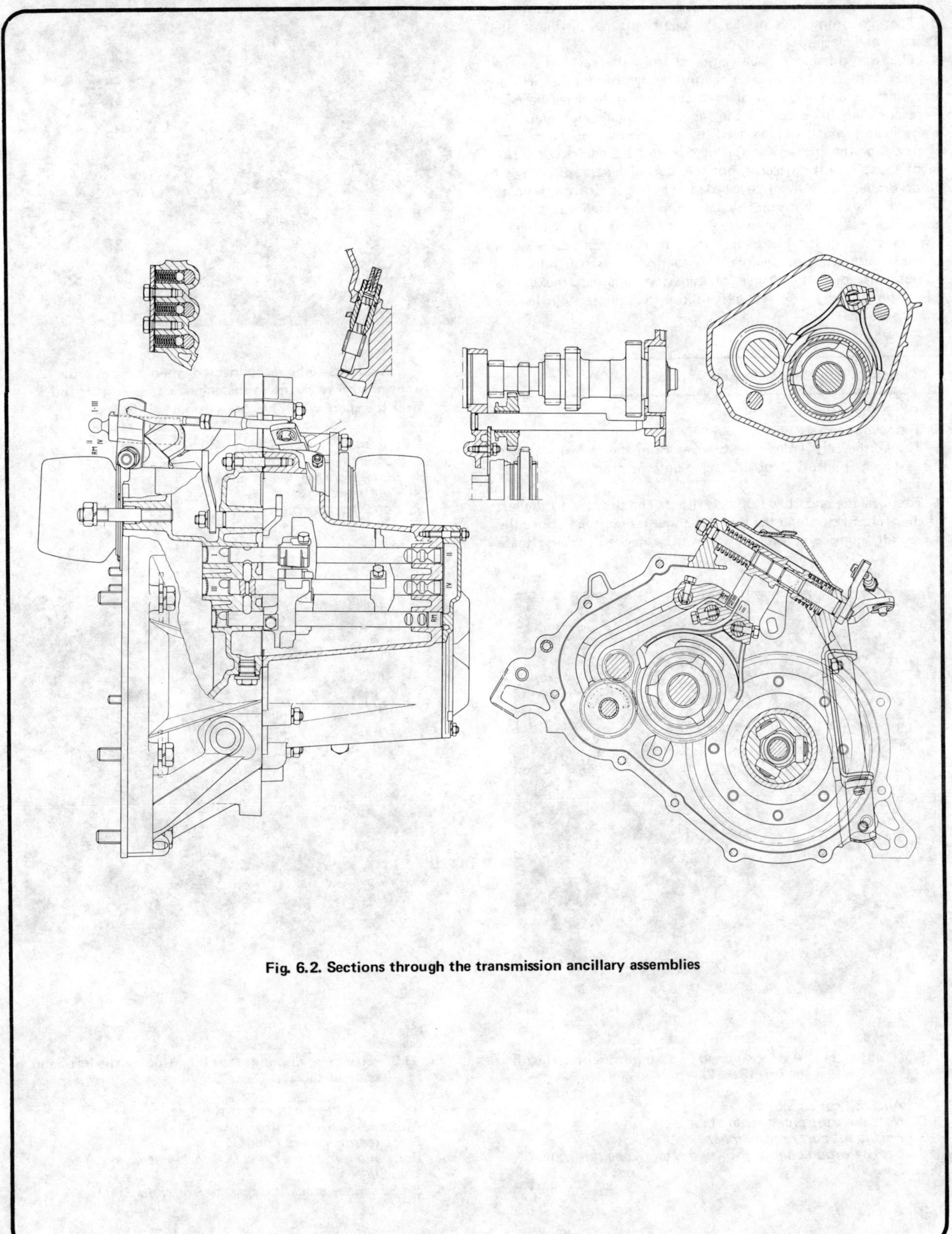

Fig. 6.2. Sections through the transmission ancillary assemblies

gearboxes in-line with the engine, the normal names, such as 'mainshaft', and 'layshaft', are not applicable. The top shaft, in line with the engine, is called the 'input shaft'. The lower one, with all the sliding dog clutches for engaging the gears, is the 'second shaft'. Below this is the final drive with differential.

3 The synchromesh is of the Porsche ring type, having a self-assisting and clash-proof action.

4 The transmission can be removed from the car leaving the engine in place. The engine cannot be removed without the transmission, and as this involves disconnecting the driveshafts, it is responsible for much of the labour of engine removal. The transmission must be removed to give access to the clutch. Removal of the transmission, unlike that of the engine, does not involve use of lifting tackle, but two good jacks, and plenty of stout timber blocks will be needed. The most likely reason to remove the transmission, apart from the obvious one of transmission failure, is an oil leak, or to overhaul the clutch. If the transmission itself is badly worn, then its overhaul is quite possible without special tools. However, the cost of individual components plus the degree of expertise required, makes the case for fitting a FIAT reconditioned unit a worthwhile consideration.

2 Transmission - removal (engine in place)

1 Drain the transmission oil.

2 Remove the spare wheel.

3 Under the car, remove the bracket from the transmission to the exhaust (photo). Undo the earthing strip from transmission to body.

4 Remove the end shield, if it is fitted. The shield is to the left of the transmission and the two bolts that retain it are above the driveshaft; considerable dexterity will be needed because these

Fig. 6.3. Parts to be disconnected when removing the transmission (working in the engine compartment and after the spare wheel has been removed) (Sec. 2)

1 Clutch release control cable
2 Speedometer cable
3 Bolts, transmission to engine
4 Bolt, reaction strut
5 Bolts, starting motor

Fig. 6.4. Parts to be disconnected (working from underneath the car) (Sec. 2)

1 Anti-roll bar supports
2 Nuts, exhaust pipe support bracket
3 Engine support crossmember
4 Gear selection control rod (two places - one not shown)
5 Shield

Fig. 6.5. Parts to be disconnected (working on the left-hand side of the car) (Sec. 2)

1 Nut, constant-velocity joint to wheel hub
2 Nut, sway bar to control arm
3 Nut, steering rod to knuckle
4 Bolts and nuts, lower shock absorber attachment to knuckle
5 Brake caliper bracket (separate from knuckle)

2.3 Transmission to exhaust support

2.5A Separate the ball and socket joint

2.5B Indicating the clip that must be removed

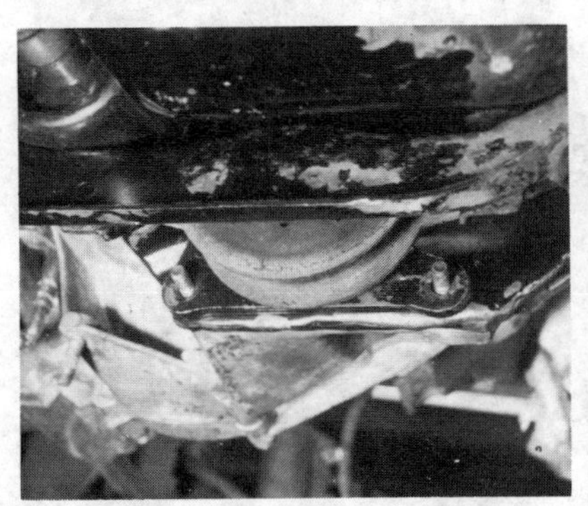
2.15 Remove the nuts from the triangular bearer bracket

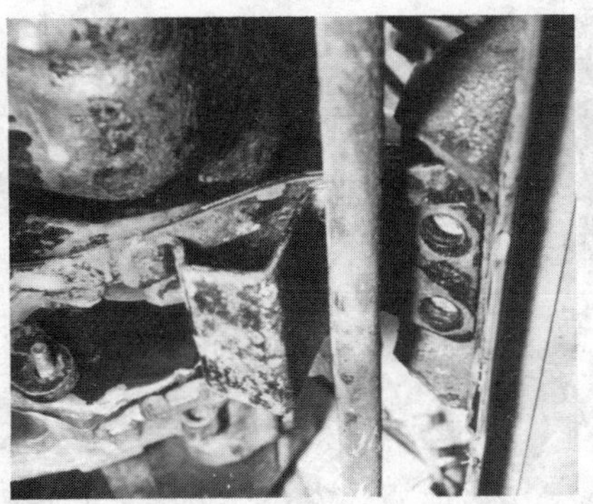
2.17A Front bolts removed from beam

2.17B Rear bolts removed from beam

bolts are fairly inaccessible.

5 Disconnect the gearchange linkage at the two points under the car, just above the support brackets. At the first point it is only necessary to prise a ball and socket joint apart, while at the second point the retaining clip must be removed from the hole in the rod. Collect the washers and bush that this operation releases. (photos)

6 Remove the solenoid wire and the heavy cable to the starter motor. Unbolt the starter and take it off the engine.

7 Undo the nut and locknut on the end of the clutch cable, and disconnect it from the clutch withdrawal lever. Tuck the cable up out of the way.

8 Disconnect the speedometer from the transmission casing by unscrewing the end-cap, and withdraw the outer and inner cables. Tie the cable back so that it will not hamper transmission removal.

9 Now prepare for the removal of the driveshafts. First remove the hub cap from the wheel, and whilst the car is still on the ground, remove the wheel hub nuts from the end of the driveshafts.

10 Jack the car up until the wheels, hanging free on the suspension, are 1 ft (300 mm) clear of the ground. Put firm supports under the strong points of the body. If the ground is at all uneven, use extra blocks for safety in case the car should shift. There will now be room to work underneath, and lift the transmission clear after it has been lowered to the ground.

11 Clean the area around the inner end of the driveshafts, so that when the rubber boots are removed, no dirt will get in.

12 Remove the driveshafts, as described in Chapter 7. The method involving the dismantling of the suspension should be used, as the anti-roll bar must be removed to give room for the transmission to slide off the engine. Basically, on the left-hand side of the car, the following points must be disconnected: wheel hub nut; anti-roll bar to control arm; steering rod to knuckle; lower shock absorber to knuckle and brake caliper bracket to knuckle. On the right-hand side of the car it is only necessary to disconnect the wheel hub nut and the anti-roll bar to control arm nut.

13 Take the weight of the power unit with a jack under the transmission.

14 Remove the single bolt that secures the reaction strut.

15 Undo the three nuts holding the triangular bearer bracket to the transmission. (photo)

16 Undo the two bolts holding the bearer to the support bracket beam under the engine.

17 Take off the triangular bearer bracket and rubber mounting, and the flywheel shield. Leave the beam behind. If in difficulty due to the inaccessibility of some of the bolts, take the beam off with the bracket, but then put it back on its own afterwards.
with the bracket, but then put it back on its own afterwards.
(photos)

18 Lower the jack so the engine is resting on the beam. This is done so that the transmission will be low enough to slide clear to the left under the wheel arch, yet the engine is still held up. Leave the jack under the transmission for now.

19 Undo the nuts and bolts holding the transmission to the engine. Support the weight of the transmission, so it does not hang on the clutch. The jack can do this.

20 Withdraw the transmission to the left, and once the stud is out of the dowel to the rear, and the input shaft clear of the clutch, lower it to the ground.

21 Although it is not always essential to remove the gearchange mechanism, we are including the full technique, to be used if required.

22 The relay lever is first disconnected from the adjustable rod by removing the clip and washer; and then removed from the casing by releasing the two bolts and washers that secure it. (photos)

23 Next, disconnect the sleeve shaft lever by removing the nut and two washers and lifting it off of the squared end of the selector lever. (photo)

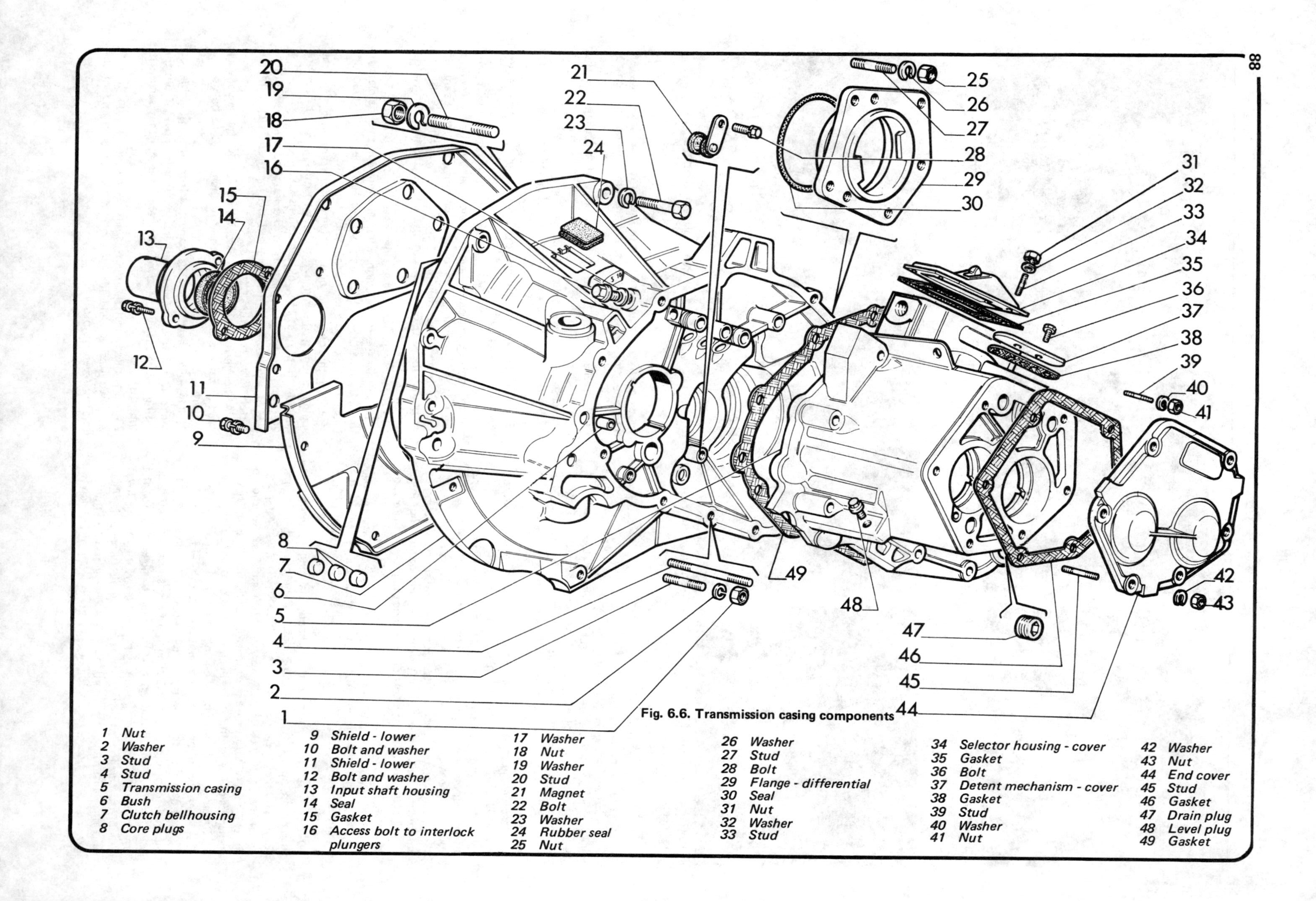

Fig. 6.6. Transmission casing components

1 *Nut*
2 *Washer*
3 *Stud*
4 *Stud*
5 *Transmission casing*
6 *Bush*
7 *Clutch bellhousing*
8 *Core plugs*
9 *Shield - lower*
10 *Bolt and washer*
11 *Shield - lower*
12 *Bolt and washer*
13 *Input shaft housing*
14 *Seal*
15 *Gasket*
16 *Access bolt to interlock plungers*
17 *Washer*
18 *Nut*
19 *Washer*
20 *Stud*
21 *Magnet*
22 *Bolt*
23 *Washer*
24 *Rubber seal*
25 *Nut*
26 *Washer*
27 *Stud*
28 *Bolt*
29 *Flange - differential*
30 *Seal*
31 *Nut*
32 *Washer*
33 *Stud*
34 *Selector housing - cover*
35 *Gasket*
36 *Bolt*
37 *Detent mechanism - cover*
38 *Gasket*
39 *Stud*
40 *Washer*
41 *Nut*
42 *Washer*
43 *Nut*
44 *End cover*
45 *Stud*
46 *Gasket*
47 *Drain plug*
48 *Level plug*
49 *Gasket*

2.22A Clip and washer on adjustable rod

2.22B Removing relay lever and housing

2.23 Showing the squared end on the sleeve shaft lever

2.24A Relay shaft cover removed ...

2.24B ... withdrawing the shaft ...

2.24C ... and releasing the sleeve, springs and cups

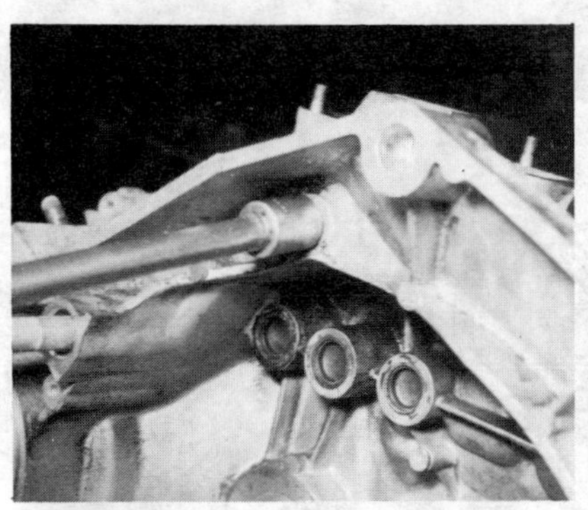
3.3 Removing clutch housing to casing nuts

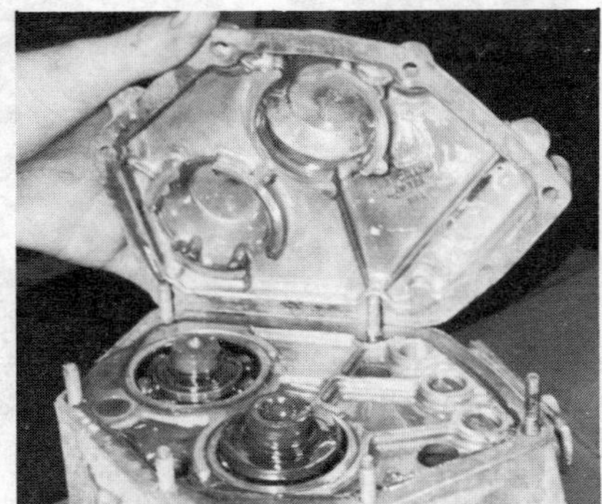
3.5 Taking off the top cover

3.6 Releasing the detent plate, springs and balls

24 Remove the four bolts that retain the cover of the control relay shaft and the single bolt and two washers that retain it in the relay shaft housing. Now, very carefully, before removing the shaft, mark its position in relation to the sleeve. This is important because it is a splined assembly. Then withdraw the shaft and collect the sleeve, springs and cups from inside the housing. (photos)

3 Transmission - dismantling

1 Clean the outside of the casing, and the inside of the clutch housing, checking in the dirt for runs indicating oil leaks, particularly from the core plug.
2 Take the clutch pull-off spring off the clutch withdrawal lever. Pivot the lever round so the withdrawal bearing is nearer, and unclip it from the fork.
3 Remove the nuts with spring washers on the studs inside the clutch housing that hold it onto the gearbox casing. (photo)
4 Stand the transmission up on end, clutch housing downwards.
5 Remove the nuts on the studs holding the end cover, now at the top of the gearbox. Take off the cover, and the old gasket. (photo)
6 Take off the two bolts holding the plate over the detents. (photo)
7 Take out the three springs. Note one is green and the two are blue. The green one is the shortest, and is for reverse.
8 Remove the three detent balls.
9 Take off the circlips, on the end of the input and second shafts. Note that the one on the second shaft has behind it two Belleville washers. These put a heavy load on the circlip. If it will not readily come out of its groove, then the load must be taken off it using a clamp. This clamp will definitely be needed for

assembly. See Section 7. (photo)

10 The Belleville washers can be compressed by screwing into the threaded end of the shaft one of the bolts that holds the transmission to the engine, using a large socket spanner to press down on the Belleville washers. Between the socket and the washer must go a semicircle of steel so the socket clears the circlip, and through the gap in it the ends of the circlip can be reached.

11 Having removed the circlips from both shafts, see if the two bearings will come off the shafts and out of the casing readily, prising with screwdrivers. If they do not, leave them on the shafts for now.

12 Remove the nuts on the other outside studs holding the gearbox casing to the clutch housing.

13 Pull the gearbox casing up off the shafts and selector rods. This will probably take a bit of time and involve tapping the casing with a soft-faced hammer in the region of the shaft bearings: don't overdo this or you might damage the casing. Eventually the casing and bearings will come up off the shafts. (photo)

15 Take out the reverse shaft and remove the gear and its bush.

16 Remove the bolts securing the selector forks and the selector levers to the three selector rods.

17 Pull the reverse selector rod up, out of the housing, and lift off the reverse fork. (photos)

14 Remove the nut holding the locking plate for the reverse idler shaft, on the face at present uppermost of the gearbox side of the clutch housing. (photo)

3.9 Showing the circlips on the ends of the shafts

3.13 Lifting off the casing

3.14 Removing reverse idler locking plate

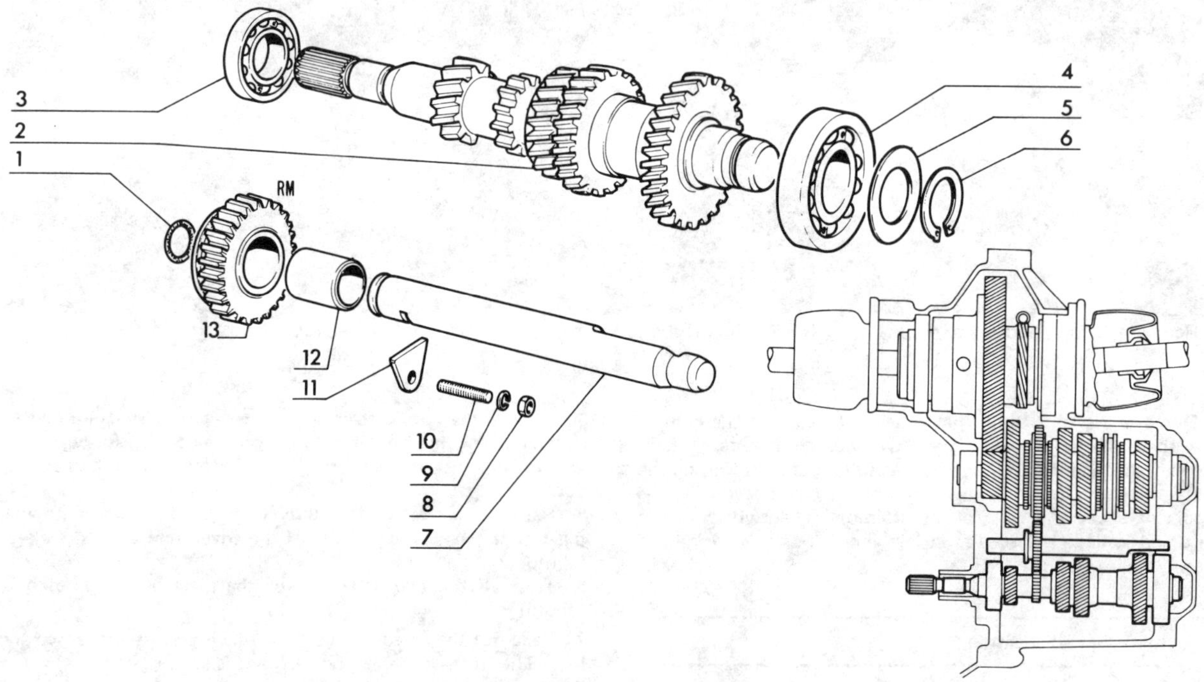

Fig. 6.7. Input and reverse idler gears

1 Seal
2 Input shaft
3 Bearing
4 Bearing
5 Thrust washer
6 Circlip
7 Reverse idler shaft
8 Nut
9 Washer
10 Stud
11 Lockplate
12 Bush
13 Reverse idler gear

18 Pull the first/second rod out of its fork, leaving the fork in position. This is more easily released after the 3rd/4th rod is removed. (photos)

19 Pull up the 3rd/4th, the centre of the three, selector rod, leaving the fork and lever still engaged, until the rod is clear of its seat, then the rod with the fork and lever can be lifted out. Keep the rods with their levers and forks. Though they cannot be fitted incorrectly, it keeps the items together. (photo)

20 Remove the bolt that gives access to the interlock plungers, and using a long thin screwdriver, remove the two interlock plungers from the passages each side of the central, 3rd/4th rod. Remove from the 3rd/4th rod the thin interlock plunger. Put these, and the detent balls in a small box so they cannot get lost. (photo)

21 Lift out the input and second shafts. They come out most easily as a pair, because of the gears being in mesh (photo).

22 Lift out the final drive/differential. (photo)

23 The two bearings in the bottom of the casing can now be

3.17A Removing the reverse selector shaft ...

3.17B ... and fork

3.18A Lifting out the 1st/2nd selector rod ...

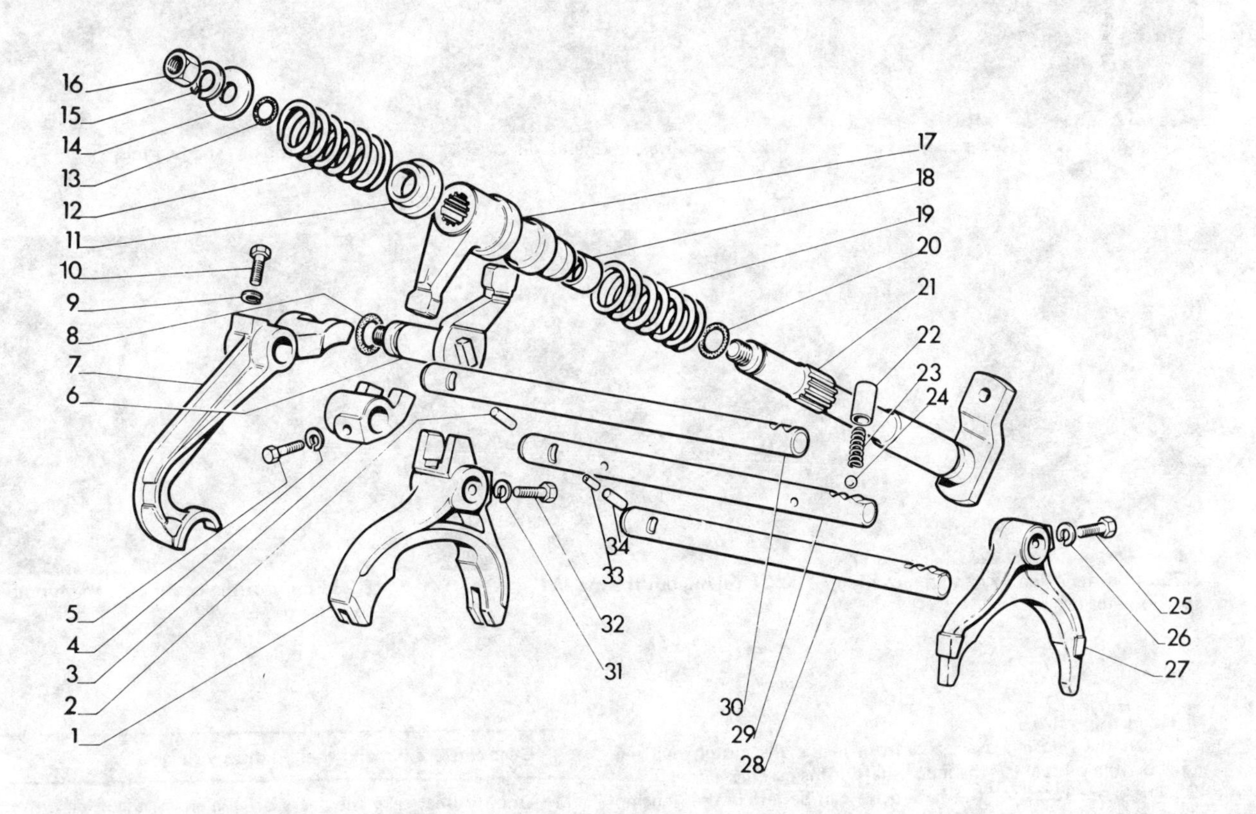

Fig. 6.8. Gearbox selector components

1 First/second selector fork and dog
2 Interlock plunger
3 Dog
4 Washer
5 Bolt
6 Selector lever
7 Reverse selector fork and dog
8 'O' ring seal
9 Washer
10 Bolt
11 Spring cup
12 Spring
13 'O' ring seal
14 Washer
15 Washer
16 Nut
17 Splined lever
18 Bush
19 Spring
20 'O' ring seal
21 Relay shaft
22 Bush
23 Detent spring
24 Detent ball
25 Bolt
26 Washer
27 3rd/4th selector fork
28 1st/2nd selector rod
29 3rd/4th selector rod
30 Reverse selector rod
31 Washer
32 Bolt
33 Interlock plunger
34 Interlock plunger

3.18B ... and its fork after the 3rd/4th rod is removed

3.19 Removing the 3rd/4th rod

3.20 Removing this bolt gives access to the interlock plungers

3.21 Lifting out the two gearshafts as a pair

3.22 Removing the final drive

3.23 Easing out the bearings

3.26 Undo the Allen key and remove the speedometer drive

3.27 Taking out the magnet

3.28 Press out the bearings in the top of the casing

lifted out. (photo)

24 Turn the housing over, and from inside the clutch housing take off the oil seal for the input shaft, in its carrier.

25 The outer races for the final drive can be left in the housing and casing unless it is decided to replace them. Unless they need replacing, the preload plate on the casing, with behind it a shim, need not be removed.

26 Remove the speedometer drive by undoing the Allen key and pulling it out. (photo)

27 Remove the nut holding the magnet at the bottom of the casing and lift out the magnet. (photo)

28 Finally, if the casing was removed with the bearings in position, these can now be pressed out, or if necessary, drifted out using a hammer and a piece of wood, just a little smaller than the diameter of the bearing. (photo)

4 Gear clusters (second shaft) - dismantling

1 One by one, take the gears off the shaft, together with their bushes and the gear-engaging dog-clutches. Keep them in the correct sequence. (photos)

2 If the synchromesh was in good order, do not dismantle it, as reassembly is difficult due to the strength of the circlip.

3 To strip the synchromesh, take the circlip off the gear wheel side. Note how all the synchronizers, springs and blocker bars are fitted and then lift them out. (photos)

4 Keep all parts laid out in order. This will greatly facilitate replacement.

5 The rebuild should present no problems providing the components are kept in the correct sequence. If in any

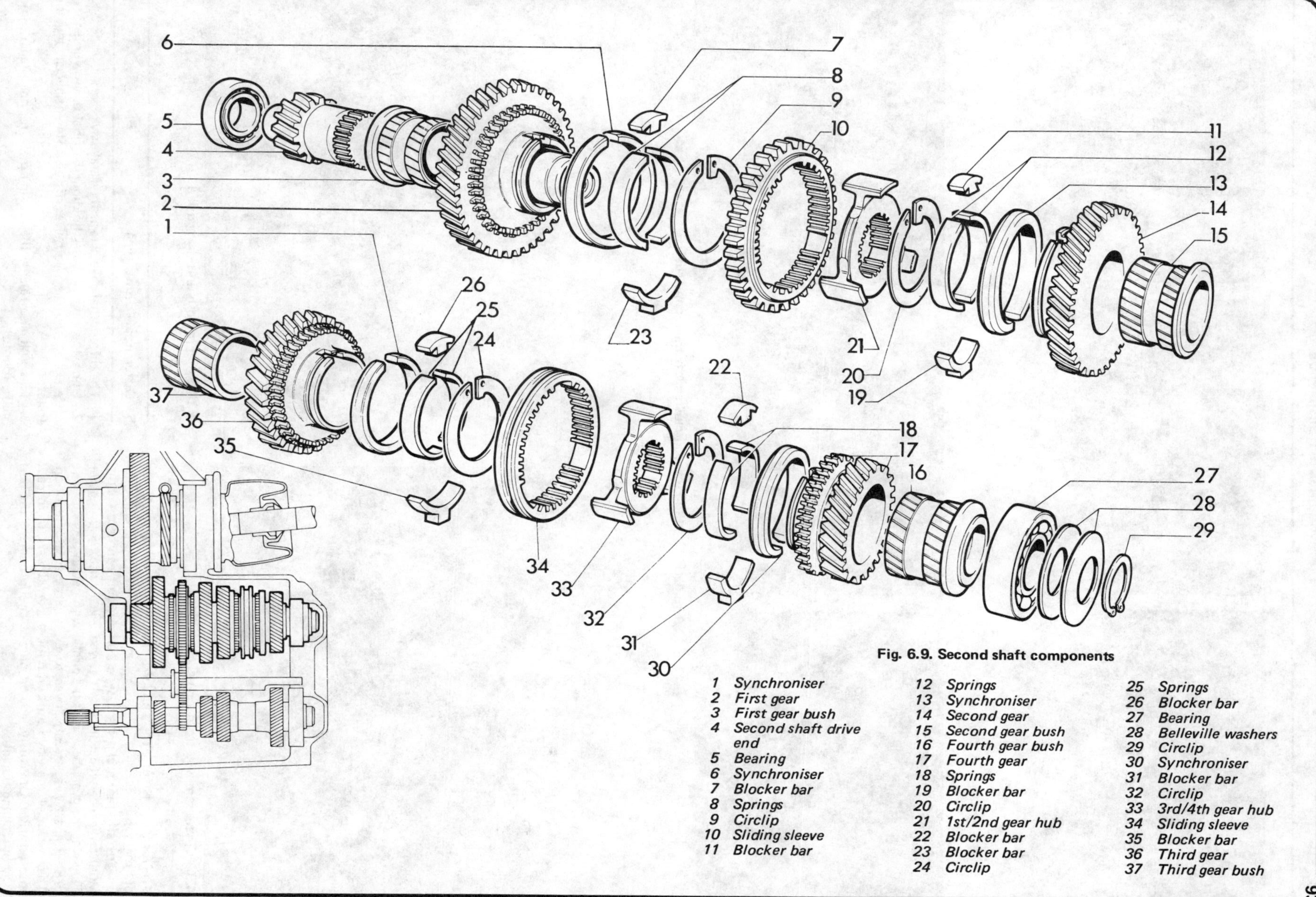

Fig. 6.9. Second shaft components

1 *Synchroniser*
2 *First gear*
3 *First gear bush*
4 *Second shaft drive end*
5 *Bearing*
6 *Synchroniser*
7 *Blocker bar*
8 *Springs*
9 *Circlip*
10 *Sliding sleeve*
11 *Blocker bar*
12 *Springs*
13 *Synchroniser*
14 *Second gear*
15 *Second gear bush*
16 *Fourth gear bush*
17 *Fourth gear*
18 *Springs*
19 *Blocker bar*
20 *Circlip*
21 *1st/2nd gear hub*
22 *Blocker bar*
23 *Blocker bar*
24 *Circlip*
25 *Springs*
26 *Blocker bar*
27 *Bearing*
28 *Belleville washers*
29 *Circlip*
30 *Synchroniser*
31 *Blocker bar*
32 *Circlip*
33 *3rd/4th gear hub*
34 *Sliding sleeve*
35 *Blocker bar*
36 *Third gear*
37 *Third gear bush*

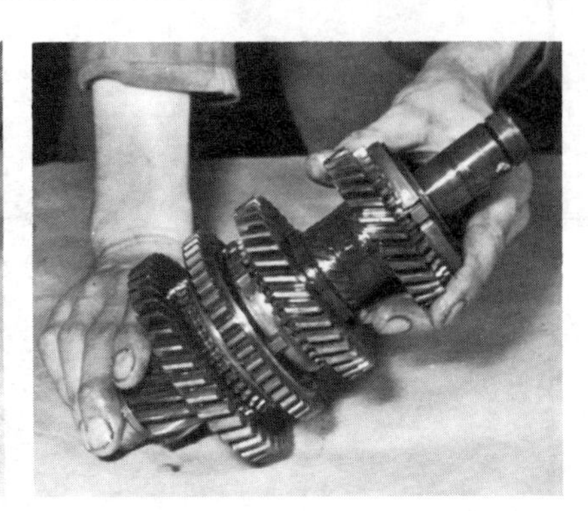

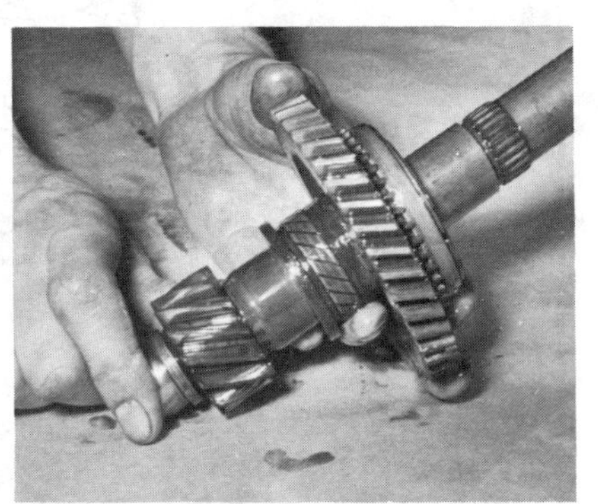

4.1A to 4.1F — Pulling the gear clusters off the second shaft

4.3A Circlip removed from a synchromesh unit

4.3B Synchromesh unit dismantled

difficulty, refer to the exploded illustration and the photographs.

6 To refit the circlip on a synchromesh unit can be difficult. Mount the gearwheel in a soft jawed, clean vice, gripping the flanks of the gearwheel. Part the circlip with circlip pliers, or else prise it apart with a screwdriver, to get it part way into position. Then work all the way round with more screwdrivers, to force it into position. Obviously, a good pair of circlip pliers is the best way of tackling the job.

5 Differential - dismantling

1 The differential cage acts as the final drive carrier.

2 Undo the ring of bolts holding the final drive wheel to the differential, and the two halves of the differential cage together. (photo)

3 This will also release the locking plate for the planet pinion shaft. (photo)

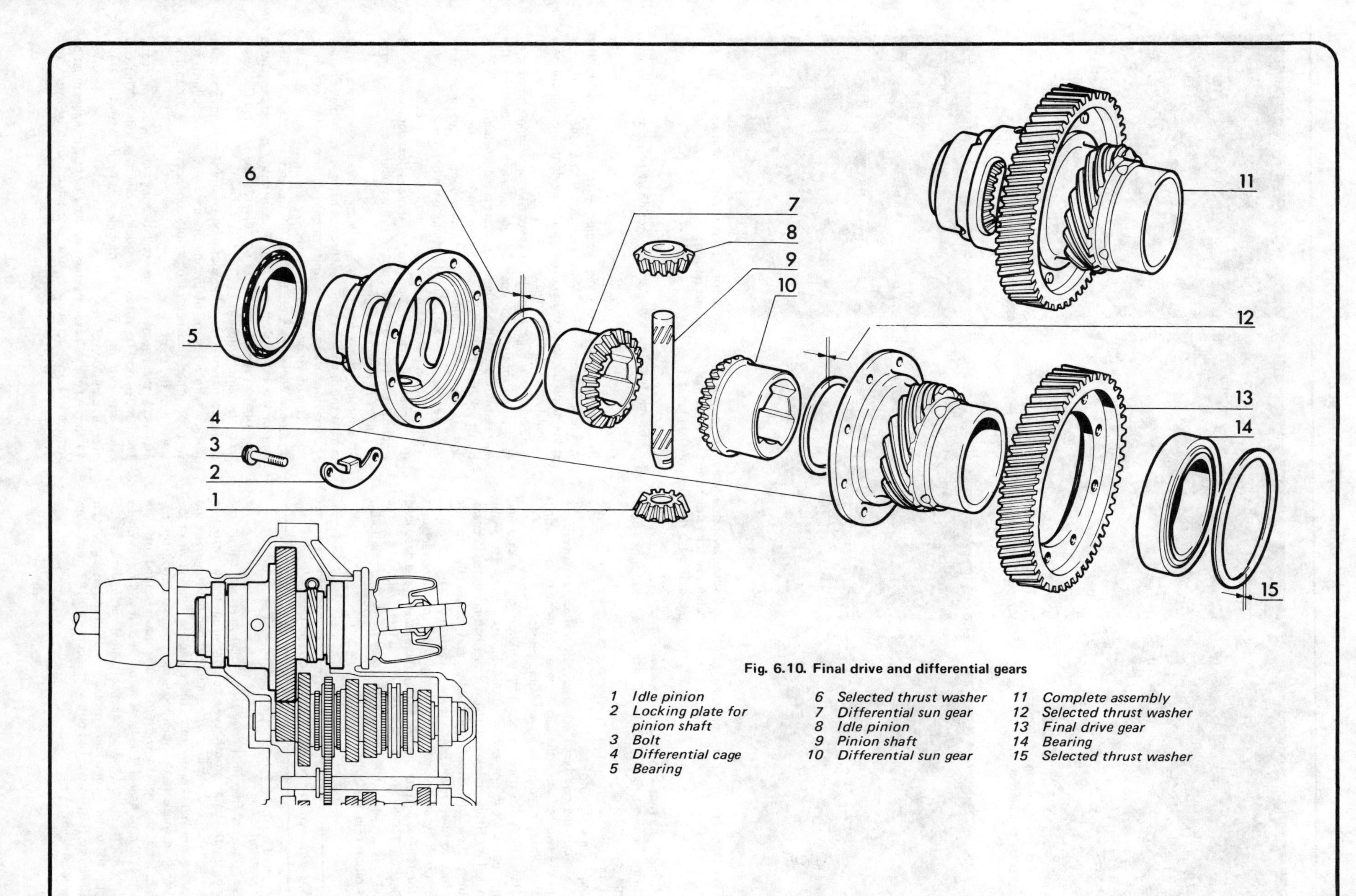

Fig. 6.10. Final drive and differential gears

1 Idle pinion
2 Locking plate for pinion shaft
3 Bolt
4 Differential cage
5 Bearing
6 Selected thrust washer
7 Differential sun gear
8 Idle pinion
9 Pinion shaft
10 Differential sun gear
11 Complete assembly
12 Selected thrust washer
13 Final drive gear
14 Bearing
15 Selected thrust washer

5.2 The differential cage is clamped together by this ring of bolts

5.3 Two of the bolts hold the lockplate for the planet pinions' shaft

5.4 Undoing them allows the whole assembly to come apart

5.6 The planet pinions come off their shaft

5.7 Both bevel side gears have a washer behind to take side thrust, and set the backlash with the teeth of the planet pinions

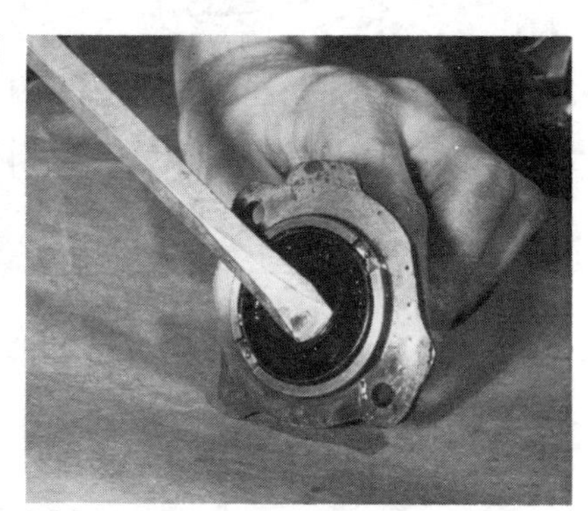

6.8 Prising out the old seal from the input shaft housing

4 Lift off the final drive wheel. (photo)

5 Mark the two halves of the differential cage, with punch dots, to aid reassembly.

6 Prise the two halves apart. Remove the shaft and the two planet pinions. (photo)

7 Take out the two bevel side gears, with their thrust washers, from the two halves of the cage. Leave the taper roller bearings in place unless they are being renewed. (photo)

8 Reassembly is the reverse. The backlash of the side and planet gears must not exceed 0.004 in. (0.1 mm). It can be adjusted by fitting thicker thrust washers behind the side gears. Tighten the bolts evenly, gradually, and diagonally, to the specified torque. If new bearings are being fitted, drive them into place carefully and evenly, only doing so on the inner race, not on the rollers. Setting the preload of these bearings is described in Section 7. The selection of thrust washers available for the bevel gears, and the bearings is shown in the exploded illustration (Fig. 6.10).

6 Transmission components - inspection and renewal

Differential

1 The races of the final drive taper roller bearings, still in the casing, should be a smooth, even colour without any pitting or other evidence of deterioration. Should either the rollers or the races be marked at all, then the complete bearing must be replaced. In this case the outer races must be extracted from the casing, and the rollers with the inner races pulled off the differential cage halves. The latter is robust so it will be simple to pull them off. But the casing being soft, great care must be taken to pull the outer races using a proper tool, and in this context, it is obviously best to try and obtain the correct FIAT tool from your local agent.

2 Check the thrust washers for scoring or other signs of wear.

Gears

3 Carefully examine all components, starting with the synchromesh assemblies. Make sure that the various parts do not show signs of wear or damage, but if evident, a complete new assembly must be obtained.

4 Examine the gearwheels for excessive wear and chipping of teeth. If a tooth is damaged on one shaft, it is likely that the corresponding tooth on the other shaft is damaged also.

5 Check the ball bearings for wear by holding the inner track and turning the outer track. Any roughness in movement indicates wear. Next, again holding the inner track, check for sideways movement. Obtain a new bearing if wear is evident.

6 Inspect all transmission casing threads for damage and, if evident, the hole will have to be drilled oversize, a new thread cut and a new bolt obtained.

7 Examine the ends of the selector forks where they rub against the channels in the synchronizer ring. If possible compare the selector forks with new units to help determine the wear that has occurred. Renew them if worn.

8 Check the splines on the end of the input shaft. Fit a new seal in the input shaft housing. This is easily prised out and a new one pressed in with the lip towards the bearing (ie; away from the clutch). (photo)

9 Inspect the reverse idler shaft and ensure that it has a smooth surface with no signs of pitting or scoring. Check the bush for wear or other evidence of deterioration and renew the seal on the end of the shaft.

10 Check the detent springs, ball bearings and safety plungers. They must not be pitted, or their movement tight. This can cause difficult or automatic disengagement in both drive and overrun conditions.

11 Inspect the selector mechanism. Any sloppy linkages must be replaced. Look particularly for elongated holes or worn rod

ends. Check the splines on the relay shaft and the condition of the associated springs. Replace the seals on the ends of the relay shaft.

12 Clean the magnet of all metallic particles and any other deposits.

7 Transmission - reassembly

1 Refit all the differential components and the final drive gear wheel to the differential cage, so it is ready as a sub-assembly. Use appropriate thrust washers behind the bevel side gears to get the correct backlash with the planet gears.

2 Fit all the gears, with their synchromesh units already assembled, to the shaft. The correct sequence can be checked against the exploded illustrations and the photographs in the dismantling sequence covered in Section 4. (photo)

3 Refit the magnet to the tapped hole in the bottom of the casing. (photo)

4 Press the two shaft lower bearings into the housings in the casing; lightly smear the tracks and inner surface with grease. (photo)

5 Refit the input shaft housing to the clutch side of the casing. With this in position the bearing upper surface should be level with the casing. (photo)

6 Replace the complete differential assembly in the casing; the bearing outer track should be in position and the assembly can be stood up on this. (The inner bearing track should be on the differential assembly, and lightly smeared with grease). (photo)

7 Fit the two gearshafts to the casing as a pair; the gears should be meshed and then the two assemblies stood up in their bearings, which should be already in the casing. (photo)

8 Refit the reverse idler gear and its bush to the shaft and place in its housing in the casing; lock it in position with the lockplate and nut. (photo)

9 Fit the 1st/2nd gear selector fork to its rod and then place the rod in its housing. (photo)

10 Position the gearchange locking pins, pushing them through the bolt hole in the casing with a suitable long drift. Firstly, one of the thicker locking pins is pushed right through to the 1st/2nd rod orifice, then the thin pin is positioned in the orifice in the 3rd/4th selector rod. The selector fork and lever can next be fitted to the rod and the assembly, complete with small lockpin, eased into its housing. Ensure the selector fork engages with the groove on the 2nd/3rd gear. (photos)

11 Now push the remaining thick locking pin into place between the 3rd/4th rod and the reverse rod position. (photo)

12 The reverse selector fork can now be engaged on the reverse idler gear and the selector rod eased through it into its housing. Bolt the fork into position. (photos)

13 Fit the selector lever to its housing alongside the 1st/2nd selector rod. A new seal should be fitted to the groove at the end of the shaft. (photo)

14 The upper half of the casing can now be eased into position over the gear assemblies. Clean the mating faces of the upper and lower casings and fit a new gasket before positioning the upper half. Finally, tighten the nuts and bolts that join the two halves, not forgetting those on the clutch side of the housing. If the differential bearings have been renewed the casing should be fitted with the bearing plate off; this will be fitted later and the necessary adjustment described (paragraph 21 refers). (photos)

15 Replace the detent balls and springs and refit the retaining plate. (photos)

16 The two gearshaft upper bearings can now be pressed into position in the top of the casing. (photo)

17 The input shaft is only retained by one thrust washer and circlip; these can be easily clipped into position, or if necessary, use a hammer and socket to get the necessary pressure. (photo)

18 The second shaft is not such an easy proposition due to the fact that there are two Belleville washers to go under the circlip

7.2 The complete gearshafts should look like this

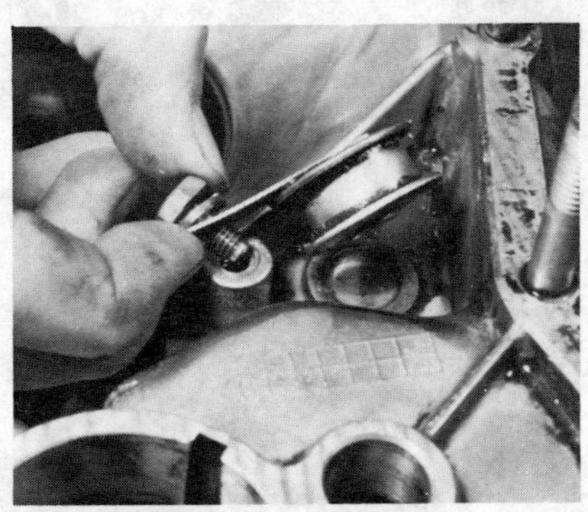

7.3 The cleaned magnet is refitted ...

7.4 ... and then the bearings in the clutch housing

7.5 The input shaft housing fits on the clutch side of the casing

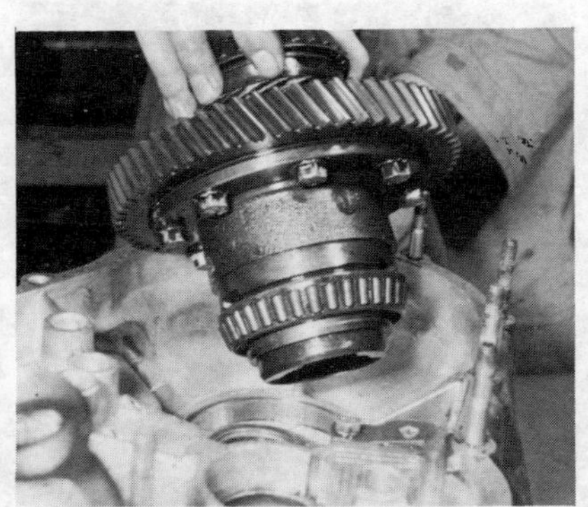

7.6 The differential assembly is then replaced

7.7 The two main gearshafts are fitted in unison

7.8 Locking the reverse idler gearshaft

7.9 Positioning the 1st/2nd selector fork and rod

7.10A Pushing into position the first thick locking pin ...

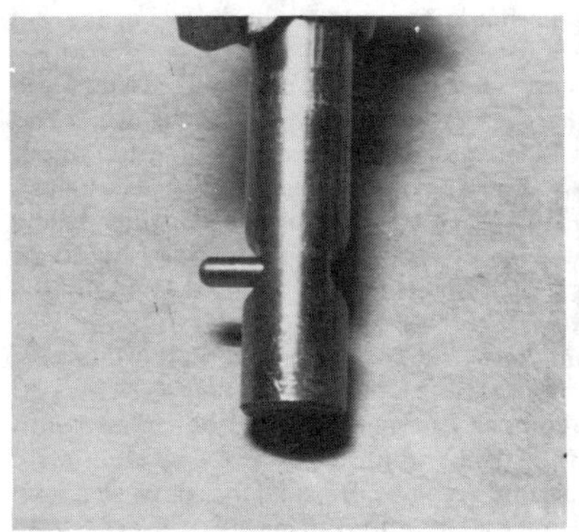

7.10B ... then the thin locking pin in the 3rd/4th rod ...

7.11 ... and then the last thick locking pin

7.12A Engaging the reverse fork ...

7.12B ... and then the rod

7.13 Inserting the selector lever

7.14A Lowering the upper half of the casing over the gears. Note the new gasket in position

7.14B If the differential bearings have been renewed, the casing should be fitted with the bearing plate off

7.15A Replace the detent balls and springs ...

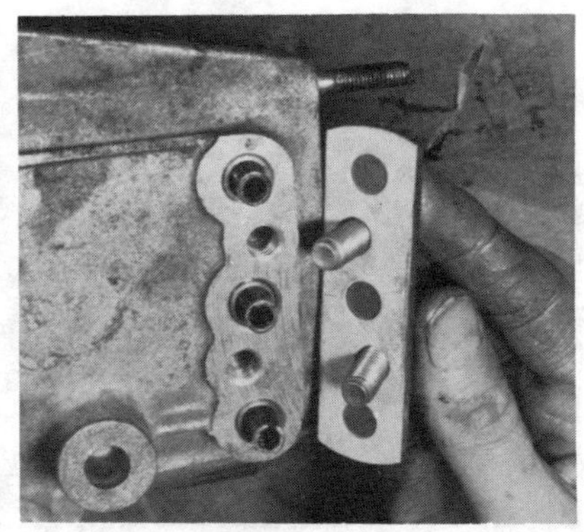

7.15B ... and then the locking plate

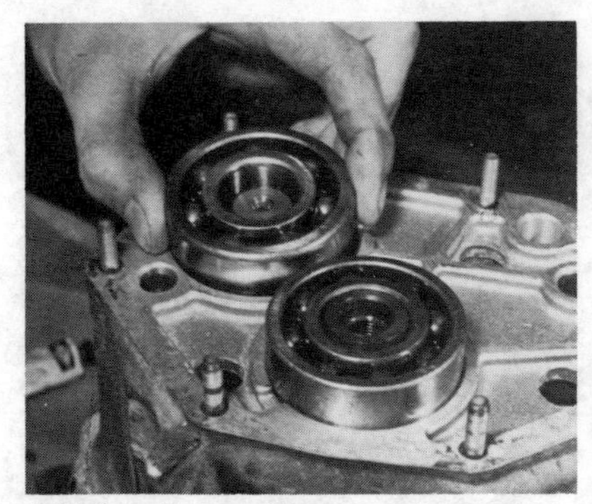
7.16 Press into position the casing top two bearings

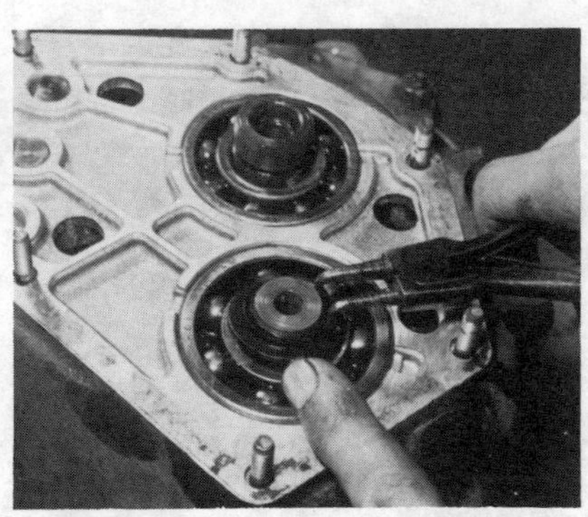
7.17A The input shaft washer and circlip should present no problems ...

7.17B ... but use a hammer and socket if at all difficult

7.19 Place two Belleville washers and the circlip in position

7.20 Method of compressing the Belleville washers and circlip

7.21 Measure the thickness of shim needed to give specified preload. Then fit it under the plate

and considerable pressure is required.

19 Fit the two Belleville washers, their outer rims next to each other, on the end of the shaft. (photo)

20 Fit the circlip. For this to get into its groove against the considerable pressure of the pair of washers, the latter must be compressed by a clamp. The end of the shaft is internally threaded, and the wheel bolts are of the same size. In order to compress the washers it is necessary to make up a local clamp, using a wheel bolt, of the type shown in the associated photograph, although anything that does a similar job will suffice.

21 If the differential bearings have been removed, now set the bearing preload. If the bearings were not removed then the remainder of this paragraph can be disregarded. Put two shims above the bearing race then fit the bearing plate, and tighten it gently down to push the bearing into place. This will be helped if a gear is engaged, and the gears rotated by the input shaft, to allow everything to roll into place. As there are two shims fitted the flange of the plate should not go right down. Release the nuts, so the plate is still in contact through the shims with the bearings, by its own weight only, and not strained. Measure the gap between the flange and its seat on the casing. This can be done by inserting feelers between the two (photo). Remove the plate, and measure the total thickness of the shims in use. Select a shim that will leave a gap of 0.003 in. (0.08 mm) between the plate and casing with no load on the fixing nuts. When the nuts are tightened, this interference will give the correct preload on the bearings, as the plate is pulled down onto the casing.

22 Fit a new gasket to the end of the casing and then fit the endplate and tighten on the studs with nuts and lockwashers. (photo)

23 Refit the speedometer drive assembly, pushing it into position and locking it in place with the Allen bolt. (photo)

24 Refit the selector mechanism; remember that the relay shaft was marked in relation to the splined sleeve and it must be refitted in exactly the same relative position. Before inserting the components ensure the tongue of the operating lever is protruding into the selector housing from its pivot point inside the casing. This tongue must fit into the recess on the splined sleeve, when in position.

25 The springs, sleeve and other components can be lined-up, using a screwdriver before inserting the relay shaft; when all components are positioned, secure with a nut and lockwasher. (photos)

26 Finally reconnect the remaining external linkages, exactly as removed. Replace the selector housing cover complete with new gasket and secure with nuts and lockwashers. (photos)

27 Clip the clutch withdrawal race into place on its fork, putting a trace of grease on its seats.

8 Transmission - refitting (engine in place)

1 The transmission having been assembled, check nothing is left to be done. The driveshafts, with their inner and outer joints, and rubber boots should be ready, but not fitted to the transmission. The oil should be put in later too.

2 The clutch must have been centralised, as described in Chapter 5.

3 Offer up the transmission to the engine, holding it square. Slide the stud on the rear into the tubular dowel, and the shaft through the clutch. An assistant can help by turning the engine over slightly by a spanner on the crankshaft pulley nut, to allow the splines to line up and engage.

4 As the transmission goes into place, keep the weight supported, so it does not hang on the clutch.

5 Fit the bolts and the nut holding the engine and transmission together. Tighten the top one first so that it takes the weight of the transmission.

6 Put a jack under the transmission, and lift the left end of the power unit to get it level.

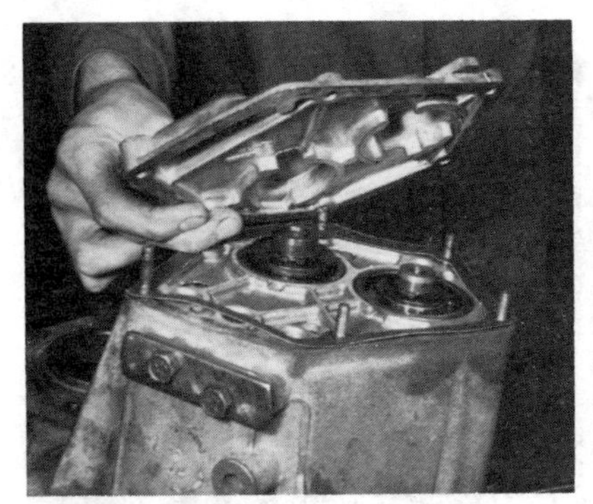
7.22 Fit a new gasket and then the cover

7.23 Insert the speedometer drive and then lock with Allen key

7.25A Line up the sleeve, springs and cups ...

7.25B ... and then fit the relay shaft. Note the tongue of the selector lever fitting between the sleeve raised ends (arrowed)

7.26A Replace the relay housing ...

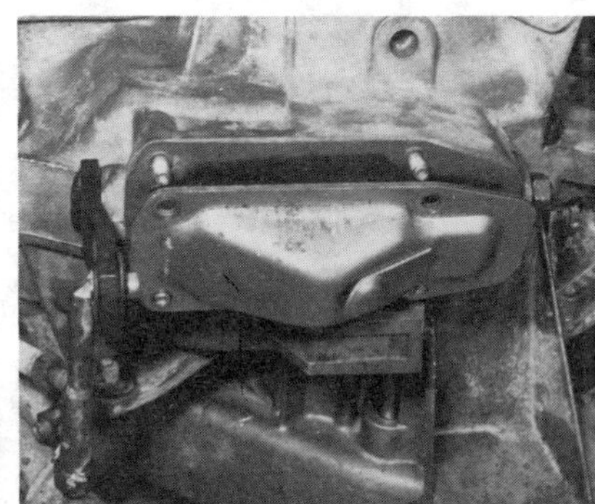
7.26B ... and then the cover

7 Fit the shield to the bottom of the flywheel housing, and then the bearer bracket to both power unit and the beam.
8 Lower the jack. Check the power unit is sitting properly in its bearers, and all are properly secured.
9 Refit the driveshafts and front suspension, as described in Chapter 7.
10 Refit the exhaust bracket. Reconnect the earthing strip.
11 Lower the car to the ground.
12 Fill the transmission with oil.
13 Refit the starter motor. Connect its heavy cable, and solenoid wire.
14 Refit the speedometer cable.
15 Connect up the clutch cable.
16 Adjust the clutch cable free-play as given in Chapter 5, to 1 in. (25 mm) free-movement at the pedal. Refit the return spring.
17 Check all bolts for tightness, particularly the mountings and those holding the transmission to the engine. Check for leaks. Check the oil level. When all is correct, refit the shield underneath.
18 If the gearchange linkage has not been maladjusted, but simply replaced as removed, then all that is necessary is to reconnect the two points under the vehicle, one being a ball and socket press joint and the other a simple rod and clip.
19 If it is necessary to adjust the mechanism, refer to Fig. 6.11 and proceed as follows:
20 With the gearlever in neutral, check that the surface plane 'P1' of the lower lever (4) is parallel with the outer plane 'P2' of the support (5).
21 If not, disconnect the adjustable rod (7) from the lever (8) and set the lever (8) in the neutral position. Position the lever (4) parallel to the plane of support (5).
22 Adjust the length of rod (7) so that it can be reconnected to lever (8) without changing the adjustments achieved in the previous paragraph.
23 Finally, check the distance between the bolt (11) and the gear lever (2) when in the neutral position; it should be 19.25 in. $\pm$ 0.16 in. (489 $\pm$ 4 mm). Then check the distance between the knob (1) of the gearlever (2) when in neutral, and the instrument panel lower padding; it should be 11.61 in. $\pm$ 0.59 in. (295 $\pm$ 15 mm).

9 Lubrication

1 The oil should be changed 500 miles after the transmission has been dismantled. This will remove any metal particles that have resulted from wear in new components as they bed in, and any bits of dirt inadvertently left in the transmission.
2 The oil should be drained after a journey, so it flows easily, and any sediment will be stirred up. Remove the plug on the left of the final drive part of the casing. Allow the oil to drain fully. Clean the plug before removal, and again before refitting.
3 Fill to the level of the plug on the front.
4 The official FIAT oil for world wide use is an SAE 90 oil. One of EP, extreme pressure, should not be used. As there is not a bevel final drive, the EP quality is not necessary, and so the side effects of the EP additives on bearings can be avoided. FIAT England recommend the use of engine oil in the transmission. This is common practice. It will run slightly cooler, compared with the SAE 90 oil.

10 Fault diagnosis - transmission

1 There will often be noises from the transmission. The important thing is to be able to decide whether a noise is normal or not: whether the transmission needs dismantling, or is safe for continued use.
2 If subdued whining comes on gradually, there is a good

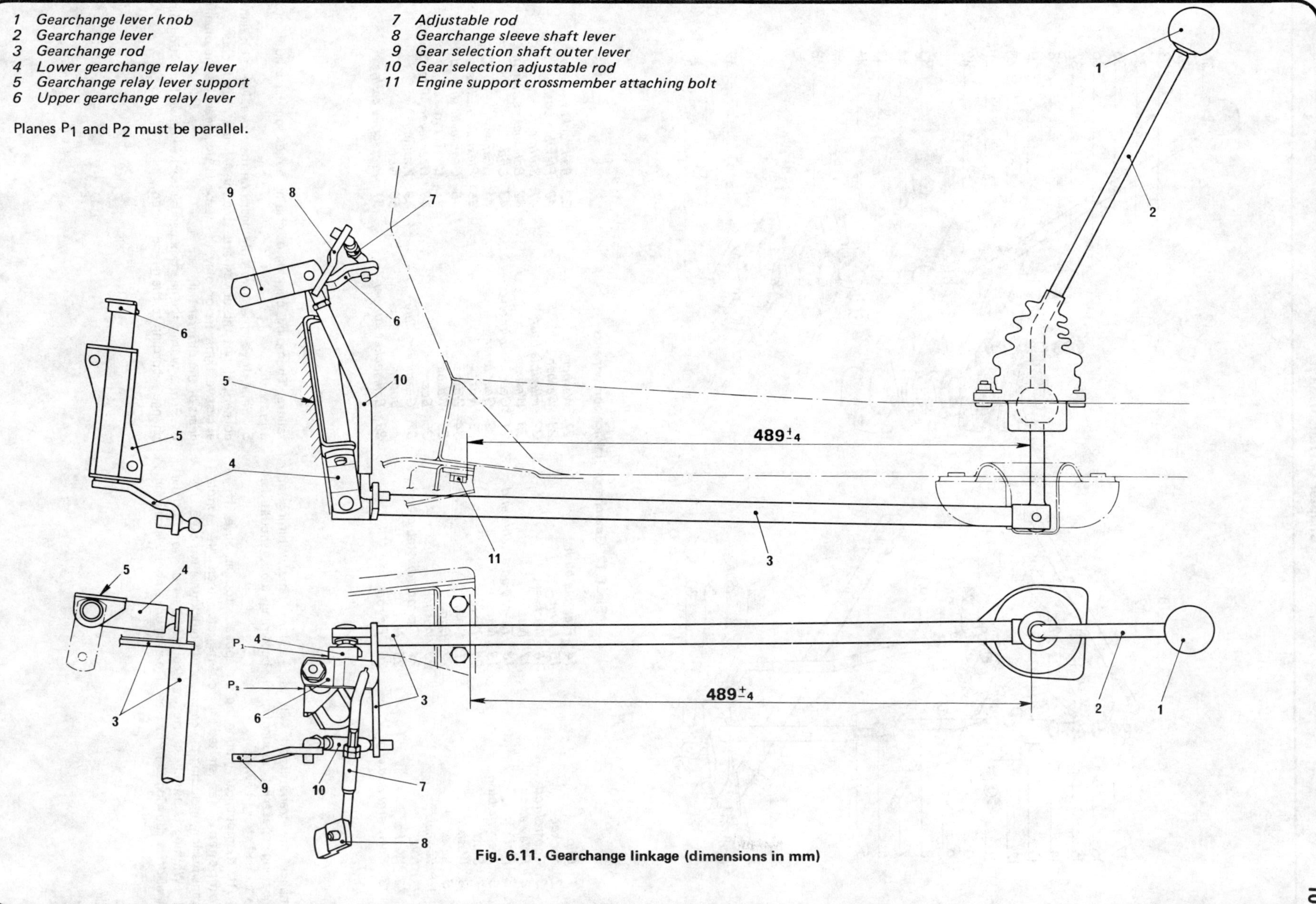

Fig. 6.11. Gearchange linkage (dimensions in mm)

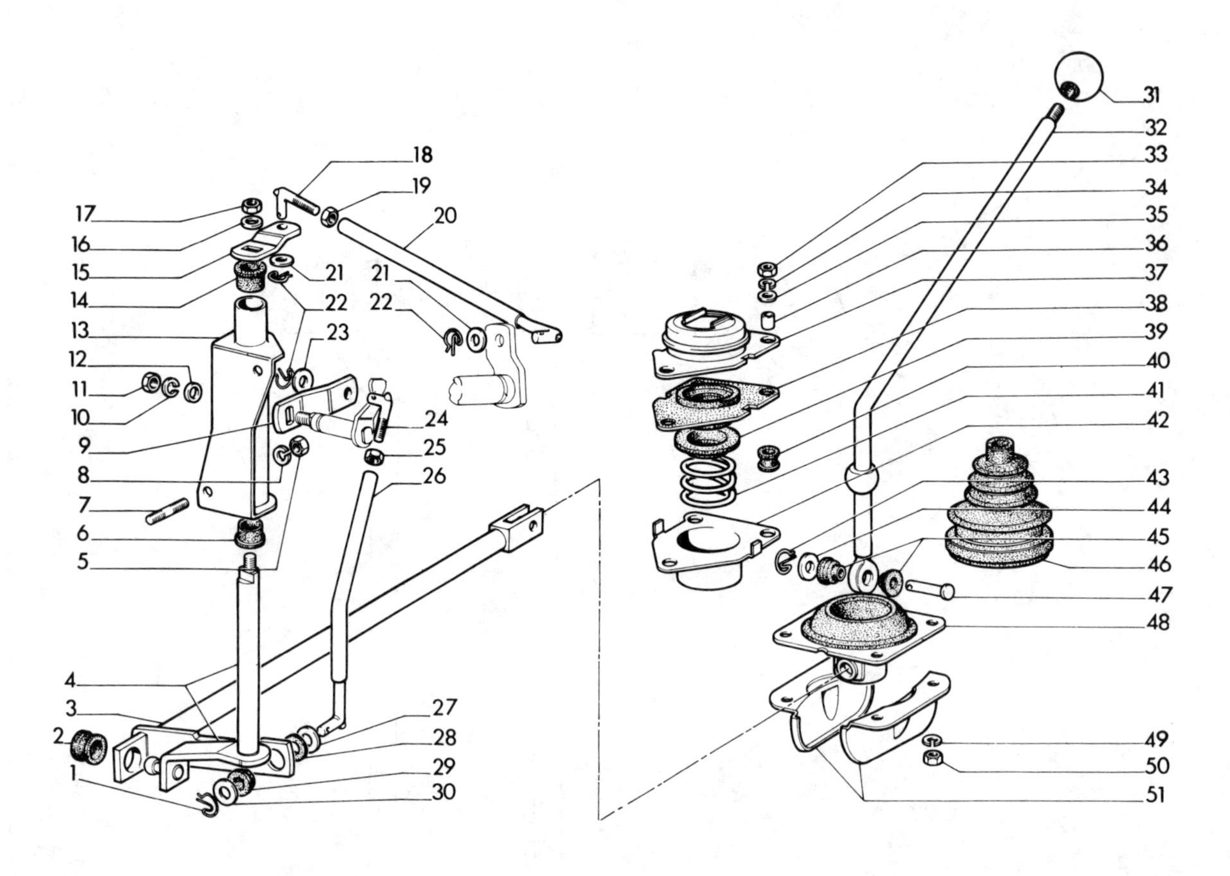

Fig. 6.12. Gearchange linkage components

1 *Clip*
2 *Socket*
3 *Control rod*
4 *Relay arm*
5 *Nut*
6 *Flexible bush*
7 *Stud*
8 *Washer*
9 *Lever*
10 *Washer*
11 *Nut*
12 *Washer*
13 *Relay arm housing*
14 *Flexible bush*
15 *Lever*
16 *Washer*
17 *Nut*
18 *Tie-rod adjustable end*
19 *Nut*
20 *Tie-rod*
21 *Washer*
22 *Clip*
23 *Washer*
24 *Tie-rod adjustable end*
25 *Nut*
26 *Tie-rod*
27 *Washer*
28 *Washer*
29 *Flexible bush*
30 *Washer*
31 *Gear lever knob*
32 *Gear lever*
33 *Nut*
34 *Washer*
35 *Washer*
36 *Bush*
37 *Gear lever plate - upper*
38 *Ball socket - upper*
39 *Ball socket - lower*
40 *Flexible bush*
41 *Spring*
42 *Ball socket housing*
43 *Clip*
44 *Washer*
45 *Flexible bushes*
46 *Gear lever shroud*
47 *Clevis pin*
48 *Gear lever plate - lower*
49 *Washer*
50 *Nut*
51 *Protective casing*

chance the transmission will last a long time. Whining or moaning appearing suddenly, or becoming loud, should be examined straight away.

3 If thumping or grating noises appear, stop at once. If a component has started to disintegrate, then the whole transmission, including the casing, could quickly be wrecked.

4 Whining is usually gearteeth incorrectly set or worn, or the final drive showing signs of wear. Roaring or rumbling is bearing failure. Thumping, clicking or grating noises suggest a hunk out of a tooth.

5 Final drive noises are only heard on the move. Noise on tight corners, if not the driveshafts (the most likely cause), could be excessive tightness or play in the bevel side gears or the planet gears of the differential.

6 Faults within the gearbox can be tracked down by noting the change according to the gear selected.

Chapter 7 Driveshafts

For modifications, and information applicable to later models, see Supplement at end of manual

Contents

Specifications

Shaft joints

Outer end	Constant velocity, Rzeppa type
Inner end	Sliding, Tripode type

Lubricant

Constant velocity joints	Molybdenum disulphide grease 3.3 oz (95 g) (Duckhams LBM 10)
Tripode joints	Transmission oil

Torque wrench settings

	lb f ft	kg f m
Hub nut, outer end of shaft	101	14
Inner end oil retainer boot to casing	7	1

1 General description

1 From the final drive and differential under the engine the drive is taken to the two front wheels by a pair of driveshafts. Because the transmission is to the left of the engine, a longer shaft is needed on the right side; it is also thinner in order to equalise the torsional flexibility of the two shafts.

2 At the inner ends of the shafts are universal joints of the 'Tripode' type. These allow for axial movement as well as misalignment as the wheels move on the suspension.

3 At the outer ends are constant velocity joints of the 'Rzeppa' type. These allow for the movement up and down of the wheel on the suspension, and the swivelling of the steering.

4 The main part of the shaft fits into the joints at the two extremities by splines, held by circlips. Rubber boots keep in the lubricant, and the road dirt out.

5 The 'Tripode' joints work in the bevel side gears of the differential, and are lubricated by the transmission oil. The constant velocity joints are packed for life with molybdenum disulphide grease.

6 Provided the rubber boots keep out the dirt the constant

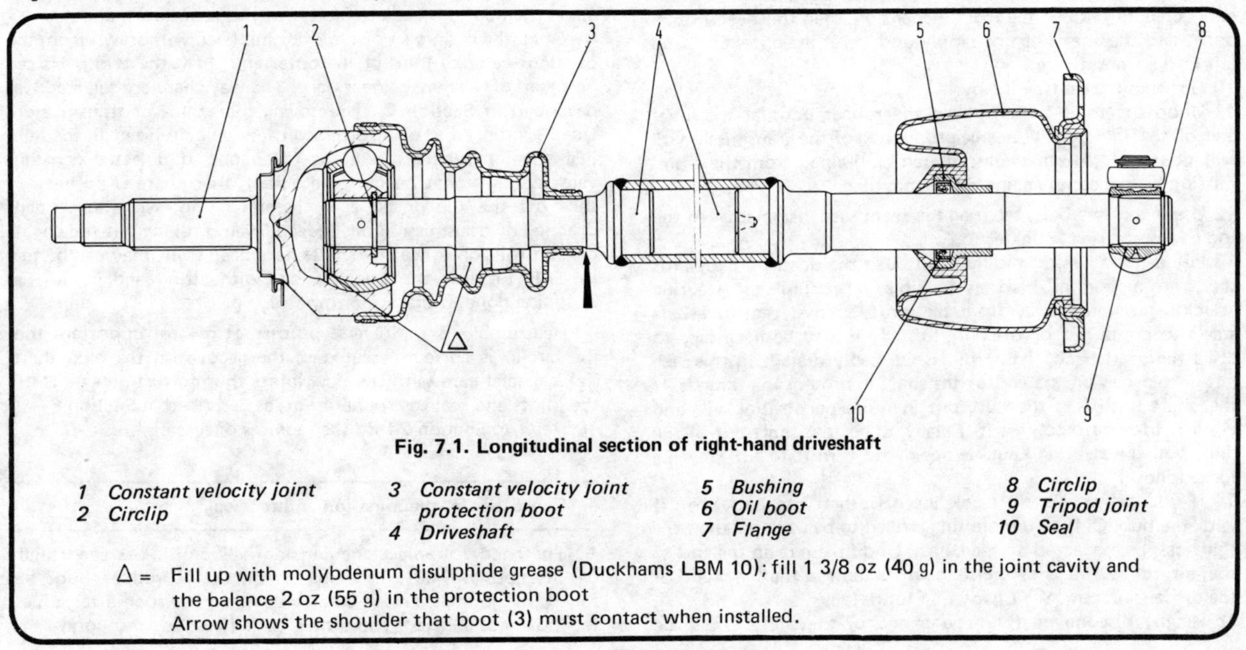

Fig. 7.1. Longitudinal section of right-hand driveshaft

1 Constant velocity joint
2 Circlip
3 Constant velocity joint protection boot
4 Driveshaft
5 Bushing
6 Oil boot
7 Flange
8 Circlip
9 Tripod joint
10 Seal

Δ = Fill up with molybdenum disulphide grease (Duckhams LBM 10); fill 1 3/8 oz (40 g) in the joint cavity and the balance 2 oz (55 g) in the protection boot
Arrow shows the shoulder that boot (3) must contact when installed.

velocity joints last well, though on cars used in towns or hills, with a high proportion of driving hard in low gears, they are unlikely to last as long as the rest of the transmission. The 'Tripode' joints should last the life of the transmission.
7 The driveshafts have to be removed to take out the engine or transmission. There are two ways of removing these shafts. Either they can be removed complete with the constant velocity joint, or taken out of the joint. The former method is lengthy, but usually easier, particularly for the first time.

2 Driveshafts (complete) - removal

1 This method involves taking the driveshafts off with the constant velocity joint still on the outer end. This avoids having to open its dirt excluding boot, and is particularly appropriate if there is nothing wrong with the shafts, but they have to be removed (eg; when the transmission is being taken out). The penalty is that quite a lot of dismantling of the suspension is involved. For the person who has not worked on the car before it is the most straightforward method, as that involving the removal of the shaft from the constant velocity joint is difficult if it is not known exactly what to look for inside the joint.
2 The suspension has to be partially dismantled. In this Section is described only the minimum necessary to remove the shafts. However, if other jobs then have to be done on the suspension, one is already part way there.
3 With the front wheel still on the ground and handbrake applied, remove the hub caps, and slacken the nuts on the outer ends of the driveshafts. Spray the shaft splines in the hub from an aerosol of rust solvent, which can be soaking in whilst work proceeds. (photo)
4 Jack-up both front wheels, and support the car firmly on stands under the reinforced points of the body just behind the front wheels. It must be very firm and at a good height so it is safe and comfortable to work underneath.
5 Remove the relevant front wheel.
6 Remove the brake caliper from the hub by pulling out the two spring clips and wedges. See Chapter 8 for details. Tie the caliper up so the flexible hose will not get strained.
7 Remove the steering ball joint from the steering arm. Refer to Chapter 10 for the method.
8 Take the nut completely off the end of the driveshaft, also its flat washer. Give another squirt of the rust solvent to the splines.
9 Under the car, clean the area around the seat for the rubber boot at the inner end of the driveshaft, on the transmission casing. But note that if the driveshafts are only being disconnected to remove the front suspension, then this need not be done, and the shafts can be simply undone at the outer end, and left in the transmission.
10 Get an oil drain tray handy.
11 Undo the three bolts holding the rubber boot at the inner end of the shaft. Quite a large proportion of the transmission oil will pour out. Alternatively, oil can be drained from the transmission before commencing the dismantling.
12 Undo the two bolts securing the front suspension strut to the top of the steering knuckle.
13 Put a hand behind the hub and push the driveshaft towards the transmission as hard as possible. Then pull the steering knuckle outwards away from the strut, so that the driveshaft starts to come out of the hub. Initially it may be necessary to drive the shaft back, if the splines are badly rusted. In this case put an old nut on the end of the shaft to protect the threads. If the shaft is stiff in the hub, put in more penetrating oil, and work it to-and-fro to ease it. Finally to get the shaft out of the hub, pull the steering knuckle down and turn it to full steering lock. (photo)
14 If you get absolutely stuck the driveshaft can be taken off with the hub. Only two more things need to be undone; then the shaft can be extracted on the bench. Undo the nut on the end of the anti-roll bar, and take out the pivot bolt at the inner end of the suspension arm. See Chapter 11 for details.
15 Having disconnected the outer end of the shaft, it can be drawn out of the transmission. Do this gently, as sometimes, particularly on early cars, the circlips on the 'Tripode' joints come off, and the needle roller bearings fall out and into the transmission. Also keep the 'Tripode' joint out of the dirt. (photos)
16 Reassembly is the reverse process. Grease the outer end of the driveshaft, the suspension bolts, and the steering ball joint pin, so another time they will be easier to remove.
17 If an assistant is available, the nuts can be tightened on the ends of the shafts by applying the foot brake and letting him torque up the nut(s). Otherwise they must be done when the car is back on the ground. New nuts should be used, and tightened to a torque of 101 lb f ft (14 kg f m). Then stake the nut by hammering its collar into the groove in the shaft with a punch (or blunt screwdriver).
18 Refill the transmission with oil when the car is back on level ground.

3 Driveshafts - removal (in two sections)

1 If the constant velocity joints are going to be replaced, then there is no harm in risking getting dirt in them, and it is then possible to take off the integral shafts without dismantling the suspension. However, a circlip has to be removed which is difficult to see, as it is smothered in black grease, and it is very difficult to get at. This method is useful if the splines are seized in the hub.
2 All the preliminary arrangements are as in Section 2, but do not dismantle the brakes, steering ball joint, or suspension.
3 Turn the steering fully inwards. Clean all round the boot at the outer end of the shaft. Take off the clip holding the boot to the joint. Peel the boot back down the shaft. Scrape all the grease off the constant velocity joint. (photo)
4 Locate the cut-out of the inner part of the constant velocity joint, and the ends of the circlip. The hub may need rotating to bring it round. (photo)
5 Part the circlip. This is almost impossible without circlip pliers, particularly working under the car.
6 Holding the circlip parted, pull the shaft towards the transmission to get the groove away from the circlip. Once it has moved, ease the circlip. (photo)
7 Check the steering is still on full lock, then pull the shaft towards the transmission, and out of the inner race of the constant velocity joint. There has been the case where the shaft would not come out, as there was not room, and the two bolts at the bottom of the suspension strut still had to be removed, to allow the hub to move out a bit. But this is not normal.
8 Once the main part of the shaft has been withdrawn from the constant velocity joint at the outer end, then the two parts can be taken off as may be necessary. The way these are released was described in Section 2. If removing the engine or transmission, there is no need to disturb the constant velocity joint in the hub, just the inner part should be taken out. If it is the constant velocity joint that needs replacement, then there is no need to take off the rubber boot at the inner end, with consequent draining of transmission oil. Indeed, even if taking out the engine or transmission, once the shafts have been split, they can be tied inwards towards the transmission, with string and no further stripping done, if preferred. (photo)
9 Reassembly is the reverse process. It is most important that the circlip is correctly seated in the groove on the axle shaft. Take special care with the new joints that no dirt gets in. Refill the joints and seal the rubber boot, as described in Section 5.
10 Refer to Section 2 for other apsects of reassembly.

4 Components - examination and renewal

1 The need to replace the constant velocity joints can usually be assessed without removing them. If the rubber boot has failed, the joint is almost certain to do so too soon after, unless the hole was spotted, and a new boot fitted in very good time.

2.3 The large nut in the centre of the front hub holds the driveshaft. Both sides are normal right-hand threads

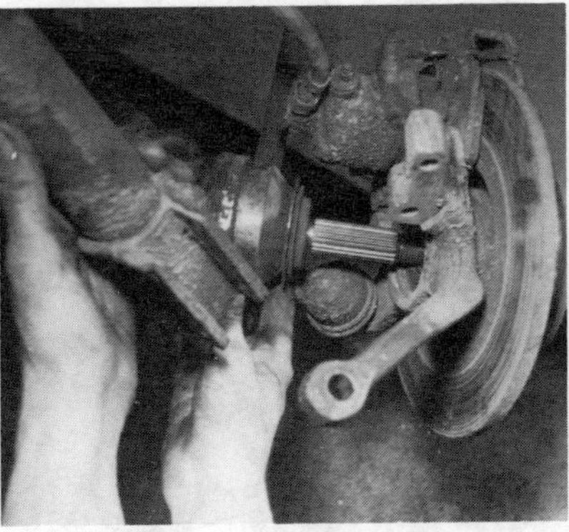

2.13 After undoing the suspension strut and steering balljoint, the knuckle can be turned enough to pull the shaft out

2.15A At the inner end clean the dirt off before undoing the boot, and put a can to catch the oil. Refill the transmission oil afterwards

2.15B Sometimes the inner, 'Tripode' joint comes apart and spill out its needle rollers, if taken out without care

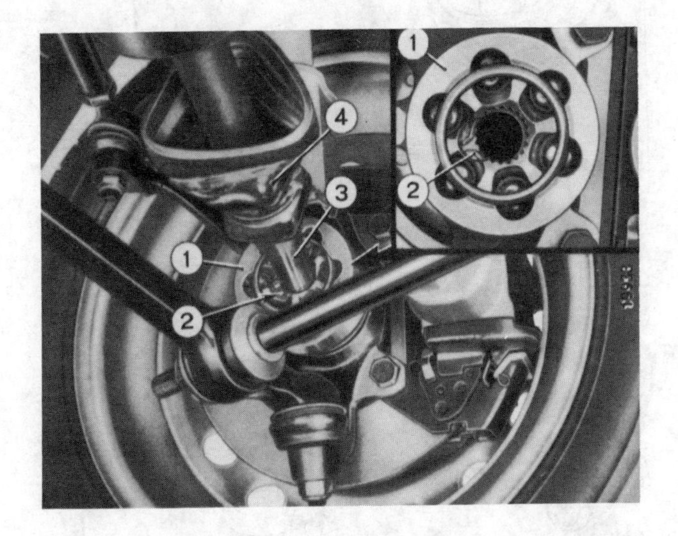

Fig. 7.2. Removing the shafts from the constant velocity joints without dismantling the suspension (Sec. 3)

1 *CV joint*
2 *Circlip*
3 *Driveshaft*
4 *Boot pulled back*

3.3 As an alternative method, after very thorough cleaning, the rubber boot is undone and peeled back ...

3.4 ... and the gap in the circlip found, which is not as easy to see when it is still on the car

3.6 The circlip can be parted and the shaft pulled out of the constant velocity joint

3.8 Then the joint can be pulled out without having to disconnect the front suspension

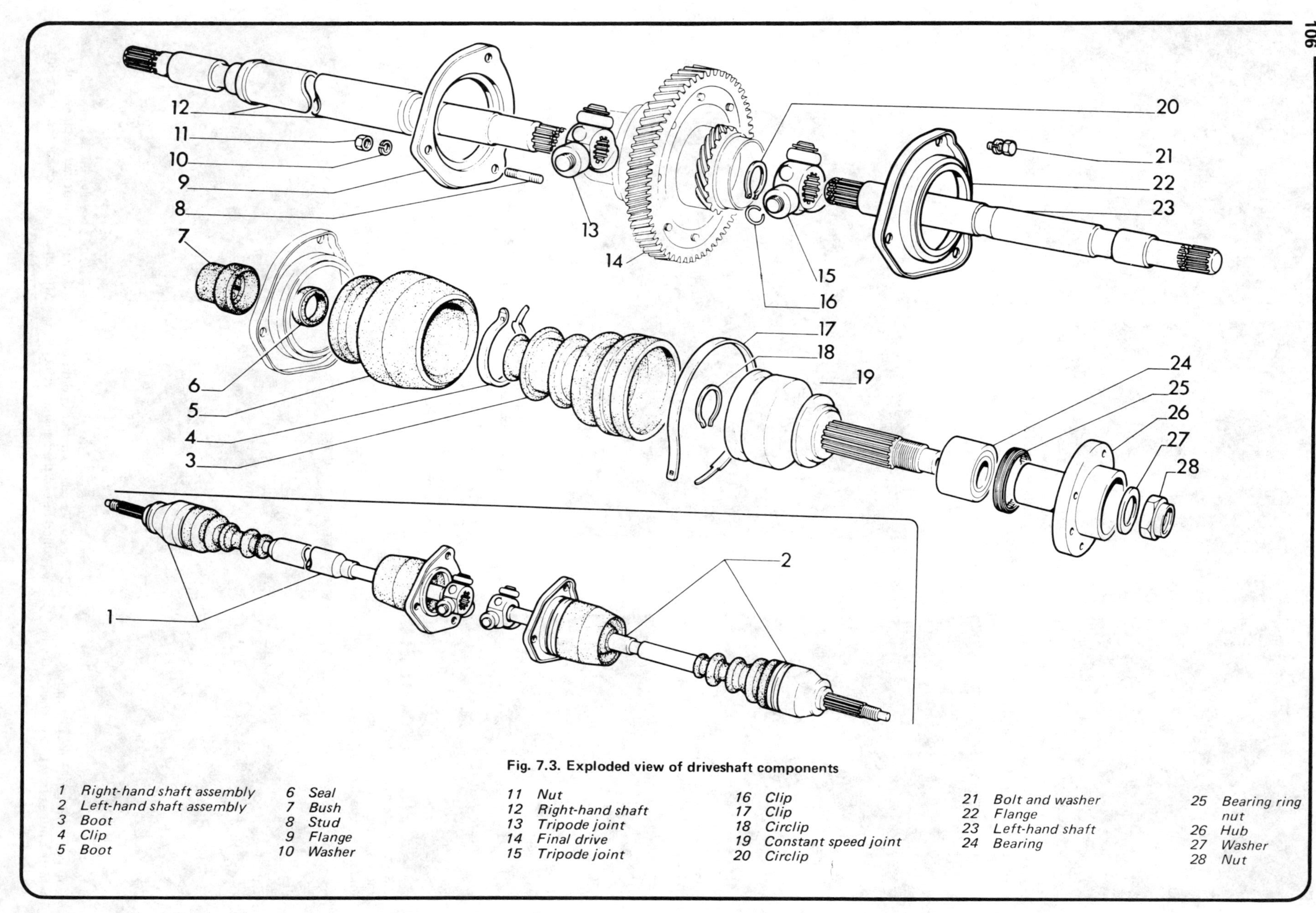

Fig. 7.3. Exploded view of driveshaft components

1 *Right-hand shaft assembly*
2 *Left-hand shaft assembly*
3 *Boot*
4 *Clip*
5 *Boot*
6 *Seal*
7 *Bush*
8 *Stud*
9 *Flange*
10 *Washer*
11 *Nut*
12 *Right-hand shaft*
13 *Tripode joint*
14 *Final drive*
15 *Tripode joint*
16 *Clip*
17 *Clip*
18 *Circlip*
19 *Constant speed joint*
20 *Circlip*
21 *Bolt and washer*
22 *Flange*
23 *Left-hand shaft*
24 *Bearing*
25 *Bearing ring nut*
26 *Hub*
27 *Washer*
28 *Nut*

By failure, it is not meant that the joint breaks. But they will become noisy, particularly at full lock. A clicking noise, later becoming a tearing noise, is a sign of excessive wear inside the joint.
2 The shaft splines and the inner 'Tripode' joint should outlast perhaps two constant velocity joints. Their condition should be examined when taken out for other reasons.
3 The shafts are not serviced as complete assemblies, so individual parts must be ordered as required. The constant velocity joints do come complete, though without the rubber boot. It is not practicable, nor are the parts available, to renovate the constant velocity joints.
4 The shafts and joints will last well provided the remarks in the next Section on the care of them are heeded.

5 Lubrication and rubber boots

1 No specific maintenance tasks have to be done on the drive-shafts or their joints. But the visual examination of the condition of the rubber boots is an important item of the check underneath the car in the 12,000 mile (20,000 km) task.
2 The rubber boot on the constant velocity joint at the outer end of the shaft must be examined to ensure there are not tears or slits in it, and that it is properly secured to the joint and the shaft, by clips at its fat and thin ends.
3 The constant velocity joint is lubricated with a molybdenum disulphide grease, such as Duckhams LBM 10. Normally this joint will not need renewing. If the rubber boot is removed, either for replacement, or when splitting the driveshaft at the constant velocity joint, then the old grease should be wiped out with clean rag to take out any dirt. Ensure complete cleanliness when doing this, or more dirt will get in, and the joint will not last long. Then pack in new grease, pushing it well in around the joint balls, and leaving a coating on the outside on the splines. Then put on the rubber boot, with grease inside it and on the shaft. Do not fill the boot with grease. Too much grease will prevent the boot from working as a concertina as the steering turns. FIAT recommend 1 3/8 oz. (40 g) of grease in the joint cavity around the balls, and 2 oz. (55 g) of grease in the protection boot.
4 Secure the boot to the rim of the joint, making sure the moulded groove of the boot is in the groove of the joint. The small end of the boot should butt against the shoulder on the shaft. For preference use the metal tape straps. If these are not available, put round two turns of soft copper wire, and twist the ends together with pliers to draw the wire firmly round the boot, but not so tight it cuts into the rubber, or is under such tension it will break after a short time. Bend the twisted ends backwards, the way they will tend to go due to shaft rotation, so they lie out of the way.
5 The 'Tripode' joint at the inner end of the shaft is lubricated by the transmission oil, so no attention need be paid to that. The rubber boot at the inner end must seal the transmission oil. Two aspects of the boot are therefore important. The rubber envelope must be free from cuts. The oil seal in which the shaft rotates must be in good condition. In order to renew the oil seal it is necessary to release the circlip that retains the tripode joint and pull the joint off of the shaft. (photos)
6 The shaft can then be eased out of the boot, the flange removed and the bush that holds the seal can be pressed out of the boot. (photos)
7 The seal is then simply prised out of the bush and a new one pressed in position. Lightly grease the seal, before pressing in position, using a silicon grease. (photo)
8 New rubber boots should be fitted to the inner ends of the shafts whenever the transmission is dismantled. Fitting them is also advisable whenever the driveshafts are removed, and there is the slightest suspicion that the boots are near the end of their life, otherwise the labour of removing the shafts may soon have to be repeated.

5.5A Releasing the circlip ...

5.5B ... and pulling the tripode joint off of the shaft

5.6A Easing the shaft out of the boot ...

5.6B ... removing the flange ...

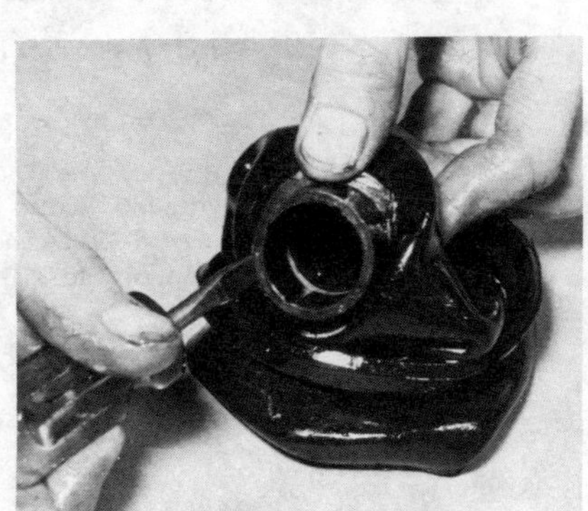
5.6C ... and prising out the bush

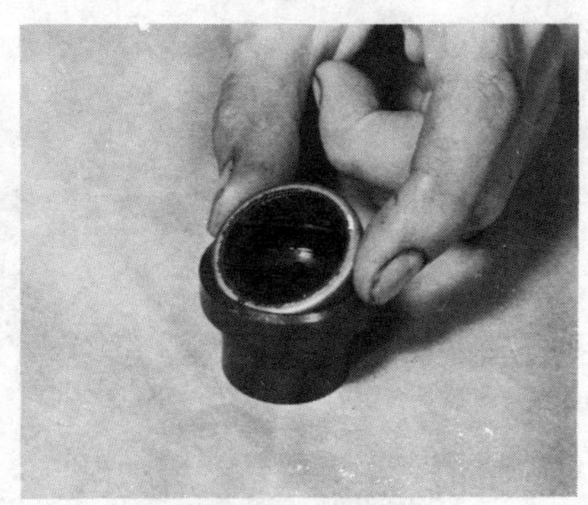
5.7 Showing the seal in the bush

Chapter 8 Braking system

For modifications, and information applicable to later models, see Supplement at end of manual

Contents

Specifications

Front brakes

Discs:

Diameter	8.937 in (227 mm)
Nominal thickness	0.392 to 0.400 in (9.95 to 10.15 mm)
Minimum allowable thickness after refacing	0.368 in (9.35 mm)
Minimum allowable thickness from wear	0.354 in (9 mm)
Maximum allowable run-out	0.006 in (0.15 mm)
Brake calipers	Floating, single cylinder type
Minimum allowable thickness of:	
Brake pad	0.08 in (2 mm)
Caliper cylinder bore	1.890 in (48 mm)
Braking lining clearance	Automatic adjustment

Rear brakes

Drum diameter	7.2930 to 7.304 in (185.24 to 185.53 mm)
Refacing drums: maximum depth of machining	0.03 in (0.8 mm)
Maximum allowable diameter (wear limit)	7.355 in (186.83 mm)
Brake linings:	
Length	7.086 in (180 mm)
Width	1.181 in (30 mm)
Thickness:	
New	0.165 to 0.177 in (4.2 to 4.5 mm)
Minimum allowable	0.06 in (1.5 mm)

Wheel cylinder bore	¾ in. (19.05 mm)
Master cylinder bore	¾ in. (19.05 mm)
Handbrake	Acting mechanically on rear brake shoes
Brake pressure regulator	Acting on rear wheels - For settings see text.

Brake fluid

Type	Hydraulic fluid to FMVSS 116 DOT 3 or SAE J1703 (Duckhams Universal Brake and Clutch Fluid)
Capacity	½ Imp. pint (0.315 litre)

Torque wrench settings

	lb f ft	kg f m
Bolt, handbrake support	11	1.5
Bolt, lock plate to front wheel brake calipers	7	1
Brake hoses to calipers	22	3
Bolt, wheel cylinder to brake backing plate	7	1
Brake hoses to wheel cylinders (rear)	14½	2
Bolt, rear brake regulator to mounting bracket	14½	2
Nut, brake master cylinder to mounting bolt	18	2.5
Rear brake backing plate bolts	18	2.5

1 General description

1 The brakes use discs at the front and drums at the rear. The hydraulic system is dual, a tandem master cylinder controlling separate circuits for the front and rear brakes.
2 A pressure regulator valve at the rear is connected to the rear suspension, and limits the braking effort at the rear when the back of the car is high, either due to light load, or heavy braking. This ensures maximum retardation and control by preventing the rear wheels locking under hard braking.
3 The brakes do not call for much actual work, but there are a number of important checks. If these checks are made any deterioration of the brakes can be noted before danger is caused. Maybe it is because the brakes are so reliable, but it is a sobering fact that many apparently enthusuastic owners allow the brakes to get in an unsafe condition.
4 Checking the hydraulic fluid is a 300 mile, or weekly, task. The level gradually falls as the brakes wear, and comes sharply up as new pads are fitted. But this is a small variation compared with the size of the reservoir. Normally the level will always be correct. But it must be checked all the same. One day it may be low, indicating a leak.
5 At 3,000 miles (5,000 km) during the check underneath the car, all pipes, particularly the flexible ones, are checked for leaks and damage, and the front pads inspected. At 6,000 miles (10,000 km) the rear drums are removed for a check inside. At such checks the signs to look for are freedom from seepage of hydraulic fluid, enough lining material left to last until the next inspection, and a good rubbing surface on drum or disc. This last is needed for good braking: it is also indicative that the brakes are working properly.
6 Poor braking can grow gradually; the driver must be alert for a fall in performance. Normal braking is quite gently. At intervals the opportunity must be made to find an empty road, and do a hard brake test.
7 The brake fluid has to stand high temperature, be non-corrosive to the brake metal parts, and not soften those of rubber. It is vital that only fluid to the correct specification is used. Avoid using different makes. Changing the fluid is a 24,000 mile (40,000 km) task. The fluid absorbs moisture from the atmosphere, so in time its quality deteriorates. Never re-use old fluid. When working on hydraulics, let all fluid that wants to drip out do so, rather than plugging the system, as this is of additional benefit in keeping the system clean. If it is ever suspected that the wrong type of fluid has been put in, then the whole system must be flushed out, and new rubber parts fitted throughout; master cylinder, all wheel cylinders, the flexible pipes, pressure regulator and pressure failure warning switch. It is recommended anyway that this is done at 60,000 miles (100,000 km). Never wash brake parts in anything other than brake fluid or methylated spirit.
8 If there is an hydraulic failure of one component, then at least the others must be suspect too.

2 Brakes (front) - checking and inspection

1 Whilst doing the 6,000 mile (10,000 km) routine maintenance task, the front brakes must be checked.
2 Once familiar with the car this can be done fairly well through the holes in the wheel, but to get a good view these should be removed.
3 Check the thickness of pad lining remaining. If this is 0.08 in (2 mm) or less, the pads must be replaced. See the next Section.
4 Check the hydraulic pipes have no seepage of fluid, and also for any signs of damage, particularly of rubbing, or weathering of the flexible pipe. Check there are no signs of fluid seeping out of the brake caliper, past the piston.
5 Check the surface of the disc. This should be shiny and smooth. Though other signs of brake inefficiency would have been apparent, if the discs have not been polished up by the pads, then the brakes cannot have been working properly.

3 Brake pads (front) - renewal

1 Jack-up and remove the front wheel.
2 Pull out the spring clips that hold the two caliper locking blocks. (photo)
3 Pull out the locking blocks. (photo)
4 Lift off the caliper, and lodge it on the suspension so its weight will not hang on the flexible pipe. (photo)
5 Take out the old pads, noting the anti-chatter springs. (photos)
6 Wipe the caliper clean, especially the flanks of the piston that are sticking out due to the thinness of the old pads. If very dirty, wash the caliper with ordinary detergent and water or methylated spirit, and dry it. Do not use petrol or other solvents, though brake fluid is very satisfactory, if messy.
7 Push the hydraulic piston back into the caliper. Watch the level in the hydraulic reservoir beside the master cylinder. The level will rise; check it does not overflow.
8 If there is risk of the reservoir overflowing, hold the caliper with the bleed nipple uppermost. Slacken the bleed nipple, and push the piston home, expelling the excess fluid out of the nipple. Retighten the nipple whilst it is still awash with escaping fluid, so no air can get in.
9 If the piston is stubborn, push it in by using a carpenter's clamp. Take care the rim of the piston is not burred.
10 Reassemble using the new pads, and checking that the anti-chatter springs are in position.
11 Press the brake pedal to bring the piston and pads up the disc.
12 Check the hydraulic fluid level.
13 Repeat for the other side.
14 Note that the pad lining type used is denoted by colour code daubs of paint. Both pads on both sides must all have the same lining. More than one type is supplied by FIAT. Ensure that the

3.2 Showing the spring clips that hold the locking blocks

3.3 Pulling out the locking blocks

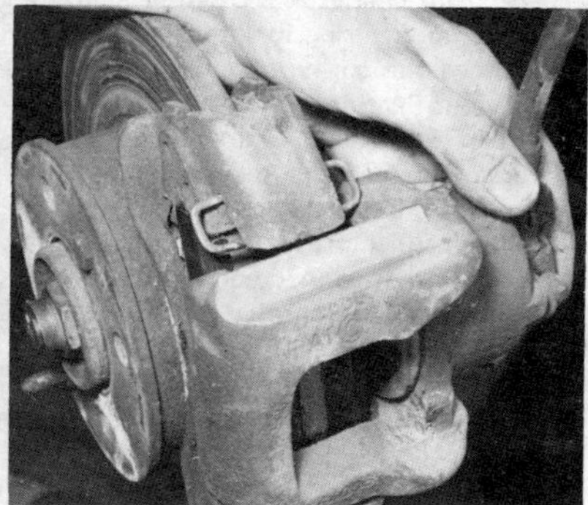
3.4 Lifting off the caliper

3.5A Removing the pads

3.5B Showing the anti-chatter springs

4.6 The support bracket can be removed from the steering knuckle

4.7 By removing just two bolts, the complete caliper assembly can be removed

lining used is approved by FIAT.

15 Use the brakes gently for at least a dozen applications so the pads can bed in.

4 Calipers (front) - removal and replacement

1 The front caliper will often need to be taken off to clear the suspension or steering for other work.

2 Remove the spring clips and locking blocks as described in the previous Section.

3 Lift off the caliper. Tie it up where it cannot strain the hose if there is no need to take it right off, as for example, when working on the front suspension.

4 If the caliper needs to be taken right off, then the flexible hose must be disconnected. As this work is presumably to overhaul the front caliper, refer to the next Section.

5 With the caliper off, remove the pads, and the anti-chatter springs.

6 Undo the two bolts that hold the caliper support bracket to the steering knuckle. (photo)

7 Once familiar with the car, if removing the caliper and support bracket for work on the suspension, it may be preferable to take off the caliper pads, and support bracket as an assembly by simply taking out the two bolts at the back. (photo)

8 On reassembly, reverse the dismantling procedures. But ensure the mounting surface for the support bracket on the steering knuckle is clean, and if the bolts are rusty, grease them. Note that as the brake pads are normally in light contact with the disc, the caliper must be carefully fitted to slide back in.

5 Calipers (front) - overhaul

1 Jack-up and remove the front wheel.

2 Wash thoroughly with water and household detergent or methylated spirit the caliper assembly, and the flexible pipe, particularly the fixing bracket and union at the car end of the flexible pipe.

3 Have ready a tin to catch the brake fluid, and sheets of clean newspaper on which to put components.

4 Take out the spring clips and locking blocks, and take the caliper off the support bracket. On models with separate cylinder and housing components, depress the retaining dowel and slide the components apart.

5 Get an assistant to press gently and slowly on the brake pedal, so as to push the piston out of the caliper cylinder. Have the tin or jar ready to catch the brake fluid.

6 Catch the piston and its dirt shield, and put them on a clean

surface (the newspaper). On some calipers it may not be possible to remove the piston, in which case a faulty caliper will have to be renewed.

7 Lodge the caliper back on the support bracket where it cannot get dirty, or strain the flexible pipe.

8 Disconnect the flexible pipe. This is best done at the car, or inboard end. Undo the union between the two parts of the pipe, and then the flexible one from the bracket. In this way the pipe is firm whilst being undone, and also the flexible pipe can be taken away for inspection with the caliper.

9 Clean all the parts very thoroughly. Do not use solvents other than alcohol (methylated spirits) or brake fluid.

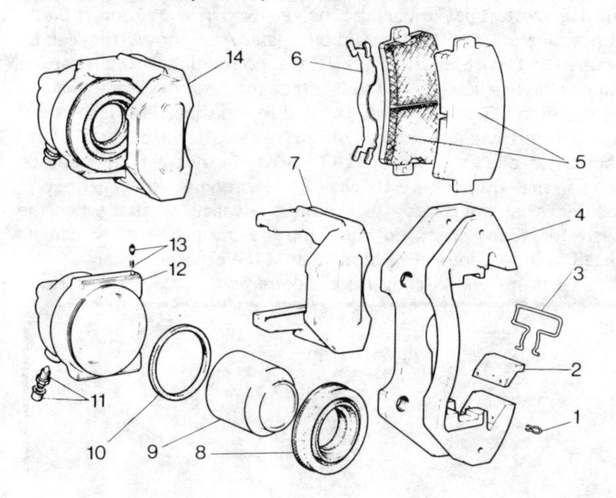

Fig. 8.1. Front brake - caliper components (Sec. 5)

1 Clip	*8 Dust boot*
2 Caliper locking block	*9 Piston*
3 Spring	*10 Seal*
4 Caliper support bracket	*11 Bleeder screw and dust cap*
5 Shoe and lining	*12 Cylinder*
6 Shoe retainer spring	*13 Spring and dowel*
7 Cylinder housing	*14 Complete caliper*

Some calipers have the cylinder and housing as a single unit

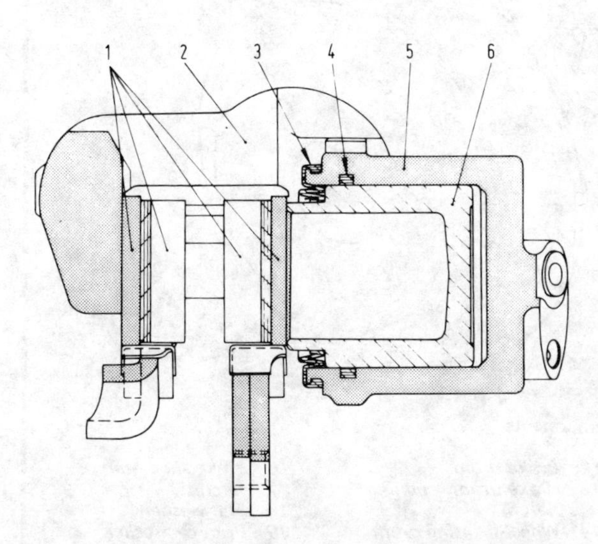

Fig. 8.2. Section through front brake caliper (Sec. 5)

1 Shoes and linings	*4 Seal*
2 Cylinder housing	*5 Cylinder*
3 Dust boot	*6 Piston*

10 Check that the caliper cylinder and the piston are free from scores. If not, they must be replaced. Ease the rubber boot from the caliper body and piston.

11 Carefully ease the 'O' ring seal from the inside of the cylinder using a non-metal pointed rod. Great care must be taken so as not to score the walls of the cylinder. This seal is shown in Figs. 8.1 and 8.2.

12 Inspect the cylinder bore and the piston for signs of scoring, and if evident, a complete new caliper assembly must be obtained.

13 To reassemble, first wet the cylinder 'O' ring seal with fresh hydraulic fluid and carefully insert it into its bore in the cylinder. Next wet the piston and place it in the cylinder bore. Push the piston down the cylinder bore until the piston rubber boot can be fitted to the piston and caliper body.

14 Reassemble the pads and caliper to the support bracket, taking care no dirt is scattered in the pipe unions.

15 Fit the flexible pipe. Make sure it is tightened with the wheel straight-ahead and without any twists or bends. Check the steering can move from lock to lock without risk of damage to the hose.

16 Fill the master cylinder reservoir, and pump up the fluid. Check for leaks. Bleed the system (see Section 16).

17 To make easier bleeding the hydraulic system, the caliper should be filled with fresh hydraulic fluid by removing the bleed nipple and pouring the fluid with the caliper in a tilted position. Refit the bleed nipple.

6 Brake disc (front) - removal and replacement

1 If a disc is badly scored, pad wear will be rapid, so the disc should be replaced or resurfaced.

2 If the disc surface is cracked or chipped, then it must be replaced at once, as braking is liable to be dangerous.

3 Remove the complete brake caliper and support bracket as described in Section 4.

4 Remove the two bolts holding the disc to the hub, one being a long wheel locating bolt. (photo)

5 Take off the flange, followed by the disc.

6 If the disc is not too badly scored, it is possible to machine it smooth, within the limits given in the specification. But the cost may be as much as a new disc.

7 Fit the new disc on the hub, washing off any preservative first.

8 Spin the hub, and measure the run out of the rim of the disc. If it is not true within the specification, take it off and refit it again. In another position the errors of hub and disc could cancel each other out.

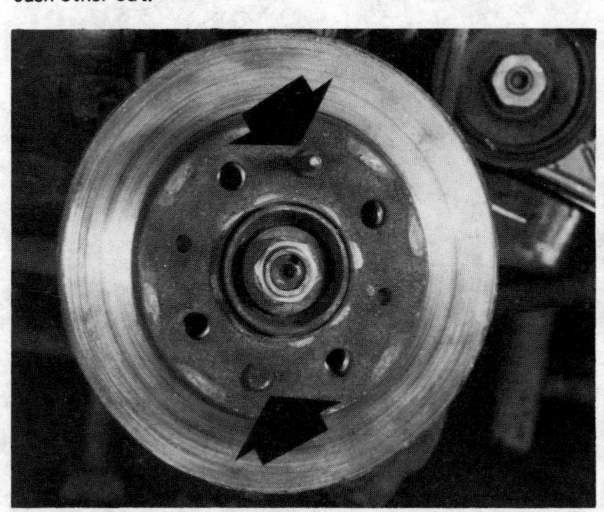

6.4 With the caliper and support bracket off, just two bolts need to be taken off to remove the disc

7 Brakes (rear) - checking and inspection

1 Part of the 6,000 mile (10,000 km) maintenance task is to check inside the rear brake drums. Later cars have inspection holes allowing the lining thickness to be seen without removing the drum, this is not really thorough enough.
2 Chock the car's other wheels, put it in gear, and release the handbrake.
3 Jack-up and remove the rear wheel.
4 Take out the two bolts holding the drum to the hub. One is the long wheel locating bolt. (photo)
5 Pull off the brake drum. (photo)
6 Check the thickness of lining remaining on the shoes. It will not be the same all round the shoe. The leading shoe also wears faster than the trailing one. If the thickness is less than 1/16 in. (1.5 mm) anywhere, the shoes must be renewed (See the next Section). (photo)
7 If the trailing shoe is still all right, it need not be changed. But what is done to one side of the car must be done to the other, to keep even braking.
8 Blow all dust out from around the hydraulic cylinder, and behind the shoes. Check hydraulic fluid is not seeping from the cylinder. If it is, overhaul the cylinder as described in Section 9. If fluid or hub grease is found on the brake linings, they must be renewed. See Section 8.
9 Wipe all dust out of the brake drum. Check its surface for scoring or cracks. If badly scored lining wear will be rapid. If cracked, the drum should be renewed. If within the limits in the specification, scoring can be removed by machining.

8 Brake shoes (rear) - removal and replacement

1 The rear brake shoes must be renewed when the lining is worn to the limit given in the specification (approximately 1/16 in. 1.5 mm). They must also be renewed if the lining is badly contaminated with brake fluid or grease. A minor wetting can be washed off with petrol, but if the contamination is deep, it cannot be eradicated, and will affect the braking characteristics of the lining. The linings are bonded to the shoe. This is a special process outside the scope of an owner. Replacement shoes should be bought from a FIAT agent. Do not get non-genuine spares, that might have linings of the wrong grade. It is recommended that the shoes are bought in advance, so that when the need for them arises during a routine inspection they can be fitted then, without need to dismantle the car again later.
2 The drum will already have been removed as described in the

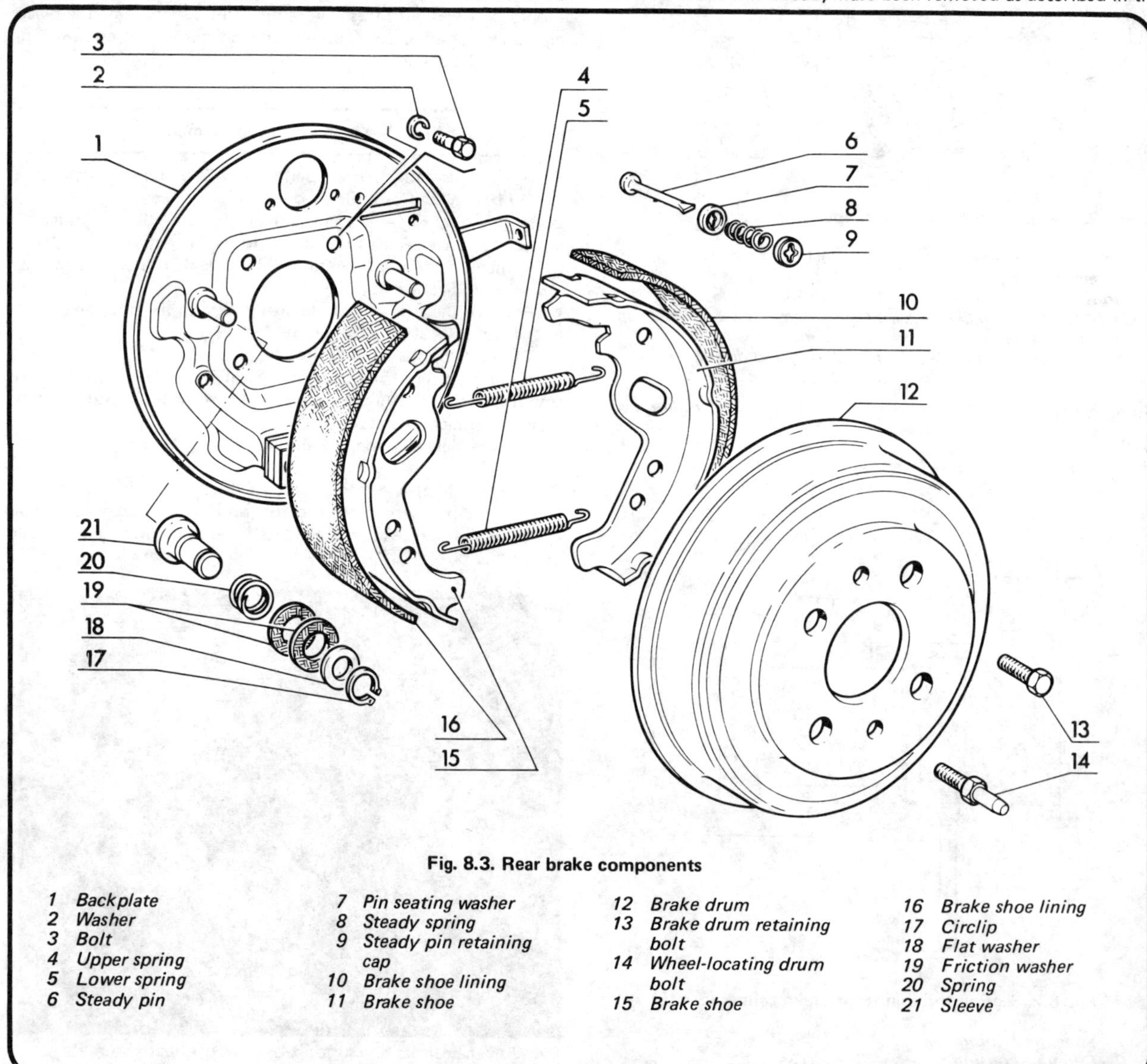

Fig. 8.3. Rear brake components

1 Backplate
2 Washer
3 Bolt
4 Upper spring
5 Lower spring
6 Steady pin
7 Pin seating washer
8 Steady spring
9 Steady pin retaining cap
10 Brake shoe lining
11 Brake shoe
12 Brake drum
13 Brake drum retaining bolt
14 Wheel-locating drum bolt
15 Brake shoe
16 Brake shoe lining
17 Circlip
18 Flat washer
19 Friction washer
20 Spring
21 Sleeve

7.4 Showing the two bolts fixing the drum to the wheel

7.5 Pulling off the drum

7.6 Brake shoes open to inspection

8.4 Depressing and a quarter turn releases the steady spring

8.5 The shoe must be worked up the self-adjuster post till it can be turned

8.8 The self-adjusters must be transferred to the new shoes

previous Section.

3 It is possible to remove the shoes with the wheel's hub in place. The flange has semi-circular sectors cut out to give room for the shoes to come off. However, there is a knack in removing the shoes, and it is much easier if the hub is first taken off. So remove the nut on the end of the stub axle. Pull off the hub (see Chapter 11/12 for details).

4 Undo the steady springs, by depressing and giving their caps a half turn to disengage the slot from the pin (photo).

5 Pull a shoe (if a right handed person, the right shoe) out from its seat on the wheel cylinder, and pivot at the bottom and work it up the post of the self adjuster. (photo)

6 Once off the self adjuster post, the pull-off spring tension is eased, as the shoe can move towards the other, so the springs can be unhooked. Before taking the springs off, note the way they are fitted and mark on the shoe, with a screwdriver the spring positions.

7 Take off the other shoe.

8 The self adjusters must now be transferred to the new shoes. To undo them, the spring must be compressed to allow the circlip to be undone. The spring is strong. If possible, get the FIAT agent to make the transfer when buying the new shoes from them. Otherwise a clamp must be organised. A vice, a carpenter's 'G' clamp, or a valve spring compressor are all possibilities. Small bits of steel must be put between the clamp and the adjuster washer so the circlip is not trapped. (photo) (Fig. 8.4)

9 After high mileages the self-adjusters need replacement. The washers wear, also the hole in the bush gets bigger, so the adjustment provided is not close enough.

10 If the hydraulic cylinder needs overhaul, now is the time with the shoes out of the way.

11 When reassembling the new shoes with the adjusters, and the shoes to the brakes, put a slight smear of brake grease on all the working surfaces for the adjusters, the bottom shoe pivot and the steady springs; but not where the shoe sits on the hydraulic cylinder piston.

12 Fit the first shoe (the left one if right handed) into place with the springs fitted in the identical position they were removed from the old shoe. Hook the second shoe onto the springs, and then pull it across until the self-adjuster can be fitted over its post. Then wriggle the shoe down into position and work the ends of the shoe into place. Make sure the handbrake linkage is in place.

13 Fit the pins for the steady springs from behind the back plate, and hold them from there whilst fitting the spring and cap from the front.

14 Some people prefer to take off and fit the pull-off springs using pliers with the brake shoes in place. This demands strong hands. If the pliers slip, the hands can get cut. If it can be done this way, then the shoe will slip on and off the self adjuster post must more easily, and all this can be done without removing the hub.

15 When refitting the hub, tighten the nut on the stub axle to the specified torque load (Chapter 11), and then stake the nut to the axle by hammering its collar into the groove.

9 Wheel cylinders (rear brakes) - overhaul

1 If there is seepage of fluid out of the ends of the wheel cylinder or if the wrong fluid has been put in the system, or if other rubber components cast suspicion on the state of the cylinder seals (cups) the cylinder should be overhauled.

2 Remove the brake shoes, as described in the previous Section.

3 Clean all round the wheel cylinder, both on the shoe side, and the outside of the brake back plate. If there is much dirt, wash with water and household detergent or methylated spirit. Do not use petrol or other solvents.

4 Spread a sheet of newspaper beside the car, with the new

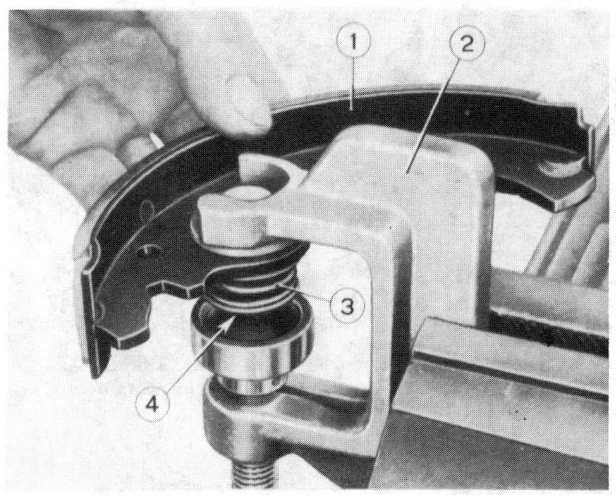

Fig. 8.4. Installing rear brake automatic adjuster using a special clamp (right - fitting the circlip) (Sec. 8)

1 Brake shoe and lining
2 Tool 'A 72246' for installing and removing brake adjuster
3 Spring
4 Sleeve

rubber seals ready. Then the work can go on steadily, even though there may be some seepage of fluid out of the end of the pipe. New seals will definitely be needed. It is pointless reassembling the cylinder with the old seals.

5 Take the rubber boots off both ends of the cylinder.

6 Push the piston in at one end and expel out of the other the train of pistons, rubber seals, spring and its two backing washers.

7 Discard the rubber parts.

8 Wash the metal parts and cylinder with brake fluid, and wipe them.

9 Examine the pistons and the cylinder bores. They should be free from scores or pitting. If the surface is not perfect then the cylinder must be scrapped.

10 Assuming the cylinder is serviceable lubricate the seals and pistons with brake fluid and reassemble ensuring absolute cleanliness. The sharp edges of the seals must go inwards.

11 Refit the shoes and drum, refill the hydraulic reservoir, and bleed the brakes (See Section 16).

12 If the wheel cylinder or piston surfaces are not good, the assembly must be changed. Imperfections will wear the sharp edge of the rubber seal and allow air into the system and in due course allow leaks as well.

13 To change the wheel cylinder, disconnect the hydraulic pipe at the rear. Take out the two bolts holding the cylinder to the back plate. Before fitting the new cylinder, clean the seating surfaces on the backplate. Ensure no dirt gets in the pipe union. Reconnect the pipe, fit the brake shoes and drum. Fill the hydraulic reservoir, and bleed the system.

10 Handbrake - adjustment

1 The handbrake does not normally need adjustment. Normal wear of the rear brake shoes will be taken care of by their self-adjusters. However, in due course the cable stretches, and there is wear in the handbrake linkage.

2 First ensure the rear brake self-adjusters have had a chance to do their work. Drive the car both forwards and in reverse and apply the brakes hard, when going at about 10 mph. This will bring both shoes in turn, when each is acting as the leading shoe, into close contact with the drum.

3 Now apply the handbrake and count the number of 'clicks' of the ratchet before the brake is hard on. If this is in excess of four, adjust the brake.

4 Jack-up both back wheels.

5 Apply the handbrake one 'click'.

6 Under the car locate the adjustment point on the cable immediately behind the lever. Undo one nut, and tighten the other to shorten the linkage. Do this till brake drag makes it slightly difficult to rotate the back wheels.

7 Check that when the lever is applied three 'clicks' it is impossible to turn the wheels, but that they are still quite free when the lever is released. Check that the linkage system in distributing the effort equally to both wheels.

8 Whilst working on it, oil the pivots of the linkage.

9 Lock the adjustment nut again.

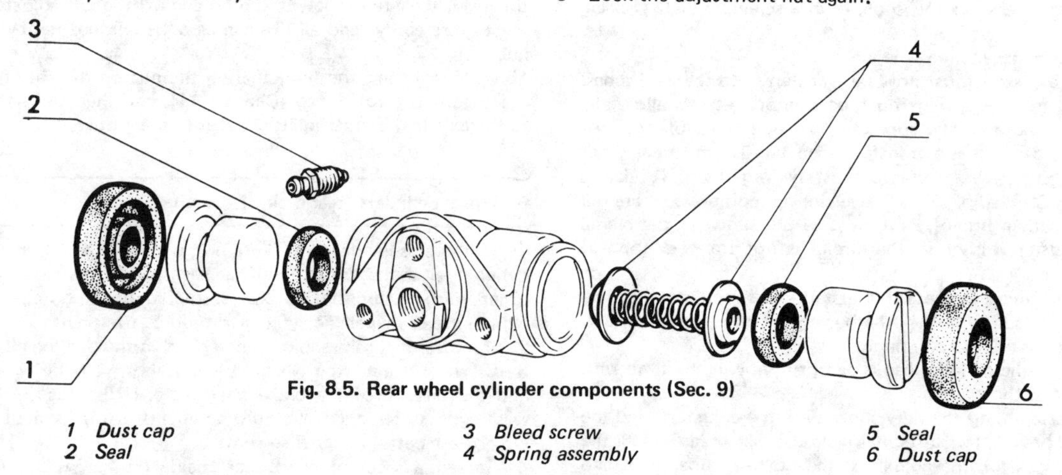

Fig. 8.5. Rear wheel cylinder components (Sec. 9)

1 Dust cap
2 Seal
3 Bleed screw
4 Spring assembly
5 Seal
6 Dust cap

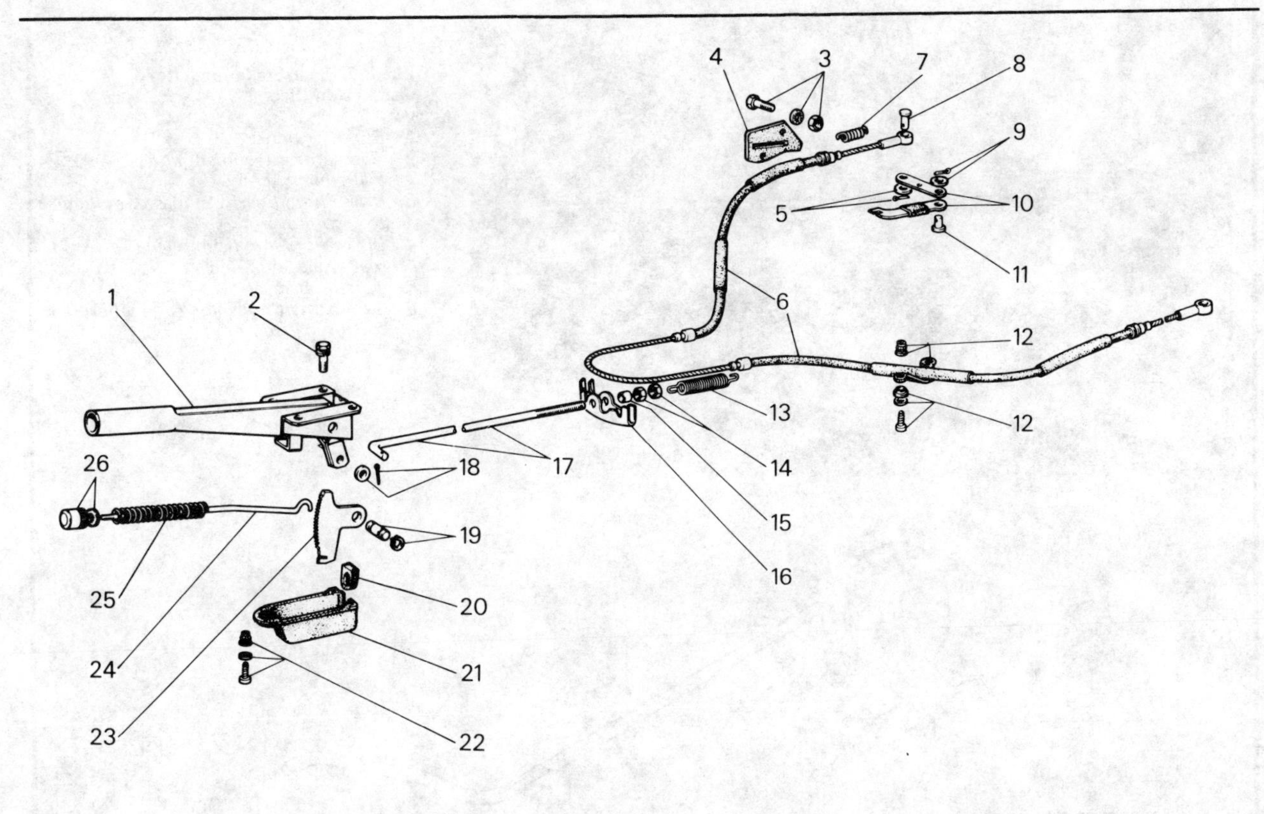

Fig. 8.6. Handbrake linkage (Sec. 10)

1 Hand lever
2 Bolt and washer
3 Bolt, nut and lockwasher
4 Pad on backing plate
5 Split pin and flat washer
6 Cable
7 Cable retracting spring
8 Pin
9 Split pin and flat washer
10 Shoe control levers
11 Pin for levers 10
12 Bolt, flat washer, cup, cable support bracket and bracket attachment block
13 Tie rod retracting spring
14 Tensioner adjusting nut and locknut
15 Spacer
16 Equaliser
17 Tie rod
18 Flat washer and split pin
19 Pin and safety ring
20 Grommet
21 Guard
22 Bolt, flat washer and block, guard to body
23 Ratchet
24 Rod
25 Spring
26 Push button and rubber ring

10 Road test the car, applying the handbrake hard at about 15 mph. It should be possible to lock both back wheels. If the adjustment is correct, but the brakes do not stop the wheels, then the inside of the drums and the condition of the linings should be examined. Drive the car about a mile and feel the temperature of the rear brake drums. They should be cold or warm. If hot, the handbrake is too tight.

11 Regulator (rear brakes) - general

1 It is important that the brakes do not lock, the rear wheels when the brakes are applied particularly hard since it causes the car to be unstable, and control may be lost. The braking force that can be applied to the rear wheels without them locking depends upon the weight upon them. So a regulator is fitted, coupled to the rear suspension and this limits the hydraulic pressure passed to the rear brakes when the rear of the car is high, either due to being unladen, or if pitching forward under heavy braking. We must stress that the regulator is not normally adjusted unless it has been removed or replaced.
2 The regulator is controlled by a rod fixed to the left suspension arm.
3 If the regulator leaks, or fails to work it should be replaced by a new one, since individual components are not serviced.
4 Under normal driving, failure of the regulator will be difficult to detect. First test the rear brakes by applying the handbrake at about 15 mph, and check it can lock the rear wheels. This proves the brake shoes are in order.
5 Now test the brakes from about 50 mph on a dry smooth level road. Wheel locking is difficult to detect but can usually be felt by an experienced driver. Other obvious indications are lack of steering, noise and tyre marks on the road. If under these conditions the rear brakes lock, it indicates that the pressure is not being limited. The ideal way of testing the regulator is with a G-meter but only an official FIAT dealer is likely to possess one.
6 To change the regulator, first wash with water, all dirt from the area, particularly the pipe unions.
7 Disconnect the link rod from the suspension.
8 Take off the hydraulic reservoir lid, and put over its mouth a sheet of polythene, then replace the lid. This will restrict the amount of fluid that leaks out. Alternatively tape over the vent hole.
9 Disconnect the pipes from the regulator, using a tin to catch fluid drips.
10 Undo the fixing bolts.
11 Fit the new regulator in position. Leave off the unions, and only put the bolts finger tight.
12 Set the link rod in position by pulling the end of the rod down until it is 1.693 ± 0.197 in (43 ± 5 mm) below the edge of the bump stop housing on the underneath of the suspension arm. (See Fig. 8.10).
13 Pull back the rubber cover of the regulator, and turn it about

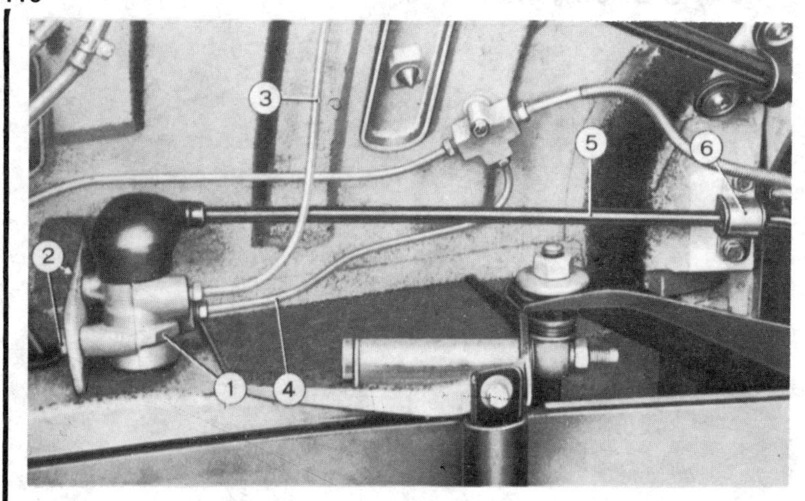

Fig. 8.7. Rear brake regulator installation (Sec. 11)

1 Brake regulator
2 Regulator attachment and adjustment screws
3 Brake fluid line from master cylinder to regulator
4 Brake fluid line from regulator to rear wheel cylinders
5 Torsion bar
6 Support with bushing for anchoring torsion bar to body

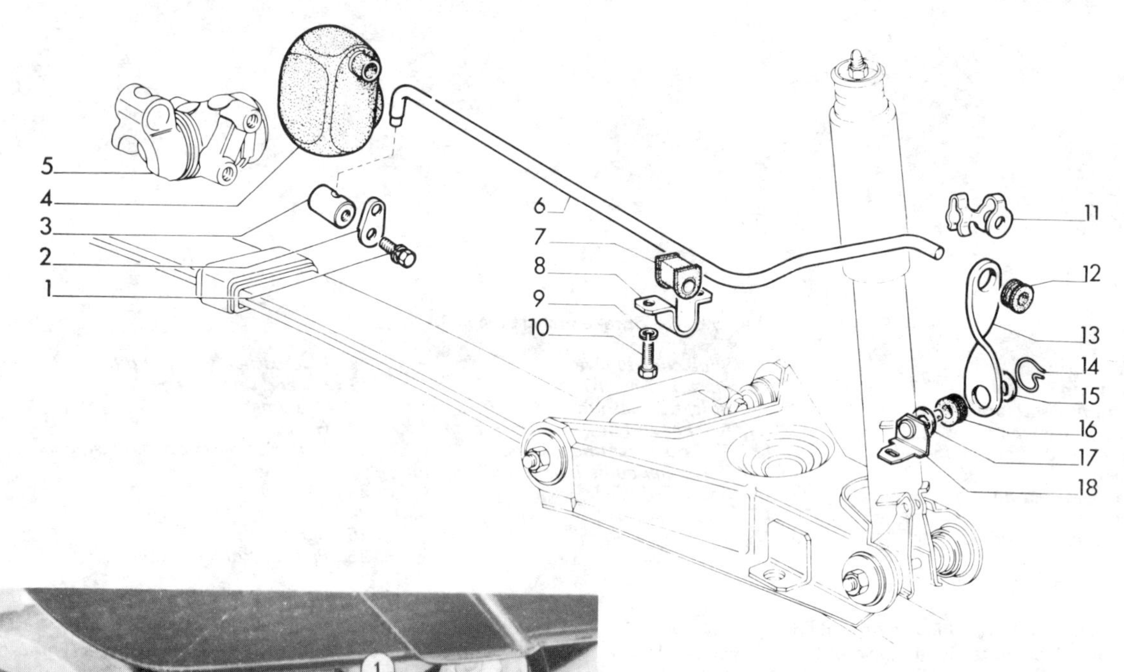

Fig. 8.8. Rear brake regulator components (Sec. 11)

1 Bolt
2 Lockplate
3 Pin
4 Rubber shroud
5 Regulator body
6 Torsion bar
7 Rubber mounting block
8 Mounting clamp
9 Washer
10 Bolt
11 Clip
12 Grommet
13 Link
14 Locking clip
15 Washer
16 Grommet
17 Washer
18 Anchor pin

Fig. 8.9. Rear brake regulator - connection to control arm (Sec. 11)

1 Buffer
2 Control arm
3 Flat washer and fastener for torsion bar link attachment to anchor pin on control arm

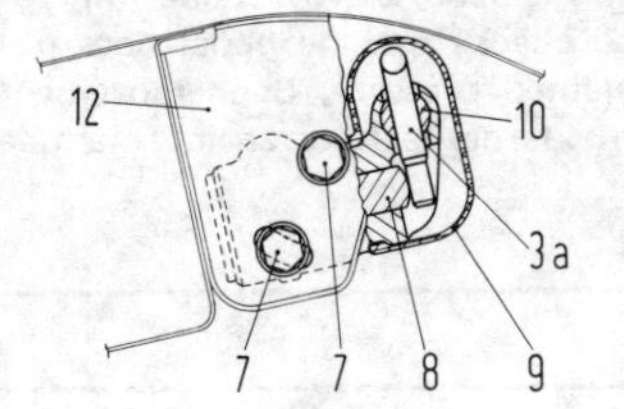

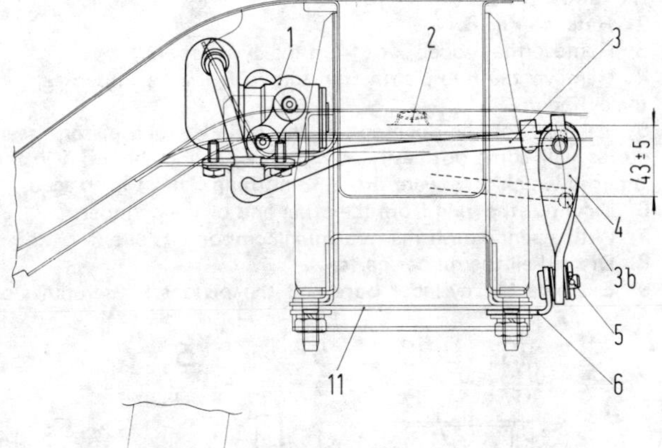

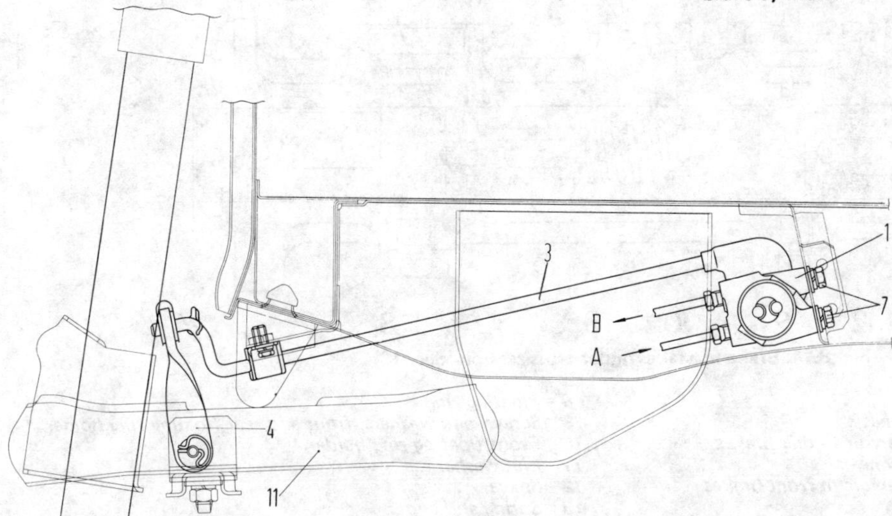

Fig. 8.10. Installation details of the rear brake regulator system (Sec. 11)

1 Brake regulator
2 Rubber buffer seat
3 Torsion bar
3a Torsion bar end, regulator side
3b Torsion bar end, link side
4 Link connecting torsion bar to control arm
5 Link anchor pin to control arm
6 Anchor pin support
7 Regulator attachment and adjustment screws
8 Regulator piston
9 Protection boot
10 Regulator pin
11 Control arm
12 Regulator support bracket

A = Brake fluid line from master cylinder to brake regulator
B = Brake fluid line from regulator to rear wheel brake cylinders

one of its mounting bolts, with the other moving in its slotted hole, till the regulator end of the rod is just touching the end of the regulator piston. Tighten the mounting bolts to a torque of 18 lb f ft (2.5 kg f m).
14 Connect the pipes to the unions.
15 Connect the link rod.
16 Refill the hydraulic reservoir, and bleed the system.

12 Master cylinder - removal and replacement

1 A likely minor defect calling for overhaul of the master cylinder is that air gets into the system calling for frequent bleeding. This could be drawn past the piston seals without there being a leak of fluid. If in slightly worse condition then fluid may leak out, showing at the master cylinder's mounting flange, or leaking down inside the car. In extreme cases, the seals may swell, so the piston is unable to return to the off position, and causes brake binding. The latter situation implies that the wrong fluid has been used, so the overhaul of the master cylinder should be done in conjunction with the complete draining of the fluid and the renewal of all the seals in the wheel cylinders, regulator, pressure warning switch and all the flexible pipes. The renewal of all such rubber components is recommended anyway as a standard procedure for brakes after 60,000 miles (100,000 km).
2 On all right-hand drive vehicles the master cylinder is mounted in a small tray that is secured to the pedal bracket assembly by two bolts, nuts and washers; this of course makes access rather difficult since one has to work underneath the dashboard. It is better in these circumstances to remove the two bolts first, release the mounting tray and then support the cylinder before the next move. (photo)
3 Disconnect the two pipes from the reservoir to the master cylinder. Use a small jar to catch the fluid after the pipes are taken off the fittings on the master cylinder.
4 Have plenty of rag below the pipes to catch any drips of fluid, as it will ruin the paint.
5 Undo the two unions for the delivery pipes to front and rear brakes.
6 Pull the master cylinder forward to disengage the actuating plunger and take it off.
7 Refitting the master cylinder is the reverse process.
8 After refitting, bleed the brakes.

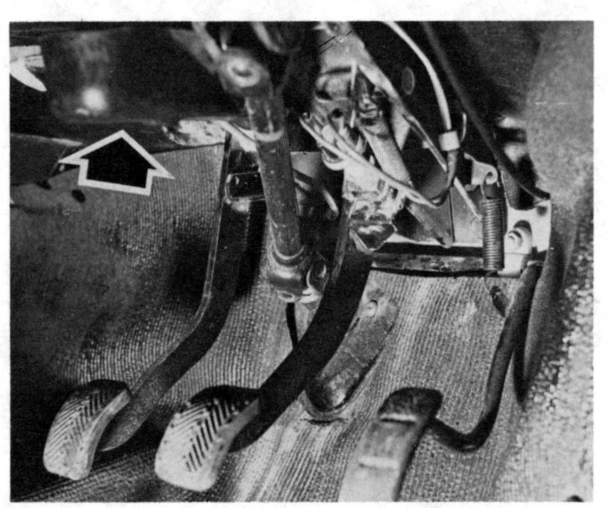
12.2 Location of master cylinder mounting tray

9 Then get an assistant to hold the pedal on whilst checking for leaks. Check visually, and also check that the pedal does not gradually go down under sustained pressure. Such movement could be allowed either by an external leak, or an internal one past the pistons.

13 Master cylinder - overhaul

1 Clean the outside of the master cylinder, but only use brake fluid or alcohol (methylated spirits). If other solvents such as petrol are used, traces will remain afterwards, and could degrade the rubber seals.
2 Refer to Fig. 8.11.
3 Remove the rubber boot from the cylinder end.
4 Remove the two piston stop screws from the underneath of the cylinder.
5 Take out of the open end of the cylinder such pistons, seals etc as will come out easily at this time, laying them out in the order in which they were fitted, so nothing can get misplaced.
6 Take out the plug from the other end of the cylinder.
7 With a pencil push the remaining components out.
8 Discard all the rubber parts.
9 Examine the cylinder bore and the pistons for scratches or

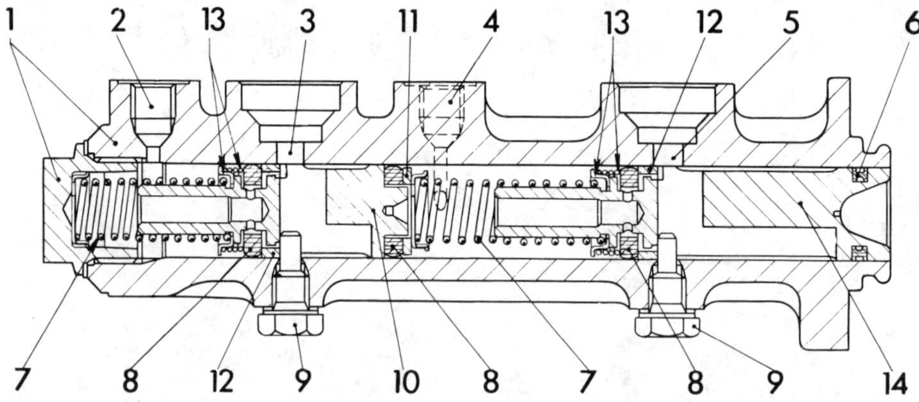

Fig. 8.11. Brake master cylinder cross-section (Sec. 13)

1 Cylinder with plug
2 Fluid line connector to rear wheels
3 Fluid line connector from reservoir to rear brakes
4 Fluid line connector to front wheels
5 Fluid line connector from reservoir to front brakes
6 Seal
7 Spring
8 Floating ring
9 Screws and washers, front and rear floating ring holder
10 Front floating ring holder
11 Flat washer
12 Spacer
13 Spring and cup
14 Rear floating ring holder

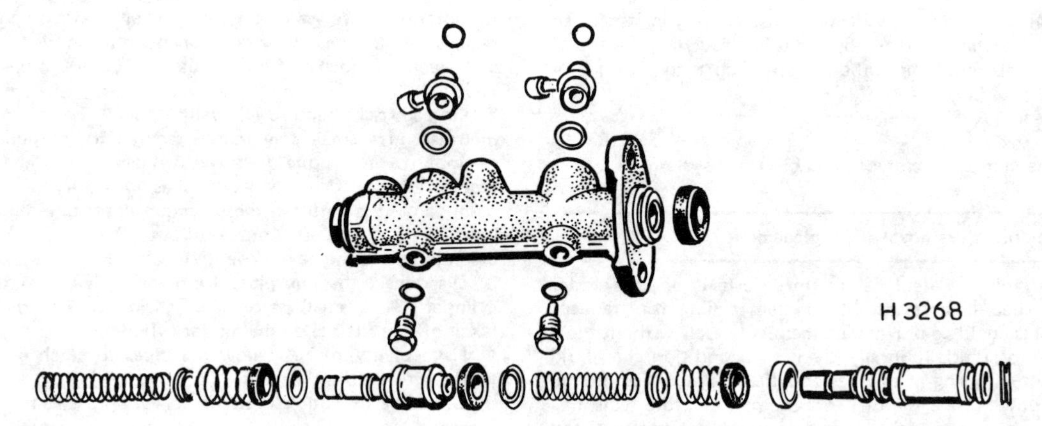

Fig. 8.12. Brake master cylinder components (Sec. 13)

Note: This view does not show the bolt-holes for fixing to the mounting tray (RHD only): the internal components are the same on LHD and RHD types.

pitting. They must have a smooth surface, otherwise there will be seepage of fluid past the seals and the latters' lips will wear rapidly. If the surface is not good, then a new cylinder is needed. In some cases it may be possible to polish out the scratches, but the clearance between bore and pistons should not exceed 0.006 in. (0.15 mm).

10 On reassembly wet all parts liberally with brake fluid.

11 Fit all the seats into place using the fingers only to push them gently into place. Seals with lips must have the lips towards the high pressure side, away from the actuating rod. Push in the secondary piston, groove **downwards,** then fit its stop screw. Check it moves freely. Then fit all the primary piston parts.

12 Check after fitting that there is a little free play between the pedal's push rod and the rear end of the piston. Otherwise the pushrod cannot return to the end of its stroke, to recoup fluid through the ports.

14 Hydraulic pipes and hoses - general

1 Periodically all brake pipes, pipe connections and unions should be completely and carefully examined.

2 First examine for signs of leakage where the pipe unions occur. Then examine the flexible hoses for signs of chafing and fraying and of course leakage. This is only a preliminary part of the flexible hose inspection, as exterior condition does not necessarily indicate the interior condition, which will be considered later.

3 The steel pipes must be examined equally carefully. They must be cleaned off and examined for any signs of dents, or other percussive damage and rust and corrosion. Rust and corrosion should be scraped off and if the depth of pitting in the pipes is significant, they will need replacement. This is particularly likely in those areas underneath the car body and along the rear axle where the pipes are exposed to full force of road and weather conditions.

4 If any section of pipe is to be taken off, first of all drain the fluid, or remove the fluid reservoir cap and line it with a piece of polythene film to make it air tight, and replace it. This will minimise the amount of fluid dripping out of the system, when pipes are removed. It is normally best to drain all the fluid, as a change is probably due anyway.

5 Rigid pipe removal is usually quite straightforward. The unions at each end are undone, the pipe and union pulled out, and the centre sections of the pipe removed from the body clips where necessary. Underneath the car, exposed unions can sometimes be very tight. As one can use only an open ended spanner and the unions are not large, burring of the floats is not uncommon when attempting to undo them. For this reason a self-locking grip wrench (Mole) is often the only way to remove a stubborn union.

6 Flexible hoses are always mounted at both ends in a rigid bracket attached to the body or a sub-assembly. To remove them it is necessary first of all to unscrew the pipe unions of the rigid pipes which go into them. The hose ends can then be unclipped from the brackets. The mounting brackets, particularly on the body frame, are not very heavy gauge and care must be taken not to wrench them off.

7 With the flexible hose removed, examine the internal bore. If it is blown through first, it should be possible to see through it. Any specks of rubber which come out, or signs of restriction in the bore, mean that the inner lining is breaking up and the pipe must be replaced.

8 Rigid pipes which need replacement can usually be purchased at any garage where they have the pipe, unions and special tools to make them up. All they need to know is the total length of the pipe, the type of flare used at each end with the union, and the length and thread of the union. FIAT is metric remember.

9 Replacement of pipes is a straightforward reversal of the removal procedure. If the rigid pipes have been made up it is best to get all the sets (bends) in them before trying to install them. Also if there are any acute bends, ask your suppliers to put these in for you on a tube bender. Otherwise you may kink the pipe and thereby restrict the bore area and fluid flow.

10 When refitting the flexible pipes check they cannot be under tension, or rub against the wheels at the limit of suspension or steering movement.

15 Brake (and clutch) pedal - removal and replacement

1 The pedals share the pivot for the clutch pedal, so must be removed as a pair.

2 Inside the car take out the split pin from its pivot, and unhook the clutch cable from the clutch pedal. Then carry out the same operation on the brake pedal to release the master cylinder pushrod.

3 Unhook the brake pedal return spring.

4 Take off the nut and washers on the right end of the pedal pivot (part of the clutch pedal).

5 Remove the screw and take off the tab holding the pedal spacer.

6 Try and withdraw the clutch pedal. It will probably be necessary to undo the electric harness bracket so the wires can be pushed out of the way.

7 Take out the clutch pedal, and then lower the brake pedal and spacer.

8 Before reassembly smear grease on the pedal pivots and the spacer.

16 Bleeding the brake hydraulic system

1 If the hydraulic system has air in it, operation will be 'spongy' and imprecise. Air will get in whenever the system is dismantled, or if the hydraulic fluid is insufficient. The latter is likely to happen as the brakes wear, and the pistons move further out in the wheel cylinders. Air can leak into the system, sometimes through a fault too slight to let fluid out. In this latter case it indicates a general overhaul of the system is needed. Bleeding is also carried out at the 36,000 miles task, to change the brake fluid.

2 You will need:

a) An assistant to pump the pedal.
b) A good supply of new hydraulic fluid.
c) An empty glass jar.
d) A plastic or rubber pipe to fit the bleed nipple.
e) A spanner for the nipple.

3 Top up the master cylinder and put fluid in the bleed jar to a depth of about ½ in (13 mm).

4 Start at the nipple furthest away from the master cylinder; ie rear brake, passenger side and work nearer.

5 Clean the nipple and put the pipe on it, the other end of which should be put in the fluid in the jar.

6 Tell your assistant to give a few quick strokes to pump up pressure, and then hold the pedal on.

7 Slacken the nipple about ½ to 1 turn, till the fluid or air begins to come out. This is usually quite apparent either as bubbles or dirt in the clean fluid in the jar. Ensure the end of the pipe always stays below the level of the fluid in the jar or air can enter the system.

8 As soon as the flow starts, tell the assistant to keep pumping. The pedal should be pushed down hard, but released slowly.

9 As soon as air has stopped coming out shut the bleed nipple as the assistant is pushing the pedal down. Do not go on too long, lest the reservoir be emptied, and more air pumped in. It is best to check the level every six pumps to be on the safe side.

10 Refill the reservoir, and repeat at the other wheels. Also keep going on the original wheel after refilling the reservoir if dirty fluid is still coming out, to get rid of all the old fluid.

11 Bleeding is greatly speeded, and can be done by one person, if spring loaded valves are fitted to the nipples. These are available from accessory shops.

12 Keep hydraulic fluid clear of the car's paint. It ruins it. Throw old fluid away. It attracts damp, so deteriorates in use.

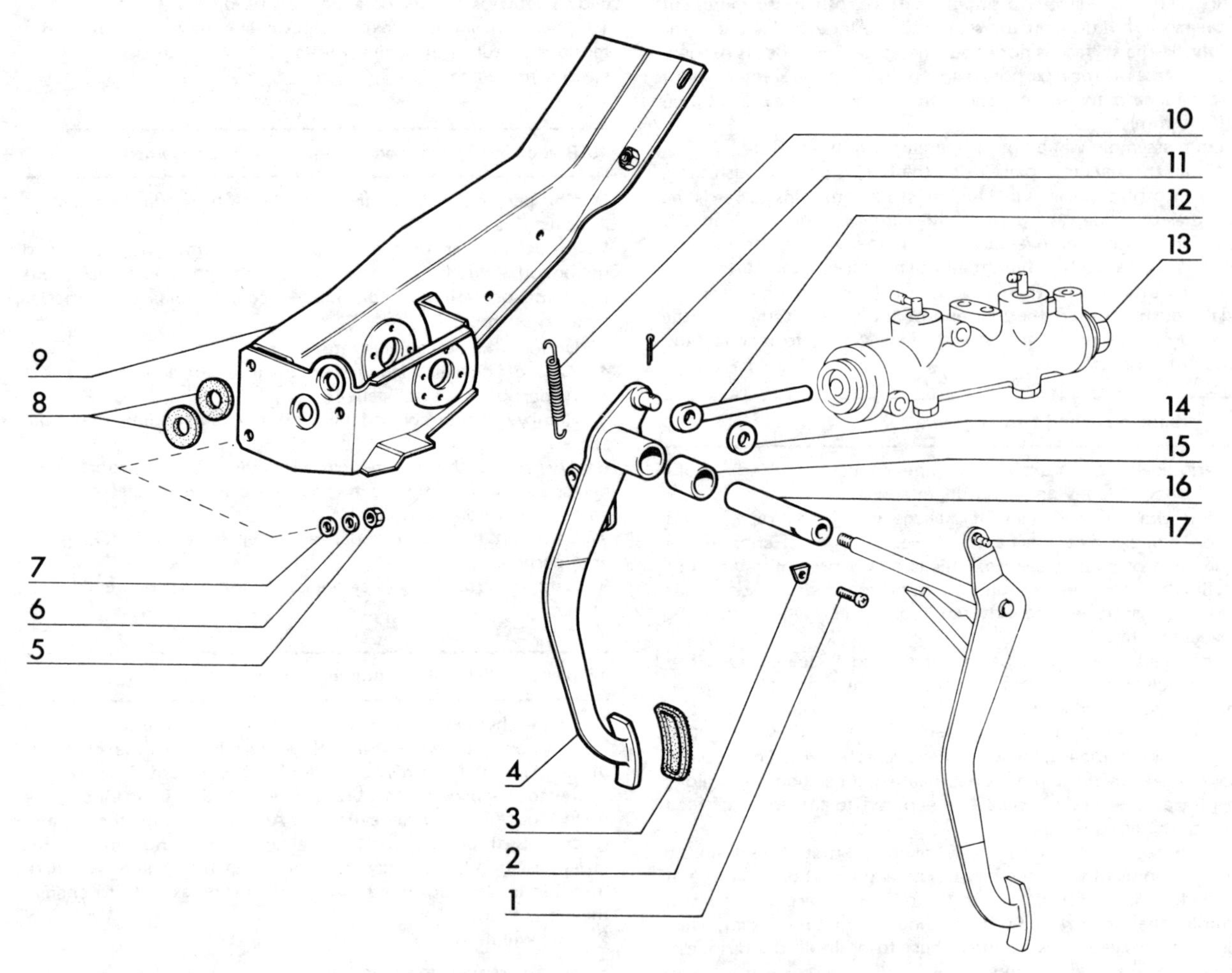

Fig. 8.13. Brake and clutch pedal assembly (Sec. 15)

1 Screw
2 Lockplate
3 Pedal rubber
4 Brake pedal
5 Nut
6 Washer
7 Washer
8 Grommets
9 Pedal bracket
10 Return spring
11 Split pin
12 Pushrod
13 Master cylinder
14 End plug
15 Bush
16 Spacer
17 Clutch cable attachment point

13 If there is difficulty in getting air out of the system, then each time the assistant depresses the pedal, close the nipple so no back flow can take place.

14 If bleeding is needed frequently, this indicates an overhaul of the master cylinder, and maybe the wheel cylinders too, is needed.

15 On Sport versions, destroy the vacuum in the servo by repeated applications of the foot pedal before starting to bleed the system.

16 Single-handed bleeding can be carried out using one of the pressure bleed kits available from accessory shops.

17 Fault diagnosis - braking system

Before diagnosing faults in the brake system ensure that any irregularities are not caused by:

1. *Uneven and incorrect tyre pressures*
2. *Incorrect 'mix' of radial and crossply tyres*
3. *Wear in the steering mechanism*
4. *Defects in the suspension and dampers*
5. *Misalignment of the bodyframe*

Symptom	Reason/s	Remedy
Pedal travels a long way before the brakes operate	Rear brake shoes set too far from the drums	Check self adjusters.
	Failure of half the hydraulic system	Check for leaks.
Stopping ability poor, even though pedal pressure is firm	Linings and/or drums badly worn or scored	Dismantle, inspect and renew as required.
	One or more wheel hydraulic cylinders seized, resulting in some brake shoes not pressing against the drums or pads against discs	Dismantle and inspect wheel cylinders. Renew as necessary.
	Failure of half of the system	Check for leaks.
	Brake linings contaminated with oil	Renew linings and repair source of oil contamination.
	Wrong type of linings fitted	Verify type of material which is correct for the car.
	Rear brake shoes wrongly assembled	Check for correct assembly.
Car veers to one side when the brakes are applied	Brake linings on one side are contaminated with oil	Renew linings and stop oil leak.
	Hydraulic wheel cylinder(s) on one side partially or fully seized	Inspect wheel cylinders for correct operation and renew as necessary.
	A mixture of lining materials fitted between sides	Standardise on types of linings fitted.
	Unequal wear between sides caused by partially seized wheel cylinders	Check wheel cylinders and renew linings and drums as required.
Pedal feels 'spongy' when the brakes are applied	Air is present in the hydraulic system	Bleed the hydraulic system and check for any signs of leakage.
Pedal feels 'springy' when the brakes are applied	Master cylinder or brake backplate mounting bolts loose	Retighten mounting bolts.
	Severe wear in brake drums causing distortion when brakes are applied	Renew drums and linings.
Pedal gradually down under sustained pressure	Small leak	Trace leak.
	Master cylinder seals failed, being by-passed	If none visible, suspect master cylinder and overhaul it.
Pedal travels right down with little or no resistance and brakes are virtually non-operative (unlikely: due to the dual system)	Leak in hydraulic system resulting in lack of pressure for operating wheel cylinders	Examine the whole of the hydraulic system and locate and repair source of leaks. Test after repairing each and every leak source.
	If no signs of leakage are apparent the master cylinder internal seals are failing to sustain pressure	Overhaul master cylinder. If indications are that seals have failed for reasons other than wear all the wheel cylinder seals should be checked also and the system completely replenished with the correct fluid.
Binding, overheating	Master cylinder faulty	Overhaul master cylinder.
	Master cylinder no free play	Check pedal clearance.
	Handbrake too tight	Readjust.
Vibration, pedal pushed up in phase with slow vibration	Discs/drums worn	Have drums skimmed. Renew discs.
Juddering	Loose back plate	Tighten.
	Dust in rear drums, or oily linings/pads or front or rear	Clean and/or reline.
	Wheels out of balance	Re-balance wheels.

Chapter 9 Electrical system

For modifications, and information applicable to later models, see Supplement at end of manual

Contents

Specifications

System voltage ... 12 volts

System polarity ... Negative earth

Battery

Capacity (standard) ...	34 amp. hour
Specific gravity at 60°F (15°C):	
Fully charged ...	1.28
Half charged ...	1.22
Nearly discharged ...	1.16
Flat ...	1.11

Dynamo

Type ...	D90/12/16/3 E
Maximum continuous output ...	16 amps (230 watts)
Maximum peak output ...	22 amps (320 watts)
Cut-in speed (12 v) (68°F - 20°C) ...	1710 - 1790 rpm
Speed for maximum continuous output ...	2550 - 2700 rpm
Speed for peak output ...	3050 - 3200 rpm
Maximum speed ...	9000 rpm
Rotation of drive ...	Clockwise
Poles ...	2
Field ...	Shunt
Engine to dynamo speed ...	1 : 1.86 (new belt)
Minimum engine speed for charge ...	970 rpm (approx)
Field winding resistance ...	8 ohms at 20°C (68°F)
Armature winding resistance ...	0.145 ohms at 20°C (68°F)

Dynamo Regulator

Type	GN 2/12/16
Cut-out closing voltage	12.2 - 13.0 volts
Reverse current maximum	16 amps
Air gap, contacts closed	0.014 in. (0.35 mm)
Points gap	0.016 - .020 in. (0.39 - 0.51 mm)
Regulator air gaps	0.039 - .044 in. (0.99 - 1.11 mm)
Setting voltage at 8 amps, warm	13.9 - 14.5 volts
Setting current at 13 volts, warm	15 - 17 amps
Regulating resistance	80 - 90 ohms
Voltage regulating resistance	16 - 18 ohms

Alternator

Type	Bosch G1 - 14V 33A 27 or Marelli A108-14V-33A-Var. 2	
Maximum output (approx.)	38 amps	40 amps
Output at 14 volts and 5,000 rpm	29 amps minimum	33 amps (at 7000 rpm)
Cut-in speed (12 v) (77°F, 25°C)	1,050 - 1,150 rpm	1,050 - 1,150 rpm
Maximum speed	14,000 rpm	14,000 rpm
Drive rotation	Clockwise	Clockwise
Engine to alternator speed	1 : 2	1 : 1.8
Field winding resistance	4 to 4.4 ohms	4 to 4.4 ohms

Alternator Regulator

Type	Bosch AD 1/14V or Marelli RTT 110 AB

Starter motor

Type	Pre-engaged by solenoid
Model	FIAT E84 - 0.8/12 Var. 3
Power	0.8 KW
Poles	4
Field	Series
No load test: 11.9 volts	25 amps or less
Solenoid winding resistance	0.39 ± 0.02 ohms
Unworn brush pressure	2.5 to 2.9 lbf (1.15 to 1.3 kgf)
Lubrication:	
Armature bushes	Engine oil (Duckhams Hypergrade)
Armature spiral grooves	SAE 10(W) oil
Free wheel splines	Multi-purpose grease (Duckhams LB 10)

Windscreen wipers

Wiper arms	Champion CCA2
Wiper blades	Champion X-3803

Heater fan

Heater fan	20 watts

Lamps

Headlamps	Dependent on market and model
Turn indicators	21 watt
Reversing light (127 Special)	21 watt
Side turn repeaters (Europe)	4 watt
Parking lights	5 watt
Licence number plate	5 watt
Interior lights	5 watt festoon type
Instrument lights	3 watt

Fuses

Fuse rating (127)	Engine fan 16 amp, remainder 8 amp
Number (127)	8 (7 x 8 amp; 1 x 16 amp)
Fuse rating (127 Special)	Engine fan and cigar lighter/rear window 16 amp, remainder 8 amp
Number (127 Special)	10 (8 x 8 amp; 2 x 16 amp)
Location	Engine compartment

See Chapter 13 for further information on fuses

1 General description

1 The electrical system is 12 volt, negative earth. On early cars the generator was DC, that is, a dynamo. Later cars have alternators. The cooling fan is electrically operated. Details are given in Chapter 2. The 'V' belt that would normally be called a 'fan belt' is at the right end of the engine, and drives the water pump and generator. It is tensioned by moving the generator on its mountings in the normal way.

The battery supplies a steady amount of current for the ignition, lighting, and other electrical circuits, and provides a reserve of electricity when the current consumed by the electrical equipment exceeds that being produced by the dynamo/alternator.

Eight fuses protect the electrical system with the exception of the starting, charging and ignition circuits.

Headlamp units of the double filament separate bulb and reflector type, are fitted to cars destined for the U.K. market. Facilities are provided for beam adjustment.

The starter motor is of the pre-engaged type.

2 Battery - removal and replacement

1 The battery is located within the engine bay on the right-hand side. It should be removed once every three months for cleaning and testing. Disconnect the negative and then the

positive leads from the battery terminals by slackening the retaining nuts and bolts. It will be noted that a shroud covers the positive terminal and that it is necessary to release the retaining clamp before the battery can be removed.
2 Remove the battery carefully and hold it vertical to ensure that none of the electrolyte is spilled.
3 Replacement is a direct reversal of the removal procedure. **Note:** Replace the positive lead before the earth lead and smear the terminals with petroleum jelly to prevent corrosion. **Never** use an ordinary grease. Don't forget to replace the shroud over the positive terminal and the clamp that retains the battery in position.

3 Battery - maintenance and inspection

1 Normal weekly battery maintenance consists of checking the electrolyte level of each cell to ensure that the separators are covered by ¼ inch (6.4 mm). If the level has fallen, top-up the battery using distilled water only. Do not overfill. If a battery is overfilled or any electrolyte spilled, immediately wipe away the excess, as electrolyte attacks and corrodes any metal it comes into contact with very rapidly.
2 As well as keeping the terminals clean and covered with petroleum jelly (vaseline), the top of the battery, and especially the top of the cells, should be kept clean and dry. This helps prevent corrosion and ensures that the battery does not become partially discharged by leakage through dampness and dirt.
3 Once every three months remove the battery and inspect the battery housing and battery leads for corrosion (white fluffy deposits on the metal which are brittle to touch). If any corrosion is found, clean off the deposit with ammonia and paint over the clean metal with an anti-rust/anti-acid paint.
4 At the same time inspect the battery case for cracks. Cracks are frequently caused to the top of the battery case by pouring in distilled water in the middle of winter **after** instead of **before** a run. This gives the water no chance to mix with the electrolyte and so the former freezes and splits the battery case.
5 If topping-up the battery becomes excessive and the case has been inspected for cracks that could cause leakage, but none found, the battery is being overcharged and the voltage regulator will have to be checked and reset.
6 With the battery on the bench at the three monthly interval check, measure the specific gravity with a hydrometer to determine the state of charge and condition of the electrolyte. There should be very little variation between the different cells

Fig. 9.1. Measuring the specific gravity of the electrolyte (Sec. 3)

and if a variation in excess of 0.025 is present it will be due to either:

a) Loss of electrolyte from the battery at some time caused by spillage or a leak, resulting in a drop in the specific gravity of the electrolyte when the deficiency was replaced with distilled water instead of fresh electrolyte.

b) An internal short circuit caused by buckling of the plates or a similar malady pointing to the likelihood of total battery failure in the near future.

7 The specific gravity of the electrolyte for fully charged conditions at the electrolyte temperature indicated, is listed in Table A. The specific gravity of a fully discharged battery at various temperatures of the electrolyte is given in Table B.

Table A

Specific gravity — battery fully charged

1.268 at	*100°F or 38°C*	*electrolyte temperature*
1.272 at	*90°F or 32°C*	" "
1.276 at	*80°F or 27°C*	" "
1.280 at	*70°F or 21°C*	" "
1.284 at	*60°F or 16°C*	" "
1.288 at	*50°F or 10°C*	" "
1.292 at	*40°F or 4°C*	" "
1.296 at	*30°F or -1.5°C*	" "

Table B

Specific gravity — battery fully discharged

1.098 at	*100°F or 38°C*	*electrolyte temperature*
1.102 at	*90°F or 32°C*	" "
1.106 at	*80°F or 27°C*	" "
1.110 at	*70°F or 21°C*	" "
1.114 at	*60°F or 16°C*	" "
1.118 at	*50°F or 10°C*	" "
1.122 at	*40°F or 4°C*	" "
1.126 at	*30°F or -1.5°C*	" "

4 Battery - electrolyte replenishment

1 If the battery is in a fully charged state and one of the cells maintains a specific gravity reading which is 0.025 or more lower than the others, and a check of each cell has been made with a voltage meter for short circuits (a four to seven second test should give a steady reading of between 1.2 and 1.8 volts), then it is likely that electrolyte has been lost at some time from the cell with the low reading.
2 Top-up the cell with a solution of 1 part sulphuric acid to 2.5 parts of water. If the cell is already fully topped-up, draw some electrolyte out of it with a hydrometer.
3 When mixing the sulphuric acid and water **never add water to sulphuric acid** — always pour the acid slowly onto the water in a glass container. **If water is added to sulphuric acid it will explode.**
4 Continue to top-up the cell with the freshly made electrolyte and then recharge the battery and check the hydrometer readings.

5 Battery - charging

1 When heavy demand is placed upon the battery such as when starting from cold, and much electrical equipment is continually in use, have the battery fully charged occasionally from an external source at the rate of 3.5 to 4 amps.
2 Continue to charge the battery at this rate until no further rise in specific gravity is noted over a four hour period.
3 Alternatively a trickle charger, charging at the rate of 1.5 amps can be safely used overnight.
4 Specially rapid 'boost' charges which are claimed to restore the power of the battery in 1 to 2 hours are most dangerous as they can cause serious damage to the battery plates through overheating.
5 While charging the battery note that the temperature of the electrolyte should never exceed 100°F (37.8°C).

6 Dynamo/alternator - removal

1 The generator will need to be removed for routine cleaning of the commutator or slip rings, be it dynamo or alternator, or after the tests carried out in position with the engine running, as described in subsequent Sections, indicate an internal fault.
2 Remove the air cleaner complete, and cover over the carburettor air intake.
3 Disconnect the leads from the generator. The two terminals of a dynamo are different sizes, so cannot be muddled. On alternators note that one of the leads to the alternator plug is live, so it is sensible to disconnect the battery negative earth terminal.
4 Slacken the pivot bolts underneath the generator, and the adjuster bolt working in the slotted bracket on top. Move the generator to slacken the 'V' belt, and take that off (photo).
5 Now completely undo and take off the adjustment and pivot nuts supporting the generator with a hand underneath. This is not as easy as it sounds, as access is difficult. In order to remove the generator it is necessary to press it tight up against the wing in order to get it off of the studs. (photo)
6 Finally, remove the two bolts that secure the manacle bracket halves together and ease the dynamo out. This operation will not apply when an alternator is fitted.
7 Replace the generator in reverse sequence. If the belt is worn, fit a new one, and adjust it to give ½ in. (12 mm) sag between pulleys when pushed hard by one finger.
8 If just the V-belt is being replaced, an alternative method of removing the belt is to loosen the generator pivot and securing nuts through the access panel inside the adjacent wing. The access panel is easily removed after the retaining nuts have been loosened (photo).

7 Dynamo - testing in position

1 If, with the engine running, no charge comes from the dynamo, or the charge is very low, first check that the drivebelt is in place and is not slipping. Then check that the leads from the control box to the dynamo are firmly attached and that one has not come loose from its terminal.
2 The generator positive terminal should be connected to terminal 51 and generator negative terminal to generator regulator terminal 67.
3 Make sure that all lights and electrical accessories are switched off. Disconnect the two cables from the rear of the dynamo. Connect the two terminals of the dynamo with a piece of short but thick wire. Start the engine and run at a normal idle speed.
4 Fit the negative lead of a 0–20 volt moving coil voltmeter to one dynamo terminal and the other lead to a good earth on the dynamo body or mounting.
5 Gradually raise the engine speed to approximately 1000 rpm. The voltmeter should rise steadily and without any fluctuations. Do not allow the voltmeter reading to exceed 20 volts or it will be damaged and do not raise the engine speed above 1,000 rpm to obtain a higher reading.
6 If no reading is obtained, dismantle the dynamo and check the brush gear. Should a reading of between 0.5 and 1 volt be obtained the field windings should be suspect. However, if a reading of between 4 and 5 volts is obtained the armature should be suspect.
7 If the dynamo is in good operating order, leave the temporary wire in position between the two dynamo terminals but remake the original connections.
8 Remove the positive lead from the dynamo regulator (terminal 51) and connect the voltmeter between this lead and the dynamo body or mounting bracket.
9 Start the engine and run as before. The readings obtained should be the same as those previously obtained. No reading indicates a fault in the cable to generator regulator terminal '51' from the dynamo.
10 Repeat the previous test for the negative lead (terminal 67).
11 Remove the temporary wire link from the rear of the dynamo. If readings are correct the control box must be checked as described in Sections 11 - 13.

8 Dynamo - dismantling and inspection

1 Mount the dynamo in a vice and unscrew and remove the two through bolts from the commutator end bracket.
2 Mark the commutator end bracket and the dynamo casing so the end bracket can be replaced in its original position. Pull the end bracket off the armature shaft. (photo)
3 Lift the two brush springs and draw the brushes out of the brush holders.
4 Measure the brushes and if worn down to 9/32 inch (7.14 mm) or less, undo the screws holding the brush leads to the end bracket. Take off the brushes complete with leads.
5 Check the condition of the ball bearing in the drive end plate by firmly holding the plate and noting if there is visible side movement of the armature shaft in relation to the end plate. If play is present, the armature assembly must be separated from the end plate. If the bearing is sound there is no need to carry out the work described in the following three paragraphs. (photo)
6 Undo and remove the larger nut and spring washer holding the pulley wheel and fan onto the armature. For this it will be necessary to hold the armature between soft faces in a bench vice.
7 Next remove the Woodruff key from its slot in the armature shaft.
8 Place the drive end bracket across the jaws of a vice with the armature downwards and gently tap the armature shaft from the bearing in the end plate with the aid of a suitable drift.
9 Carefully inspect the armature and check it for open or short circuited windings. It is a good indication of an open circuited

6.4 Removing the drivebelt for the generator and water pump

6.5 The alternator has to be moved as close as possible to the wing in order to get it off of the studs

6.8 Removing the access panel to get at the generator retaining nuts and the drivebelt

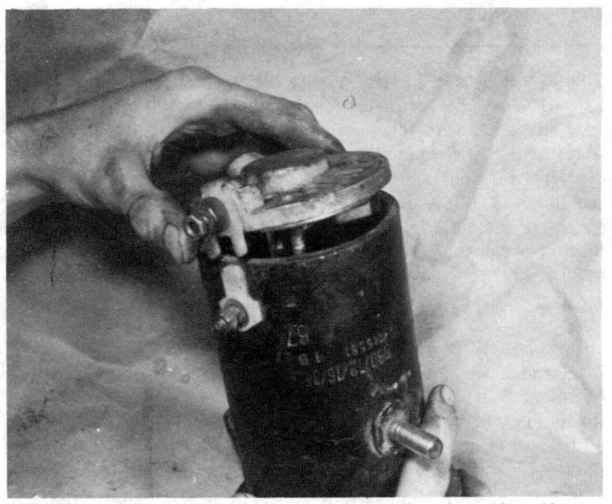

8.2 Having taken off the two nuts the endplate just lifts off

8.5 Then the yoke can be lifted off the armature and drive endplate

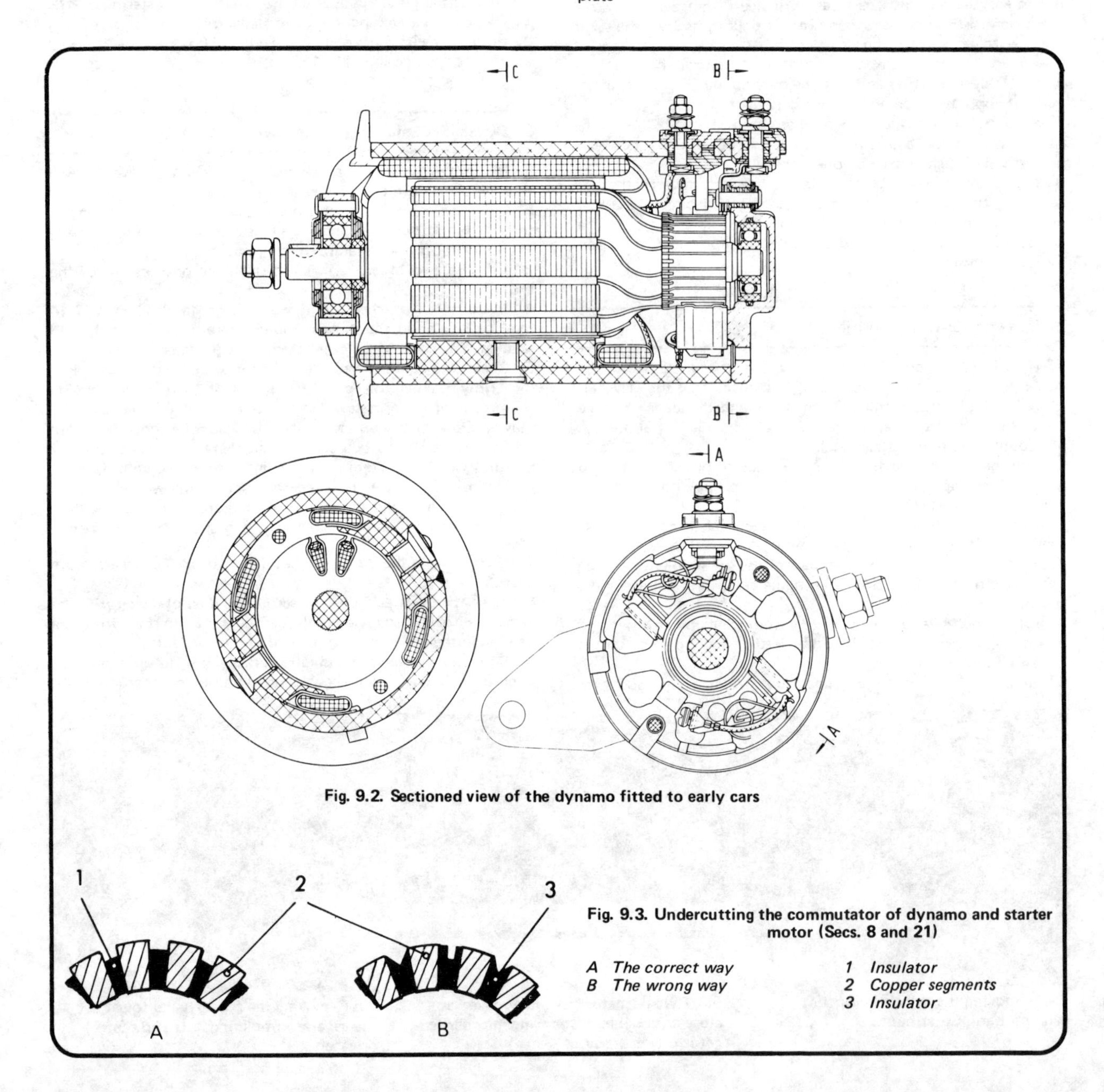

Fig. 9.2. Sectioned view of the dynamo fitted to early cars

Fig. 9.3. Undercutting the commutator of dynamo and starter motor (Secs. 8 and 21)

A The correct way
B The wrong way

1 Insulator
2 Copper segments
3 Insulator

armature when the commutator segments are burnt. If the armature has short circuited, the commutator segments will be very badly burnt, and the overheated armature windings badly discoloured. If open or short circuits are suspected, then test by substituting a new armature for the suspect one.
10 Check the resistance of the field coils. To do this, connect an ohmmeter between the field terminal and the yoke and note the reading on the ohmmeter which should be about 7.7 to 8.1 ohms. If the ohmmeter reading is infinity this indicates an open circuit in the field winding. If the ohmmeter reading is below 6 ohms this indicates that one of the field coils is faulty and must be replaced.
11 Field coil replacement involves the use of a wheel operated screwdriver and this operation is considered to be beyond the scope of most owners. Therefore if the field coils are at fault, either purchase a rebuilt dynamo or take the body to a FIAT garage or auto-electrician for new field coils to be fitted.
12 Next check the condition of the commutator. If it is dirty and blackened, clean it with a petrol dampened rag. If the commutator is in good condition the surface will be smooth and quite free from pits or burnt areas, and the insulated segments clearly defined.
13 If, after cleaning, the commutator still has pits and burnt spots wrap a strip of glass paper round it taking great care to move the commutator ¼ of a turn every ten rubs until it is thoroughly clean.
14 In extreme cases of wear the commutator can be mounted in a lathe and with the lathe turning at a high speed, a very fine cut may be taken off the commutator. Then polish the commutator with glass paper. If the commutator has worn so that the insulators between the segments are level with the top of the segments, then undercut the insulators to a depth of 1/32 inch (0.8 mm). The best tool for this purpose is half a hacksaw blade ground to a thickness of the insulator, and with the handle end of the blade covered in insulating tape to make it more comfortable to hold.
15 Check the bush bearing in the commutator end bracket for wear by noting if the armature spindle rocks when placed in it. If worn it must be renewed.
16 The bush bearing can be removed by a suitable extractor or by screwing in a tap of suitable diameter four or five turns. The tap complete with bush is then pulled out of the end bracket.
17 **Note:** The bush bearing is of the porous phosphor bronze type and before fitting a new one, it is essential that it is allowed to stand in engine oil for at least 24 hours before fitment. In an emergency the bush can be immersed in hot oil (100°C/212°F) for two hours.
18 Carefully fit the new bush into the endplate, pressing it in until the end of the bush is flush with the inner side of the endplate. If available press the bush in with a smooth shouldered mandrel the same diameter as the armature shaft.

9 Dynamo - repair and reassembly

1 To renew the ball bearing fitted to the drive end bracket, drill out the rivets which hold the bearing retainer plate to the end bracket and lift off the plate.
2 Press out the bearing from the end bracket and remove the packing washers from the bearing housing.
3 Thoroughly clean the bearing housing and the new bearing and pack with high melting point grease.
4 Place the packing washers into the bearing housing and then using a suitable diameter drift refit the bearing.
5 Replace the bearing plate and fit three new rivets.
6 Open the rivet ends with the aid of a suitable cold chisel and lock them in position with a ball hammer.
7 Refit the drive end bracket to the armature shaft. Do not try and force the bracket on but with the aid of a suitable socket abutting the bearing, tap the bearing on gently, so pulling the end bracket down with it.
8 Refit the Woodruff key onto the armature shaft.
9 Replace the fan and pulley wheel and then fit the spring

9.12 To put the endplate back on, hook the brush springs up so they hold back the brushes

washer and nut. Tighten the latter. The drive bracket end of the dynamo is now fully assembled.
10 If the brushes are a little worn and are to be used again then ensure that they are placed in the same holders from which they were removed. When refitting brushes either new or old, check that they move freely in their holders. If either brush sticks, clean with a petrol moistened rag and if still stiff, lightly polish the sides of the brush with a very fine file until the brush moves quite freely in its holder.
11 Tighten the two retaining screws and washers which hold the wire leads to the brushes in place.
12 It is far easier to slip the end piece with brushes over the commutator, if the brushes are raised in their holders and held in this position by the pressure of the springs resting against their flanks. (photo)
13 Refit the armature to the casing and then the commutator endplate, taking care to align the previously made marks.
14 Replace the two through bolts and tighten securely.
15 Finally hook the ends of the two springs off the flanks of the brushes and onto their heads so the brushes are forced down into contact with the armature.

10 Control box (dynamo) - general

1 The control box has three relays. One is the cut-out, another limits the dynamo voltage generated, and the third its current.
2 The cut-out disconnects the dynamo from the battery when it is no longer charging, otherwise it could run off the battery as a motor.
3 The two regulator relays by their combination of voltage and current control regulate the output to match the electrical load that might be switched on, and to suit the state of charge of the battery.
4 The current control is set to the maximum safe limit for the dynamo. The voltage regulator is set to a potential that will limit the charge given a healthy battery to a mere trickle.
5 If the control box has a complete failure the ignition warning light will come on. If there is a partial failure, unless an ammeter is fitted, there will be no warning. Undercharging may become apparent as a flat battery. Minor overcharging will give the need for frequent topping up of the battery. Gross overcharging may blow light bulbs and perhaps result in a smell of burning from the overloaded dynamo.
6 Major defects are likely to be the burning of the points on the relays, so that they never make contact and no regulation takes place.
7 Minor defects occur due to wear and general ageing altering

the voltage/current at which the cut-out or regulator work.

8 A car-type ammeter will reveal these aberrations, and is a more useful fitting for the long term. But for fault-finding more accurate instruments are needed. Unless you have some experience of such things, and of the instruments, it is suggested you do not tamper with the control box. If done incorrectly a new control box and a new dynamo may be needed.

9 In any work on the control box it is important that the leads are not fitted to wrong terminals or the unit will be ruined.

11 Cut-out (dynamo) - checking and adjustment

1 The cut-out is the relay on the right as you look at them with the terminals at the bottom. It is the only one of the three with points open when the engine is switched off.

2 The dynamo will have been proved satisfactory by earlier fault-finding.

3 Take off the control box cover.

4 Check the voltage at the terminal from the dynamo, number '51' and the output from the cut-out, number '30'. This will prove that the defect is at the cut-out.

5 Assuming there is 12/15 volts at the terminal '51', but none at '30', check the operation of the relay of the cut-out. With the engine running fast enough to charge, try pushing it with a finger.

6 If the relay does not hold down there is a fault in the wiring.

7 If it stays down but there is still no voltage at the output terminal '30', points appear to need cleaning. If the push made it work it may need resetting.

8 To reset the cut-out wire a voltmeter from terminal '51' (the dynamo connection) to earth. Start up: Warm up the engine for 15 minutes. Increase engine speed gradually, watching both the voltmeter and the cut-out. Note the voltage at which the cut-out closes: there will be a little kick of the voltmeter. It should close at 12.6 volts. This voltage should be set after the cut-out is warmed up by about 15 minutes running.

9 Adjust by bending the arm on which the spring of the contact rests, increasing the spring tension to raise the operating voltage.

10 On slowing down the cut-out should 'drop-off', that is cut out, when the dynamo stops charging. The reverse current should never be high: The official maximum is 16 amps, which is high as such things go. To improve drop-off bend the fixed contacts so that the moving one cannot be drawn so close to the armature. A car ammeter will show this negative current before 'drop-off'.

12 Voltage regulator (dynamo) - checking and adjustment

1 The circuit needed for testing must not have any current flow through the regulator. See Fig. 9.5.

2 Slide a piece of paper in between the cut-out points and connect the voltmeter from the dynamo connection terminal 51 on the left of the control box and to earth. This must be an instrument accurate to 0.3 volt.

3 Start up, and run the engine at 3,000 rpm. The regulator should limit the voltage to 15.5 volts. It is important to take the reading quickly to avoid temperature effects.

4 The reading should be steady. Fluctuations imply the contact points need cleaning. This should be done with fine glass paper, and all dust removed.

5 If adjustment is needed, bend the arm onto which the spring blade rests, increasing the spring tension to raise the regulated voltage. The voltage regulator is the one farthest from the cut-out. (Nearest the front of the car).

6 The FIAT setting procedure is fairly complicated. The setting figure of 15.5 volts quoted is a compromise. It is valid at an ambient temperature of 10°C (50°F). At 20°C (68°F) set to 15 volts. These voltages are the maximum. Do not exceed this. If the car is used extensively on long journeys a voltage lower by 0.5 volts should be used.

13 Current regulator (dynamo) - checking and adjustment

1 The current control is set using an ammeter accurate to 0.5 amp. See Fig. 9.5. The ammeter is wired between the control box output terminal number '30' and the leads that are normally connected to it.

2 Wedge cardboard between the voltage regulator armature (the one furthest from the cut-out) and its arm to hold the contacts closed, so no voltage regulation can take place.

3 Start up the engine.

4 Turn on the headlamps to load the dynamo.

5 Speed up the engine to a speed of about 3,000 rpm.

6 The maximum current should be 16 amps.

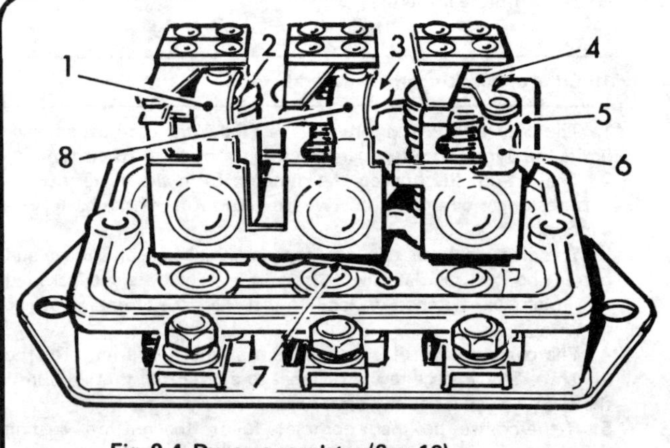

Fig. 9.4. Dynamo regulator (Sec. 10)

1 *Voltage regulator stationary contact carrier*
2 *Voltage regulator armature*
3 *Current regulator armature*
4 *Cutout relay armature*
5 *Cutout relay armature stop*
6 *Cutout relay stationary contact carrier*
7 *Voltage regulator series resistor cable*
8 *Current regulator stationary contact carrier*

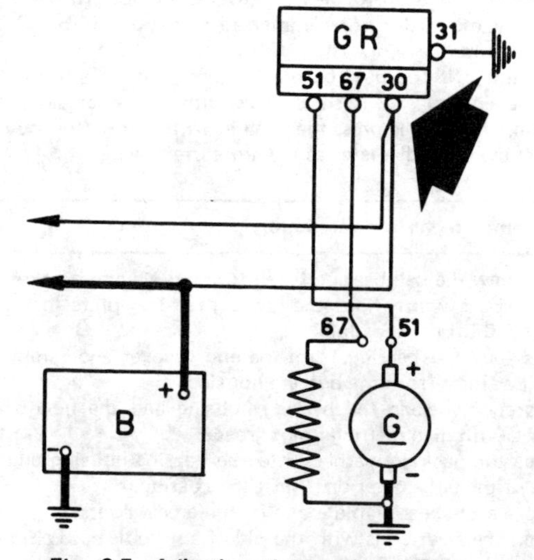

Fig. 9.5. Adjusting the regulator. The broad arrow shows where the ammeter must be inserted for setting the current. When setting the voltage, inserting paper in the cut out points has the same effect as cutting the circuit at that point

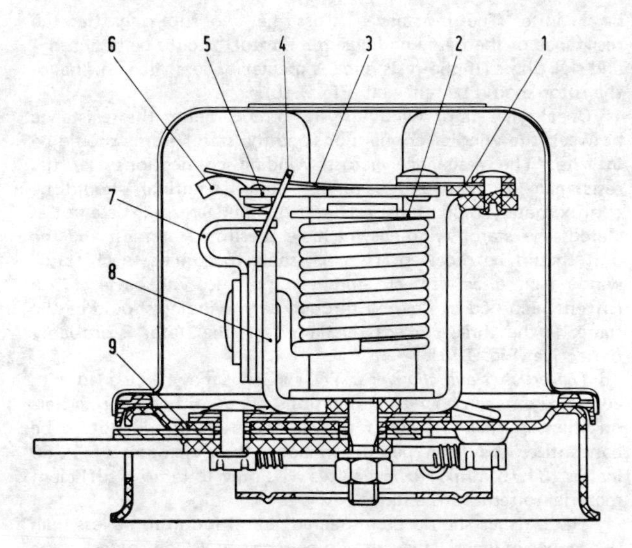

Fig. 9.6. Cut-out relay (Sec. 11)

1 *Bi-metal hinge spring*
2 *Core*
3 *Adjusting spring*
4 *Armature stop*
5 *Armature*
6 *Setting arm*
7 *Stationary contact carrier*
8 *Body*
9 *Base plate*

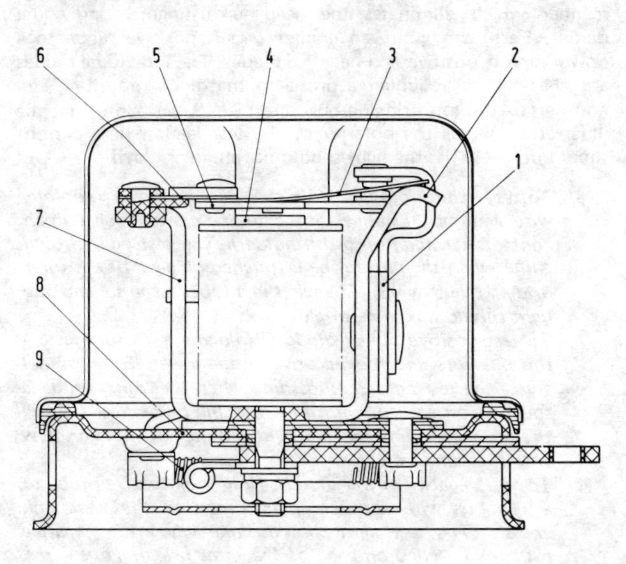

Fig. 9.7. Voltage and current regulator (Secs. 12 and 13)

1 *Stationary contact carrier*
2 *Setting arm*
3 *Adjusting spring*
4 *Core*
5 *Armature*
6 *Steel and bi-metal hinge spring*
7 *Body*
8 *Base plate*
9 *Voltage regulator series resistor*

7 The current should be steady. Fluctuations imply the contacts need cleaning: do this with fine glass paper and remove all dust.
8 If adjustment is needed bend the arm onto which the spring blade rests, increasing the spring tension to raise the controlled current. The current regulator is the centre one.

14 Alternator - safety precautions

If there are indications that the charging system is malfunctioning in any way, care must be taken to diagnose faults properly, otherwise damage of a serious and expensive nature may occur to parts which are in fact quite serviceable.

The following basic requirements must be observed at all times, therefore, if damage is to be prevented.
1 **All** alternator systems use a **negative** earth. Even the simple mistake of connecting a battery the wrong way round could burn out the alternator diodes in a few seconds.
2 Before disconnecting any wires in the system the engine and ignition circuits should be switched off. This will minimise accidental short circuits.
3 The alternator must **never** be run with the output wire disconnected.
4 Always disconnect the battery positive lead from the car's electrical system if an outside charging source is being used.
5 Do not use test wire connections that could move accidentally and short circuit against nearby terminals. Short circuits will not blow fuses - they will burn out diodes or transistors.
6 Always disconnect the battery cables and alternator output wires before any electric welding work is done on the car body.

15 Alternator - description and testing

1 The alternator develops its current in the stationary windings, the rotor carrying the field. The brushes therefore carry only a small current, so they last a long time and only simple slip rings are needed instead of a commutator.
2 The AC voltage is rectified by a bank of diodes. These also prevent battery discharge through the alternator.
3 Very little servicing is needed. Every 36,000 miles the alternator should be stripped and the brushes cleaned and checked, together with the slip rings.
4 Fault diagnosis is more a matter of confirming the fault is in the alternator, and it is probable then that a new unit will have to be fitted. However, if parts are available, component repair is possible. To fit new rectifiers or stator windings requires experience with a soldering iron, and should not be done by someone completely inexperienced in electronic or electrical servicing.
5 The adjustment of the alternator drivebelt is a 3,000 mile (5,000 km) task, and was detailed in task 2.1 of the Routine Maintenance.
6 The alternator should be tested on the car, as then it can be run up to speed by the engine, whilst under load from the car's electrical equipment.
7 Locate the alternator's output wire. This is the 'B+' terminal, which has a brown wire. It is difficult to pick it up at the alternator plug itself, but it goes to a junction point for brown and red wires, where the leads from the battery and the main switches join. Put an ammeter in this lead. That the correct brown lead is being used can be confirmed by turning on lights, ignition etc. The ammeter should show no current at this time. Also connect a voltmeter from this point to earth.
8 Start up the engine, and set it to run at 2,000 rpm. Turn on the headlights and heater fan to load the alternator. The voltage should be about 14 and the amperage about 29.
9 If the current is low the alternator appears to be faulty. If the voltage is low, then the regulator is probably malfunctioning.

16 Alternator - dismantling for inspection or repair

1 It should be stated that it is necessary to have a multimeter available or equipment capable of measuring electrical resistance, if a detailed examination is to be completed. A small (12 watt) soldering iron will be necessary if items such as brushes and the rectifier are to be replaced. The same iron will also be needed to

separate the stator from the rear end assembly. If such equipment is not available the alternator should be taken to your local auto-electrical specialist.
2 It is worth checking before the alternator is dismantled that a range of spares - rectifier pack, brushes, rotor bearings etc, are available locally. You may be in a position where the only course of action to be taken when the alternator is proved faulty is to buy a factory exchange unit.
3 Replaceable items include the rectifier pack (but not the component parts of it), bearings and brushes, rotor assemblies and stator assemblies. Detailed examination of these components begins with the removal of the alternator as described in Section 6 of this Chapter.
4 Now with the unit on a clean bench, clamp the alternator unit in a vice with the lower pivot lug. Insert a rod into the centrifugal fan adjacent to the pulley and hold the rotor shaft still whilst the shaft end nut is undone and removed. Pull the pulley and fan from the shaft and retrieve the Woodruff key used to locate them on the shaft. **Note:** This step is only necessary if the alternator is to be completely dismantled. If the alternator is being opened-up for inspection only, proceed to paragraph 6, then paragraph 8. For brush replacement only, refer to paragraphs 6 and 7.
5 Where applicable, remove the radio suppression condenser (capacitor) from the rear end frame (one screw and washer, and a plug-in connection).
6 Undo the two screws which retain the brush holder to the rear frame of the alternator, then ease the holder out of the alternator. Inspect the brushes and if worn so that there is less than 0.4 inch (10 mm) of brush protruding from the holder, both brushes should be replaced (Fig. 9.9).
7 If only the brushes are being renewed, this operation can be carried out now; if further dismantling is required, fit new brushes and install the brush holder during the final stages of reassembly. When soldering in new brushes hold the connecting wire in a pair of pliers to prevent the molten solder flowing down the wire strands. The wire insulation must be clamped in the lug next to the soldered lug.
8 Lightly centre pop the end fittings and stator exterior to show alignment before disassembly. Continue the dismantling of the alternator by undoing the three long screws which hold the two end frames together. Separate the rotor and front frame from the stator and rear end frame.
9 If further dismantling is required, unsolder the three stator winding connections (be quick because the diodes in the rectifier pack are sensitive to heat). Use the 12 watt instrument soldering iron.
10 Once the stator connections have been unsoldered, the stator may be separated from the rear end frame.
11 Turning your attention to the rear face of the rear end frame, undo the nut from the 'B+' terminal and retrieve the lockwasher, washer and insulating washer.
12 Now from the inside of the end frame, remove the two rectifier pack fixing screws and extract the pack from the end frame.
13 By now you should have the following groups/sub assemblies in front of you, for examination and testing:

(i) *Forward end frame complete with rotor shaft bearing and fixing screws.*
(ii) *Rear end frame minus rotor bearing.*
(iii) *Rotor, complete with rear bearing.*
(iv) *Stator.*
(v) *Rectifier pack plus fixing screws and 'B+' terminal parts.*
(vi) *Brush holder plus fixing screws.*
(vii) *A collection of parts including the fan, pulley and rotor end nuts.*

Note that in the rear end frame bearing location there will be a backing washer and, where applicable, a shield.
14 Check the rotor windings with a multimeter or a test light. The resistance between the slip rings and the rotor frame should be infinite. The resistance across the two slip rings (ie; the resistance of the field windings in the rotor) should be between 4 and 4.4 ohms. If the resistance is greater or less than this band, the rotor should be renewed (Fig. 9.10).
15 Check the stator windings with a multimeter; the resistance between the winding connections and the stator frame should be infinite. The resistance across winding connections (ie; the resistance of two phases of the stator winding) should be approximately 0.35 ohms. Again, if the meter readings are not as stated, the stator windings must be presumed at fault and the unit should be renewed. It is advisable to connect each stator winding in series with an ammeter and a 1.5 volt battery. The current recorded as flowing through each winding should be the same. If the three current readings vary, the stator is probably deflective (Fig. 9.11).
16 *The brushes and slip rings:* The slip rings may be cleaned with very fine emery paper and then polished. Avoid generating flats on the slip ring surface. If the slip rings are found not to be concentric, they can be turned down to a diameter of 1.250 inches (31.75 mm). However, it is wise only to remove sufficient material to remove the flats.

The brushes should be inspected, and if found to be less than 0.4 inches in length, they should be renewed. When soldering the new brushes in place, hold the connecting wire with a pair of pliers to prevent the molten solder from flowing down the wire strands from the connection. If solder were allowed onto the strands the wire would become rigid and unserviceable.
17 *Diodes:* There are three bus wires running across the diode terminals on the rectifier pack, which has been assembled in a manner which eliminates the need to distinguish individual diodes. Each wire joins an exciter circuit positive diode to a power circuit positive and negative diodes. The individual diodes are checked by touching a probe to the diode mounting and another to the top bridging bus wire (Fig. 9.12). Note that the illustration shows the polarity of the test leads where current should not flow (ie; the light should not shine or glow).

a) *To test exciter diodes: Hold the 'plus' probe to a top bus wire and the 'negative' probe to the contact rail (which contacts terminal on brush holder). The test light should shine when the plus probe is touched on each of the three bridging bus wires. Change the probes around and the light should* **not** *shine or glow.*
b) *To test positive power diodes: Hold the 'plus' probe to a top bus wire and the 'negative' probe to the 'B+' terminal stud. The test light should shine when the 'plus' probe is touched on each of the three bridging bus wires. Change the probes around and the light should* **not** *shine or glow with the same test.*
c) *To test negative power diodes: Hold the negative probe to a top bus wire and the 'plus' probe to the heat sink chassis. The test light should shine when the negative probe is touched on each of the three bridging bus wires. Change the probes around and the light should* **not** *shine or glow with the same test.*

If any diode is found faulty, the whole rectifier pack should be renewed.
18 *Reassembly of the alternator:* Reassembly of this alternator follows the reverse sequence to dismantling. Make sure that the stator winding ends are soldered quickly to the proper location. Assemble the individual alternator parts so that the marks made prior to dismantling coincide.

17 Regulator (alternator) - general

1 Because the alternator starts to charge at 1,100 rpm, which is an engine speed of 550 rpm, it charges even when the engine is idling. Also its output is much higher than the dynamo's. So setting of the regulator is not so critical. This is fortunate, as adjustments are much more likely to give an opportunity for the unwary to damage the alternator.

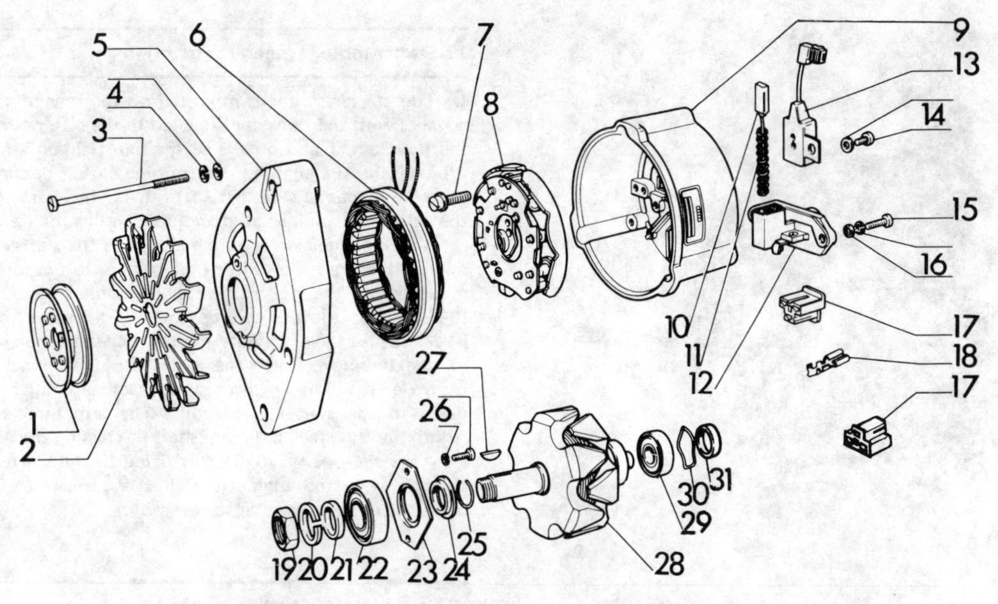

Fig. 9.8. Exploded view of alternator fitted to later models

1 Pulley
2 Fan
3 Bolts
4 Washers
5 Drive-end bracket
6 Stator windings
7 Plate screw
8 Diode plate (rectifier pack)
9 Body
10 Brush
11 Spring
12 Brush holder
13 Condenser
14 Screws and washers
15 Screws and washers
16 Screws and washers
17 Plug socket
18 Connector
19 Shaft nut
20 Spring washer
21 Thrust ring
22 Bearing
23 Retainer plate
24 Thrust ring
25 Spring washer
26 Screw and washer
27 Key
28 Rotor
29 Bearing
30 Backing washer
31 Shield (where applicable)

Fig. 9.9. Removing the brush holder (Sec. 16)

Fig. 9.10. Resistance checks of the rotor windings (Sec. 16)

Fig. 9.11. Resistance checks of the stator windings (Sec. 16)

2 If the tests given in Section 15 indicate that the regulator is at fault, this could be confirmed by a car electrician. In this case a specialist can be just as helpful as the FIAT agent, as the alternator is made by Bosch, and it and the regulator are conventional. Fitting a new regulator is quite simple, but then it ought to be checked for adjustment after fitting.

3 When fitting a new regulator, it is important that the leads are fitted exactly as removed, or the alternator may be severely damaged.

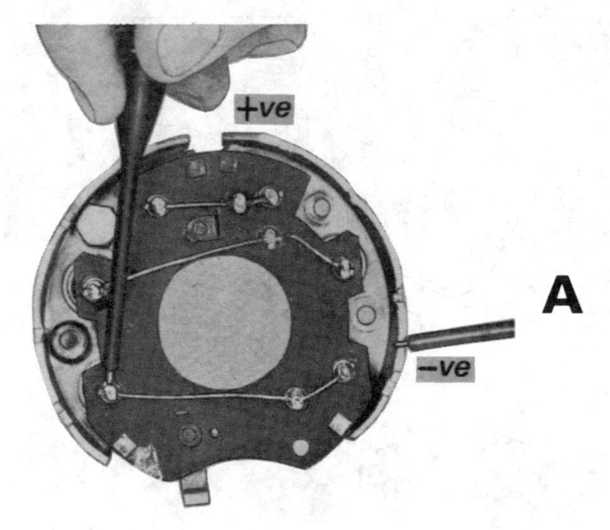

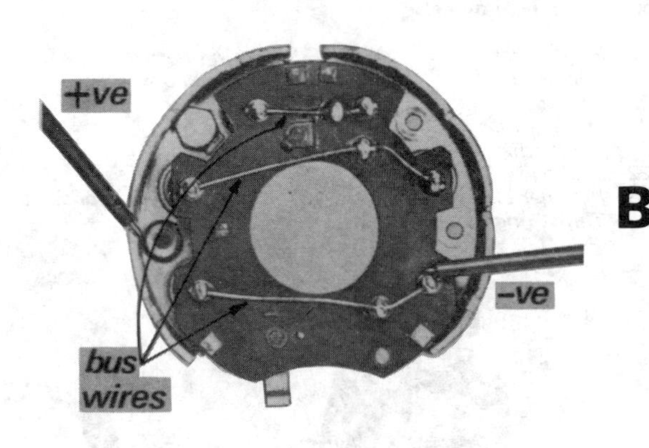

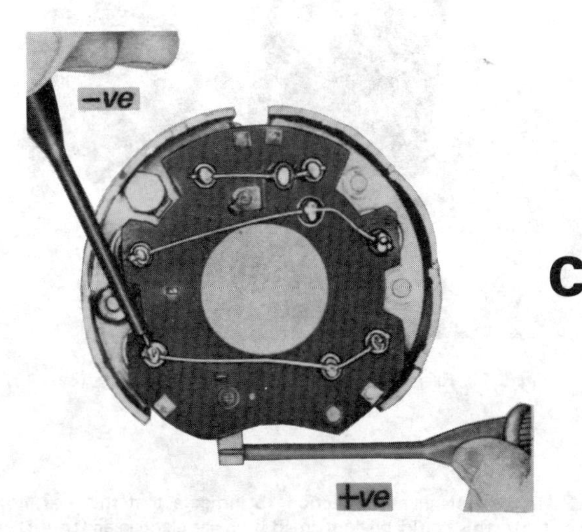

Fig. 9.12. Testing the rectifier pack diodes (Sec. 16)

a) Exciter diodes. No current should flow with this test lead polarity
b) Positive power diodes. No current should flow with this test lead polarity
c) Negative power diodes. No current should flow with this test lead polarity

18 Starter motor - general description

The starter motor is mounted on the rear of the engine and engages with the ring gear attached to the outer circumference of the flywheel. The motor is of the four field coil, four pole piece type of the pre-engaged overrunning clutch design. Four spring loaded brushes are used, two of these brushes are earthed and the other two are insulated and attached to the field coil ends.

When the ignition is switched on and the starter motor switch operated, current flows from the battery to the solenoid which is mounted on the top of the starter motor body. The plunger in the solenoid moves inwards and, via a pivoted drive engaging lever, pushes the drive pinion into mesh with the starter ring gear on the flywheel. When the solenoid plunger reaches the end of its travel it closes an internal contact and full starting current flows to the starter field coils. The armature is then able to rotate the flywheel and crankshaft so starting the engine.

A special one way clutch is fitted to the starter drive pinion so that when the engine just fires and starts to operate on its own it does not drive the starter motor.

19 Starter motor - testing on engine

1 If the starter motor fails to operate then check the condition of the battery by turning on the headlamps. If they glow brightly for several seconds and gradually dim, the battery is in an uncharged condition.
2 If the headlamps continued to glow brightly and it is obvious that the battery is in good condition, then check the tightness of the battery wiring connections (and in particular the earth lead from the battery terminal to its connection on the bodyframe). If the positive terminal on the battery becomes hot when an attempt is made to work the starter, this is a sure sign of a poor connection on the battery terminal. To rectify, remove the terminal, clean the inside of the cap and the terminal post thoroughly and reconnect. Check the tightness of the connections at the relay switch and at the starter motor.
3 If the wiring is in order then check that the starter motor is operating. To do this, press the rubber covered button in the centre of the solenoid. If it is working the starter motor will be heard to 'click' as it tries to rotate. Alternatively check it with a voltmeter.

If the battery is fully charged, the wiring in order and the switch working, and the starter motor fails to operate, then it will have to be removed from the car for examination.

20 Starter motor - removal and refitting

1 Disconnect the battery earth terminal, for safety reasons.
2 Make a note of the electrical cables at the rear of the solenoid and detach the cables. (photo)
3 Remove the upper starter motor fixing.
4 Working under the car, loosen and then remove the lower starter motor fixing taking care to support the motor so as to prevent damage to the drive components.
5 Lift the starter motor out of engagement with the flywheel ring gear and lower it from the car.
6 Replacement is a reversal of the removal procedure.

21 Starter motor and drive gear - dismantling and reassembly

1 Slacken the cover band clamping screw and slide the cover band from the end of the starter motor body.
2 Lift up the brush hold down springs using an electrician's screwdriver and retain in the raised position by placing the spring on the side of each brush. (photo)

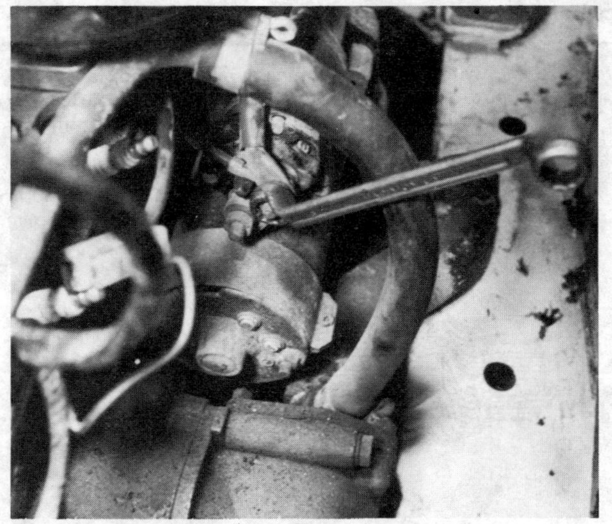

20.2 Disconnecting the main feed to the starter motor; ensure the battery lead is disconnected before attempting this

21.2 It is no use just replacing the brushes through the windows, as the commutator will need cleaning

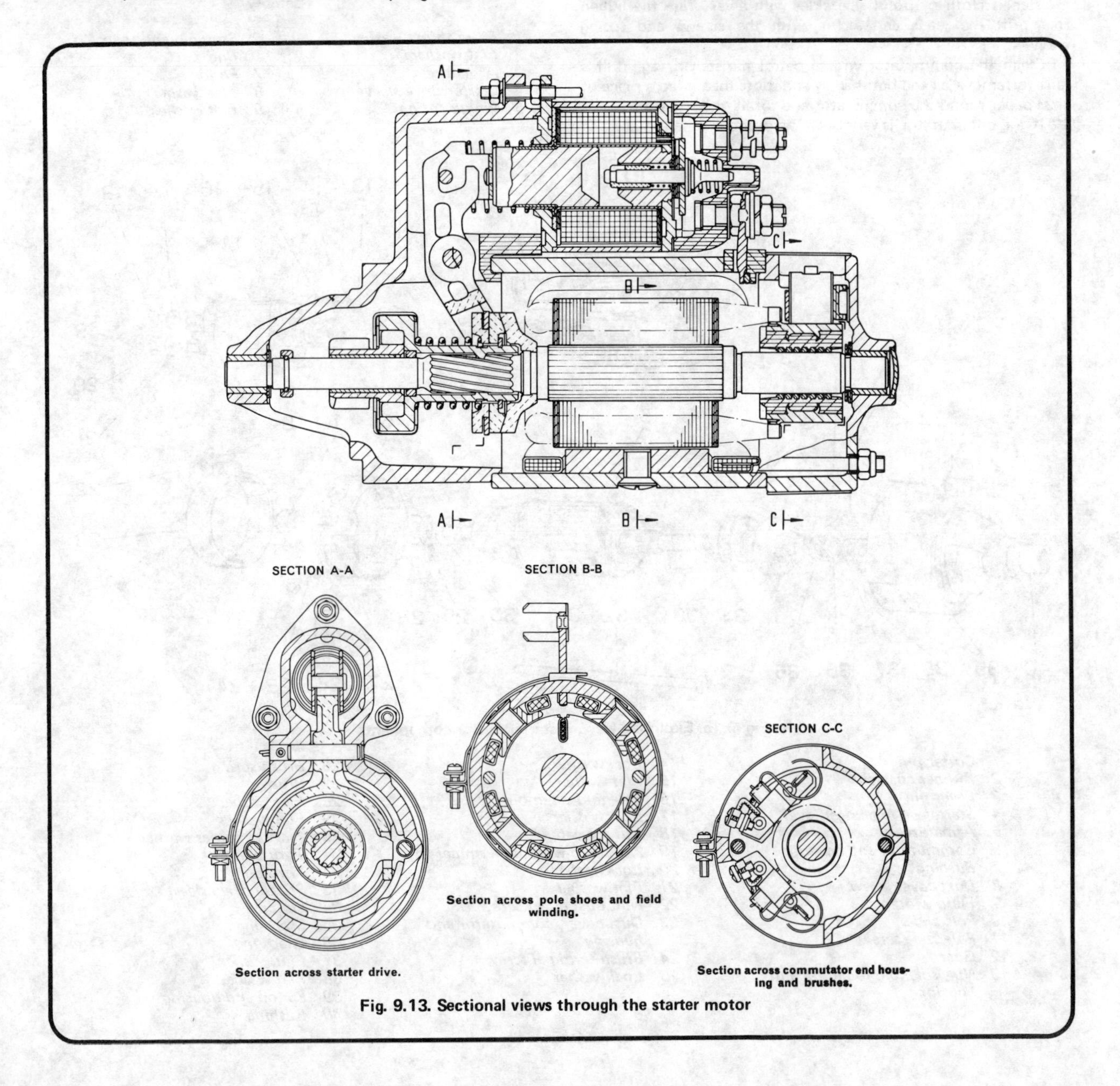

Fig. 9.13. Sectional views through the starter motor

3 Undo and remove the two through bolts, spring washers and nuts and lift away the commutator end bracket.
4 Undo and remove the nut and spring washer securing the heavy duty cable to the solenoid lower terminal. Also remove the field coil cables from the solenoid.
5 Undo and remove the nuts, spring and plain washers securing the solenoid to the front drive end housing. Carefully disengage the plunger from the operating fork and lift away the solenoid.
6 The body of the starter motor together with field coils may now be lifted from the front drive end housing.
7 To remove the operating fork, first note which way round it is fitted and then extract the split pin securing the pivot pin into the front drive end housing. Withdraw the pivot pin and lift away the operating fork.
8 Lift away the armature assembly complete with the clutch drive mechanism.
9 Inspect the brushes for wear and if worn down to 9/32 inch (7.14 mm) or less, undo the screws holding the brush leads to the end bracket. Take off the brushes complete with leads.
10 Fit new brushes to the holders and make sure that they move freely in their holders. If they do not, wipe with a petrol moistened cloth or polish the sides with a very fine file. When free refit the leads and secure with the screws and spring washers.
11 Clean the commutator with a petrol moistened rag. If this fails to remove all the burnt areas and spots then wrap a piece of glass paper round the commutator and rotate the armature.
12 If the commutator is very badly worn mount the armature in a lathe and, with the chuck turning at a high speed, take a very fine cut out of the commutator and finish the surface by polishing with glass paper. **Do not undercut the mica insulators between the commutator segments. (Fig. 9.3).**
13 With the starter motor dismantled, test the four field coils for open circuit. Connect a 12 volt battery with a 12 volt bulb in one of the leads between the field terminal post and the tapping

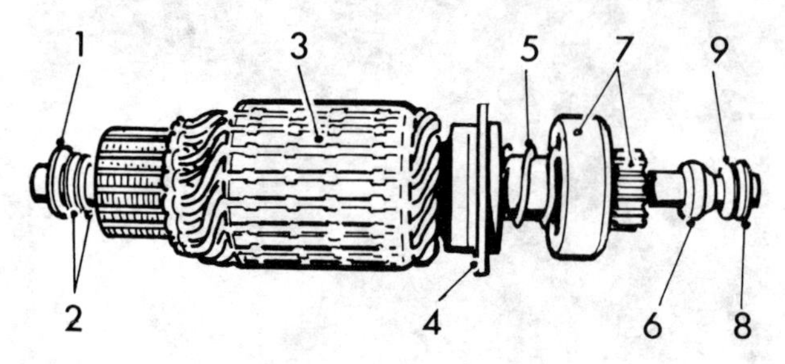

Fig. 9.14. View of armature showing the overrunning clutch assembly

1 Fibre thrust washer
2 Flat washers
3 Armature
4 Starter drive sleeve
5 Pinion spring
6 Stop pin and snap ring
7 Pinion
8 Flat washer
9 Flat washer

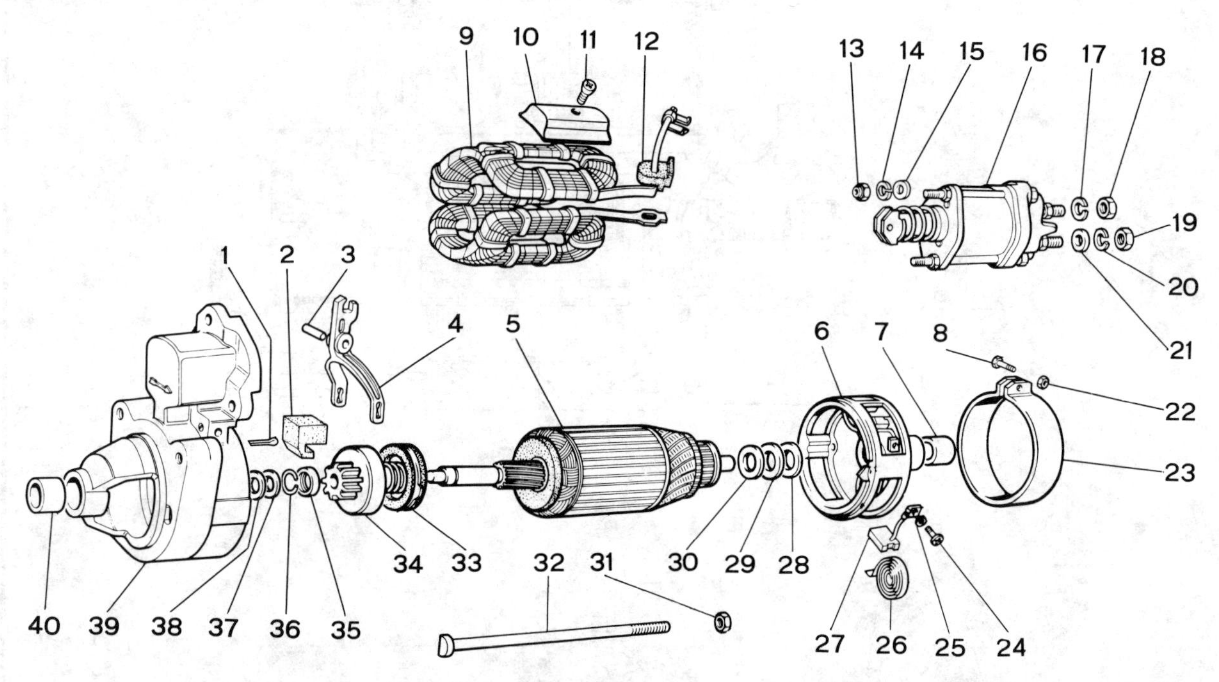

Fig. 9.15. Exploded view of starter motor components

1 Cotter pin
2 Rubber cushion
3 Lever pin
4 Starter drive plunger lever
5 Armature
6 Commutator end housing
7 Bushing
8 Dust cover screw
9 Field winding
10 Pole shoe
11 Pole shoe screw
12 Guard
13 Nut, solenoid to pinion end housing
14 Lockwasher
15 Flat washer
16 Solenoid assembly
17 Lockwasher
18 Nut for cable
19 Nut, field winding terminal
20 Lockwasher
21 Flat washer
22 Nut, dust cover screw
23 Dust cover, commutator end housing
24 Brush terminal screw
25 Lockwasher
26 Brush spring
27 Brush
28 Flat washer
29 Flat washer
30 Fibre shoulder washer
31 Thru bolt nut
32 Thru bolt
33 Starter drive sleeve
34 Pinion
35 Stop ring
36 Snap ring
37 Flat washer
38 Flat washer
39 Pinion end housing
40 Bushing

point of the field coils to which the external connections are made. An open circuit is proved by the bulb not lighting.

14 If the bulb lights, it does not necessarily mean that the field coils are in order, as there is a possibility that one of the coils will be earthing to the starter body or pole shoes. To check this, remove the lead from the external connector and place it against a clean portion of the starter yoke. If the bulb lights the field coils are earthing.

15 Replacement of the field coils calls for the use of a wheel operated screwdriver and is beyond the scope of the majority of owners. The starter body should be taken to an auto electrician or FIAT garage for new field coils to be fitted. Alternatively, purchase an exchange starter motor.

16 If the armature is damaged this will be evident after visual inspection. Look for signs of burning, discolouration and for conductors which have lifted away from the commutator.

17 Inspect the parts of the drive gear for signs of wear or damage possibly caused by accidental engagement of the starter drive gear with the engine running.

18 Reassembly is a straightforward reversal of the dismantling procedure.

22 Starter motor bushes - inspection, removal and replacement

1 With the starter motor stripped down check the condition of the bushes in the commutator end bracket and the front drive end bracket. They should be renewed when they are sufficiently worn to allow visible side movement of the armature shaft.

2 The commutator end bracket bush is simply driven out with a suitable drift.

3 The front drive end bracket bush is removed by screwing in a tap of suitable diameter four or five turns and then pulling out the tap and bush.

4 As the replacement bushes are of the porous phosphor bronze type it is essential that they are allowed to stand in engine oil for at least 24 hours before fitment. In an emergency the bush can be immersed in hot oil (100°C/212°F) for 2 hours.

5 Carefully fit the new bushes until the ends of the bushes are flush with the inner side of the housing. If available press the bushes in with a smooth shouldered mandrel the same diameter as the armature shaft.

23 Starter motor solenoid - removal and refitting

1 For safety reasons, disconnect the battery earth terminal.

2 For ease of working, remove the starter motor from the engine as described in Section 20.

23.3 The solenoid is removed by disconnecting the cable to the motor, and then undoing the three nuts

3 Make a note of the cable connections on the solenoid and disconnect these from the solenoid. (photo)

4 Undo and remove the solenoid mounting nuts and washers and lift away from the starter motor.

5 If the solenoid is faulty then a new one should be fitted as it is a partially sealed unit. Do not attempt to dismantle and overhaul.

6 Refitting is the reverse sequence to removal. Take care to ensure all cable connections are clean and tight.

24 Fuses

1 Seven 8-amp fuses and one 16-amp fuse are found in the engine compartment, under the bulkhead and behind the battery.

2 If any of the fuses blow due to a short circuit or similar trouble, trace the source of the trouble and rectify before fitting a new fuse.

3 To assist in tracing a fault in any given circuit, tables are given below to indicate the fuse protected circuits: The numbers against each fuse refer to Fig. 9.16 (w/l = warning light).

1 (8 Amps)
Oil pressure w/l
Water temperature w/l
Fuel gauge and w/l
Direction indicator and w/l
Windscreen wiper
Blower
Stop light

2 (16 Amps)
Courtesy light
Horn
Fan

3 (8 Amps)
L.H. main beam
Main beam w/l

4 (8 Amps)
R.H. main beam

5 (8 Amps)
L.H. dipped beam

6 (8 Amps)
R.H. dipped beam

7 (8 Amps)
L.H. side light
R.H. rear light
L.H. number plate light
Panel light

8 (8 Amps)
R.H. side light
L.H. rear light
R.H. number plate light
Side light w/l

4 Unprotected circuits are: charging; ignition warning light; ignition; starter; and cooling fan relay coil.

Fig. 9.16. Fusebox and other electrical components (127 models - not 127 Special)

1 8 fuses (see Section 24)
9 Turn signal flasher unit
10 Electrofan relay
11 Generator regulator

No. 2 is a 16 amp fuse while all the others are 8 amp

25 Headlamps - alignment and bulb replacement

Notes: *Holts Amber Lamp is useful for temporarily changing the headlight colour to conform with the normal usage on Continental Europe.*

1 The double-filament bulb can be reached from inside the engine compartment. With reference to Fig. 9.17 withdraw the connector plug (B), the rubber boot (C), and finally the clip (D) by depressing and turning anticlockwise. (photos)

2 Fit the clip on the new bulb and ensure that the peg engages with the associated hole. Replace the rest of the components in the reverse manner to removal.

3 The headlights may be adjusted for both vertical and horizontal beam adjustment by the two screws, 1 and 2 (Fig. 9.18. They should be set so that on full or high beam the beams are set slightly below parallel with a level road surface. Position the car on level ground with the headlights 16 ft 5 in. (5 m) from a vertical wall. Check the tyre pressures and adjust as necessary. Bounce the front and rear of the car up and down several times to settle the suspension.

4 Next, refer to Fig. 9.19 and draw two crosses on a wall corresponding to the headlamp centres. Switch on the dipped beams. Reference points 'P-P' should lie 4 3/8 in. (110 mm) below the associated headlamp centre marks on the wall. To adjust, turn screws 1 and 2 (Fig. 9.18).

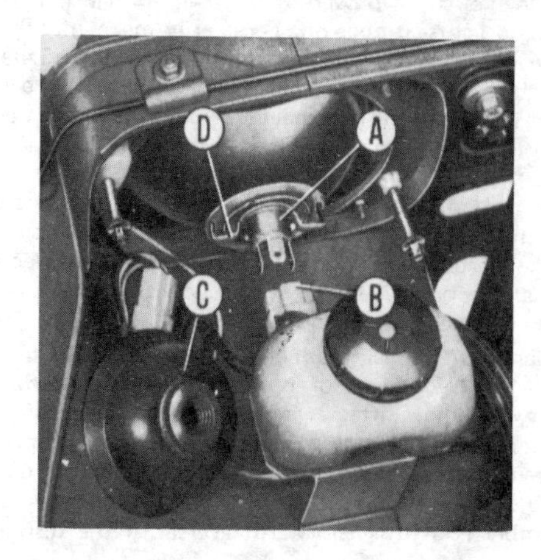

Fig. 9.17. Headlight - bulb replacement (early models) (Sec. 25)

25.1A Removing the electrical connector

25.1B View after removing the rubber boot

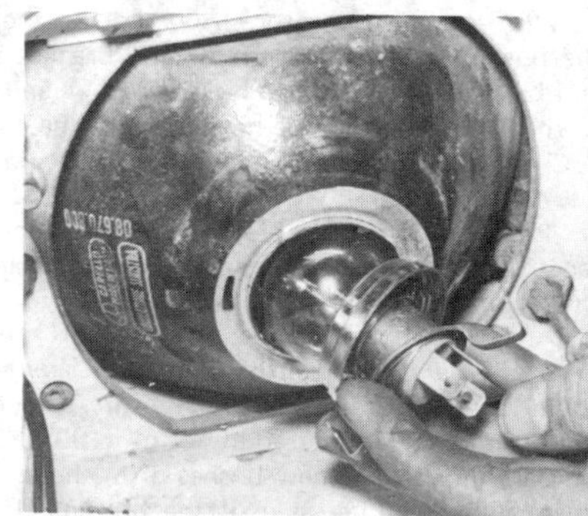

25.1C Release the bulb by depressing and twisting anticlockwise

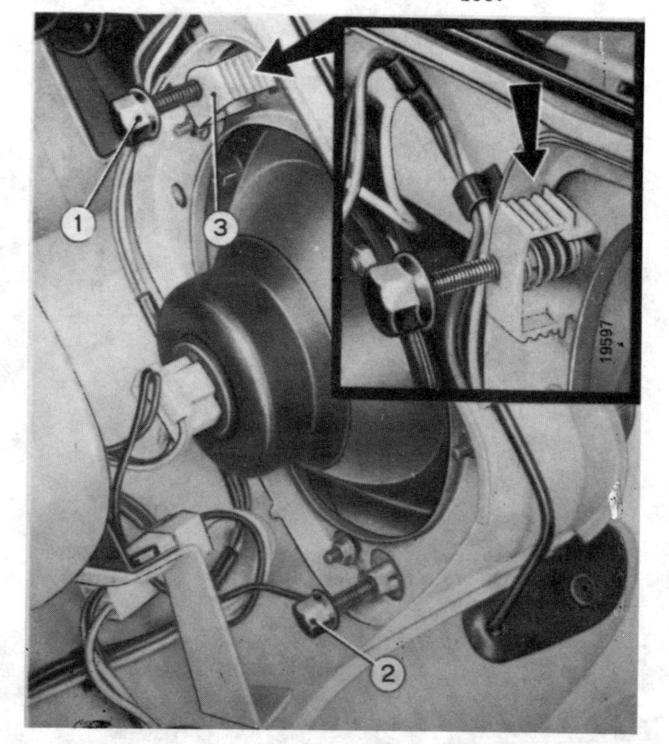

Fig. 9.18. Headlight adjustment details (early models) (Sec. 25)

1 *Screw for adjusting the low beam vertically*
2 *Screw for adjusting the low beam horizontally*
3 *Device for adjusting headlight vertical position depending on the load*

The arrow in the inset shows the manual adjuster for setting the headlight vertical position according to the load of the car

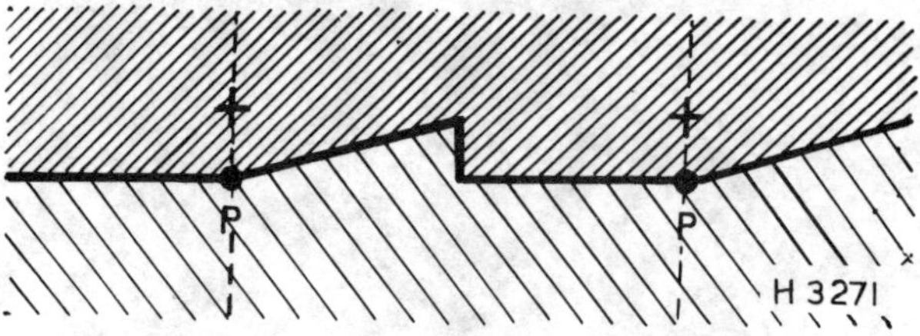

Fig. 9.19. Headlight beam aiming pattern (Sec. 25)
Note: This beam pattern applies to RH drive cars.
A reverse beam pattern should be obtained for LH drive cars

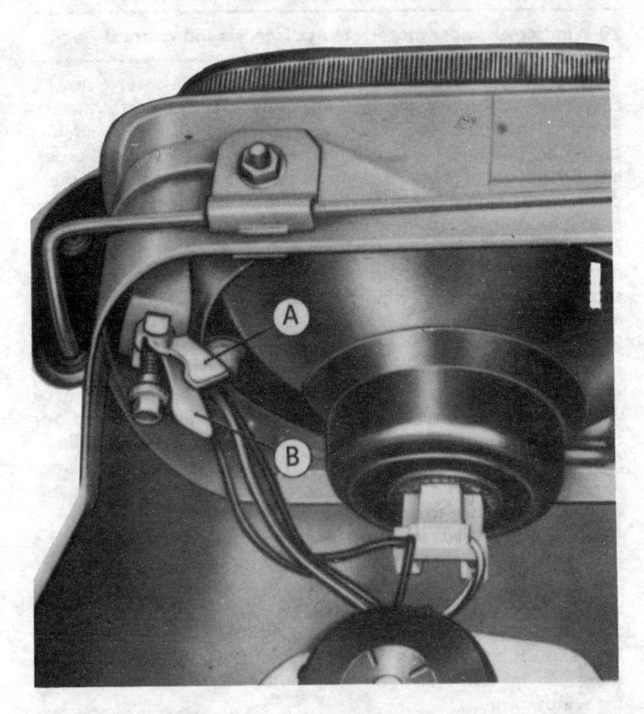

Fig. 9.20. Headlight adjustment details (later models) (Sec. 25)

A Headlight manual adjuster in position for car unloaded and at half load (aiming position)
B Headlight manual adjuster in position for fully loaded car

5 Models of the three-door type are fitted with a manual beam angle corrector to compensate for the changes in vehicle pitch under load. Refer to Fig. 9.20.

6 Use position 'A' when the car is to carry light loads; turn to position'B' before moving off with a heavy load. Ensure that the headlamp angle correctors are always set in the same position relative to one another. The headlamp alignment adjustment operation is as described in paragraph 5, except that reference points 'P-P' should lie 3 3/8 in. (85 mm) below the associated crossmarks on the wall.

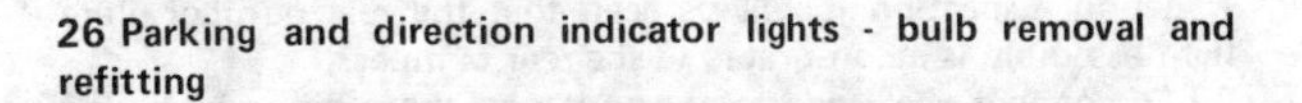

26 Parking and direction indicator lights - bulb removal and refitting

Front parking light and direction indicator light

1 Undo and remove the screws that secure the light unit lens to the light body and draw the unit forwards. (photo)

2 Remove the appropriate bulb by twisting in an anticlockwise direction. It has a bayonet type fitting.

3 Refitting is the reverse sequence to removal.

Direction repeater lights

4 The pull-out type bulb holder may be reached from the inside of the wing.

Rear light cluster

5 Undo and remove the three screws securing the light lens to the body and draw the lens forwards to give access to the bulbs. (photo)

6 Remove the appropriate bulb by twisting in an anticlockwise direction. It has a bayonet type fitting.

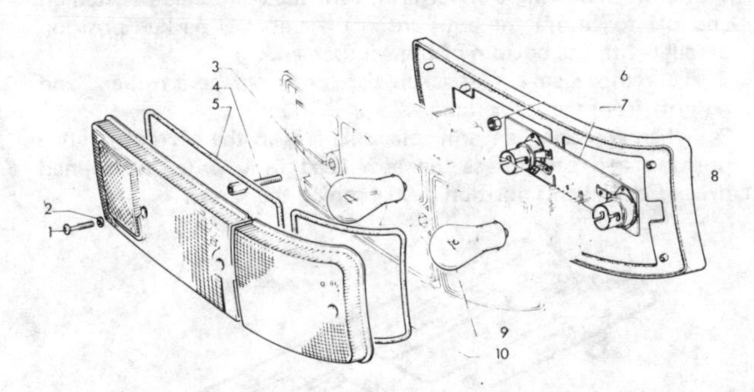

Fig. 9.21. Parking and direction indicator light fitting - rear (Sec. 26)

1 Screw - lens
2 Washer
3 Parking bulb
4 Screw - body
5 Gasket
6 Nut - body
7 Bulb holder assembly
8 Gasket
9 Indicator bulb
10 Gasket

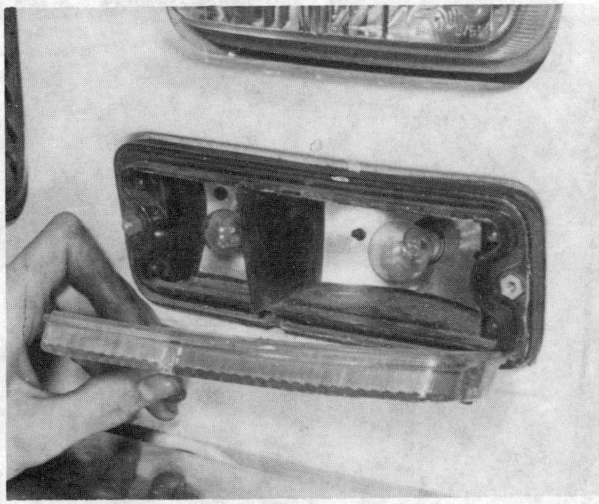

26.1 Removing the lens - front parking light

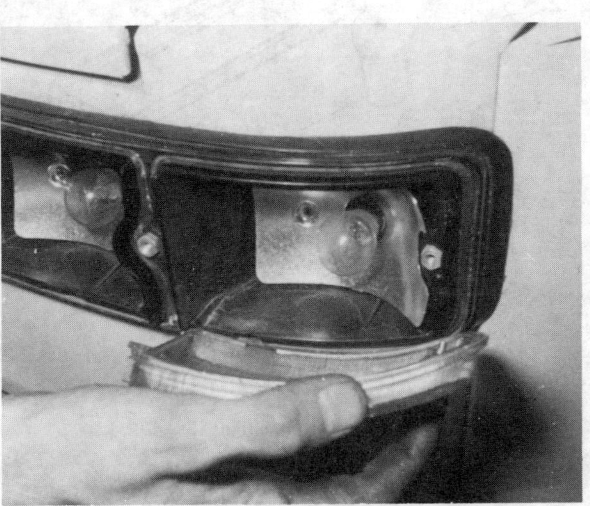

26.5 Removing the lens - rear direction indicator light

Rear number plate light

7 Upon inspection it will be seen that the rear number plate light assembly is incorporated in the rear bumper.
8 Disconnect the leads to the unit and then, by compressing the lugs of the bulb holder it is possible to pull the unit out of the bumper.
9 Remove the bulb(s); they are fitted with a bayonet type fitting.

27 Windscreen wiper mechanism - maintenance

1 Renew the windscreen wiper blades at intervals of 12,000 miles, or more frequently if necessary.
2 The linkage which operates the wiper arms from the motor should be greased every 6,000 miles. The washer round the two pivot housings should be lubricated with several drops of glycerine every 6,000 miles.

28 Windscreen wiper arms - removal and replacement

1 Before removing a wiper arm, turn the windscreen switch on and off to ensure the arms are in their normal parked position parallel with the bottom of the windscreen.
2 To remove an arm, slacken the screw on the arm head and pull off from the splined drive.
3 When replacing an arm, place so it is in the correct relative position and then press the arm head fully onto the splined drive. Lock in this position by tightening the screw.

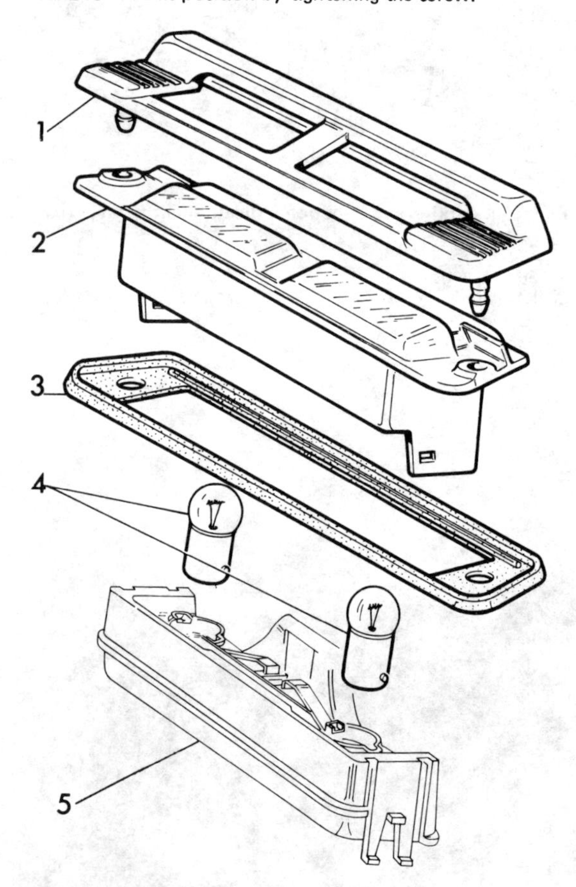

Fig. 9.22. Number plate light (Sec. 26)

1 Outer body
2 Inner body
3 Gasket
4 Bulbs
5 Removable bulb holder

29 Windscreen wiper motor - fault diagnosis and removal

1 Should the windscreen wipers fail, or work very slowly, check the 'A' fuse. If the fuse has blown, replace it after having checked the wiring of the motor and other electrical circuits served by this fuse for short circuits. If the fuse is in good order, check the wiper switch. Then check the terminals for loose connections, and make sure the insulation of the external wiring is not broken or cracked. Use a test lamp to ensure current is reaching the motor. If this is in order, check the current the motor is taking by connecting up a 1-20 ammeter in the circuit and turning on the wiper switch. Consumption should be approximately 2 amps.
2 If the wiper takes a very high current, check the wiper blades for freedom of movement. If this is satisfactory check the motor mountings and linkages for tightness. Should all appear to be well, the motor will have to be stopped and checked for tightness of the armature in its bearings or insufficient endfloat.
3 If the motor takes a very low current, ensure that the battery is fully charged. If the battery is well charged, then one is left with no option but to assume the motor is faulty.
4 Having analysed the fault the next step is to remove the motor to either obtain an exchange unit or delve into its internals. The motor location is shown in the associated photo.
5 Disconnect the plug for the leads.
6 Pull off the wiper arms.
7 Take off the fixtures on the ends of the arm spindles so that these can be withdrawn through the bushes in the body below the windscreen.
8 Undo the two bolts holding the motor mounting frame to the bracket on the bulkhead.
9 Remove the whole assembly.
10 When refitting, grease the arm spindles in their bushes, and all the other links for the connecting levers.
11 Operate the motor before fitting the blade arms. Once the motor has put the spindles in the parked position, the arms can be fitted in their correct orientation. Make sure the assembly is refitted without distortion which could make the linkage stiff.
12 If the operation of the wiper becomes very sluggish due to the need for lubrication of the interconnecting mechanism, or in heavy snow, there is risk of burning out the motor. Always switch off the wiper if it stalls, and if it cannot get back to the parked position, unplug the motor.

29.4 Showing the location of the windscreen wiper motor

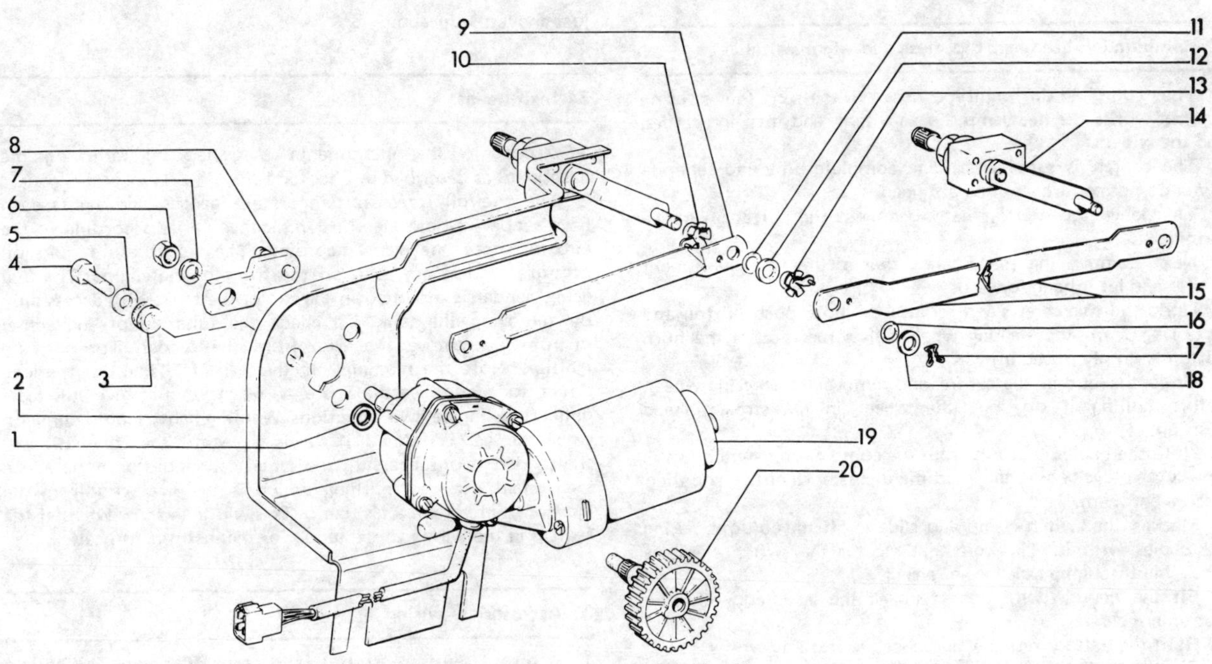

Fig. 9.23. Windscreen wiper details

1 Washer
2 Mounting frame
3 Bush
4 Bolt
5 Washer
6 Locknut
7 Lockwasher
8 Crank
9 Lever
10 Locating bush
11 Washer
12 Washer
13 Locating bush
14 Articulated head
15 Lever
16 Washer
17 Washer
18 Clip
19 Wiper motor
20 Large gear wheel (inside motor)

30 Windscreen wiper motor - checking the brushes

1 Although many owners will opt for buying an exchange unit, or one from a breaker's yard, there is no reason for not making an inspection of the motor brushes, since they are relatively accessible.
2 Unscrew the two outer through bolt nuts and lockwashers from the outside of the commutator endframe.
3 Carefully withdraw the endframe and insulation pack from the motor body noting the steel and fabric washers on the armature shaft.
4 Unscrew the two inner through bolt nuts and spring washers.
5 Lift away the brush mounting plate from the through bolts. As the plate is drawn off the commutator the two brushes will, under the action of their tensioning springs, move out of their holders.
6 Inspect the brushes and if they are worn, unsolder the brush cable tags from the field coil.
7 Clean out the brush holders and plate assembly and fit the new brushes. Make sure they are free to move in the holders and then resolder them to the field coils.
8 Reassemble the motor in the reverse manner to dismantling.

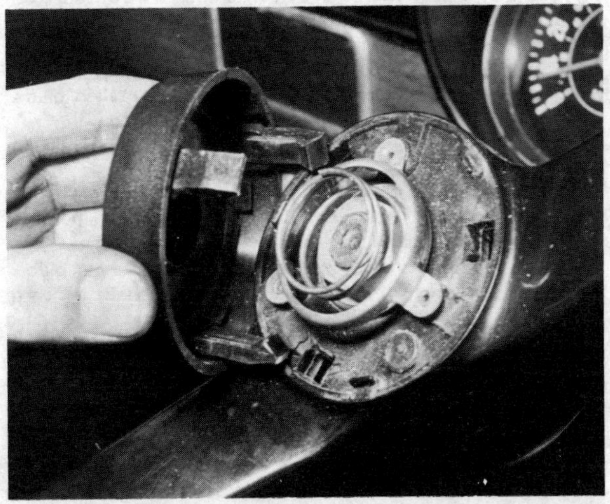

31.3 Remove the horn button by prising off the complete cap

31 Horn

1 If a horn fails it is usually a fuse or wiring fault. Horns themselves tend to be fairly robust. The fuse is easily checked as it is shared with the interior light and the cooling fan.
2 The horn button is the earth return, the black wire. Join a short length of wire from that terminal on one of the horns to an earth point and see if that makes the horn work. If it does, then the fault lies in the button or the associated wiring. If it is still faulty, check the horn live side, the violet wires, with a test bulb. If the test bulb doesn't light then the cable between the horn terminal and the fuse is faulty.
3 The horn button can be removed from the steering wheel by prising it off. Try earthing the wire to the steering column if the contacts are suspect. (photo)
4 As a last resort, if all else fails, it can be adjusted by a screw on the back. Try turning the screw in first: only a turn or so. If the horn works again, adjust the screw in or out for the best note. Afterwards put some paint on the screw to seal out water, and to lock the screw. Having proved the horn totally defective, it will have to be renewed.

32 Combination headlamp dip, flash and wiper switch

1 The combination, multi-purpose switch on the steering column works the headlamp dip and flash, and turn indicators, and the wipers.
2 The switch is supplied as one complete unit and separate minor components cannot be obtained.
3 To change the switch, first disconnect the battery negative terminal.
4 Next, remove the single screw that secures the instrument cluster and let it hang forward.
5 Undo the two screws and remove the horn push button and its spring from the steering wheel. On some models the horn button is simply prised off.
6 Undo the nut in the centre and remove the steering wheel. Before pulling it off, put the wheel in the straight-ahead position.
7 Note the cables that run from the combination switch to the connector plugs behind the instrument cluster: identify and then disconnect them.
8 Slacken the switch clamp and slide it off the column, taking the cables with it. The bolt that secures the switch clamp is accessible through a hole in the fairing.
9 Fit the new switch, connect up all the leads and refit the instrument cluster.
10 Refit the battery terminal and check everything works.
11 Fit the steering wheel in the straight position.
12 Check that the indicator self-cancelling mechanism works.
13 Refit the horn button.

33 Flashing indicator system

1 Failure of a bulb will be shown by a change in the speed of flashing, and the note of the clicking. Failure of the flasher itself is usually complete, with the lamp permanently lit or unlit.
2 If the flasher unit fails to operate, or works very slowly or very rapidly, check out the flasher circuit as detailed below, before assuming there is a fault in the unit itself.
3 Examine the direction indicator bulbs front and rear for broken filaments.
4 If the external flashers are working but the internal flasher warning light has ceased to function, check the filament of the warning bulb and replace as necessary.
5 Check all the flasher circuit connections if a flasher bulb is sound but does not work. Pay particular attention to the earth connections on the flasher units: these should be cleaned thoroughly and the connection re-made.
6 In the event of total direction indicator failure, check the 'A' fuse. This, however, will be obvious due to the failure of all the other warning lights supplied from the same fuse.
7 If all other items check out then the flasher unit itself is faulty and must be renewed.
8 The flasher unit itself is readily accessible, being mounted just above the fusebox.

34 Instruments

1 Access to the instruments is gained by removing the dashboard as described in Chapter 12.
2 The fuel/oil, pressure/temperature gauges and lights each have senders whose electrical resistance varies according to the circumstances they are recording. The temperature, and oil pressure senders are both on the front of the engine. The fuel gauge sender is mounted in the top of the tank, with access only by removing the tank. To check the function of the actual instrument, remove the wire from the sender. Turn on the ignition. Note the reading with the lead off. Then put the lead direct to a good earth. These two tests should give full scale deflections in the two directions. Warning lights should come on when the lead is earthed. If this is successful, then the fault is in the sender. If unsuccessful, the same test should be done at the instrument, first establishing which is the live wire from the ignition switch with a test lamp. This will then show whether the fault is in the wiring to the sender or the instrument itself.

35 Accessories - wiring

1 A most useful accessory would be an ammeter. This should be connected into the brown lead between the battery and all other circuits other than the starter motor. Having mounted the dial on the dashboard, lead a pair of wires to the junction of brown and red wires for the alternator and other main circuits.
2 Other items should be wired in from a fused circuit as appropriate. For instance a spot light or fog light should be wired in from fuse 8. If forgotten, it will be switched off when the lights are switched off.
3 Connect all wires firmly. Ideally use proper car type connectors, or else terminal blocks sold for household or radio fittings. Where wires pass through holes in the bodywork, fit a grommet to prevent chafing. Strap the wires to the existing harness.
4 When connecting a radio, try to position the aerial as far as possible from the distributor in order to minimise interference. It will also be necessary to fit suppression condensers to the dynamo or alternator, and the ignition circuit.
5 The last point to suppress the dynamo output would be at terminal '30' (pink cable) on the control box; this is because it is a convenient point for earthing the condenser casing on the chassis with the mounting screw. Ensure that a good earth connection is made by rubbing down to bare metal around the fixing hole and lightly smearing with vaseline. The ignition can be suppressed in a similar manner by connecting the condenser to the negative (black cable) terminal on the ignition coil.
6 If you are at all uncertain about any of the above points, do take an auto-electricians advice, since its only too easy to cause yourself trouble and expense through faulty wiring.

36 Fault diagnosis - electrical system

Symptom	Reason/s	Remedy
Starter motor fails to turn engine		
No DC supply at starter motor	Battery discharged	Charge battery.
	Battery defective internally	Fit new battery.
	Battery terminal leads loose or earth lead not securely attached to body	Check and tighten leads.
	Loose or broken connections in starter motor circuit	Check all connections and tighten any that are loose.
	Starter motor switch or solenoid faulty	Test and replace faulty components with new.
DC supply at starter motor: faulty motor	Starter brushes badly worn, sticking, or brush wires loose	Examine brushes, replace as necessary, tighten brush wires.
	Commutator dirty, worn, or burnt	Clean commutator, recut if badly burnt.
	Starter motor armature faulty	Overhaul starter motor, fit new armature.
	Field coils earthed	Overhaul starter motor.
Starter motor turns engine very slowly		
Electrical defects	Battery in discharged condition	Charge battery.
	Starter brushes badly worn, sticking, or brush wires loose	Examine brushes, replace as necessary, tighten brush wires.
	Loose wires in starter motor circuit	Check wiring and tighten as necessary.
Starter motor operates without turning engine		
Free wheel in drive faulty	Seized or stuck	Remove, examine, replace as necessary.
Mechanical damage	Pinion or ring gear teeth broken or worn	Fit new gear ring, and new pinion to starter motor drive.
Starter motor noisy or excessively rough engagement		
Lack of attention or mechanical damage	Pinion or ring gear teeth broken or worn	Fit new ring gear, or new pinion to starter motor drive.
	Starter motor retaining bolts loose	Tighten starter motor securing bolts. Fit new spring washer if necessary.
Battery will not hold charge for more than a few days		
Wear or damage	Battery defective internally	Remove and fit new battery.
	Electrolyte level too low or electrolyte too weak due to leakage	Top-up electrolyte level to just above plates.
	Drive belt slipping	Check belt for wear, replace if necessary, and tighten.
	Short in lighting circuit causing continual battery drain	Trace and rectify.
	Regulator unit not working correctly	Check setting, clean, and replace if defective.
Generator not charging - battery runs flat in a few days		
Charge warning light not extinguishing	Drive belt loose and slipping, or broken	Check, replace, and tighten as necessary.
	Brushes worn, sticking, broken or dirty commutator dirty	Examine, clean, or replace brushes as necessary.
	Brush springs weak or broken	Examine and test. Replace as necessary.
	Alternator only - Stator coils faulty	Fit new item or reconditioned generator.
	Rotor windings faulty	Fit new or reconditioned item or generator.
	Slip rings dirty	Clean slip rings.
Wipers		
Wiper motor fails to work	Blown fuse	Check and replace fuse if necessary.
	Wire connections loose, disconnected, or broken	Check wiper wiring. Tighten loose connections.
	Brushes badly worn	Remove and fit new brushes.
	Armature worn or faulty	Remove and overhaul and fit replacement armature.
	Field coils faulty	Purchase reconditioned wiper motor.
Wiper motor works very slowly and takes excessive current	Commutator dirty, greasy or burnt	Clean commutator thoroughly.
	Drive linkage bent or unlubricated	Examine drive and straighten out curvature. Lubricate.
	Wiper arm spindle binding or damaged	Remove, overhaul, or fit replacement.
	Armature bearings dry or unaligned	Replace with new bearings correctly aligned.
	Armature badly worn or faulty	Remove, overhaul, or fit replacement armature.

Wiper motor works slowly and takes little current	Brushes badly worn Commutator dirty, greasy, or burnt Armature badly worn or faulty	Remove and fit new brushes. Clean commutator thoroughly. Remove and overhaul armature or fit replacement.
Wiper motor works but wiper blades remain static	Driving linkage disengaged or faulty Wiper motor gearbox parts badly worn	Examine and if faulty, replace. Overhaul or fit new wiper motor.
Lights		
Lights do not come on	If engine not running, battery discharged Battery earth connection faulty Wire connections loose, disconnected or broken Light switch faulty	Charge battery. Make tight corrosion-free connection. Check all connections for tightness and cable for breaks. By-pass light switch to ascertain if fault is in switch and fit new switch as appropriate.
Lights come on but fade out	If engine not running, battery discharging	Push start car, and charge battery.
Lights give very poor illumination	Lamp glasses dirty Reflector tarnished or dirty Lamps badly out of adjustment Incorrect bulb with too low wattage fitted Existing bulbs old and badly discoloured	Clean glasses. Fit new light unit. Adjust lamps correctly. Remove bulb and replace with correct grade. Renew bulb.

Key to Wiring Diagram (see page 143)

Note: Diagram shows dynamo. Wiring for models fitted with an alternator will be found in Chapter 13.

1 Front turn signal lights
2 Front parking lights
3 Headlight high-low beam
4 Radiator electrofan
5 Thermal switch, electrofan
6 Horn
7 Turn signal side repeaters
8 Relay, electrofan
9 Starting motor
10 Dynamo
11 Generator regulator
12 Sending unit, water temperature
13 Ignition coil
14 Sending unit, low oil pressure
15 Spark plugs
16 Ignition distributor
17 Battery
18 Turn signal flasher
19 Stop light press switch
20 Windscreen wiper motor
21 Fuses
22 Instrument cluster lights
23 Junction block
24 Parking light indicator (green)
25 Turn signal indicator (green)
26 High beam indicator (blue)
27 High water temperature warning light (red)
28 No charge warning light (red)
29 Low fuel warning light (red)
30 Fuel gauge
31 Low oil pressure warning light (red)
32 Ignition and starting key switch
33 Exterior lighting and instrument cluster light switch
34 Door courtesy light switches
35 Switch, electrofan
36 Interior ventilation electrofan
37 Windshield wiper lever switch
38 Headlight high-low beam change-over switch and low beam flasher
39 Turn signal selector switch
40 Horn button
41 Interior light with incorporated switch
42 Fuel gauge sending unit
43 Rear turn signal lights
44 Rear parking and stop lights
45 License plate lights

Wiring colour code:

Azzurro = Light blue
Giallo = Yellow
Nero = Black
Grigio = Grey
Verde = Green
Marrone = Brown
Viola = Violet
Rosa = Pink
Bianco = White
Rosso = Red
Blu = Dark blue
Arancio = Amber

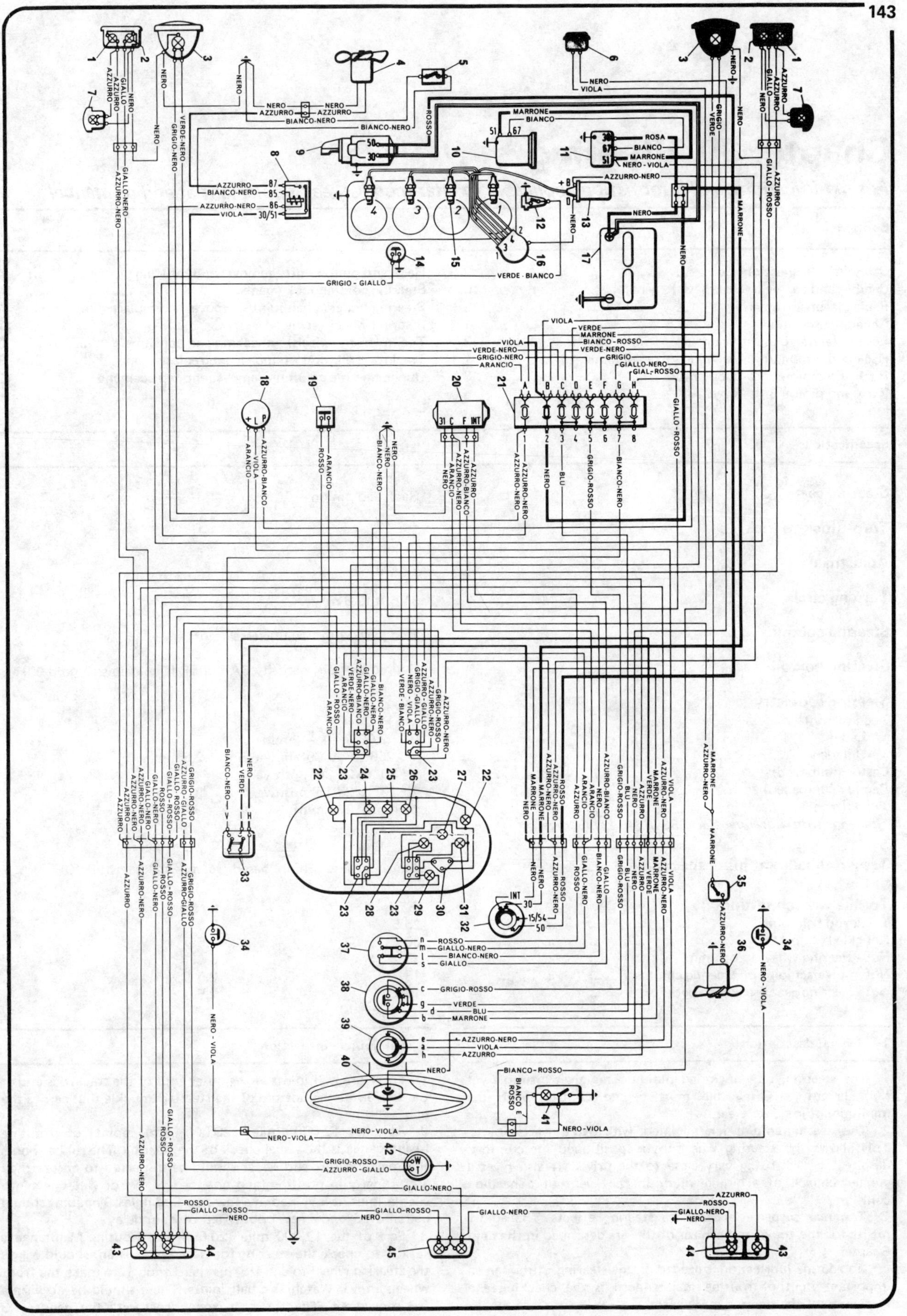
NERO
AZZURRO
BIANCO-NERO
ROSSO
MARRONE
BIANCO
ROSA
NERO-VIOLA
AZZURRO-NERO
VERDE-NERO
GRIGIO-NERO
GIALLO-NERO
GIALLO-ROSSO
VIOLA
VERDE
GRIGIO
ARANCIO
BLU
GRIGIO-ROSSO
GRIGIO-GIALLO
VERDE-BIANCO
AZZURRO-BIANCO
AZZURRO-GIALLO
BIANCO-ROSSO
BIANCO E ROSSO
GIAL.-ROSSO
INT
15/54
30/51
87
85
86
50
30
51
67
31 C F INT
A B C D E F G H

Chapter 10 Steering

For modifications, and information applicable to later models, see Supplement at end of manual

Contents

Specifications

Steering gear	Rack and pinion
Turns (lock-to-lock)	3½
Adjustment	By shims
Turning circle	31.5 ft (9.6 m)
Steering column	Two part, with two universal joints
Steering box oil	Hypoid gear oil, viscosity SAE 90EP (Duckhams Hypoid 90)

Steering geometry

Toe-in (front):	
Laden*	0 ± 0.08 in (0 ± 2 mm)
Unladen	−0.256 to −0.098 in (−6.5 to −2.5 mm)
Castor (unloaded)	1° 30' to 2° 30' positive
Camber (unloaded)	1° 15' to 2° 15' positive
Track (front)	50.4 in (1280 mm)

**Laden is with 4 occupants and 88 lb (40 kg) luggage*

Trackrod balljoint lubricant type	Multi-purpose lithium based grease (Duckhams LB 10)

Torque wrench settings

	lb f ft	kg fm
Track rod ball joint	25.5	3.5
Knuckle ball joint	36	3.5
Nut - steering wheel to column	36	5
Nut - universal joint on steering column	18	2.5
Nut - steering gear bracket to body	18	2.5

1 General description

1 The steering is by rack and pinion. This gives a simple yet rigid layout, ensuring the most precise control, with the minimum of joints to wear.
2 The steering column is articulated, with two universal joints. This allows the steering wheel to be positioned for comfort, despite the rack being very close to the driver. It also gives a greater chance of avoiding injury in the event of a head-on collision.
3 The rear suspension can have steering effects. The adjustments for the toe-in and camber of this are described in the next Chapter.
4 No routine lubrication is needed on the steering. Although an important item of routine maintenance is the checking and inspection of the steering mechanism.

2 Ball joints - inspection

1 There are ball joints at the outer ends of the trackrods and at the swivels at the bottom of the steering knuckle that carries the hub.
2 At the 12,000 mile (20,000 km) point of Routine Maintenance is the visual check of the steering. The rubber boots that exclude dirt and water should be inspected to ensure they are properly in position, and not torn. If dirt or water gets into such a joint, it is ruined in a few hundred miles. The joint should be removed, and a new boot fitted without delay.
3 Part of the 12,000 mile (20,000 km) Routine Maintenance task is to check the steering for wear. An assistant should wiggle the steering wheel to-and-fro, just hard enough to make the front wheels move. Watch the ball joints. There should be no visible free-movement. Then grasp a front wheel with the hands at 3

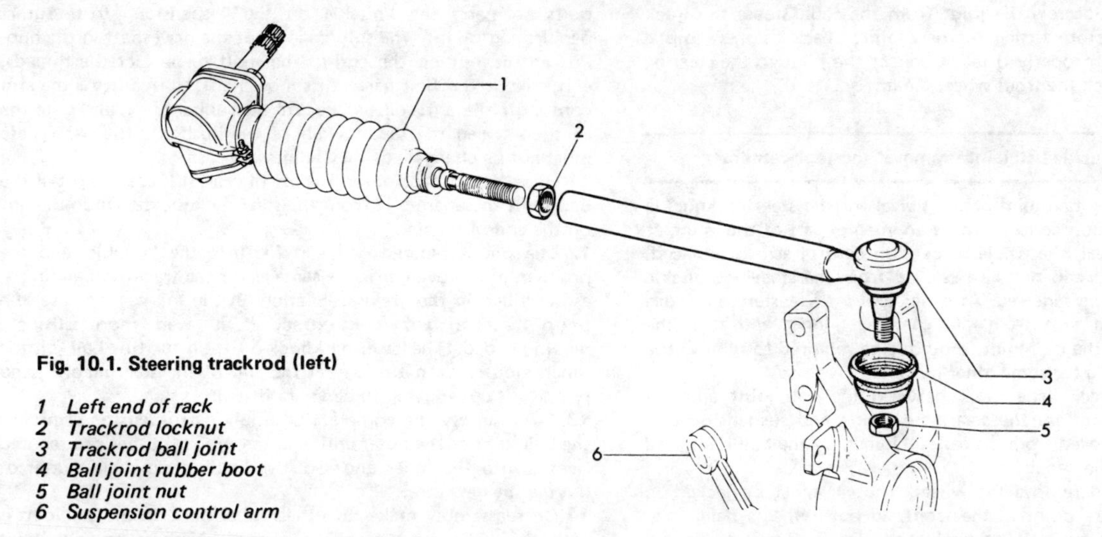

Fig. 10.1. Steering trackrod (left)

1 Left end of rack
2 Trackrod locknut
3 Trackrod ball joint
4 Ball joint rubber boot
5 Ball joint nut
6 Suspension control arm

and 9 o'clock on the wheel. Work at the wheel hard, rocking it in the horizontal plane. The rocking should shift the steering wheel but no lost motion should be felt.
4 There are also ball joints at the inner end of the trackrods, where they join the rack. But these are so well shielded and lubricated that no wear should develop there for considerable mileages.
5 The steering knuckle ball joint is more difficult to check. Again grip the wheel, but this time with both hands at the top. Rock to-and-fro vigorously, while an assistant watches the joint.
6 If any free-movement is seen or felt on a ball joint, it must be replaced.

3 Trackrod ball joint - removal and replacement

1 The ball joints on the outer end of the trackrod fit into the steering arm with a tapered pin, held by a nut at the bottom. They have to be removed to disconnect the steering when the suspension is being worked on, as in removing a driveshaft. They have to be taken off the steering arm to fit a new rubber boot. Finally, they have to be taken off for renewal.
2 Jack-up the car and remove the front wheel.
3 Undo the self-locking nut on the bottom of the tapered pin under the steering arm.
4 Now extract the taper from the steering arm. This is the difficult bit. It is best to buy a universal steering ball joint separator. This is a small extractor, and over the years will earn its keep, fitting all makes of cars, as ball joints are pretty well standardised. The separator presses the taper out of its seating. Without a separator, the taper must be jumped out by impact. Steady the steering arm with a heavy hammer beside the ball joint. Then hit the steering arm, on the part through which the ball joint is fitted. The hit must be a hard blow with a heavy hammer. Note: this hit is from the side. Do not try to drive the ball joint taper out by direct hammer blows, as the threads at the end will be burred over. If the ball joint proves hard to get out either using a separator or a hammer, allow some rust solvent penetrating oil to soak in, then try again. (photo)
5 When reconnecting the tapered ball joint pin to the steering arm, do not grease it.
6 If fitting a new rubber boot to the existing ball joint, wipe away all dirt and old grease. Then smear some molybdenum disulphide grease over the inside of the joint, fill the rubber boot with it, then fit the boot, squeezing it down tight to force out excess grease and get the boot close in to the joint.
7 If fitting a new ball joint, hold the trackrod, undo the

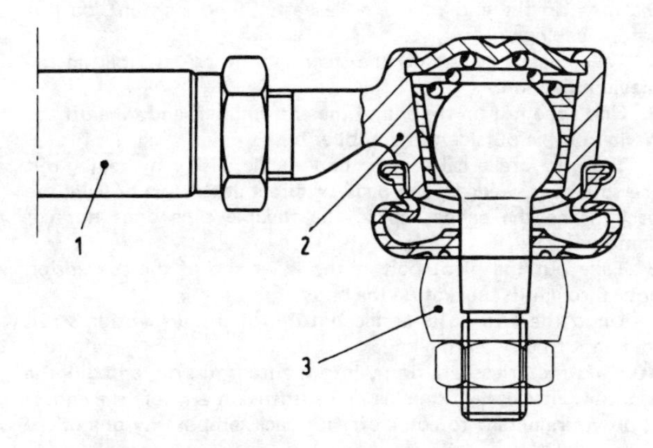

Fig. 10.2. Trackrod ball joint (Sec. 3)

1 Trackrod
2 Ball joint body
3 Steering arm

3.4 Preferably press the taper out of the steering arm. Otherwise it must be given a hard hammer blow on the side

locknut, and unscrew the joint from the rod. Grease the track rod threads before fitting the new joint. Check the new joint's rubber boot is properly fitted. Connect the joint to the steering arm. Then check the front wheel toe-in.

4 Steering knuckle ball joint - removal and replacement

1 The steering ball joint at the swivel for the steering knuckle on the suspension control arm is an integral part of the latter. If it needs renewal, the whole suspension control arm is replaced. The ball joint need not be removed from the steering knuckle unless it is being renewed. At other times it is easier to remove the suspension arm from the car at its inner end than the knuckle from the ball joint, should it be required to remove the steering knuckle for some other purpose.
2 There is very little room between the ball joint and the driveshaft. Therefore the steering knuckle and suspension arm should be removed from the car still joined by the ball joint, and separated on the bench.
3 Jack-up and remove the wheel. The car must be jacked-up from the centre point at the front, so both wheels hang free, otherwise the anti-roll bar will be twisted. Support the car securely under the strong points behind the wheel arches.
4 Get an assistant to press hard on the brake pedal, and undo the nuts on the end of the driveshaft, (if no assistant, do this before jacking-up).
5 Take the ball joint off the steering arm, as described in the previous Section.
6 Undo the nut on the end of the anti-roll bar, and take off the washer on the outside of the rubber bush.
7 Take the brake caliper off the knuckle, either by taking out the split pins, wedges, and pads, or direct at the two bolts at the back. Hang the caliper up so the flexible pipe does not get damaged.
8 Take out the pivot bolt at the inner end of the suspension arm, through its bracket on the body.
9 Undo the two bolts at the bottom of the suspension strut, joining it to the steering knuckle.
10 Hold the driveshaft firmly into the transmission, and pull the assembly of steering knuckle and suspension arm off the end. If it proves reluctant to come off, put back temporarily one of the bolts to hang the knuckle on the suspension strut. Squirt penetrating oil into the splined joint of the driveshaft in the hub. Put an old nut on the end of the shaft to protect the threads, and drive it the first part of its movement, to free it. As the arm comes off the anti-roll bar, secure the rubber bush, and note the shims between the second half of the bush and the bar. These must not be changed, as they adjust the caster.
11 Once the suspension arm is off the car, it is easy to get at the ball joint to separate it from the steering knuckle. Undo the nut on the end of its pin.
12 The pin is tapered where it fits into the knuckle, and the problem of removing it is the same as for the trackrod ball joints, as described in the previous Section. But in this case the use of a proper separator to extract it is even more strongly recommended. The steering knuckle fitting for this ball joint is much stouter than the one at the end of the steering arm, and jumping it out is going to be very difficult.
13 FIAT supply the complete assembly of suspension arm with the ball joint at the outer end, and the rubber bushes pressed and riveted into the inner end. So 'overhaul' is simply a matter of buying the new arm.
14 On reassembly make sure the rubber boot on the ball joint is in position. Grease the splines of the driveshaft, and the pivot bolt at the inner end of the suspension arm, so they all fit easily, and will come off easily another time.
15 Ensure no grease gets on the rubber bushes of the arm and the anti-roll bar. Tighten their nuts after the jack has been lowered, so they are clamped up in the loaded position.
16 Recheck the tightness of all nuts after a road test.
17 Reset the front wheel toe-in.

5 Rack and pinion unit - removal and refitting

1 Inside the car remove the pinch bolt through the bottom end of the steering column, where it joins the pinion.
2 Jack-up the front of the car, and remove both wheels. Make sure the car is secure, as you will be working underneath. Check also there is plenty of room between the car and the garage wall on the right, to bring the rack out sideways.
3 Remove the spare wheel as a general aid to vision.
4 Remove the shield over the transmission (not fitted to some

Fig. 10.3. Steering components on the wheel

1 Nut, tie-rod end joint to steering arm
2 Steering arm
3 Tie-rod

Fig. 10.4. Disconnection points under the car to aid steering gear removal (Sec. 5)

1 Steering gear housing and dust and oil boot
2 Bolts, steering gear housing to body
3 Bolts, crossmember rear end to underbody
4 Gearshifting tie-rod

Fig. 10.5. Steering gear disconnection points inside the car (Sec. 5)

1 *Bolt and nut, lower steering shaft universal joint to pinion shaft*
2 *Gasket cover*
3 *Pinion shaft*

cars).

5 Undo the ball joints at the ends of the trackrods from the steering arms, as described in Section 3.

6 Take out the two bolts from the two 'U' clamps that hold the rack to the car body. Then disconnect the rear end of the engine crossmember and support and lower the engine sufficiently to allow the pinion shaft to emerge from the hole in the bulkhead.

7 Pull the rack towards the engine to disengage the pinion from the steering column. Once it is clear, pass it out to the right side of the car.

8 Before refitting the rack, fill it with oil and check the adjustment of the damping yoke as described in subsequent Sections.

9 Inside the car, turn the steering wheel to the straight-ahead position. Get the rack to its midway position by turning it fully one way, then counting the turns of the pinion, wind it fully to the other extreme. Then bring it back half the number of turns. This must be done accurately, counting to about a 1/16th of a turn. The pinion can be turned by padding its splines, and snapping on a selfgrip wrench of the 'Mole' type.

10 Refit the rack from the right-hand side, and mount it in the clamps, making sure the rubber packing pads are in good condition. Note the clamps and pads are handed left and right.

11 Before reconnecting the trackrods, turn the steering wheel from lock-to-lock, to recheck the column is correctly aligned. Then put back in the pinch bolt at the bottom of the column.

12 Reconnect the engine crossmember support.

6 Rack and pinion unit - overhaul

1 The rack and pinion unit should last for a high mileage, particularly if the rubber boots at the end are never damaged. The most likely need is to reset the damping yoke after a high

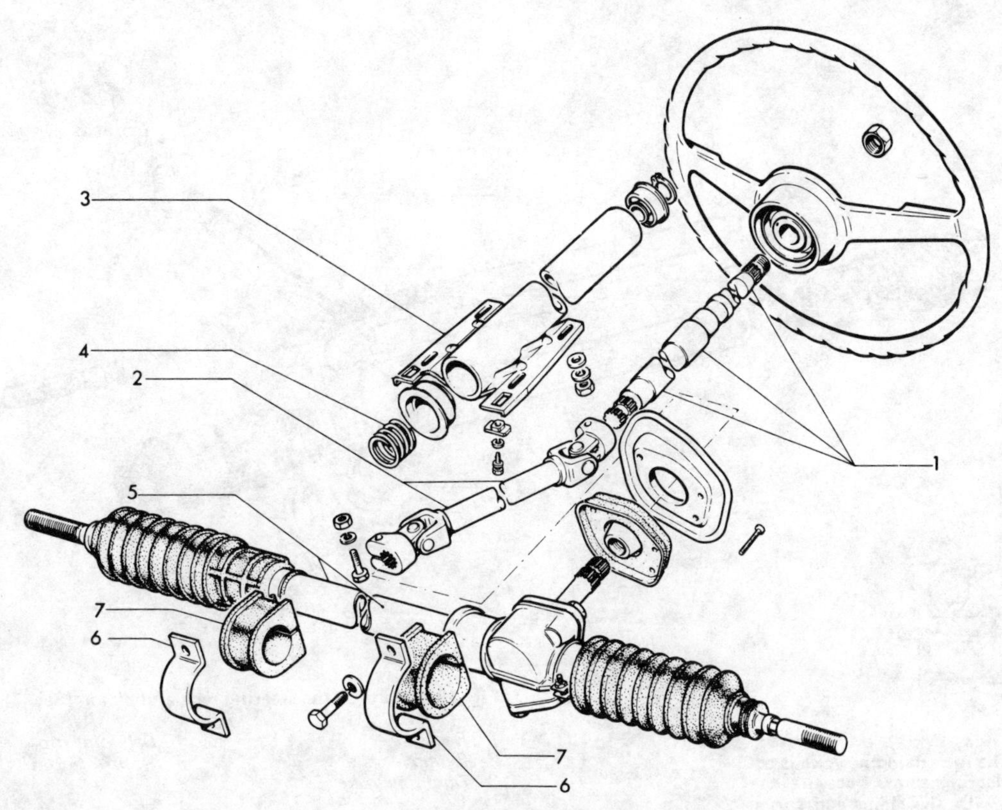

Fig. 10.6. The steering gear

1 *Upper column*
2 *Lower column*
3 *Upper column outer tube*
4 *Bearing spring*
5 *Rack and pinion*
6 *Rack mounting clamps left and right*
7 *Mounting padding, left and right*

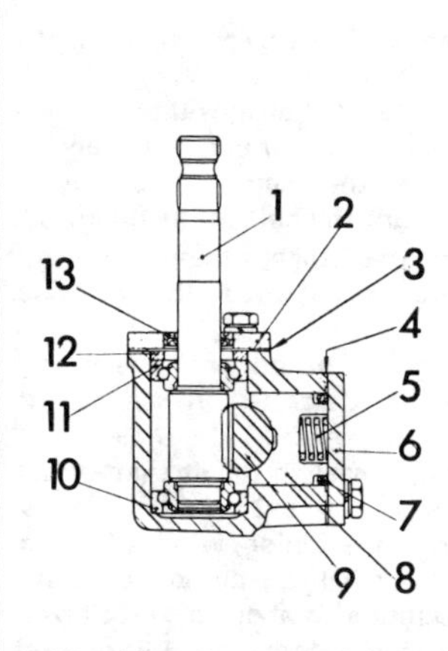

Fig. 10.7. Section through steering rack at pinion and damping yoke (Sec. 6)

1 *Pinion*
2 *Cover*
3 *Gasket*
4 *Yoke shims*
5 *Spring*
6 *Yoke cover*
7 *Seal*
8 *Yoke*
9 *Rack*
10 *Lower bearing*
11 *Top bearing*
12 *Pinion shim*
13 *Pinion seal*

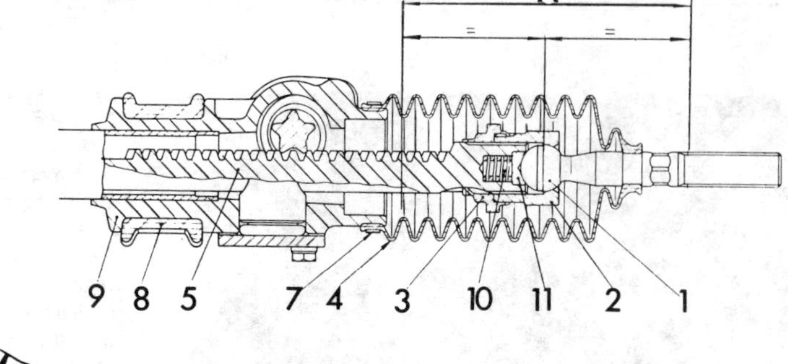

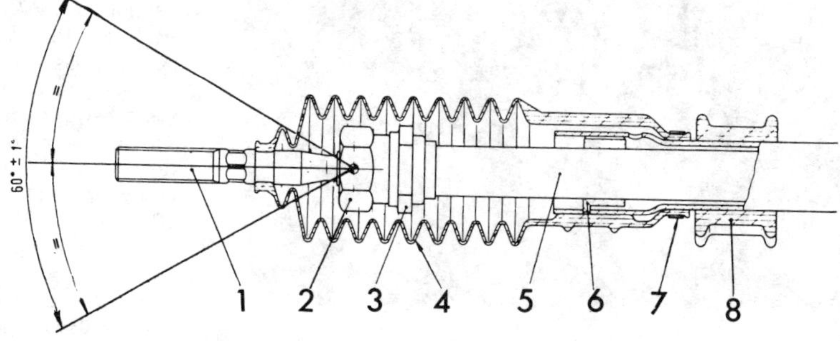

Fig. 10.8. Sections of rack and pinion (Sec. 6)

1 *Ball joint rods*
2 *Adjustable ball joint heads*
3 *Locknuts*
4 *Boots*
5 *Rack*
6 *Rack bushes*
7 *Boot clip*
8 *Rubber mounting pads*
9 *Housing*
10 *Ball joint spring*
11 *Ball seat. Rack travel = A = 5.118 in (130 mm). Ball joint rod solid angle = 60°*

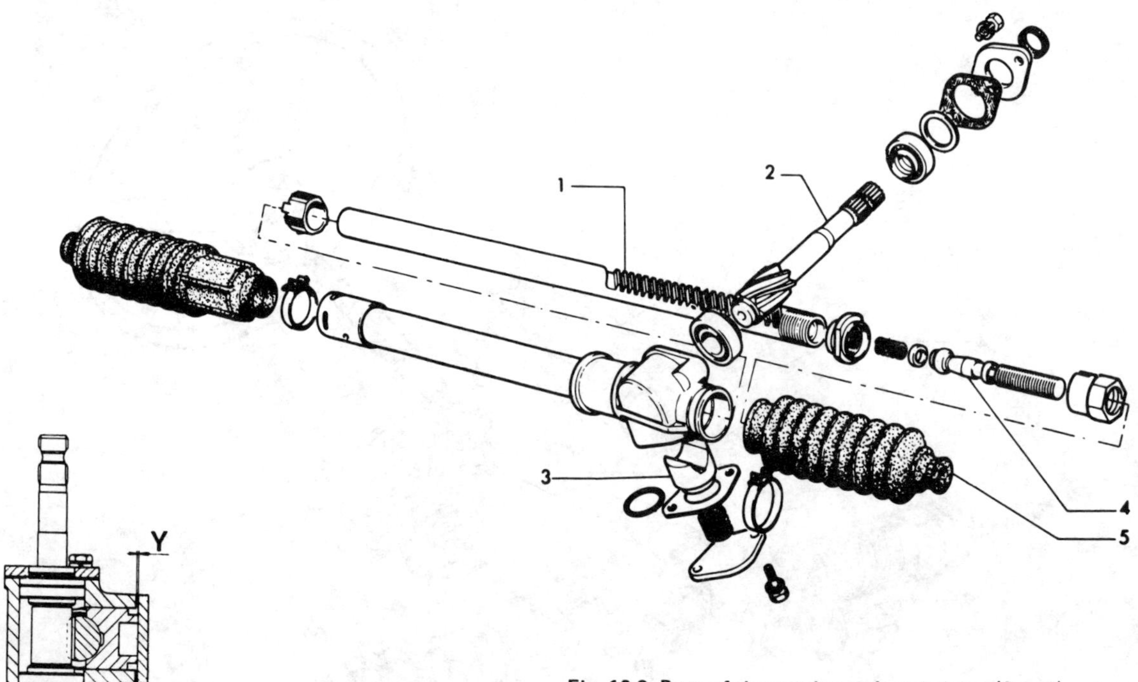

Fig. 10.9. Parts of the steering rack and pinion (Sec. 7)

1 *Rack*
2 *Pinion*
3 *Damping yoke*
4 *Ball joint rod*
5 *Rubber boot (left-hand)*

Fig. 10.9A. The rack damping yoke is adjusted by selecting shims; measure the distance 'Y' with the shims and spring removed. The shim thickness should be Y+ 0.002 to 0.005 in (Y+ 0.05 to 0.13 mm).

mileage. This is described in the next Section. If the rack and pinion and the ball joints at the inner end all seem worn, then it is worth considering replacing the complete assembly. This is certainly worthwhile if there is accident damage.
2 Dismantle the rack by first removing the trackrods from the inner ball joints rods of the rack. Note that there are flats on the latter so they can be held by an open-jawed spanner.
3 Remove the clamps holding the rubber boots to the rack housing.
4 Take off the rubber boots.
5 Take off the clamps that held the rack to the car, and their rubber packing.
6 Take off its cover plate, and remove the rack damping yoke with spring, and shims, and oil sealing ring. (See Fig. 10.7).
7 Remove the two bolts holding the pinion bearing plate to the housing. Take out the pinion, with oil seal, gasket, shim and top ball bearing.
8 Mount the rack housing in a vice, but be careful not to crush it.
9 Unstake the lock for the locknuts for the adjustable head for the ball joints at the ends of the rack. The locknut is the inner one. Undo the locknuts themselves. (See Fig. 10.8).
10 Take off the ends of the rack housing the adjustable heads for the ball joints, bringing with them the ball joints, and their sockets and springs.
11 Slide the rack out of the housing.
12 Take out the lower bearing for the pinion.
13 Check all parts for signs of excessive wear or damage, or corrosion. New gaskets, a new seal for the pinion shaft, and new rubber boots should be fitted.
14 As all parts are refitted, lubricate them thoroughly with an SAE 90 EP oil. Take care the rack teeth do not scrape the housing bearings as they are slid in.
15 Fit the pinion bearing plate with too many shims, and screw in its bolts gently to settle bearings. Release the bolts until they are finger-tight. Measure the gap between the housing and plate. Measure with a micrometer the thickness of the gasket. Calculate the thickness of shims needed. The shims should be thicker than the gap between housing and plate, allowing for the gasket, by 0.001 - 0.005 in. (0.025 - 0.13 mm). Shims are available in four thicknesses from 0.12 - 2.5 mm. The pinion should be able to turn freely, without jerks, and without stiffness exceeding 0.3 lb f ft (0.04 kg fm). There should not be any endfloat.
16 Reassemble the ball joints to the ends of the rack. Tighten the adjustable heads till a torque of 1.5 to 3.25 lb f ft (0.2 to 0.5 kg fm) is needed to turn the ball joints in their seats. Check that the trackrods can rotate through the solid angle of a cone of 60°. Then lock the heads with the locknuts, and stake them in position.
17 Adjust the damping yoke, and refill with oil as described in subsequent Sections.

7 Rack and pinion unit - damping yoke adjustment

1 The yoke in the rack housing presses the rack into mesh with the pinion. This cuts out any backlash between the gears. Also due to its pressure it introduces some stiffness into the rack, which cuts out excessive reaction from the road to the steering wheel.
2 In due course, wear reduces the pressure exerted by the damping yoke. The pressure is controlled by the yoke cover plate and a spring.
3 The yoke setting should be reset if the rack has been dismantled for overhaul.
4 The need for resetting of the yoke if the car has run a long mileage, but the rack is not being dismantled, is not easy to detect. On bumpy roads the shock induced through the steering will give a feeling of play, and sometimes faint clonking can be heard. In extreme cases free-play in the steering may be felt, though this is rare. If the steering is compared with that of a new rack on another car, the lack of friction damping is quite apparent in the ease of movement of the steering wheel of the worn one.
5 Access to the yoke with the rack fitted is a little difficult, and if the car is dirty underneath, there is risk of this getting into the rack. So it is much easier done on the bench, removing the race as described in Section 5.
6 Turn the steering to the straight-ahead position.
7 Take the cover plate off the damping yoke, remove the spring and shims, and refit it. Refit the bolts, but only tighten them enough to hold the yoke firmly against the rack.
8 Turn the pinion through 180° either way to settle the rack.
9 Measure the gap between the cover plate and the rack housing. (See Fig. 10.9A).
10 Select shims to a thickness 0.002 to 0.005 in. (0.05 to 0.13 mm) more than the measured gap. (Shims are available 0.1 and 0.15 mm thick).
11 Remove the cover plate again, and refit the spring.
12 Smear each shim with soft setting gasket compound, and fit them and the cover plate.

8 Rack and pinion unit - lubrication

1 The steering rack and pinion unit is lubricated for 'life'. If it is dismantled, it must be reassembled with a new supply.
2 The lubricant is gear oil of SAE 90 EP, such as Duckhams Hypoid 90. The amount is 0.14 litre (0.24 Imp. pint, 0.3 US pint, 4¾ fl oz).
3 The oil is inserted into the rack at the pinion end, that is the driver's end. If this is being done on the bench before refitting the rack to the car, fit the rubber boot to the other end, pour in the oil at the pinion end, then fit that boot. Turn the rack to-and-fro to spread the oil and check the boots are not being pressurised by excess oil.
4 If there has been a leak, and it is required to refill the rack on the car, undo the rubber boots and allow any old oil left to drain out, so there will not be too much in it after refilling.
5 Refit the boot on the passenger's side. Jack-up the car very high on the driver's side.
6 Pour the oil into the end of the rack and the boot, and quickly refit the boot to prevent any dribbling out.
7 Turn the rack from lock-to-lock to distribute the oil and check the rubber boots are relaxed, and showing no signs of the rack being over full.
8 Refit the clamping clips to the large ends of the rubber boots. Position the tightening screws, if doing this on the bench, so their heads can be reached from below after the rack has been refitted to the car.

9 Toe-in (front wheels) - checking and adjustment

1 Garages have accurate gauges to measure all aspects of steering geometry. Even so, there is possibility of error. Measuring at home is far less accurate, but time is at less of a premium, so it can be done often enough to get several consistent readings to know that the setting is correct.
2 If the car is old, and you have not set its steering yourself, the steering wheel and trackrods may be off centre. It must be established that the steering rack is in the centre of its travel when the wheels are straight, and that then the steering wheel is 'straight' too.
3 Make or obtain an alignment gauge. One can be made up from a length of tubing or bar, cranked to clear the engine/transmission and having a screw and locknut at one end.
4 Use the gauge to measure the distance between the inner roadwheel rims at hub height at the rear of the wheels.
5 Push the car forward so that the wheels rotate through 180° (half a turn) and, using the gauge, measure the distance between the inner rims again at hub height, at the front of the wheels. This last measurement should differ from the first and represents the toe-in or toe-out given in Specifications.
6 Where new components have been fitted or the front wheel alignment is grossly out of adjustment, set the front roadwheels

parallel with the trackrods of exactly equal length before checking with the gauge. With the steering centred in this way, check that the steering wheel is in the 'straight ahead' position. If not, remove the wheel (Section 11) and move it round a few splines until it is correctly positioned.

7 Adjustment is made by screwing the rod of the ball joint on the end of the rack into or out of the trackrod. Hold the trackrod to prevent it turning, and undo the locknut. Put a spanner on the rack ball joint rod, and turn that. If the rubber boot wrinkles up, turn this back to its original position. Carefully tighten the locknut so the setting is not disturbed.

8 Mention has been made of ensuring the steering wheel is 'straight' before adjusting the toe-in. It will probably prove impossible to get it quite straight. Also, different road cambers, or side winds, will affect the amount of steering applied. A careful note should be made of the steering wheel angle when driving on an absolutely straight and camber free road, without any wind. Then if a kerb is nudged, and the toe-in perhaps altered, the steering wheel angle can be checked to give a guide as whether there is any misalignment.

10 Camber and castor (front wheels) - general

1 It is not practical for the owner to check front wheel camber. Camber is the amount the top of the wheel slopes out. It is not adjustable, and will normally only be upset by accident damage. Realignment of the body and suspension mounting points must then be done by a firm with the jigs, measuring equipment, and experience of such work.

2 Again, it is not practicable for the owner to do anything about castor. Castor is the arrangement whereby the ground contact point of the wheel is behind its centre of pivot for steering, so the wheel is self-centering. Castor is adjusted by shifting the lower mounting forward or back, in relation to the top mounting in the mudguard. This shift is achieved by altering the shims at the end of the anti-roll bar. Such an adjustment is easily made. The difficulty with castor is the actual measurement of the angle. Small variations of castor are of minor consequence. The owner can limit these by always ensuring the same shims are kept on the end of the anti-roll bar. Again, after accident damage, the FIAT agent's skill and equipment should be used.

11 Steering wheel - removal

1 Undo the two screws on the underside of the steering wheel. The horn button and cover should now be loose. On some models the horn button and cover are simply prised off.

2 Unclip the cover from the button and then remove each, separately.

3 Undo the nut at the centre of the steering wheel.

4 Pull the wheel off the column. Under no circumstances use force to drive or tap the wheel from the column. Use a suitable puller if necessary.

12 Upper steering column - removal and replacement

1 The steering column is in two sections. The top part is mounted in two ball bearings. The lower part has a pair of universal joints and connects the upper column to the steering pinion. On later models the ball bearings in the upper shaft are replaced by rubber bushes, and the upper shaft, support plate and housing are very slightly modified.

2 To remove the upper column, first disconnect a battery lead. Then remove the steering wheel, as described in the previous Section.

3 Release the screw that secures the instrument panel cluster, lift away the cluster and disconnect the electrical connectors and speedometer cable behind it.

4 Remove the two screws that secure the upper and lower column shrouds and the single screw that locates the turn signal selector unit.

5 Ease the turn signal selector unit off of the upper column and then remove the upper and lower shrouds.

6 Remove the pinch bolt through the universal joint where the two parts of the column meet.

7 Remove the nuts and bolts holding the column supporting bracket, holding the upper column by hand.

8 Move the upper column away, and pull it off the lower column. Catch the spring and ring below the bottom bearing. These are not fitted on later models that have bushes in lieu of bearings.

9 On the bench, take off the split ring at the top end of the upper column. Then push the column first one way then the other within the outer column, to get the bearings out. An alternative is to prise the bearings out; this also applies to the later models fitted with bushes.

10 Before reassembly, on early models, grease the bearings. On all models it is necessary to lubricate the splined coupling between the steering wheel and upper shaft with graphited oil.

11 Ensure the wheels are in the straight-ahead position, then engage the splines on the lower column with the wheel straight. If it is found difficult to reassemble due to the spring at the bottom end of the upper shaft, leave off the circlip at the top end until after the upper and lower columns are rejoined.

13 Lower steering column - removal

1 Once the upper steering column has been removed, the lower follows easily.

2 Take out the pinch bolt securing the bottom universal joint to the pinion shaft of the steering gear, then pull the shaft off.

14 Steering column lock - general

1 On some models the ignition switch is combined with a steering lock.

Fig. 10.10. Disconnection points to commence removing steering column upper shaft (Sec. 12)

1 Nut, steering wheel to shaft
2 Screw, instrument cluster to panel
3 Screw, upper, steering column
4 Screw, turn signal selector unit
5 Screw, lower, steering column

2 If the lock jams, try turning the column gently, to take the load off the locking pawl, whilst working the key. Give the lock and column a squirt of oil from an aerosol of easing oil.

3 If the lock has to be changed, difficulty will be experienced getting out the screws, as these are 'burglar proof'. When the switch is fitted, the centre of three screws is tightened until its head shears off. To remove it, a hole must be drilled down the bolt, and then an 'Easiout' inserted, to screw it off. If no 'Easiout' is available, the old bolt must be completely drilled away.

Fig. 10.11. Removing the turn signal indicator switch (Sec. 12)

1 Electric connections, located under the cluster
2 Turn signal selector unit

Fig. 10.12. Removing the upper steering shaft support (Sec. 12)

1 Nuts and bolts; steering shaft support to instrument panel
2 Nut and bolt; upper steering shaft to upper universal joint yoke

SECTION A-A

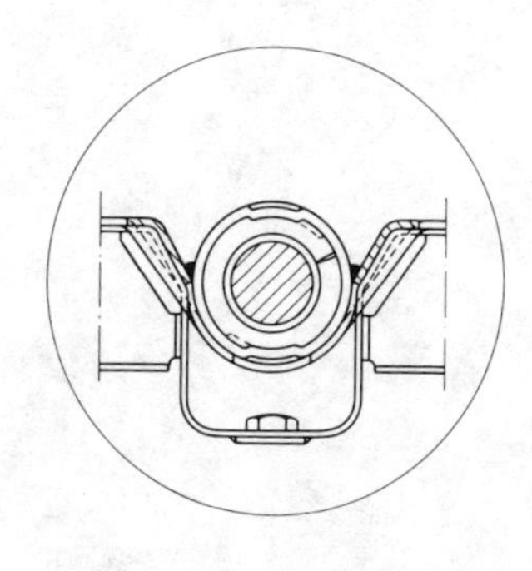

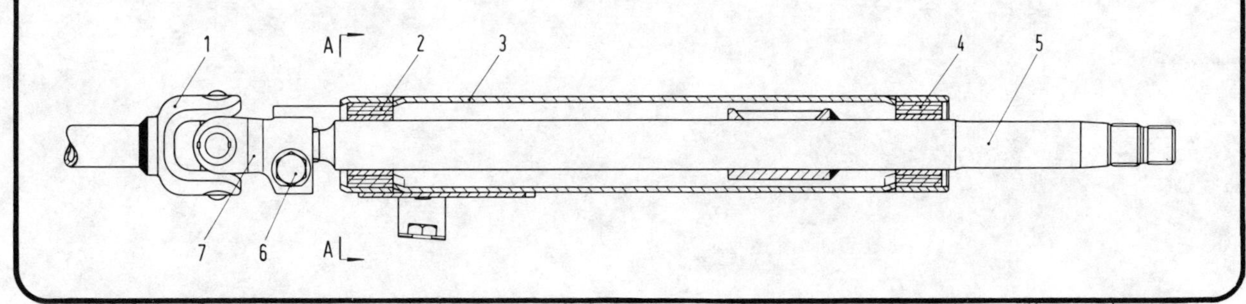

Fig. 10.13 Section through the upper steering shaft

1 Lower steering shaft
2 Lower rubber bushing
3 Column
4 Upper rubber bushing
5 Upper steering shaft
6 Bolt and nut, upper steering shaft to upper universal joint yoke
7 Upper universal joint yoke

15 Fault diagnosis - steering

Before assuming that the steering mechanism is at fault when mishandling is experienced make sure that the trouble is not caused by:

1 Binding brakes
2 Incorrect mix of radial and crossply tyres
3 Incorrect tyre pressures
4 Misalignment of the body and rear suspension

Symptom	Reason/s	Remedy
Steering wheel can be moved before any sign of movement of the wheels is apparent	Wear in the steering linkage, gear and column coupling	Check movement in all joints and steering gear and overhaul and renew as required.
Vehicle difficult to steer in a consistent straight line; wandering: unstable on corners	As above	As above.
	Wheel alignment incorrect (indicated by excessive or uneven tyre wear)	Check toe-in.
	Rear suspension toe-in camber wrong	See Chapter 11.
	Front wheel hub bearings loose or worn	Adjust or renew as necessary.
	Worn ball joints, trackrods or suspension arms	Renew as necessary.
Steering stiff and heavy	Incorrect wheel alignment (indicated by excessive or uneven tyre wear)	Check wheel alignment.
	Excessive wear or seizure in one or more of the joints in the steering linkage or suspension arm ball joints	Renew as necessary.
	Excessive wear in the steering gear unit	Dismantle, check, relubricate the rack and pinion.
Wheel wobble and vibration	Roadwheels out of balance	Balance wheels. (See Chapter 11).
	Roadwheels buckled	Check for damage.
Excessive pitching and rolling on corners and during braking	Defective shock absorbers	Check and renew as necessary. (See Chapter 11).

Chapter 11 Suspension

For modifications, and information applicable to later models, see Supplement at end of manual

Contents

Specifications

Front suspension

Type	Independent: Mac Pherson struts and anti-roll bar
Spring type	Coil
Spring length:	
Yellow daubed springs	9.252 in. (235 mm) or more loaded at 551 lb (250 kg)
Green daubed springs	9.252 in. (235 mm) or less loaded at 551 lb (250 kg)
Minimum load to achieve above length	507 lb (230 kg)
Shock absorber type	Telescopic, double acting (within struts)

Rear suspension

Type	Independent: Transverse leaf spring
Spring type	Two leaf with interspaced rubber blocks
Shock absorber type	Telescopic, double acting
Rear hub bearing endfloat (maximum)	0.0016 inch (0.04 mm)

Wheels (front)

Hub bearing lubricant type	Multi-purpose lithium based grease (Duckhams LB 10)

Wheels (rear)

Camber:	
Laden car*	-2° 30' to -3° 30'
Unladen car	0° to 1°
Toe-in:	
Laden car*	0.157 to 0.315 in (4 to 8 mm)
Unladen car	0.118 to 0.275 in (3 to 7 mm)
Track (rear)	51.0 in (1295 mm)

**Laden is with 4 occupants and 88 lb (40 kg) luggage*

Wheels (all)

Type	4.00 x 13: Pressed steel disc

Tyres (all)

Size and type	135 SR-13: Radial ply
Tyre pressures (2-door):	
Front	24.2 psi (1.7 kg/cm^2)
Rear	27 psi (1.9 kg/cm^2), heavy loads 31 psi (2.2 kg/cm^2)
Tyre pressures (3-door):	
Front	24.2 psi (1.7 kg/cm^2)
Rear †	27 psi (1.9 kg/cm^2), heavy loads 31 psi (2.2 kg/cm^2)

† With one occupant + 330 kg (728 lbs) the rear figure should be 31 psi (2.2 kg/sq cm).

Torque wrench settings	lbf ft	kgf m
Wheel nuts	51	7
Front and rear hub nuts:		
M20 x 1.5	116	16
M18 x 1.5	101	14
Front suspension		
Front suspension arm to body	18	2.5
Front suspension balljoint/steering knuckle	58	8
Anti-roll bar/front suspension arm	43	6
Anti-roll bar/body	22	3
Front strut spindle (M10 nut)	18	2.5
Front suspension strut/steering knuckle	43	6
Front suspension strut mounting (M6 nuts)	7	1
Front wheel bearing ring nut	43	6
Steering balljoint/track rod	36	5
Balljoint/steering arm	25	3.5
Brake caliper/steering knuckle	36	5
Rear suspension		
Rear suspension strut spindle nut	18	2.5
Rubber bush/rear suspension arm pin nut	32.5	4.5
Rear suspension arm/hub spindle	58	8
Rear suspension arm pivot pin/body	36	5
Rear suspension strut/body upper mounting	18	2.5
Rear suspension strut/hub upper mounting	43	6
Leaf spring/control arm insulator nuts	22	3

1 General description

1 The suspension is independent all round. There is an anti-roll bar at the front, which also acts as the fore and aft tie-rod for the suspension. The rear suspension uses one transverse leaf spring. This is not mounted at its centre, but in two clamps to each side. This gives an anti-roll action to the spring.
2 The rear suspension should seldom need attention. But when it does, the setting of the camber and toe-in is important, otherwise the steering effect of the rear wheels will affect the handling, and wear might be bad. Dismantling of the front suspension is sometimes needed, not because it wears, but to remove the driveshafts, or allow the transmission to be taken off the engine.
3 The handling of the car, and its reaction to various road surfaces, have been specifically designed for the use of radial ply tyres. Tyres of other types **should not** be used.

2 Suspension - inspection

1 Checking the suspension is part of the 6,000 mile (10,000 km) task.
2 Inspect the outside of the shock absorbers for leaks and check their operation by bouncing.
3 With the car parked on level ground check that it sits level from left to right, and does not appear to be drooping at one end, particularly down at the back.
4 Examine all the rubber bushes of the suspension arms and rear spring mountings. The rubber should be firm, not softened by oil or cracked by weathering. The pin pivoted in the bush should be held central, and not able to make metal to metal contact.
5 Check the outside of the springs. If rusting, they should be sprayed with oil.
6 Check the tightness of all nuts, particularly those holding the front suspension strut to the steering knuckle, and the rear shock absorber to the hub.
7 Grip the top of each wheel, in turn, and rock vigorously. Any looseness in the bearings, or the suspension can be felt. Failed rubber bushes giving metal-to-metal contact can be heard.

3 Suspension (front) - dismantling and reassembly

1 Sometimes the front suspension will need taking off to release the driveshafts. In that case there is no need to remove the front suspension strut. At other times, the way work is tackled will depend on whether a steering ball joint removing tool is available. If it is, the steering knuckle can be readily detached from the suspension arm. If not then it is going to prove easier to remove the arm from the car, then take the arm off the knuckle. Again, it is easier to take the anti-roll bar off the suspension arm if the arms are removed from both sides, and the anti-roll bar from the car, but this assumes both sides need attention. In tackling a particular job, only go as far as is necessary, depending on the work needed and the tools and space available.
2 Before jacking-up, slacken the nuts on the outer ends of the driveshafts.
3 Jack-up the front of the car quite high, and support it firmly under the reinforced points behind the wheel arches, and the central jacking point at the front; make sure it is secure. Remove the front wheels.
4 If a front strut is to be removed without dismantling anything else, undo the two bolts at the bottom of the suspension strut (photo), that hold it to the steering knuckle. Then undo the three nuts on the studs through the wing. Lift out the strut.
5 To completely dismantle the suspension leave the strut on until later.
6 Disconnect the steering ball joint on the outer end of the trackrod from the steering arm. See Chapter 10, Section 4 for methods of freeing these.
7 Take the brake caliper off. Either pull out the two spring clips and the wedges, or undo the two bolts holding the whole assembly to the knuckle. See Chapter 8 for details. Tie the caliper up with string so the flexible pipe will not get pulled.
8 Take out the bolt through the inner end pivot of the suspension arm, at the bracket on the body. (photo).
9 Undo the two bolts at the bottom of the suspension strut that hold it to the steering knuckle.
10 Detach the front suspension on the other side of the car in the same way.
11 Remove the two driveshafts from the hubs as described in Chapter 7. If leaving their inner ends in the transmission, tie them in. Then temporarily put back one bolt to hold the knuckle to the bottom of the suspension strut.
12 Undo the two brackets at the front of the car in which the anti-roll bar pivots. Each bracket is held only by two bolts, one of which is tapped into the chassis while the other is retained by a nut.

3.4 To remove the front strut it is only necessary to undo the two bolts at the bottom

3.8 The rest of the suspension is released by taking out the pivot bolt at the inner end of the suspension arm, and then dismounting the antiroll bar

Fig. 11.1. Left-hand front suspension - internal view

1 *Anti-roll bar threaded end*
2 *Control arm*
3 *Nut and bolt, control arm to body*
4 *Nut, steering tie rod to knuckle*
5 *Hydraulic shock absorber*
6 *Bolts, anti-roll bar to body*
7 *Anti-roll bar*
8 *Anti-roll bar support bracket*

Fig. 11.2. Left-hand front suspension - external view

1 *Nut, steering tie-rod to steering arm*
2 *Bolt and nut, control arm to body*
3 *Control arm*
4 *Anti-roll bar threaded end*

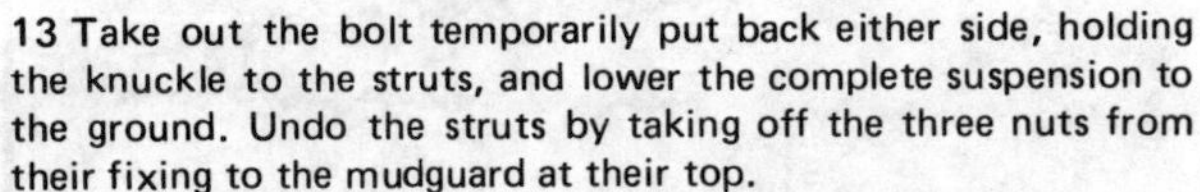

13 Take out the bolt temporarily put back either side, holding the knuckle to the struts, and lower the complete suspension to the ground. Undo the struts by taking off the three nuts from their fixing to the mudguard at their top.

14 Now the suspension arms can be taken off the anti-roll bar. Undo the nut on the end of the bar. Pull off the arm, noting the position of the washers, bushes and the shims. The shims control the castor angle and it is most important that they are replaced exactly as removed (See Chapter 10).

15 Take the steering knuckle off the arm, again referring to Chapter 10, Section 4.

16 Reassembly is the reverse process. Grease all bolts before fitting them, so they will be easier to strip next time. If any of the rubber bushes have been renewed, check the toe-in. (See Chapter 10). After road test, recheck all bolts for tightness.

17 When refitting the bolt at the inner end of the suspension arm, do not tighten it until the jack has been lowered, so the bush will be clamped in its relaxed position at the static height and the degree of twist in it will be minimised.

4 Suspension spring or shock absorber (front) - renewal

1 The front suspension strut consists of a concentric spring and shock absorber. As an assembly the strut is simplicity itself to remove, as described in the last Section. Only the bolts at the bottom and the top have to be removed, not the centre spindle nut, but those at the top are for its mounting to the mudguard. When removed from the car, the shock absorber will be held at full extension by the spring.

2 To release the spring the load must be taken off the shock absorber, then the nut can be removed from the shock absorber's central rod and the spring allowed to expand. This must be done in a special spring compressor. Without the use of one of these, the nut must not be taken off the shock absorber rod, lest the spring fly out uncontrollably. Many makes of car have MacPherson strut suspension, so if a FIAT agent is not available, some other agent should be able to provide one. It has been known for an over-enthusiastic owner to attempt to restrain the springs with wire ties. Don't do this; it really is fraught with danger.
3 The most common reason for needing to separate the two will be to replace the shock absorber, as this has a shorter life than the spring. Even so, the spring should be carefully examined when free, as should all the components of the bush at the top, and the spring seat at the bottom.
4 Springs come in two grades of length. It will usually be found that the one on the driver's side will have settled more than the other.
5 When replacing a faulty spring, ideally both springs should be replaced simultaneously, but in any case the colour codes of replacement springs must match.

5 Suspension arm bushes (front) - renewal

1 The rubber bushes at the inner end of the suspension control arm should be renewed if they perish or soften, which will show in the way they curl round at the ends, and the way the arm is not held centrally on the bolt.
2 The metal central tube and end washers are peened into position.
3 This peening is done in a powerful press with a load of 1 ton. Unless this is done properly the bush as a whole will be too long, and it will prove difficult, or impossible, to get it into its mounting. The FIAT agent should therefore be given the job of putting in the new bushes.
4 However, if the rubber bushes need renewal, the condition of the ball joint at the steering knuckle must be suspect and it may well be worth replacing that too. The complete assembly of arm, knuckle, with the rubber bushes pressed in properly, is serviced as a spare by FIAT. See Section 4 in the previous Chapter.

6 Hub bearings (front) - overhaul

1 Normally the front wheels should be silent, have negligible rim-lock, and turn smoothly. Incipient failure of the bearings is evidenced by noise, excessive play and even bad steering if they are particularly worn. The bearings are sealed for life, are a press fit in the hub, and need such force to extract them that this should only be attempted when the bearings are going to be renewed, since they are surê to be damaged when being pressed out.
2 The construction of the hub is shown in Fig. 11.5.
3 First dismantle the suspension, as described in Section 3 to remove the steering knuckle from the car.
4 Then take the brake disc off the hub by undoing the standard bolt and long-headed wheel aligning bolt.
5 Now the hub must be pressed out of the bearings. It is a tight

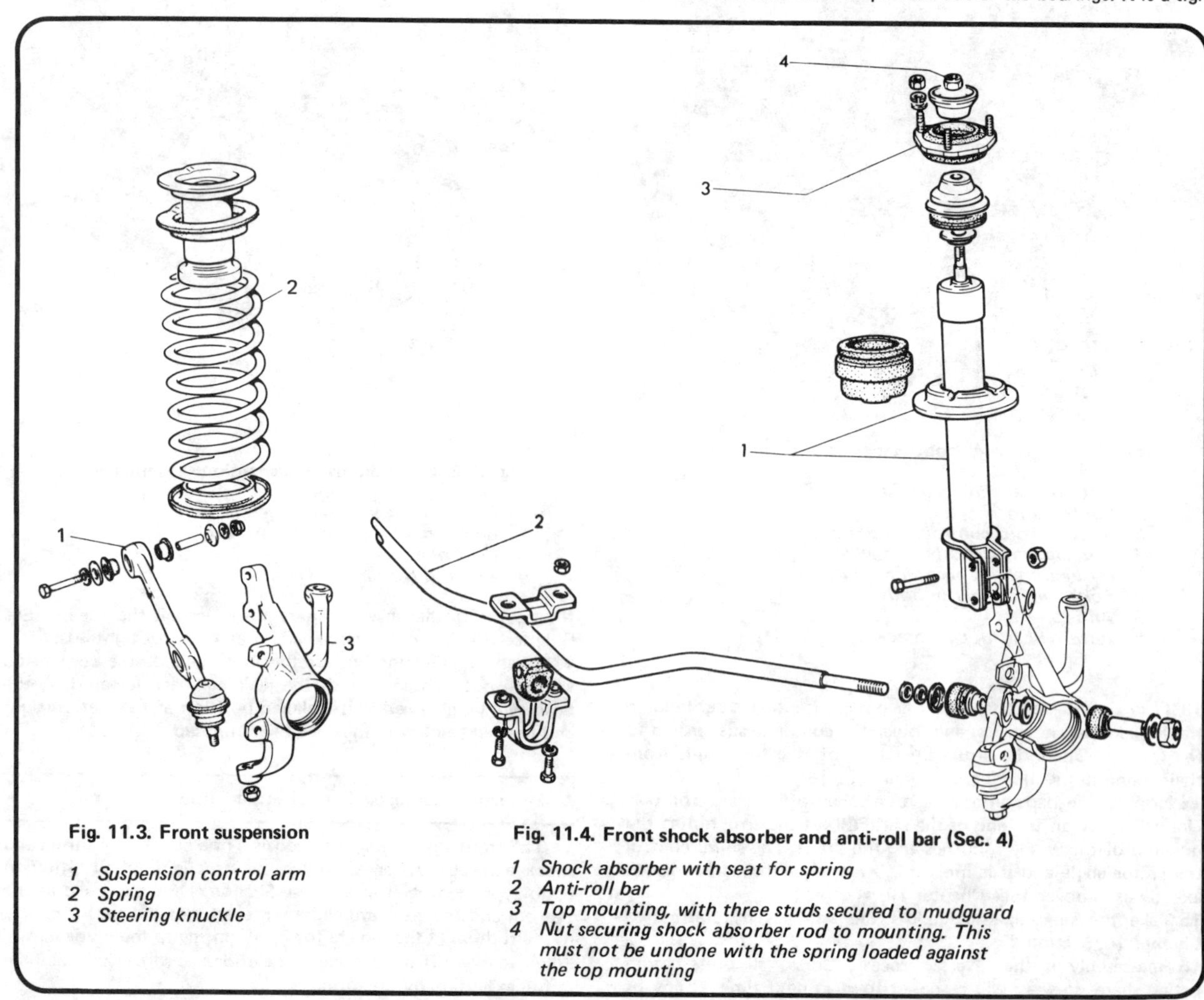

Fig. 11.3. Front suspension

1 *Suspension control arm*
2 *Spring*
3 *Steering knuckle*

Fig. 11.4. Front shock absorber and anti-roll bar (Sec. 4)

1 *Shock absorber with seat for spring*
2 *Anti-roll bar*
3 *Top mounting, with three studs secured to mudguard*
4 *Nut securing shock absorber rod to mounting. This must not be undone with the spring loaded against the top mounting*

fit, and may have rusted in. A good soak in penetrating oil will help. The steering knuckle must be well supported whilst the hub is pressed from the inside. The press mandrel must only rest on the hub, not the bearing inner race. Again, the help of a FIAT agent is recommended, as they have the special press tools.

6 Once the hub is out, the bearing locking collar on the outside can be reached. This is screwed in, and should have been staked in position with a punch. The staking should be cut away, and the nut undone. Again, a press is needed to get out the old bearings.

7 Clean and rub off all rust from the knuckle bearing housing, and the hub.

8 Press in the new bearings, only doing so on the outer race. Use a new bearing nut, and stake it in position.

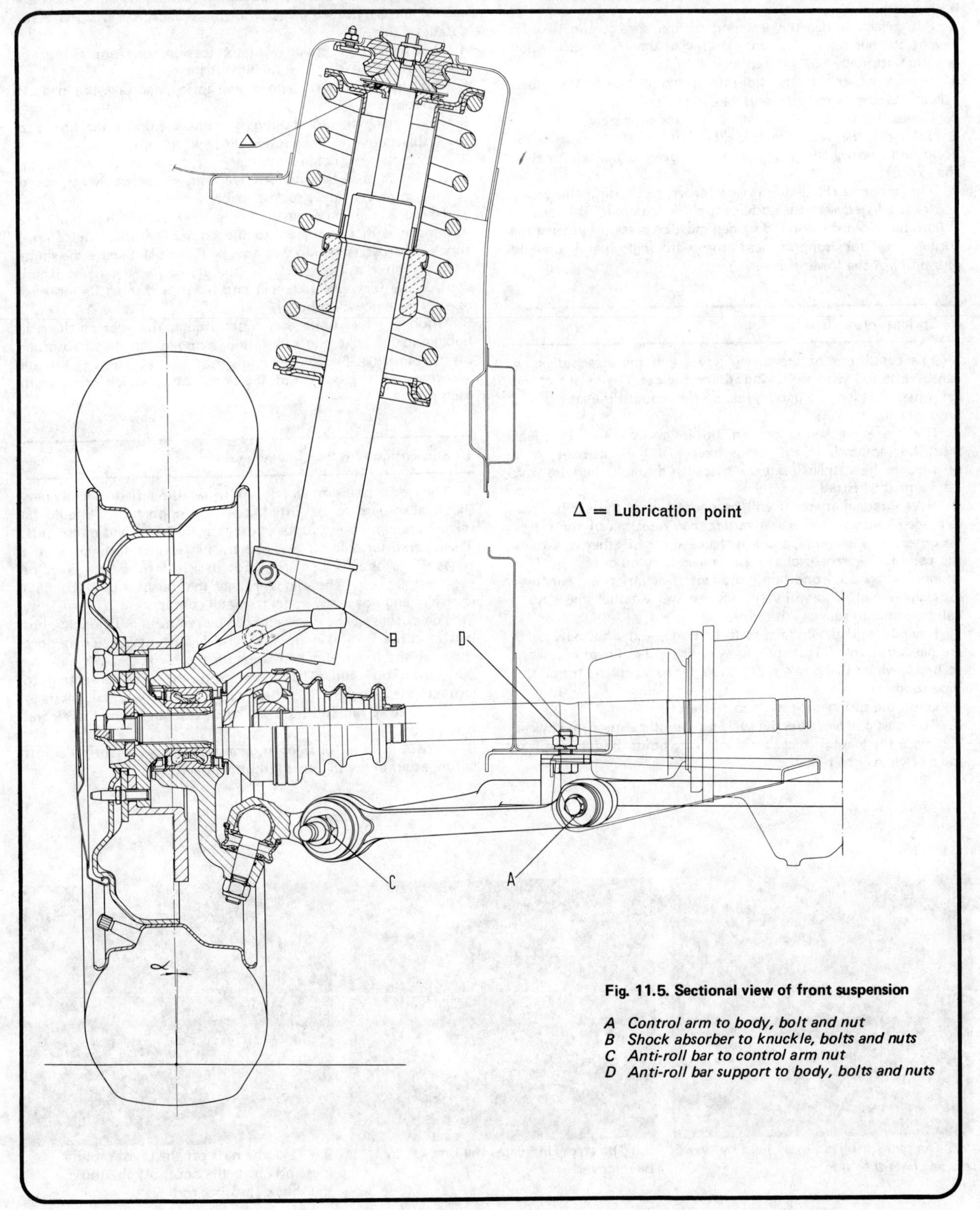

Fig. 11.5. Sectional view of front suspension

A Control arm to body, bolt and nut
B Shock absorber to knuckle, bolts and nuts
C Anti-roll bar to control arm nut
D Anti-roll bar support to body, bolts and nuts

7 Springs (rear) - removal and refitting

1 The rear spring can be removed without disturbing the rest of the rear suspension.
2 Jack-up, and block the body firmly on stands under the body strong points to support it safely so that work is comfortable underneath.
3 Put a jack under the left end of the spring, and take its weight so that no load is on the bracket under the suspension arm that normally holds the spring.
4 Undo the two bolts that hold the spring bracket to the underside of the suspension arm, and take it off. (photo)
5 Repeat for the bracket on the other side of the car.
6 Take off the two inner brackets holding the spring to the body and remove the spring, passing it over to the left to clear the exhaust.
7 The spring is refitted in reverse order, but ensure the spring centre is aligned with the body centreline. Maximum tolerance is ± 0.08 in. (± 2 mm). An 0.8 in. gap must be present between the spring insulator support and the right-angle bend on the extremity of the lower leaf spring.

8 Springs (rear) - overhaul

1 The conditions of the spring leaves and the interleaf lining can be assessed with the spring still on the car. The height of the car when unladen is a good guide of the amount the leaves may have settled.
2 If a leaf is cracked or broken, this is easy to see, and the leaf must be replaced. If the leaves have lost their camber, it is possible to have them reset by specialist firms. Otherwise new leaves must be fitted.
3 New interleaf material will be needed after the car has run extended mileage, somewhat earlier than resetting of the leaves becomes really essential, and it is the need for this that will make the complete overhaul of the spring necessary on older cars. If it is not in good condition, the interleaf friction becomes excessively high, and will vary between wet weather and dry; it will be prone to squeak when dry.
4 Having decided what needs to be done, and what new parts are needed, don't forget some new spring clips. These are likely to break whilst being taken off. Also get new rubbers for all the mountings.
5 Undo the spring clips and part the leaves.
6 The spring leaves' rubbing surfaces should be smooth. If wear has made the leaves noticeably thin at any point the relevant leaf must obviously be replaced.

9 Control arms and shock absorbers (rear suspension) - removal

1 Jack-up the rear of the car and support it on blocks under the body strong points.
2 Remove the wheel.
3 Disconnect the brake flexible pipe at the body end. Undo the end of the handbrake cable from the brakes. See Chapter 8 for details of this.
4 Disconnect the brake regulator torsion bar from its bracket on the left suspension arm taking out the spring pin.
5 Put a jack under the end of the spring, and take the load off the suspension.
6 Undo the two nuts holding the bracket under the spring to the suspension arm. Then lower the jack. (photo)
7 Inside the car, behind the rear seat, take the nut off the top of the shock absorber rod. An assistant will probably be needed outside to prevent the shock absorber from turning.
8 Remove the nuts on the top two studs that hold the suspension arm pivot shaft to the body. The assembly of arm, shock absorber, and hub can now be lifted off. Secure the shims that are on the two studs on the body for the pivot: these control the rear wheel camber and toe-in and must be replaced exactly as removed. (photo)
9 Undo the two bolts that pass through the eyes on the rear hub knuckle. Take the suspension arm and the shock absorber off the knuckle. Secure the shims that were between the shock absorber and the bushes of the arm, noting which came from each side.

10 Suspension arm (rear) - overhaul

1 The rear suspension will normally be dismantled to allow new shock absorbers to be fitted. Such an opportunity should be taken to examine the rubber bushes at the inner and outer ends. Their condition can be assessed by the rubber showing at the edges. This should be firm: not perished by weathering, or softened by oil. The pin through the bush should be held centrally and not allow metal to metal contact.
2 The bushes at the outer end can be removed individually, but will be quite firmly in position and need pressing. To remove those at the inner end first press on the pivot to push out one bush from one end. Then press from the other direction to extract the second. The first one should only be partially extracted so it can still hold the pivot straight whilst the second is removed.
3 Inspect the arm for signs of distortion, or bad rusting. Paint it before assembly with a rust inhibiting paint.

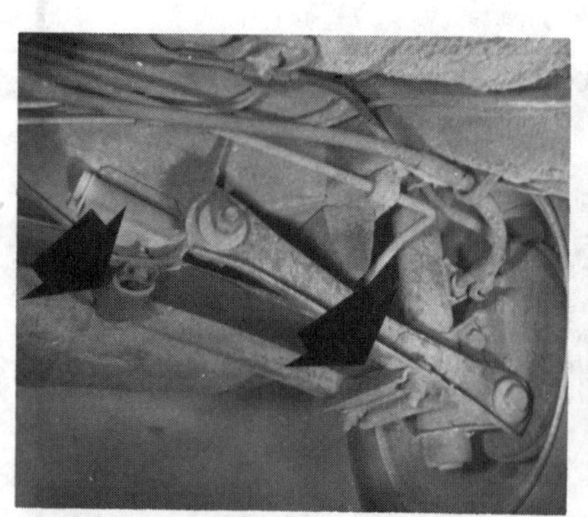

7.4 The rear spring is simply held by two clamps on either side

9.6 The arrow indicates the bracket to be removed

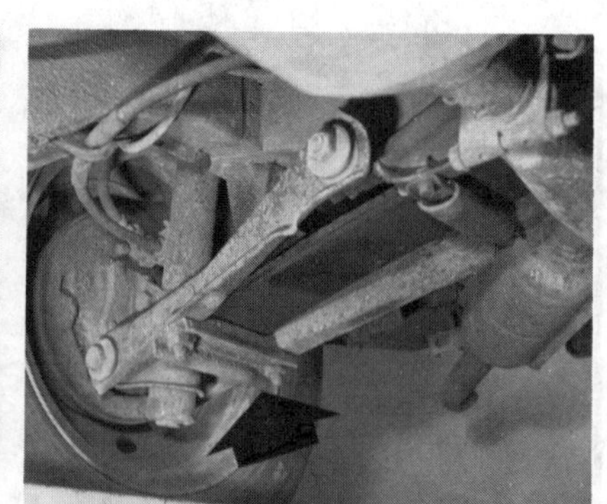

9.8 Take the nuts off the pivot mountings and inside the boot, off the top of the shock absorber rod

Fig. 11.6. Left-hand rear suspension components

1 *Nut, control arm pin to body*
2 *Nut, pin to control arm*
3 *Shock absorber (suspension strut)*
4 *Upper screw, shock absorber lower attachment to wheel spindle*
5 *Lower screw, shock absorber lower attachment to control arm and spindle*
6 *Shims*
7 *Nuts, leaf spring to control arm*
8 *Insulator*

4 Press in the new bushes. They need good support whilst being inserted, so they are not distorted. It may be wise to get your FIAT agent to do this unless you have the equipment that makes it easy. If the job is found difficult, then damage to the bushes is likely.

5 Once the bushes have been fitted, the shims must be selected to give the correct fit with the shock absorber (strut) at the outer end of the arm. (See Fig.11.7).

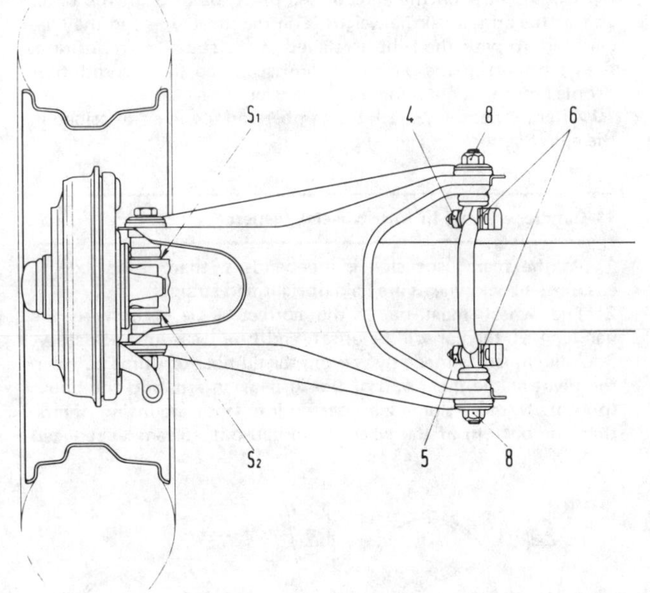

Fig. 11.7. Rear suspension adjustment (Sec. 10)

S1 & S2 Shims adjusting fit of shock absorber (strut) to arm
4 & 5 Nuts securing arm pivot to body
6 Shims adjusting camber and toe-in
8 Nuts clamping bushes on pivot

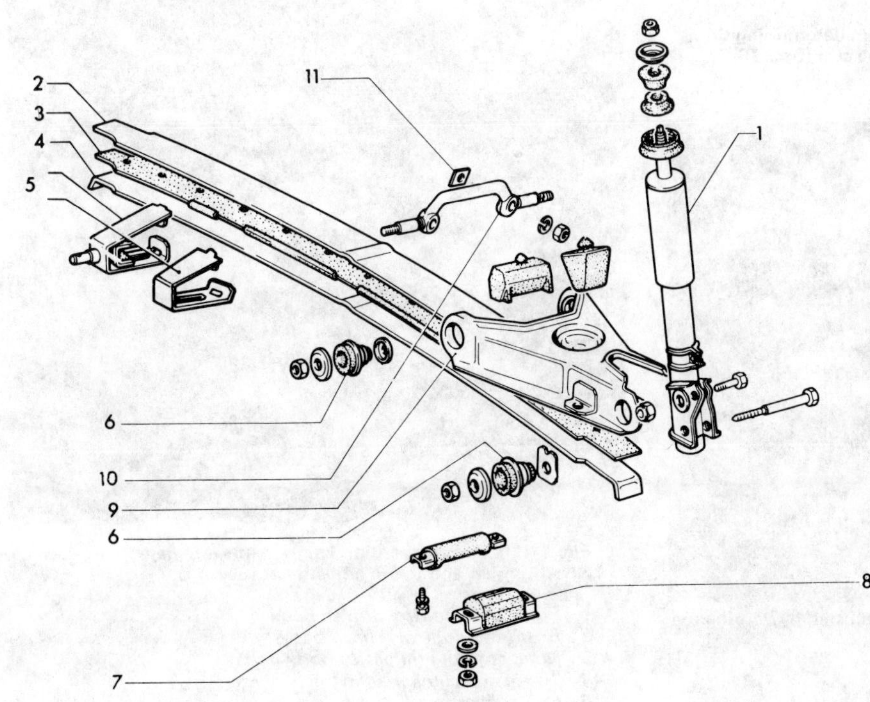

Fig. 11.8. Rear suspension components

1 *Shock absorber (suspension strut)*
2 *Spring top leaf*
3 *Interleaf lining*
4 *Main leaf*
5 *Spring clips*
6 *Arm bush*
7 *Spring securing brackets*
8 *Spring securing brackets*
9 *Suspension control arm*
10 *Pivot for arm bushes*
11 *Shims for adjusting toe-in and camber*

6 Measure the width of the gap between the bushes in the suspension arm. This = a: see Fig. 11.9.
7 Measure the width of the shock absorber (strut) across its mounting lugs. This = b.
8 Shims must go either side. Their total thickness = s.

s = (a-b) + 0.118 inch (3.0 mm).

The thickness 's' must be divided so that the shimming ahead of the suspension strut is 0.020 inch (0.5 mm) thicker than that behind it.
9 Reassemble the new components to the car, but do not tighten the nuts on the ends of the pivot, or through the outer end of the arm, until the weight is on the suspension, so they are clamped up with the rubber relaxed at the static laden position, then the twist in the bush is minimised. Load the car with four people before retightening the suspension.
10 After reassembly, check the camber and toe-in as described in the next Section.

11 Camber and toe-in (rear wheels) - general

1 As the rear suspension is independent, there is no axle to ensure the back wheels are held upright and straight.
2 The wheel must be at the correct angle, otherwise the handling of the car will be upset, and tyre wear unsatisfactory.
3 Adjustment is made by varying the number of shims between the pivot at the inner end of the suspension arm and the body. (photo). If more shims are inserted into both mounting points, then the bottom of the wheel is pushed out, so camber reduced (or made more negative). If shims are taken out of the front point and transferred to the rear, the camber stays the same, but the toe-in will be increased. (See Figs. 11.7 and 11.11).
4 The toe-in can be measured by the system detailed in Chapter 10. It must be ensured that this toe-in is given equally to each wheel. This can be done by looking along toe-in setting bars when they are laid along the wheel, and seeing where the bars, if extended, would touch the front wheels. Of course, it is important the front wheels are set truly 'straight-ahead' too. It will take some care to ensure this is correctly done.
5 The camber is quoted in the specification for the laden and unladen condition. Laden camber is difficult to achieve and people's weights vary. Anyway, the unladen camber is easier to measure, because zero camber is within the specification. The wheel can therefore be adjusted to be upright, and the only gauge needed to measure it is a builder's plumb line. Even a nut hung on a piece of string will make a crude plumb line.
6 These two adjustments have been described separately. But toe-in and camber should be done concurrently so that the number of times the suspension has to be dismantled to insert shims is minimised.
7 Having given the relevant techniques, we should in all fairness state that unless you are a very competent and experienced mechanic, this type of tricky adjustment is best left to your FIAT agent.

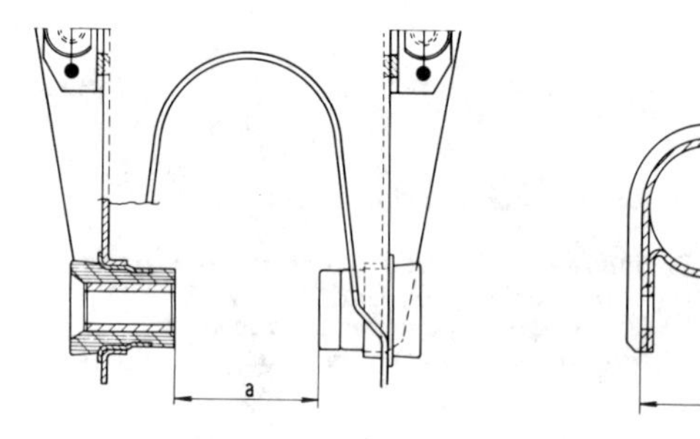

Fig. 11.9. Measurements 'a' and 'b' used in determining shims in mounting the shock absorber to the arm (Sec. 10)

11.3 The rear wheel toe-in and camber is adjusted by varying the number of shims behind the pivot mountings

12 Hubs and wheel bearings (rear) - removal and refitment

1 The rear hub bearings are sealed for life and serviced as a complete assembly.
2 Failure of the bearings will be brought to notice by the test in Section 2. But it will also be convenient to remove the hubs when changing the rear brake shoes, as it makes the job much easier, particularly for someone who is not used to the job.
3 Jack-up and remove the wheel.
4 Remove the brake drum by undoing the one bolt and the long headed wheel aligning bolt that holds it to the hub flange.

Fig. 11.10. Rear suspension components associated with rear wheel camber and toe-in adjustment (Sec. 11)

1 *Front bolt, control arm to body*
2 *Rear bolt, control arm to body*
3 *Nuts, control arm pin to body bolts*
4 *Toe-in and camber adjustment shims*
5 *Control arm pin*

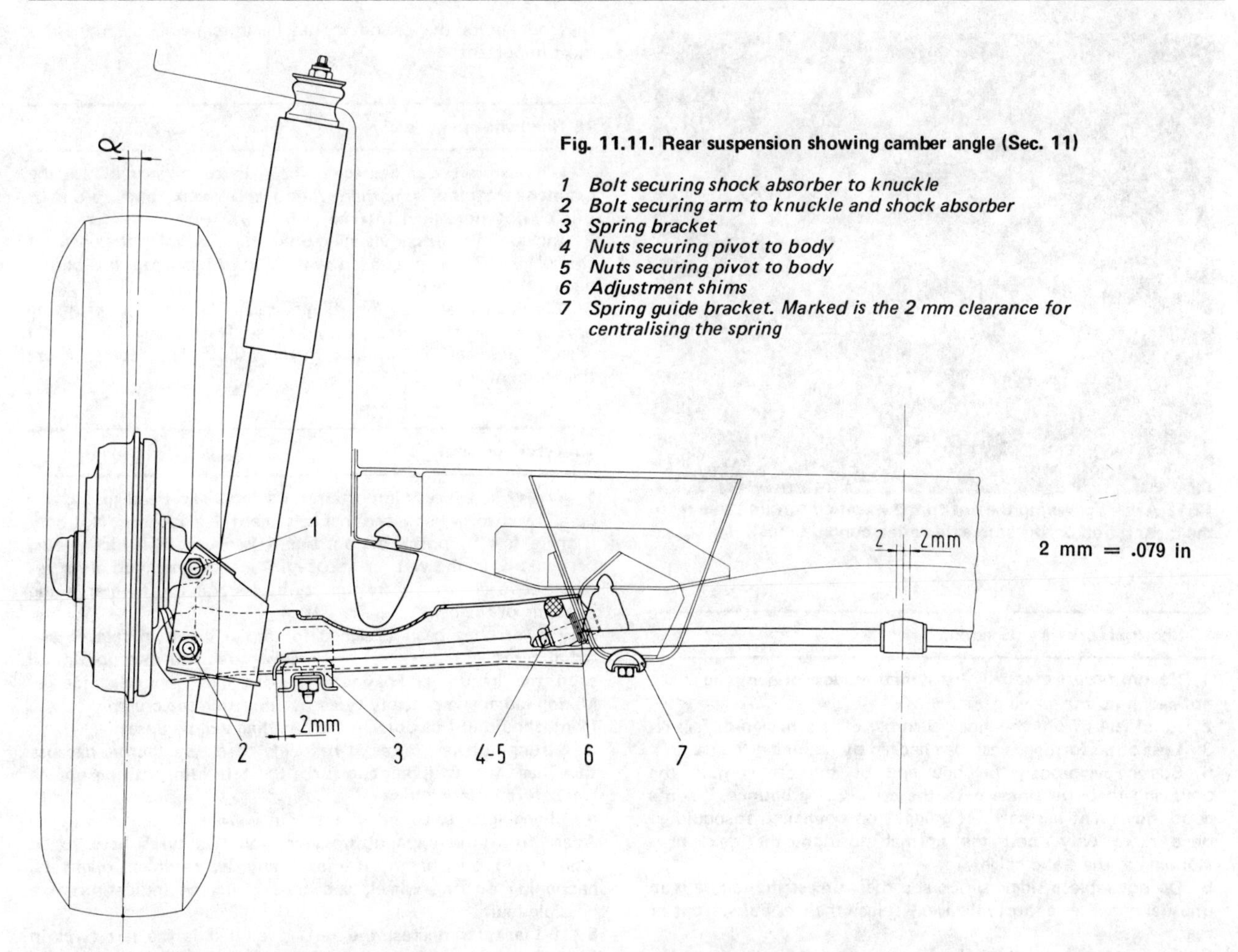

Fig. 11.11. Rear suspension showing camber angle (Sec. 11)

1 Bolt securing shock absorber to knuckle
2 Bolt securing arm to knuckle and shock absorber
3 Spring bracket
4 Nuts securing pivot to body
5 Nuts securing pivot to body
6 Adjustment shims
7 Spring guide bracket. Marked is the 2 mm clearance for centralising the spring

5 Prise off the bearing cap. This is soft and readily deformed if not removed carefully. It can be started by prising with a screwdriver between its flange and the hub. Make sure it is kept square, or it will jam. Hammer blows from the direction of the brake shoes can also help. (photo)
6 Undo the collared nut on the stub axle and take off its washer. (photo)
7 Pull off the hub from the stub axle. A conventional hub puller may be needed. (photo)
8 The hub will have to be replaced if the bearings no longer move smoothly and silently, or if they have developed free-play.
9 Before reassembly, smear a little grease on the stub axle. If the old hub is being reused, do not immerse it in cleaning fluid, as this will get past the bearing seals, but cannot be dried out, or new grease put in.
10 Put the hub onto the stub axle, and fit the washer and a new collared nut.
11 Tighten the nut to the specified torque load and lock it by staking it to the groove in the stub axle with a punch (photo).
12 Smear a little grease over the bearing cap, and tap it gently and squarely into place. There is no need to drive it hard in to get it fully home, as long as it has gone on far enough to be secure.
13 Reassemble the drum and wheel. Then check the new hub turns freely, smoothly, and without rim rock.

12.5 The bearing cap must be carefully prised and tapped off to give access to the sealed bearings

12.6 Remove the nuts and washer

12.7 Then pull off the hub. This is replaced with its bearings as a unit

12.11 After tightening the nut to the specified torque stake it to the indentation on the stub axle with a punch, to lock it

13 Shock absorbers - inspection

1 Failure occurs gradually, so sometimes loss of damping is not noticed in normal driving.

2 Fluid leaks from the shock absorber are a sure sign of failure.

3 Loss of performance can be checked by bouncing the car.

4 Bounce vigorously on one end of the car, timing the bouncing to be in phase with the car's spring bounce. When a good movement has been set going, stop bouncing. It should go once more only, past the normal position, the next time stopping at the static height.

5 Do not replace single shock absorbers unless the failure is an unusual one after a short mileage. Replace them as pairs, front or rear.

6 To do the replacement, the suspension must be dismantled as described in Section 10 for the rear suspension.

7 The front suspension strut is very easily removed, as described in Section 3. But then the spring and shock absorber must be separated, as described in Section 4.

8 It is probable that the FIAT agent will stock the shock absorber in its outer casing. But the inside working parts are serviced as a sub-assembly. To extract them from the casing, the shock absorber should be extended. Then the gland fitting where the rod enters the casing should be unscrewed. Cleanlines˒ is most important.

14 Roadwheels - general

1 Wheels can be damaged if a kerb is struck. During maintenance tasks spin them round and check they are out of true by not more than :/16 in. (1.5 mm).

2 Include in your car cleaning programme washing the backs of the wheels. These get caked with mud. Always do this before having them balanced.

3 The wheels require repainting occasionally, either after fitting new tyres, or after some two years use. Rusting is liable to start where the wheel rim is joined to the disc. If this becomes severe the wheel may fail.

15 Tyres - general

1 The tyres are very important for safety. The tread should not be allowed to get so worn that it is near the legal limit. Not only is there fear of police action, but it is exceedingly dangerous, particularly in the wet, or at speed. The recommended pressures must be used to ensure the tyres last, and to preserve the handling of the car.

2 The car has been designed for radial tyres (in fact special radial tyres at that). Use of crossply tyres will give noise, and spoil the handling. They also have a much shorter mileage. Mixing radial and crossply tyres, even putting the crossply on the front, should not be done as the handling will be upset.

3 Fitting tubeless tyres is best left to a tyre factor, because they have the equipment to force the tyre into position on the rim before inflating it.

4 Changing tyres round to even out wear is not recommended. Apart from the expenditure when all five tyres have to be replaced in one batch, it masks any wear which might be happening on one wheel, and thus giving an indication of a possible fault.

5 If the tyres nearest the kerb side, that is the left tyres in Great Britain, wear faster than the ones on the opposite side, this indicates the wheels are toeing-in too much, and vice versa.

6 If the tread wear in the centre is worse than that at the edges it indicates over-inflation, and vice versa.

7 It will pay to have the wheels balanced. Wheel out-of-balance causes vibration in the steering, which is unpleasant, and causes much wear to the suspension and tyres. In bad cases it will spoil the roadholding and handling of the car and cause severe vibration at certain speeds.

Chapter 12 Bodywork and fittings

For modifications, and information applicable to later models, see Supplement at end of manual

Contents

1 General description

The combined bodyshell and underframe is an all-welded unitary structure of sheet steel. There are two basic two-door body types; one is provided with a tailgate or third door as some term it, while the other is of the conventional two-door with luggage boot concept.

Extreme care has gone into providing the maximum of space in a rather small body. This has been achieved by mounting the combined engine and transmission transversely, and by a long wheelbase that places the rear wheels far enough back to avoid intrusion of the rear wheel arches into the rear seat width. The unusual MacPherson strut rear suspension design allows an abnormally deep boot for a vehicle of this size.

Strong points are provided at the front and rear of the vehicle for jacking it up, and in the engine bay for carrying the engine supports. Needless to say, these points should be particularly examined for corrosion. The remainder of the vehicle is of a conventional nature, apart from the spare wheel being mounted under the bonnet, which is front-hinged and retained in the raised position by a torsion bar.

2 Maintenance - bodywork and underframe

The general condition of a vehicle's bodywork is the one thing that significantly affects its value. Maintenance is easy but needs to be regular. Neglect, particularly after minor damage, can lead quickly to further deterioration and costly repair bills. It is important also to keep watch on those parts of the vehicle not immediately visible, for instance the underside, inside all the wheel arches and the lower part of the engine compartment.

The basic maintenance routine for the bodywork is washing - preferably with a lot of water, from a hose. This will remove all the loose solids which may have stuck to the vehicle. It is important to flush these off in such a way as to prevent grit from scratching the finish. The wheel arches and underframe need washing in the same way to remove any accumulated mud which will retain moisture and tend to encourage rust. Paradoxically enough, the best time to clean the underframe and wheel arches is in wet weather when the mud is thoroughly wet and soft. In very wet weather the underframe is usually cleaned of large accumulations automatically and this is a good time for inspection.

Periodically, except on vehicles with a wax-based underbody

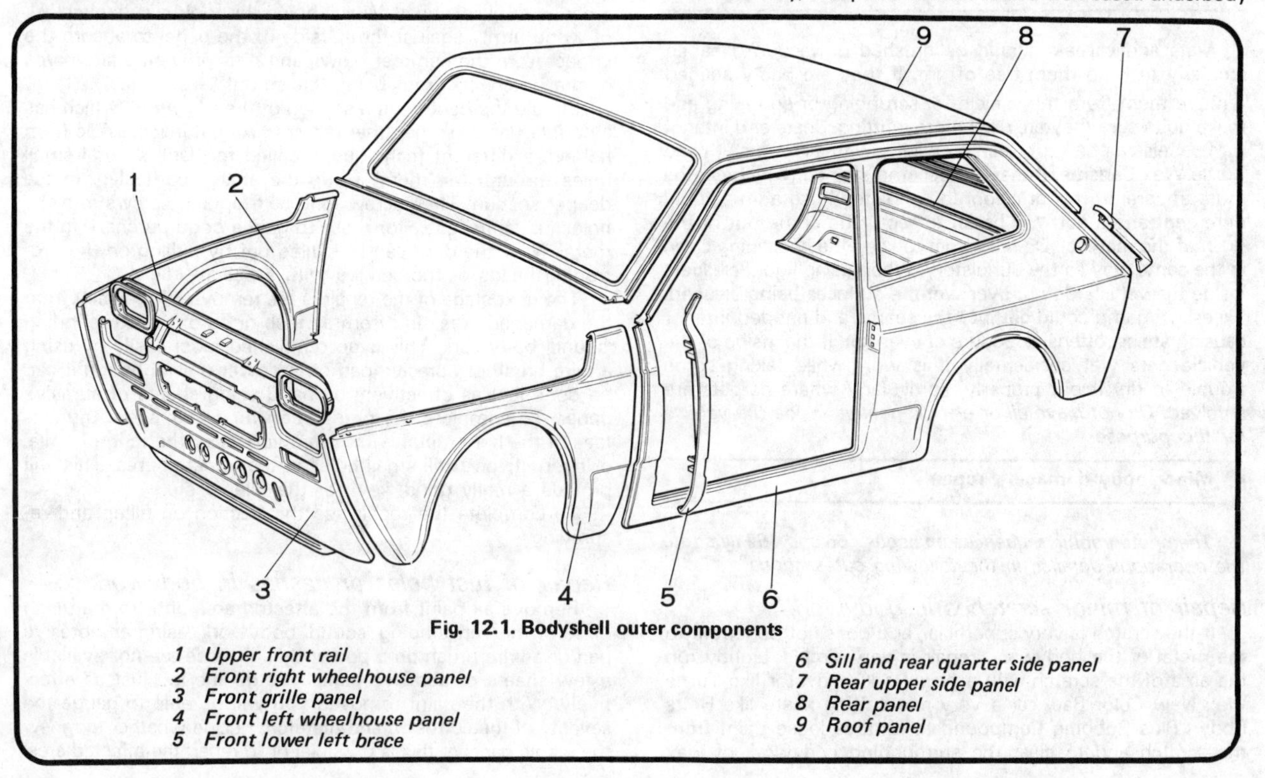

Fig. 12.1. Bodyshell outer components

1 *Upper front rail*
2 *Front right wheelhouse panel*
3 *Front grille panel*
4 *Front left wheelhouse panel*
5 *Front lower left brace*
6 *Sill and rear quarter side panel*
7 *Rear upper side panel*
8 *Rear panel*
9 *Roof panel*

protective coating, it is a good idea to have the whole of the underframe of the vehicle steam cleaned, engine compartment included, so that a thorough inspection can be carried out to see what minor repairs and renovations are necessary. Steam cleaning is available at many garages and is necessary for removal of the accumulation of oily grime which sometimes is allowed to become thick in certain areas. If steam cleaning facilities are not available, there are one or two excellent grease solvents available, such as Holts Engine Cleaner or Holts Foambrite, which can be brush applied. The dirt can then be simply hosed off. Note that these methods should not be used on vehicles with wax-based underbody protective coating or the coating will be removed. Such vehicles should be inspected annually, preferably just prior to winter, when the underbody should be washed down and any damage to the wax coating repaired using Holts Undershield. Ideally, a completely fresh coat should be applied. It would also be worth considering the use of such wax-based protection for injection into door panels, sills, box sections, etc, as an additional safeguard against rust damage where such protection is not provided by the vehicle manufacturer.

After washing paintwork, wipe off with a chamois leather to give an unspotted clear finish. A coat of clear protective wax polish, like the many excellent Turtle Wax polishes, will give added protection against chemical pollutants in the air. If the paintwork sheen has dulled or oxidised, use a cleaner/polisher combination such as Turtle Extra to restore the brilliance of the shine. This requires a little effort, but such dulling is usually caused because regular washing has been neglected. Care needs to be taken with metallic paintwork, as special non-abrasive cleaner/polisher is required to avoid damage to the finish. Always check that the door and ventilator opening drain holes and pipes are completely clear so that water can be drained out. Bright work should be treated in the same way as paint work. Windscreens and windows can be kept clear of the smeary film which often appears by the use of a proprietary glass cleaner like Holts Mixra. Never use any form of wax or other body or chromium polish on glass.

3 Maintenance – upholstery and carpets

Mats and carpets should be brushed or vacuum cleaned regularly to keep them free of grit. If they are badly stained remove them from the vehicle for scrubbing or sponging and make quite sure they are dry before refitting. Seats and interior trim panels can be kept clean by wiping with a damp cloth and Turtle Wax Carisma. If they do become stained (which can be more apparent on light coloured upholstery) use a little liquid detergent and a soft nail brush to scour the grime out of the grain of the material. Do not forget to keep the headlining clean in the same way as the upholstery. When using liquid cleaners inside the vehicle do not over-wet the surfaces being cleaned. Excessive damp could get into the seams and padded interior causing stains, offensive odours or even rot. If the inside of the vehicle gets wet accidentally it is worthwhile taking some trouble to dry it out properly, particularly where carpets are involved. *Do not leave oil or electric heaters inside the vehicle for this purpose.*

4 Minor body damage – repair

The photographic sequences on pages 166 and 167 illustrate the operations detailed in the following sub-sections.

Repair of minor scratches in bodywork

If the scratch is very superficial, and does not penetrate to the metal of the bodywork, repair is very simple. Lightly rub the area of the scratch with a paintwork renovator like Turtle Wax New Color Back, or a very fine cutting paste like Holts Body + Plus Rubbing Compound to remove loose paint from the scratch and to clear the surrounding bodywork of wax polish. Rinse the area with clean water.

Apply touch-up paint, such as Holts Dupli-Color Color Touch or a paint film like Holts Autofilm, to the scratch using a fine paint brush; continue to apply fine layers of paint until the surface of the paint in the scratch is level with the surrounding paintwork. Allow the new paint at least two weeks to harden: then blend it into the surrounding paintwork by rubbing the scratch area with a paintwork renovator or a very fine cutting paste, such as Holts Body + Plus Rubbing Compound or Turtle Wax New Color Back. Finally, apply wax polish from one of the Turtle Wax range of wax polishes.

Where the scratch has penetrated right through to the metal of the bodywork, causing the metal to rust, a different repair technique is required. Remove any loose rust from the bottom of the scratch with a penknife, then apply rust inhibiting paint, such as Turtle Wax Rust Master, to prevent the formation of rust in the future. Using a rubber or nylon applicator fill the scratch with bodystopper paste like Holts Body + Plus Knifing Putty. If required, this paste can be mixed with cellulose thinners, such as Holts Body + Plus Cellulose Thinners, to provide a very thin paste which is ideal for filling narrow scratches. Before the stopper-paste in the scratch hardens, wrap a piece of smooth cotton rag around the top of a finger. Dip the finger in cellulose thinners, such as Holts Body + Plus Cellulose Thinners, and then quickly sweep it across the surface of the stopper-paste in the scratch; this will ensure that the surface of the stopper-paste is slightly hollowed. The scratch can now be painted over as described earlier in this Section.

Repair of dents in bodywork

When deep denting of the vehicle's bodywork has taken place, the first task is to pull the dent out, until the affected bodywork almost attains its original shape. There is little point in trying to restore the original shape completely, as the metal in the damaged area will have stretched on impact and cannot be reshaped fully to its original contour. It is better to bring the level of the dent up to a point which is about $\frac{1}{8}$ in (3 mm) below the level of the surrounding bodywork. In cases where the dent is very shallow anyway, it is not worth trying to pull it out at all. If the underside of the dent is accessible, it can be hammered out gently from behind, using a mallet with a wooden or plastic head. Whilst doing this, hold a suitable block of wood firmly against the outside of the panel to absorb the impact from the hammer blows and thus prevent a large area of the bodywork from being 'belled-out'.

Should the dent be in a section of the bodywork which has a double skin or some other factor making it inaccessible from behind, a different technique is called for. Drill several small holes through the metal inside the area – particulary in the deeper section. Then screw long self-tapping screws into the holes just sufficiently for them to gain a good purchase in the metal. Now the dent can be pulled out by pulling on the protruding heads of the screws with a pair of pliers.

The next stage of the repair is the removal of the paint from the damaged area, and from an inch or so of the surrounding 'sound' bodywork. This is accomplished most easily by using a wire brush or abrasive pad on a power drill, although it can be done just as effectively by hand using sheets of abrasive paper. To complete the preparation for filling, score the surface of the bare metal with a screwdriver or the tang of a file, or alternatively, drill small holes in the affected area. This will provide a really good 'key' for the filler paste.

To complete the repair see the Section on filling and respraying.

Repair of rust holes or gashes in bodywork

Remove all paint from the affected area and from an inch or so of the surrounding 'sound' bodywork, using an abrasive pad or a wire brush on a power drill. If these are not available a few sheets of abrasive paper will do the job just as effectively. With the paint removed you will be able to gauge the severity of the corrosion and therefore decide whether to renew the whole panel (if this is possible) or to repair the affected area.

New body panels are not as expensive as most people think and it is often quicker and more satisfactory to fit a new panel than to attempt to repair large areas of corrosion.

Remove all fittings from the affected area except those which will act as a guide to the original shape of the damaged bodywork (eg headlamp shells etc). Then, using tin snips or a hacksaw blade, remove all loose metal and any other metal badly affected by corrosion. Hammer the edges of the hole inwards in order to create a slight depression for the filler paste.

Wire brush the affected area to remove the powdery rust from the surface of the remaining metal. Paint the affected area with rust inhibiting paint like Turtle Wax Rust Master; if the back of the rusted area is accessible treat this also.

Before filling can take place it will be necessary to block the hole in some way. This can be achieved by the use of aluminium or plastic mesh, or aluminium tape.

Aluminium or plastic mesh or glass fibre matting, such as the Holts Body + Plus Glass Fibre Matting, is probably the best material to use for a large hole. Cut a piece to the approximate size and shape of the hole to be filled, then position it in the hole so that its edges are below the level of the surrounding bodywork. It can be retained in position by several blobs of filler paste around its periphery.

Aluminium tape should be used for small or very narrow holes. Pull a piece off the roll and trim it to the approximate size and shape required, then pull off the backing paper (if used) and stick the tape over the hole; it can be overlapped if the thickness of one piece is insufficient. Burnish down the edges of the tape with the handle of a screwdriver or similar, to ensure that the tape is securely attached to the metal underneath.

Bodywork repairs – filling and re-spraying

Before using this Section, see the Sections on dent, deep scratch, rust holes and gash repairs.

Many types of bodyfiller are available, but generally speaking those proprietary kits which contain a tin of filler paste and a tube of resin hardener are best for this type of repair, like Holts Body + Plus or Holts No Mix which can be used directly from the tube. A wide, flexible plastic or nylon applicator will be found invaluable for imparting a smooth and well contoured finish to the surface of the filler.

Mix up a little filler on a clean piece of card or board – measure the hardener carefully (follow the maker's instructions on the pack) otherwise the filler will set too rapidly or too slowly. Alternatively, Holts No Mix can be used straight from the tube without mixing, but daylight is required to cure it. Using the applicator apply the filler paste to the prepared area; draw the applicator across the surface of the filler to achieve the correct contour and to level the filler surface. As soon as a contour that approximates to the correct one is achieved, stop working the paste – if you carry on too long the paste will become sticky and begin to 'pick up' on the applicator. Continue to add thin layers of filler paste at twenty-minute intervals until the level of the filler is just proud of the surrounding bodywork.

Once the filler has hardened, excess can be removed using a metal plane or file. From then on, progressively finer grades of abrasive paper should be used, starting with a 40 grade production paper and finishing with 400 grade wet-and-dry paper. Always wrap the abrasive paper around a flat rubber, cork, or wooden block – otherwise the surface of the filler will not be completely flat. During the smoothing of the filler surface the wet-and-dry paper should be periodically rinsed in water. This will ensure that a very smooth finish is imparted to the filler at the final stage.

At this stage the 'dent' should be surrounded by a ring of bare metal, which in turn should be encircled by the finely 'feathered' edge of the good paintwork. Rinse the repair area with clean water, until all of the dust produced by the rubbing-down operation has gone.

Spray the whole repair area with a light coat of primer, either Holts Body + Plus Grey or Red Oxide Primer – this will show up any imperfections in the surface of the filler. Repair these imperfections with fresh filler paste or bodystopper, and once more smooth the surface with abrasive paper. If bodystopper is used, it can be mixed with cellulose thinners to form a really thin paste which is ideal for filling small holes. Repeat this spray and repair procedure until you are satisfied that the surface of the filler, and the feathered edge of the paintwork are perfect. Clean the repair area with clean water and allow to dry fully.

The repair area is now ready for final spraying. Paint spraying must be carried out in a warm, dry, windless and dust free atmosphere. This condition can be created artificially if you have access to a large indoor working area, but if you are forced to work in the open, you will have to pick your day very carefully. If you are working indoors, dousing the floor in the work area with water will help to settle the dust which would otherwise be in the atmosphere. If the repair area is confined to one body panel, mask off the surrounding panels; this will help to minimise the effects of a slight mis-match in paint colours. Bodywork fittings (eg chrome strips, door handles etc) will also need to be masked off. Use genuine masking tape and several thicknesses of newspaper for the masking operations.

Before commencing to spray, agitate the aerosol can thoroughly, then spray a test area (an old tin, or similar) until the technique is mastered. Cover the repair area with a thick coat of primer; the thickness should be built up using several thin layers of paint rather than one thick one. Using 400 grade wet-and-dry paper, rub down the surface of the primer until it is really smooth. While doing this, the work area should be thoroughly doused with water, and the wet-and-dry paper periodically rinsed in water. Allow to dry before spraying on more paint.

Spray on the top coat using Holts Dupli-Color Autospray, again building up the thickness by using several thin layers of paint. Start spraying in the centre of the repair area and then, with a single side-to-side motion, work outwards until the whole repair area and about 2 inches of the surrounding original paintwork is covered. Remove all masking material 10 to 15 minutes after spraying on the final coat of paint.

Allow the new paint at least two weeks to harden, then, using a paintwork renovator or a very fine cutting paste such as Turtle Wax New Color Back or Holts Body + Plus Rubbing Compound, blend the edges of the paint into the existing paintwork. Finally, apply wax polish.

5 Major body damage - repair

1 Where serious damage has occurred or large areas need renewal due to neglect, new sections or panels will need welding in and this is best left to professionals. If the damage is due to impact it will also be necessary to check the alignment of the body structure.

2 If a body is left misaligned it is first of all dangerous as the car will not handle properly - and secondly, uneven stresses will be imposed on the steering, engine and transmission, causing abnormal wear or complete failure. Tyre wear will also be excessive.

6 Body corrosion

1 The ultimate scrapping of a car is usually due to rust, rather than it becoming uneconomic to renew mechanical parts.

2 The rust grows from two origins: from the underneath unprotected after the paint was blasted off by the road grit: from inside, where damp has collected inside hollow body sections, without a chance to drain, and no rust protection to the metal.

3 The corrosion is particularly prone to start at welding joints in the body, as there are stresses left in after the heat of welding has cooled. These seams are also traps for the damp, and difficult to rustproof.

4 Salt on the roads in winter promotes this horror. It is hygroscopic; it attracts moisture, so the car stays damp even in a garage. It is also an electrolyte when wet, so promotes violent

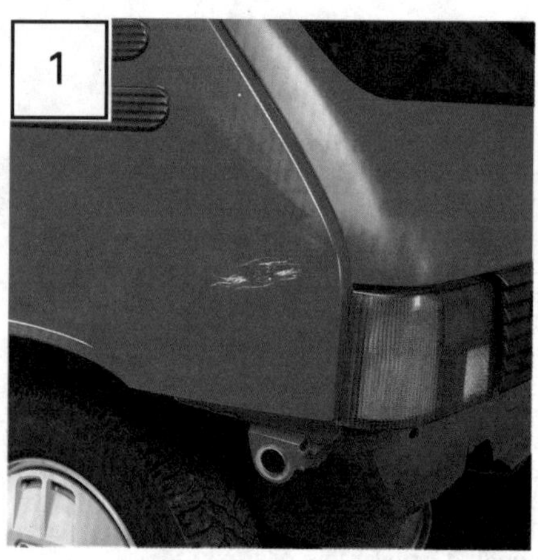

This photographic sequence shows the steps taken to repair the dent and paintwork damage shown above. In general, the procedure for repairing a hole will be similar; where there are substantial differences, the procedure is clearly described and shown in a separate photograph.

First remove any trim around the dent, then hammer out the dent where access is possible. This will minimise filling. Here, after the large dent has been hammered out, the damaged area is being made slightly concave.

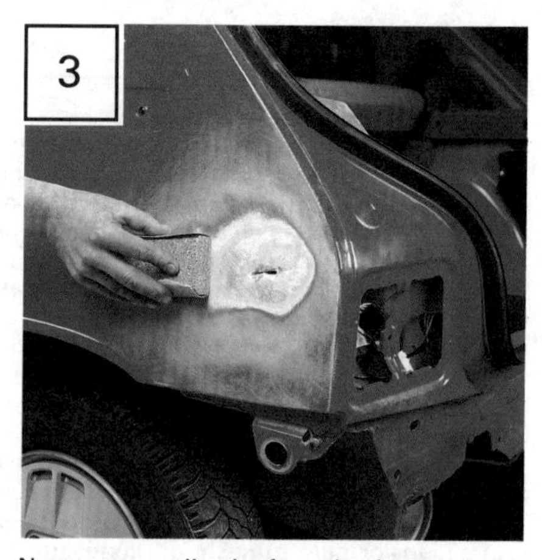

Next, remove all paint from the damaged area by rubbing with coarse abrasive paper or using a power drill fitted with a wire brush or abrasive pad. 'Feather' the edge of the boundary with good paintwork using a finer grade of abrasive paper.

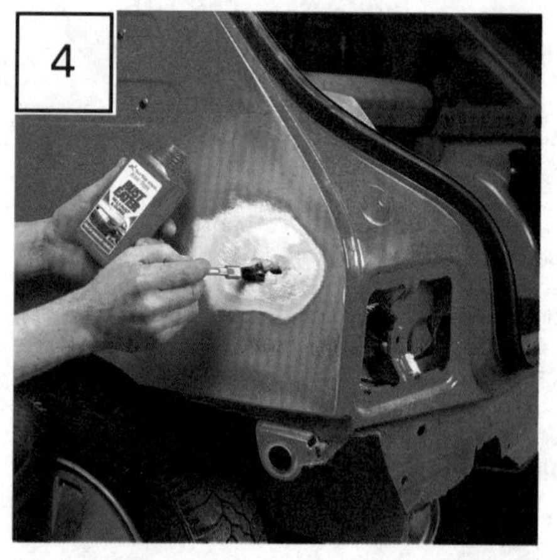

Where there are holes or other damage, the sheet metal should be cut away before proceeding further. The damaged area and any signs of rust should be treated with Turtle Wax Hi-Tech Rust Eater, which will also inhibit further rust formation.

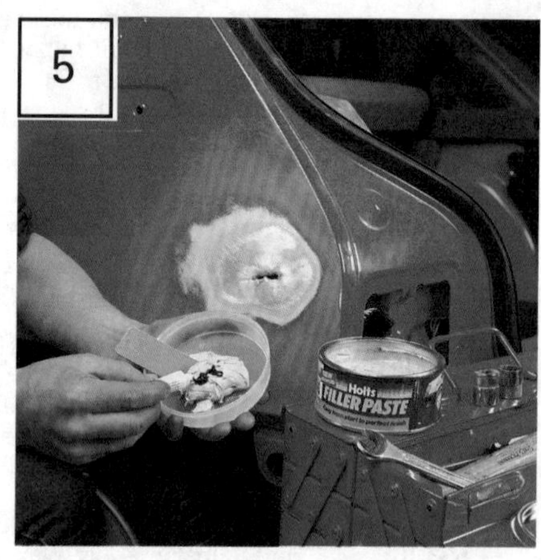

For a large dent or hole mix Holts Body Plus Resin and Hardener according to the manufacturer's instructions and apply around the edge of the repair. Press Glass Fibre Matting over the repair area and leave for 20-30 minutes to harden. Then ...

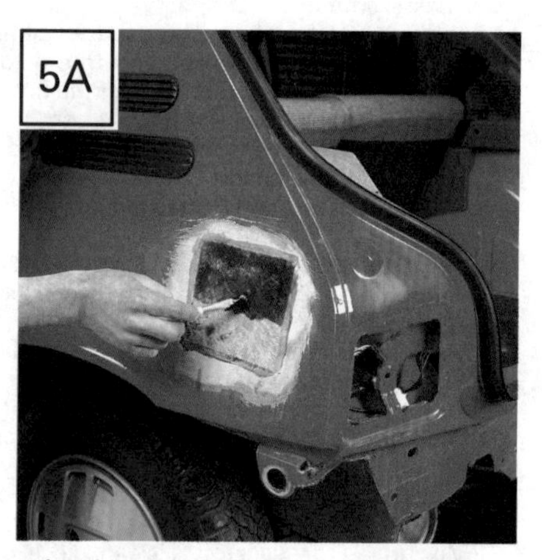

... brush more Holts Body Plus Resin and Hardener onto the matting and leave to harden. Repeat the sequence with two or three layers of matting, checking that the final layer is lower than the surrounding area. Apply Holts Body Plus Filler Paste as shown in Step 5B.

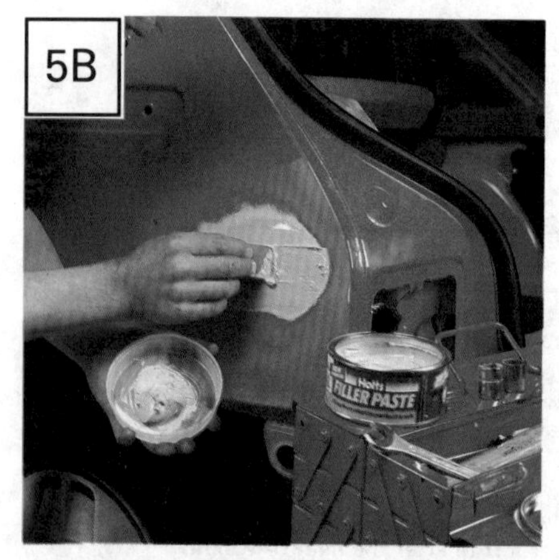

For a medium dent, mix Holts Body Plus Filler Paste and Hardener according to the manufacturer's instructions and apply it with a flexible applicator. Apply thin layers of filler at 20-minute intervals, until the filler surface is slightly proud of the surrounding bodywork.

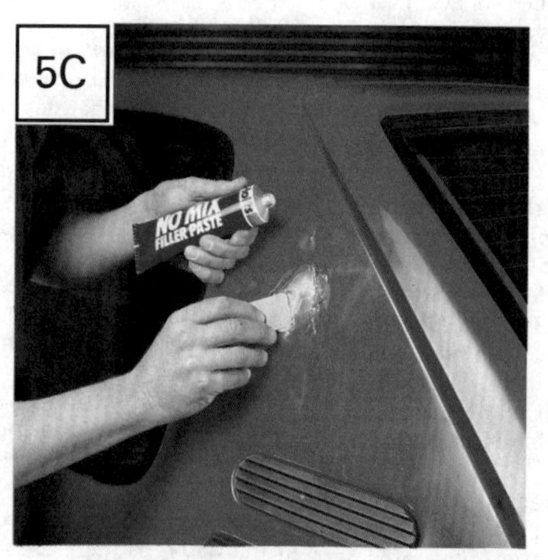

For small dents and scratches use Holts No Mix Filler Paste straight from the tube. Apply it according to the instructions in thin layers, using the spatula provided. It will harden in minutes if applied outdoors and may then be used as its own knifing putty.

Use a plane or file for initial shaping. Then, using progressively finer grades of wet-and-dry paper, wrapped round a sanding block, and copious amounts of clean water, rub down the filler until glass smooth. 'Feather' the edges of adjoining paintwork.

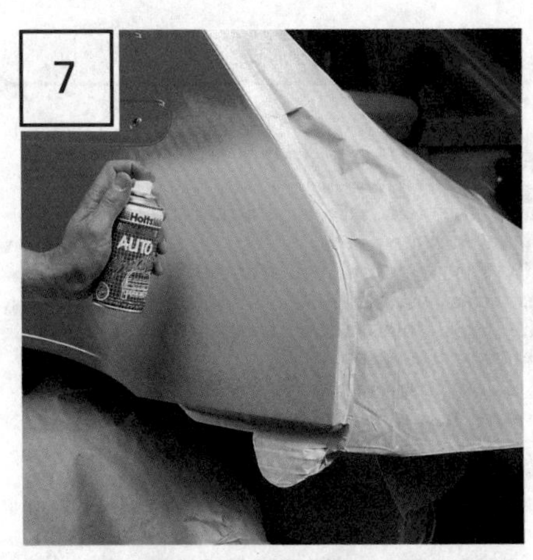

Protect adjoining areas before spraying the whole repair area and at least one inch of the surrounding sound paintwork with Holts Dupli-Color primer.

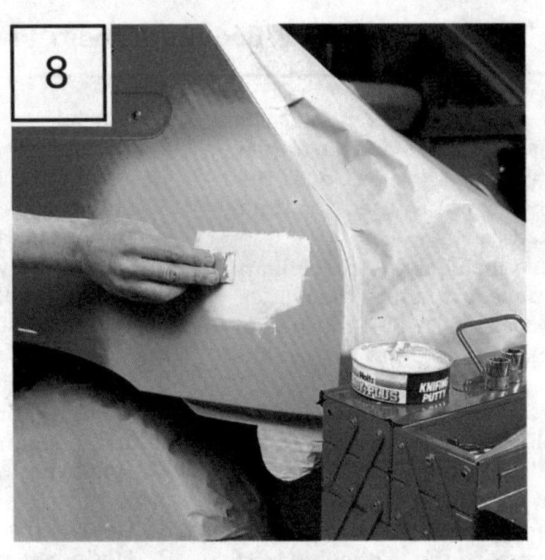

Fill any imperfections in the filler surface with a small amount of Holts Body Plus Knifing Putty. Using plenty of clean water, rub down the surface with a fine grade wet-and-dry paper – 400 grade is recommended – until it is really smooth.

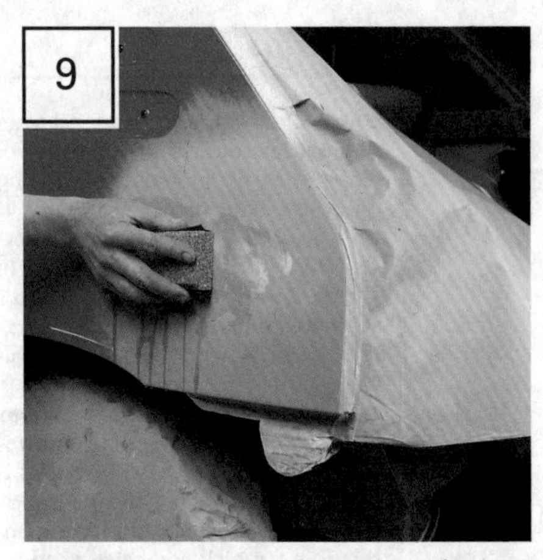

Carefully fill any remaining imperfections with knifing putty before applying the last coat of primer. Then rub down the surface with Holts Body Plus Rubbing Compound to ensure a really smooth surface.

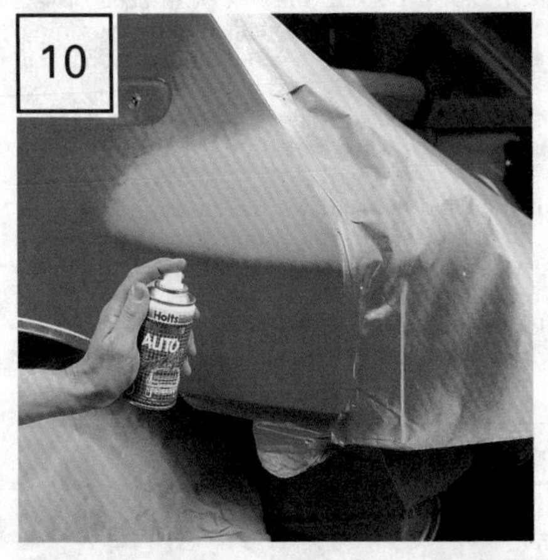

Protect surrounding areas from overspray before applying the topcoat in several thin layers. Agitate Holts Dupli-Color aerosol thoroughly. Start at the repair centre, spraying outwards with a side-to-side motion.

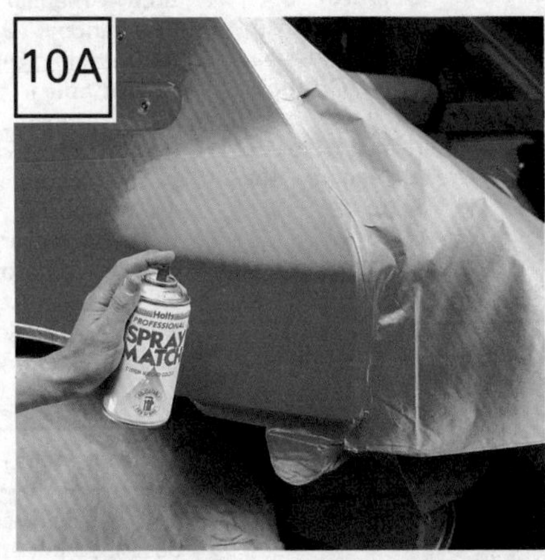

If the exact colour is not available off the shelf, local Holts Professional Spraymatch Centres will custom fill an aerosol to match perfectly.

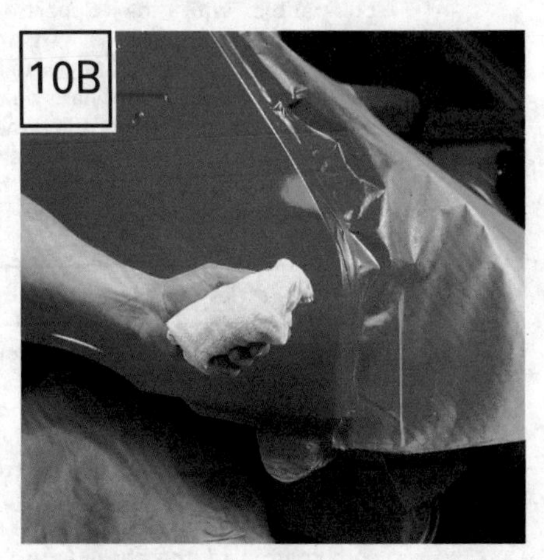

To identify whether a lacquer finish is required, rub a painted unrepaired part of the body with wax and a clean cloth.

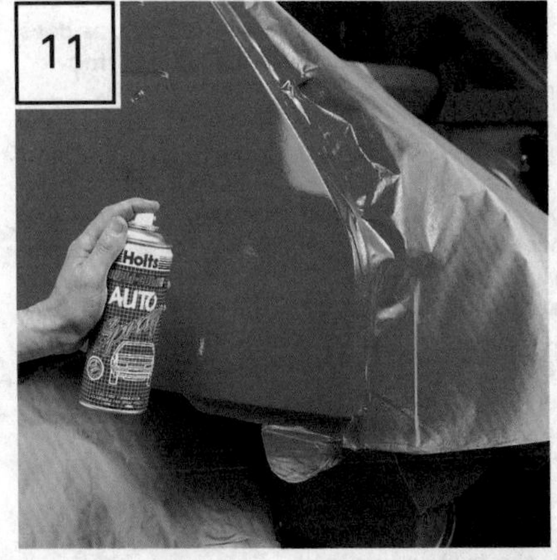

If *no* traces of paint appear on the cloth, spray Holts Dupli-Color clear lacquer over the repaired area to achieve the correct gloss level.

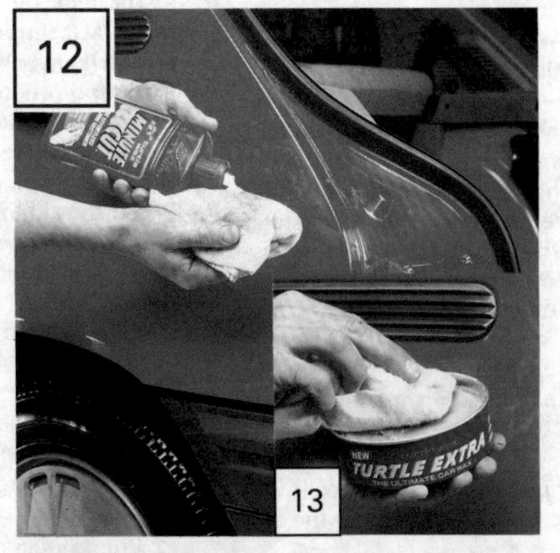

The paint will take about two weeks to harden fully. After this time it can be 'cut' with a mild cutting compound such as Turtle Wax Minute Cut prior to polishing with a final coating of Turtle Wax Extra.

When carrying out bodywork repairs, remember that the quality of the finished job is proportional to the time and effort expended.

corrosion. If the car has been used on salty roads it must be desalted as soon as possible. A couple of days rain after a thaw clears the roads, and then driving the car in the wet does this naturally. The damage is worse if the car is not used for some time after being used on wet salted roads.

5 The Routine Maintenance schedule includes the checking of the underneath for rust. It should be done just before the winter, so that the car is prepared for its ravages whilst dry and free of salt. Then it needs to be done again in the spring to remove the winter's damage.

6 The bodywork should be explored and all hollow sections found. Into these a rust inhibitor should be injected; aerosols like 'Supertrol 001', or 'Di-Nitrol 33B' are good. Easily accessible areas can be wetted by the inhibitor bought more cheaply as a liquid and applied by paintbrush. An example is the insides of the doors, which can be reached by removing the panel linings. If hollow sections are sealed, it pays to drill a hole, spray in the aerosol, and then seal with the underseal paint smeared over the hole.

7 The underneath needs painting, and where abraded by grit flung up by the wheels, this must be one of the special thick resilient paints. 'Adup bronze super seal' is recommended. It is compatible with the 'Supertrol 001' and 'Di-Nitrol 33B' inhibitors. So if these are put on first the two between them make a good job in getting into corners. The 'Adup' underseal paint can be used as ordinary paint on sheltered areas. On those showered by grit from the wheels, thick layers need to be applied. At the mudguards the underseal paint needs to be applied round the edges so that these are protected, the corners being particularly vulnerable.

7 Door rattles - tracing and rectification

1 The commonest cause of door rattles is a misaligned, loose, or worn striker plate, but other causes may be:

a) Loose door handles, window winder handles or door hinges

b) Loose, worn or misaligned door lock components

c) Loose or worn remote control mechanism

2 It is quite possible for door rattles to be the result of a combination of the above faults so a careful examination must be made to determine the causes of the fault.

3 If the nose of the striker plate is worn and as a result the door rattles, renew it and adjust the plate.

4 If the nose of the door lock wedge is badly worn and the door rattles as a result, then fit a new lock.

5 Should the hinges be badly worn, then they must be renewed.

8 Door trim pad - removal and refitting

1 Unscrew the two screws holding the door armrest.

2 Prise off the escutcheon round the interior door handle.

3 Remove the window winder. With one hand push the winder surrounding escutcheon towards the door so the spring clip holding the handle to the shaft is visible. Insert a thin screwdriver to push out the clip. Normally the clip will have been inserted in such a way that this push must come from the axis of the handle.

4 Carefully prise out the trim pad from the door, starting at the bottom. Insert a screwdriver close to a clip. Then pull out the next clip.

5 When refitting everything, put the winder handle halfway on its shaft. Then put the spring clip in position, in as far as it will go. Then push the handle fully home onto its shaft, and the clip will spring into its groove.

9 Door interior - dismantling and reassembly

1 Refer to Section 8, and remove the door interior handles and trim panel.

2 Refer to Fig.12.4 and undo the fixings securing the door glass run channel to the interior panel.

3 Undo and remove the two screws securing the ventilator window frame to the main door frame.

4 Temporarily refit the window regulator handle and wind down the window as far as possible.

5 Carefully pull the ventilator window rearwards and lift away the outer edge of the door frame.

6 Refer to Fig.12.2 and unscrew the fixings securing the door lock and remote control to the door inner panel. Lift away the complete assembly.

7 Remove the screws which secure the glass to the control cable.

8 Detach the window regulator cable from the jockey wheels.

9 Undo and remove the window regulator screws and then lift away the complete regulator assembly.

10 The door glass may now be lifted away from the main door frame.

11 Reassembling the door is the reverse sequence to dismantling but there are several additional points that should be noted:

12 Wind the control cable onto the window regulator grooves in such a manner that the turns do not interfere with each other. The window regulator cable tension can be adjusted by loosening its retaining nut and shifting the movable pulley along the slotted hole in the mounting bracket.

13 Position the glass so that it abuts against the buffer and then, without moving the cable, secure the glass to the cable by means of the lockplate.

14 Temporarily refit the window regulator handle and make sure that full movement of the door glass is obtained without strain on the regulator. If necessary re-adjust the position of the door glass relative to the cable.

15 Refit the interior trim panel and handles, as described in Section 8. Do not forget to well lubricate all moving parts.

10 Door locks - general

1 To gain access to the door locks, remove the trim panel, as described in Section 8.

2 Wind the window up. The lock can then be reached through the access hole in the inner panel.

11 Door - adjustment

1 The door position is adjustable at the hinges and the striker at the lock by the use of mounting points with large holes. The hinge attachment points to the body have adjustment, as does the striker, achieved by slackening their fixing screws, moving them and retightening.

2 Before making any adjustment, mark round the edge of the striker, or hinge, with pencil, so that the original position and the amount moved can be seen.

3 It should not be necessary to move the hinge. The striker may need to be moved in, or up, to compensate for wear in the lock. After long periods the hinges wear, so more weight is taken by the striker. This makes the door difficult to close. But the striker must not be lowered, or the bottom of the door will rub on the body.

4 The hinges should be oiled regularly. The key locks to the doors should also be oiled, particularly in winter to keep out the damp and prevent them freezing.

12 Windscreen glass - removal and refitting

1 If the windscreen shatters, fitting a replacement windscreen is one of the few jobs which the average owner is advised to leave to the professional. For the owner who wishes to do the job himself the following instructions are given:

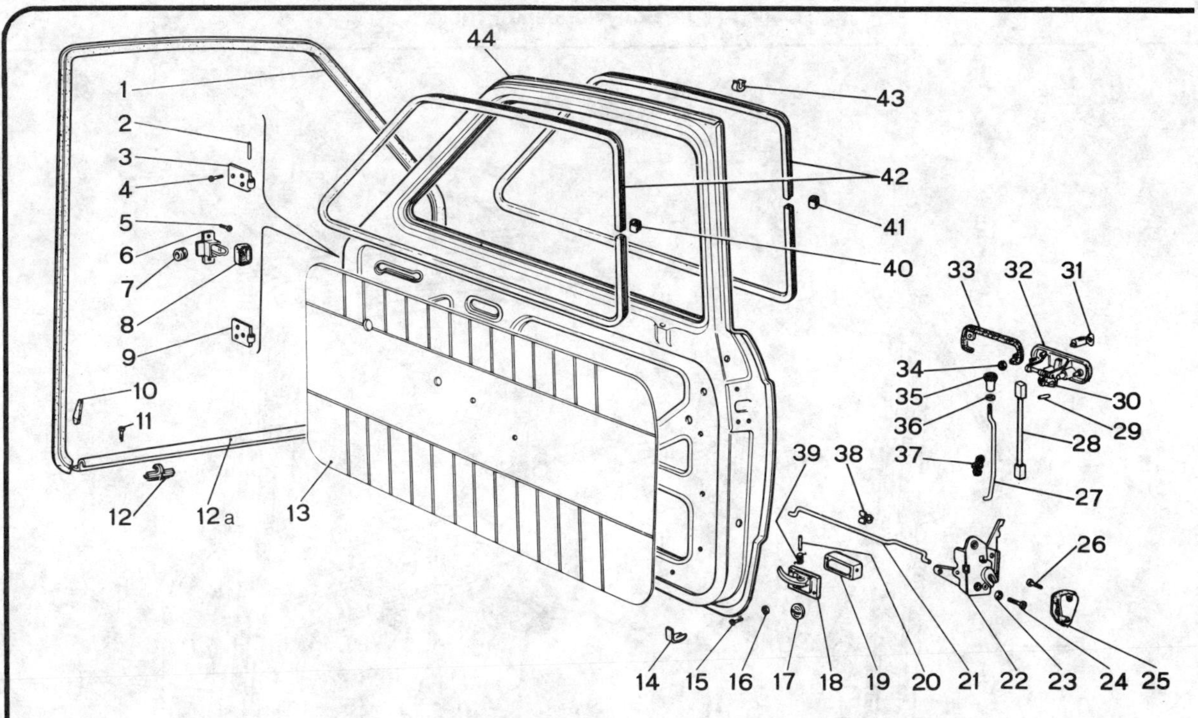

Fig. 12.2. Exploded view of right door components (handles and locks)

1 *Weatherstrip*
2 *Hinge pin*
3 *Upper fixed hinge*
4 *Hinge screw*
5 *Check band screw*
6 *Check band*
7 *Pulley*
8 *Pad*
9 *Lower fixed hinge*
10 *Block*
11 *Scuff plate screw*
12 *Clip*
12a *Scuff plate*
13 *Trim panel*
14 *Clip*
15 *Screw*
16 *Lockwasher*
17 *Escutcheon*
18 *Remote control*
19 *Cap*
20 *Remote control pin*
21 *Remote control link*
22 *Lock*
23 *Lockwasher*
24 *Lock screw*
25 *Striker*
26 *Striker screw*
27 *Safety lock link*
28 *Safety lock outside control rod*
29 *Pin*
30 *Safety lock control on handle*
31 *Handle lock cylinder*
32 *Outside door handle*
33 *Bezel*
34 *Handle nut*
35 *Button*
36 *Rubber ring*
37 *Button rod retainer*
38 *Clip*
39 *Remote control handle return spring*
40 *Butt cover*
41 *Butt cover*
42 *Window frames, inner and outer*
43 *Retainer*
44 *Door*

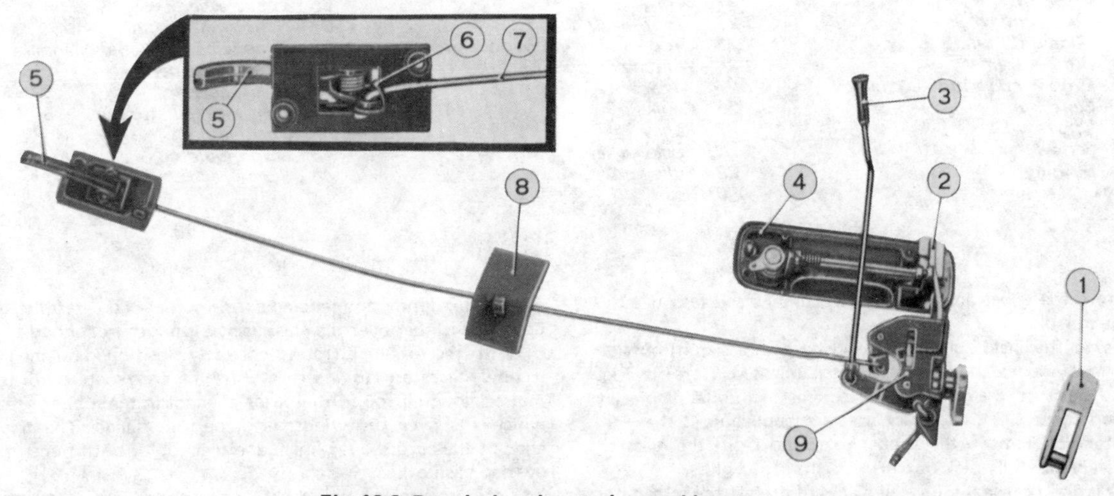

Fig. 12.3. Door lock and controls assembly

1 *Striker plate*
2 *Handle to lock link*
3 *Safety lock button and link*
4 *Outside door handle*
5 *Inside door handle*
6 *Rod fastener*
7 *Inside handle to lock rod*
8 *Rod sound deadener*
9 *Lock*

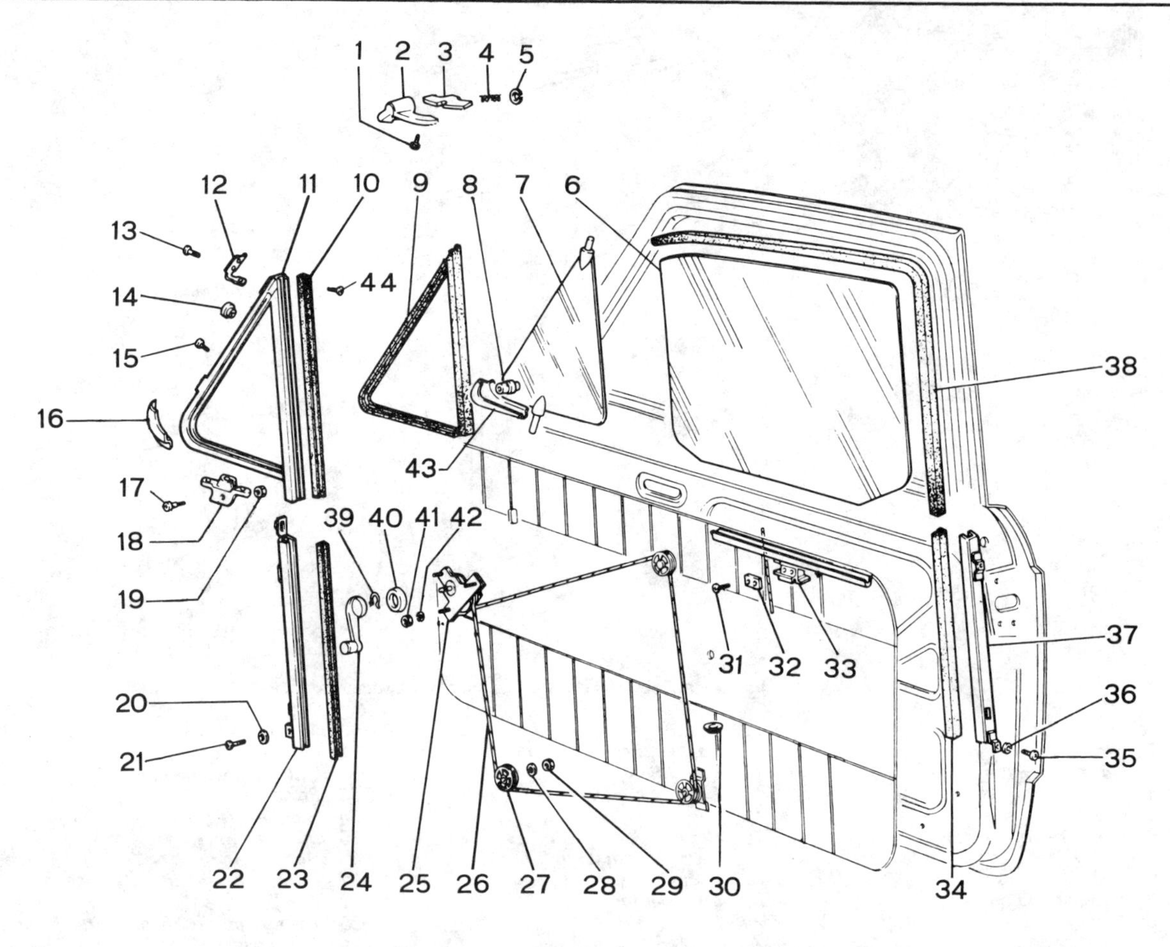

Fig. 12.4. Exploded view of right door components (glasses and controls)

1 Vent window handle screw
2 Vent window handle
3 Handle release lever
4 Spring
5 Lockwasher
6 Drop window glass
7 Vent window glass
8 Handle (2) pin
9 Rubber frame
10 Drop window rubber channel
11 Centre post
12 Upper swivel
13 Screw
14 Bushing
15 Screw
16 Vent window stop
17 Screw
18 Lower swivel support
19 Nut
20 Lockwasher
21 Screw
22 Front lower channel
23 Drop window rubber channel
24 Window regulator handle
25 Window regulator
26 Cable
27 Pulley
28 Lockwasher
29 Pulley nut
30 Stop
31 Screw
32 Plate
33 Glass bottom channel
34 Drop window rubber channel
35 Screw
36 Lockwasher
37 Rear lower channel
38 Drop window rubber channel
39 Handle retainer
40 Escutcheon
41 Window regulator nut
42 Lockwasher
43 Drip moulding
44 Screw

2 Remove the wiper arms from their spindles and take out the rear view mirror.
3 Move to the inside of the car and have an assistant outside the car ready to catch the glass as it is released. Refer to Fig. 12.8 and push on the glass with the palms of the hands placed at the top corners. This is, of course, not applicable if the glass has shattered. Remove the rubber surround from the glass or alternatively carefully pick out the remains of the glass.
4 Now is the time to remove all pieces of glass if the screen has shattered. Use a vacuum cleaner to extract as much as possible. Switch on the heater boost motor and adjust the controls to screen defrost but watch out for pieces of flying glass which might be blown out of the ducting.
5 Carefully inspect the rubber surround for signs of splitting or deterioration. Position the glass into the lower channel of the rubber surround commencing at one corner and carefully lift the lip of the rubber over the glass using a smooth flat ended tool.
6 With the rubber surround correctly positioned on the glass it is now necessary to insert a piece of cord about 15 ft (4.5 metres) long all round the outer channel in the rubber surround which fits over the windscreen aperture flange. The two free ends of the cord should finish at either top or bottom centre and overlap each other.
7 Offer the screen up to the aperture and get an assistant to press the rubber surround hard against the body flange. Slowly pull one end of the cord, moving round the windscreen and so drawing the lip over the windscreen flange on the body. To assist this operation a piece of tapered wood may be used.
8 Make sure that the rubber surround is correctly seating all round and finally refit the screen wiper blades.

13 Rear window glass - removal and refitting

The sequence for removal and refitting the rear window glass is basically identical to that for the windscreen. When removing the glass however, pressure must be applied to the lower corners of the pane and not to the top. Refer to Section 12 for full details.

14 Rear quarter window glass - removal and refitting

1 The sequence for removal and refitting the rear quarter window glass is basically identical to that for the windscreen. The trim moulding should be fitted to the slot in the rubber surround after the surround is correctly positioned in the aperture.

2 On cars with vent type quarter windows, first remove the screws that secure the handle to the rear roof pillar. Then turn the glass outwards and disengage the hinges. To install the glass, reverse the above procedure.

15 Bonnet (hood) - removal and refitting

1 Because the bonnet is hinged at the front, it gets in the way when working on the engine. Usually, FIAT mechanics are so

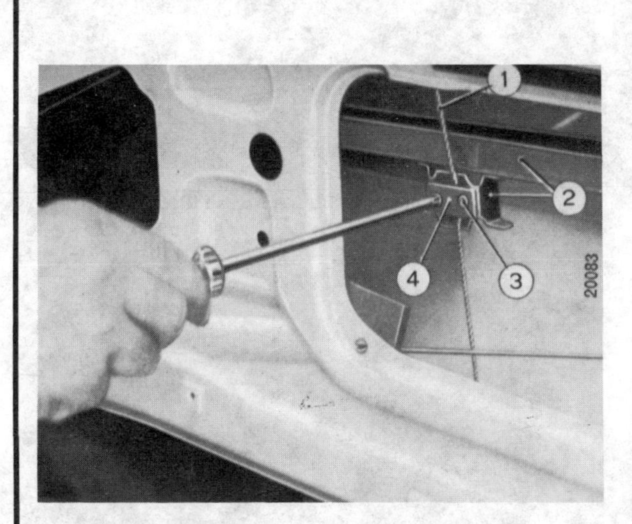

Fig. 12.5. Adjusting window travel

1 *Window regulator cable*
2 *Glass channel*
3 *Plate screws*
4 *Glass channel to cable attachment plate*

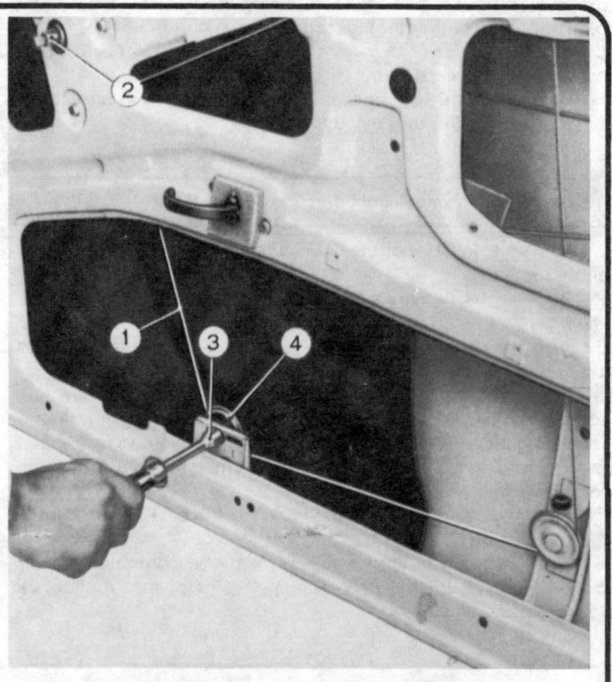

Fig. 12.6. Adjusting tension of window regulator cable

1 *Window regulator cable*
2 *Window regulator*
3 *Wrench for adjusting movable pulley*
4 *Movable pulley*

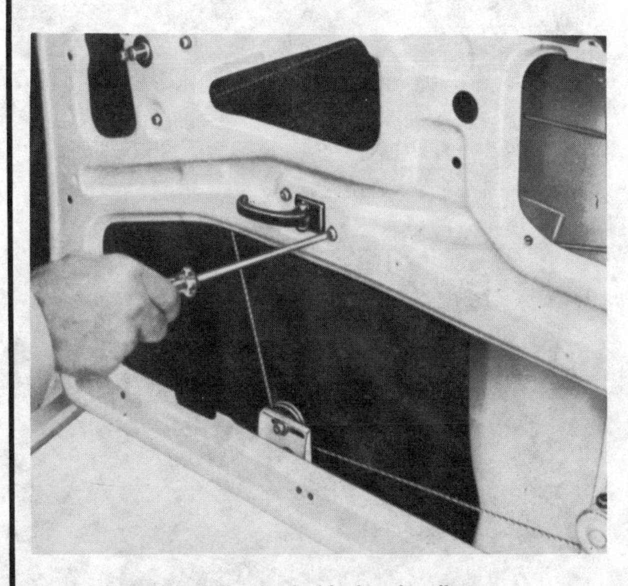

Fig. 12.7. Adjusting inside door handle

Fig. 12.8. Removing an unbroken windscreen (Sec. 12)

used to the cars, and have the benefit of special tools, they do not remove the bonnet even when taking out the engine. However its removal is recommended for such jobs as taking off the cylinder head, or taking out the engine as everything can be reached more easily.
2 Mark the fore and aft positions of the hinges on the body front panel, with pencil.
3 Undo the hinge bolts whilst an assistant holds the bonnet.
4 Tip the top of the bonnet towards the windscreen and lift it up, so the hinges can come clear of their mountings. Put the bonnet where it cannot get scratched.
5 Refitting the bonnet is the reverse. Be sure to replace the hinges in the same position so the bonnet lines up with the body. Close the bonnet gently, in case it is not straight, to check it fits properly into its opening. If necessary slacken the hinges and reposition it, utilising the elongated hinge mounting holes.
6 The bonnet lock mechanism can also be adjusted for position by virtue of oversize mounting holes. Before adjusting the bonnet lock position, ensure that the bonnet is positioned squarely in relation to the body.

16 Dashboard - removing the instrument panel

1 The panel in front of the driver that carries the instruments is easily removed.
2 Remove the fixing screw just below the centre of the face of the panel.
3 Push the panel down slightly to disengage two little hooks at the top.
4 From the engine compartment push the speedometer cable, after releasing it from its clip, through the scuttle (firewall) till inside the car the instrument panel can be pulled out far enough to reach behind, to undo the speedometer cable from the instrument.
5 Turn the panel over, till the wiring at the back is uppermost. Unplug the wiring.
6 When refitting the panel make sure the speedometer cable is pulled gently from the engine compartment so that it adopts a fair curve.

17 Dashboard - removal of the main panel pad

1 Remove the instrument panel, as described in the previous Section.
2 On some models the main panel pad is secured by nuts, which are accessible from the engine compartment. On other models the pad is secured by clips.
3 Take the ashtray out of its housing.
4 Press the pad upwards, from inside the ashtray housing (Fig. 12.13), to overcome the resistance of the clips (if fitted).
5 Lift the pad away.

18 Tailgate - adjustment, removal and refitting

1 On the three-door models, if the tailgate is not fitting correctly it is adjusted in much the same way as a door. The easiest adjustment is to simply loosen the screws that retain the striker plate and re-position it as necessary. If sufficient adjustment is not achieved by this method, then remove the top hinge covers, loosen the hinge screws and adjust the tailgate position. Retighten the screws after the tailgate is locked in the correct position.
2 To remove the tailgate, first remove the hinge covers, take out the locking pin, then withdraw the articulation pin. This needs to be done at each hinge.
3 Disconnect the tailgate prop by removing the nut that secures the prop to the top pivot point. Carefully lift the tailgate off.
4 Replacement is the reverse of removal, but it is almost certain that it will need to be re-aligned, as described previously.

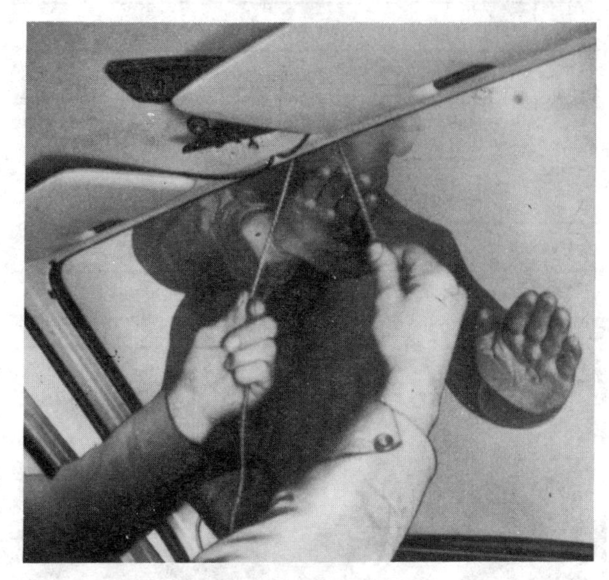

Fig. 12.9. Replacing the windscreen (Sec. 12)

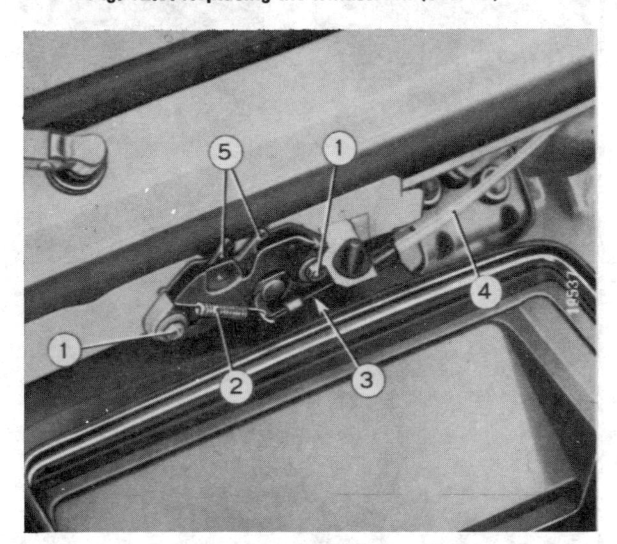

Fig. 12.10. Bonnet (hood) latch

1 Hood latch attachment bolts (bolt holes are oversized for adjustment)
2 Latch hook retracting spring
3 Latch release cable
4 Cable sheath
5 Latch hooks

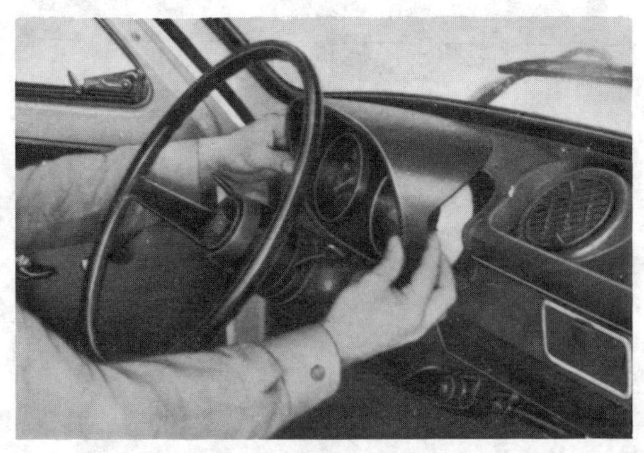

Fig. 12.11. Removing the instrument panel (Sec 16)

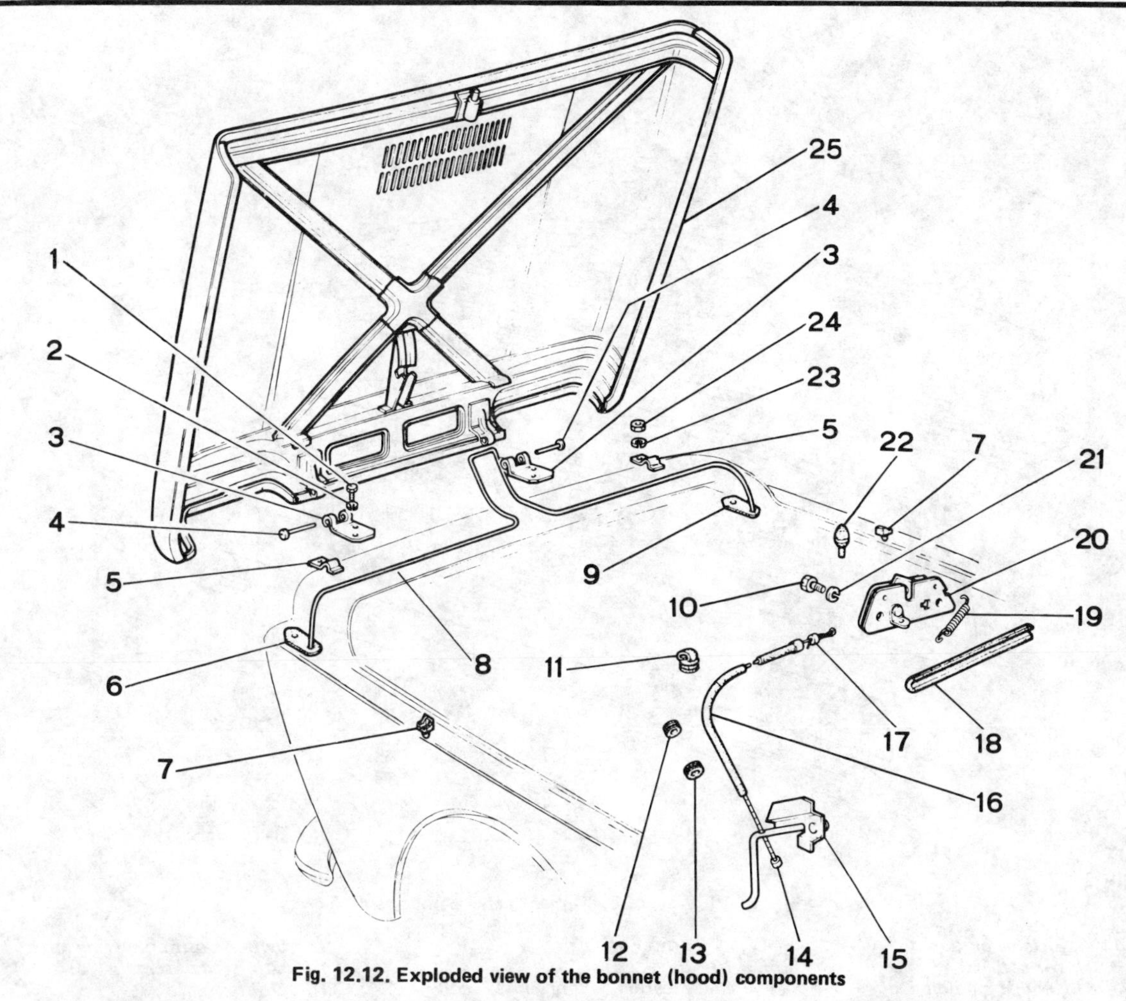

Fig. 12.12. Exploded view of the bonnet (hood) components

1 Hinge attachment bolt to front upper cross bar
2 Lockwasher
3 Fixed hinge
4 Hinge pin
5 Torsion bar bracket
6 Torsion bar left end seat
7 Side buffer
8 Torsion bar
9 Torsion bar right end seat
10 Hood latch bolt
11 Sheath clip
12 Grommet
13 Grommet
14 Latch release wire
15 Hood release control lever
16 Sheath
17 Fastener
18 Rubber strip
19 Latch hook retracting spring
20 Hood latch
21 Lockwasher
22 Hood buffer
23 Lockwasher
24 Nut
25 Hood panel

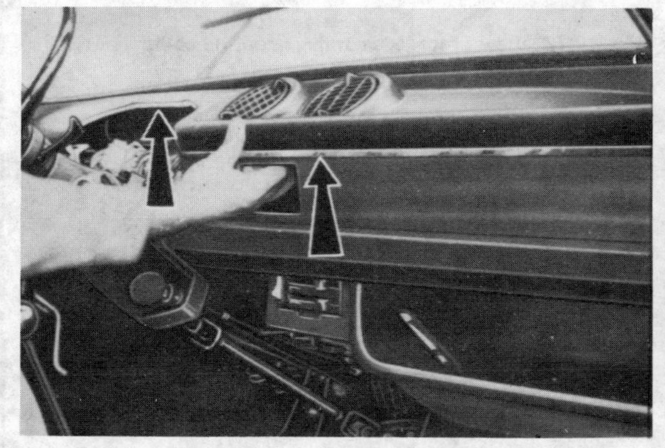

Fig. 12.13. Removing the main panel pad (Sec 17)

Fig. 12.14. View with main panel pad removed (Sec. 17)

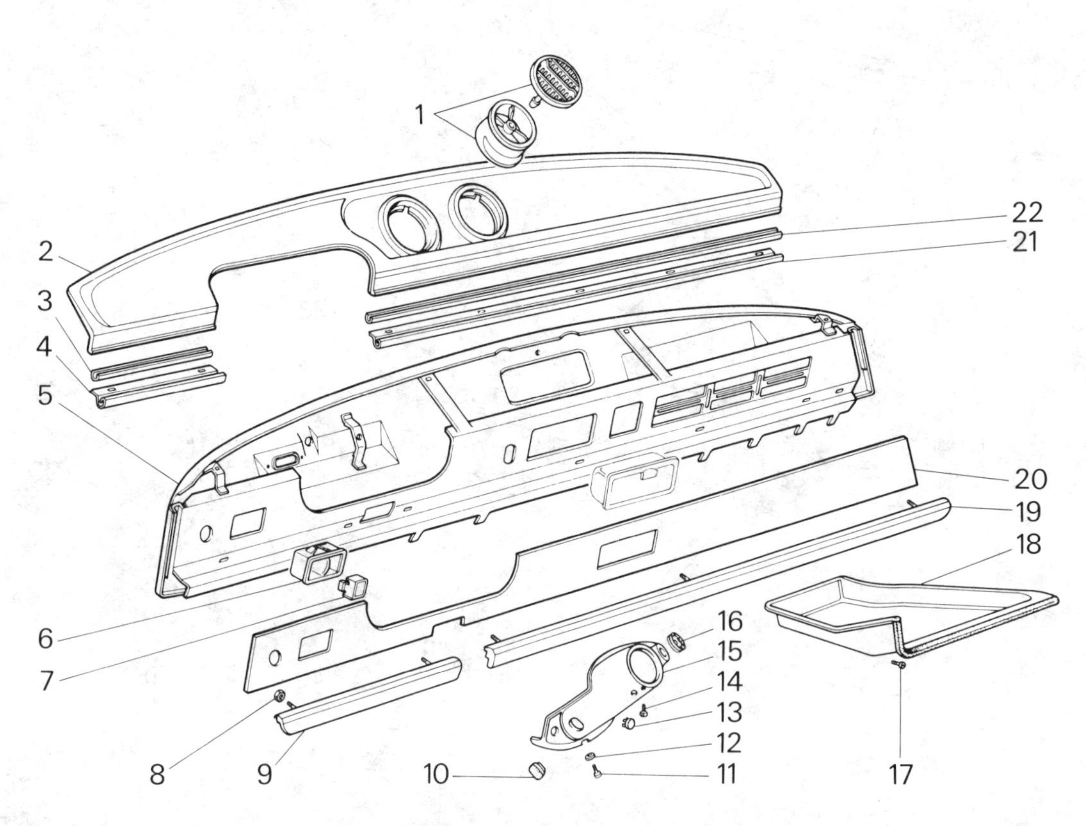

Fig. 12.15. View of instrument panel components

1 *Adjustable air outlet*
2 *Upper pad*
3 *Upper left moulding*
4 *Lower left strip*
5 *Instrument panel*
6 *Switch bezel*
7 *Plug*
8 *Lower pad nut*
9 *Lower left pad*
10 *Block*
11 *Steering column shield screw*
12 *Lockwasher*
13 *Rubber block*
14 *Steering column shield screw*
15 *Steering column shield*
16 *Switch positioning ring nut*
17 *Utility shelf screws*
18 *Utility shelf*
19 *Lower right pad*
20 *Frontal pad*
21 *Lower right strip*
22 *Upper right moulding*

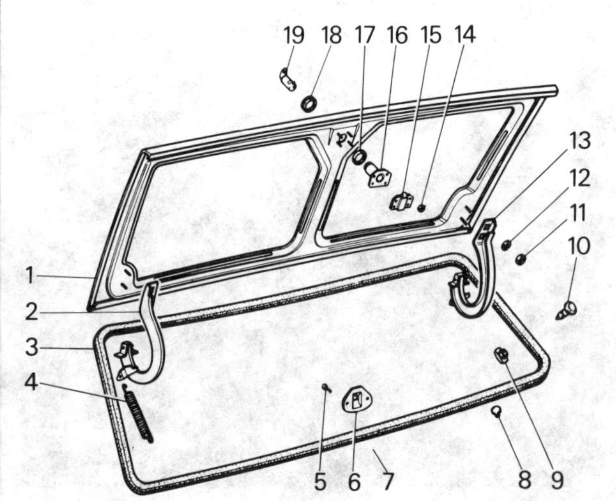

Fig. 12.16. View of luggage compartment lid components (Sec. 19)

1 *Lid*
2 *Left hinge*
3 *Luggage compartment rubber strip*
4 *Reaction spring*
5 *Striker screw*
6 *Striker*
7 *Rubber strip clip*
8 *Buffer*
9 *Snap fastener*
10 *Snap fastener*
11 *Nut, hinge to lid*
12 *Lockwasher*
13 *Right hinge*
14 *Locknut*
15 *Lock*
16 *Lock cylinder housing*
17 & 18 *Inner and outer lock cylinder rubber rings*
19 *Lock cylinder*

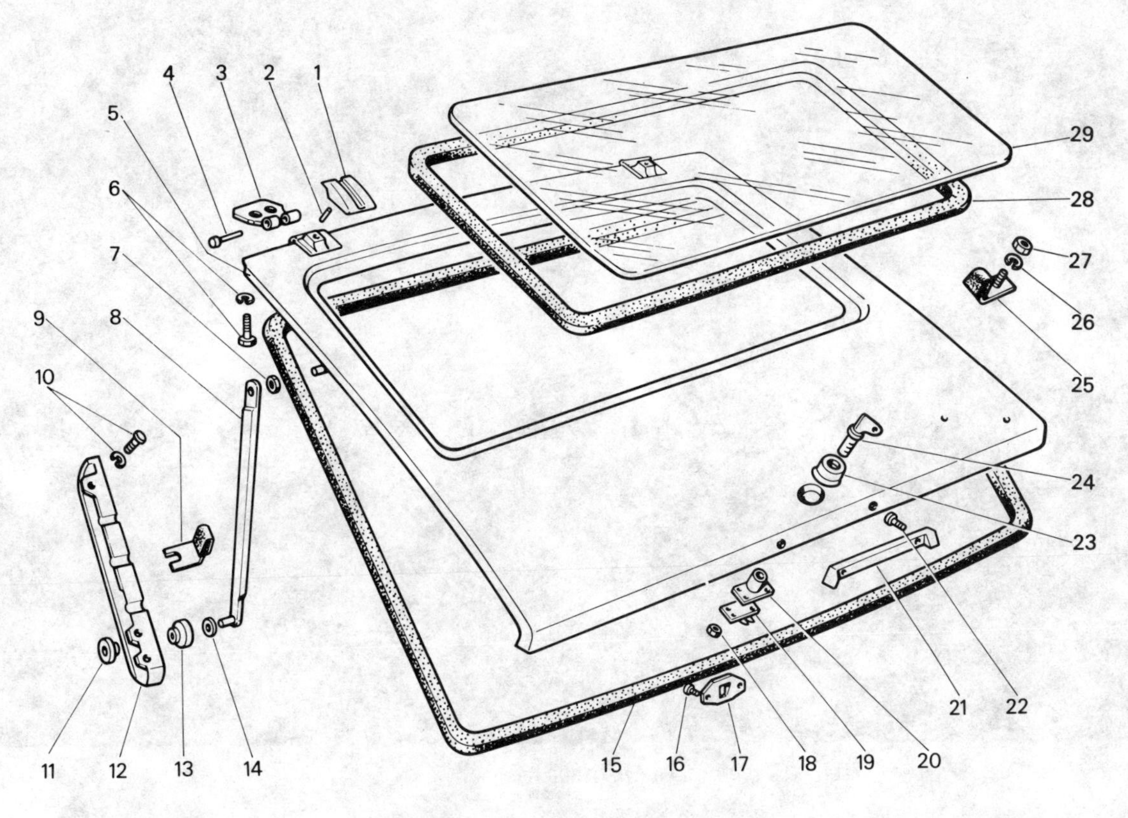

Fig. 12.17. Tailgate components (Sec. 18)

1 *Hinge cover*
2 *Pin*
3 *Hinge*
4 *Articulation pin*
5 *Tailgate*
6 *Hinge bolt and spring washer*
7 *Flat washer*
8 *Prop*
9 *LH bracket*
10 *Prop guide bolt and spring washer*
11 *Pulley*
12 *Prop guide*
13 *Pulley*
14 *Flat washer*
15 *Tailgate weatherstrip*
16 *Striker plate bolt*
17 *Striker plate*
18 *Locknut*
19 *Lock*
20 *Lock control*
21 *Handle*
22 *Handle bolt*
23 *Escutcheon*
24 *Lock cylinder*
25 *RH bracket*
26 *Spring washer*
27 *Bracket bolt*
28 *Glass rubber channel*
29 *Glass*

19 Luggage compartment lid - adjustment, removal and refitting

1 On models not fitted with a tailgate, the removal of the luggage compartment lid is straightforward and only involves removal of the nuts that secure the lid to the hinges. When replacing the lid, only tighten the screws sufficiently to retain the lid in position; it can then be manipulated to achieve the correct alignment, and then the screws fully tightened.

2 If the lid will not open or close smoothly, ie; through jamming, or failing to engage the lock, adjustment is by one of the following methods:

a) *First position the lid correctly, by use of the elongated holes on the hinges and then fully tighten the nuts.*
b) *To ensure engagement of the lock mechanism, adjust the position of the striker plate, using its elongated holes.*
c) *If the lid fails to rise after the lock has been released, this is due to the reaction spring being incorrectly positioned, or weakened. Replace if necessary.*

Fiat 127 1300 GT

Fiat 127 1050 Super

Chapter 13 Supplement: Revisions and information on later models

Contents

1 Introduction

Since its first introduction in 1971 the FIAT 127 has undergone a continuing programme of improvements concerned mainly with styling and trim alterations. However, in 1977 a new model range was introduced which included the option of a larger engine of 1049 cc capacity with a single overhead camshaft. In appearance, cars in the new range are broadly similar to the earlier models but they have a lower bonnet line, larger glass area and redesigned bumpers.

In October 1978, the FIAT 127 Sport was launched. This is a high performance version of the three-door hatchback having the 1049 cc engine with a twin choke carburettor.

In April 1982 the body styling was improved to include a new grille, rectangular front headlamps and new rear light clusters. Additionally the 1300 GT was introduced; being fitted with a larger capacity version of the 1049 cc overhead camshaft engine.

This supplement has been added to cater for some of the production changes which have been made, but principally to cover the overhead camshaft engine. The material given in Chapters 1 to 12 still applies unless superseded by information contained in this Supplement. Therefore, always read this Chapter in conjunction with the rest of the book.

2 Specifications

The Specifications listed here are revisions of, or supplementary to, the main Specifications given at the beginning of each Chapter

Dimensions, weights and capacities

Dimensions - Special, Super and 1300 GT (1982 on)

Overall length	12 ft 2 in (3711 mm)
Overall width	5 ft 1 in (1552 mm)
Overall height:	
Special and Super	4 ft 6 in (1370 mm)
1300 GT	4 ft 5½ in (1360 mm)
Track:	
Front	4 ft 2¼ in (1288 mm)
Rear	4 ft 3 in (1303 mm)
Wheelbase	7 ft 3 in (2225 mm)

Weights

Kerb weight:	
900 cc	1565 lb (710 kg)
1050 cc	1610 lb (730 kg)
1301 cc	1709 lb (775 kg)
Fully laden weight:	
900 cc	2447 lb (1110 kg)
1050 cc	2492 lb (1130 kg)
1301 cc	2591 lb (1175 kg)
Carrying capacity	As Standard, De Luxe and Special
Maximum towing capacity:	
Special and Super	1764 lb (800 kg)
1300 GT	1863 lb (845 kg)

Capacities

Engine oil (including filter):	
903 cc engine	6.5 Imp. pints (3.7 litres)
1049 cc engine	6.0 Imp. pints (3.4 litres)
1301 cc engine	6.9 Imp. pints (3.9 litres)

Coolant (including heater):	
903 cc engine	8.8 Imp. pints (5.0 litres)
1049 cc and 1301 cc engines	9.7 Imp. pints (5.5 litres)
Fuel	6.7 Imp. gal (30.5 litres) including reserve of 0.8 Imp. gal (3.5 litres)
Transmission:	
Four-speed	4.2 Imp. pints (2.4 litres)
Five-speed	4.4 Imp. pints (2.5 litres)

Engine (1049 and 1301 cc)

General

Type number:	
1049 cc	127 A 000
1301 cc	127 A 3 000
Bore:	
1049 cc	2.992 in (76.0 mm)
1301 cc	2.996 in (76.1 mm)
Stroke:	
1049 cc	2.276 in (57.8 mm)
1301 cc	2.815 in (71.5 mm)
Compression ratio:	
1049 cc:	
Sport	9.8 to 1
All models but Sport	9.3 to 1
1301 cc	9.75 to 1
Maximum HP (DIN):	
1049 cc:	
Sport	70 at 5600 rpm
All models but Sport	50 at 5600 rpm
1301 cc	75 at 5750 rpm
Maximum torque (DIN):	
1049 cc:	
Sport	61.3 lbf ft (8.5 kgf m) at 4500 rpm
All models but Sport	57.1 lbf ft (7.9 kgf m) at 3000 rpm
1301 cc	76.0 lbf ft (10.5 kgf m) at 3500 rpm

Valve mechanism

Valve timing:	**1049 cc (Sport)**	**1049 cc (all models but Sport)**	**1301 cc**
Inlet opens	6° BTDC	2° BTDC	1° BTDC
Inlet closes	46° ABDC	42° ABDC	51° ABDC
Exhaust opens	47° BBDC	42° BBDC	42° BBDC
Exhaust closes	7° ATDC	2° ATDC	12° ATDC

Valve clearances (cold):	**1049 cc (all models but Sport)**	**1049 cc (Sport) and 1301 cc**
For checking valve timing:		
Inlet	0.028 in (0.70 mm)	0.032 in (0.80 mm)
Exhaust	0.028 in (0.70 mm)	0.032 in (0.80 mm)
Adjustment for running (engine cold):		
Inlet	0.012 in (0.30 mm)	0.016 in (0.40 mm)
Exhaust	0.016 in (0.40 mm)	0.020 in (0.50 mm)

Cylinder block and connecting rods

	inches	**mm**
Cylinder bore diameter*	2.992 to 2.994	76.00 to 76.05
**Cylinder bores are graded and have a variation of 0.0004 inch (0.01 mm) between each grade*		
Main bearing housing bore diameter	2.044 to 2.045	51.921 to 51.943
Width of centre main bearing cap between thrust washers	1.088 to 1.091	27.64 to 27.70
Connecting rod big-end diameter	1.8555 to 1.8560	47.130 to 47.142
Connecting rod small-end diameter	0.8638 to 0.8646	21.94 to 21.96
Big-end bearing shell thickness	0.0605 to 0.0608	1.537 to 1.544
Big-end bearing shell undersize range	0.01;0.02;0.03;0.04	0.254;0.508;0.762;1.016
Gudgeon pin interference fit in small-end:		
1049 cc engine	0.0004 to 0.0017	0.01 to 0.042
1301 cc engine	0.0020 to 0.0040	0.05 to 0.102
Big-end bearing running clearance:		
1049 cc engine	0.0014 to 0.0032	0.036 to 0.080
1301 cc engine	0.0014 to 0.0028	0.036 to 0.070

Pistons

	inches	**mm**
Piston-to-bore clearance	0.002 to 0.003	0.06 to 0.08
Gudgeon pin clearance in piston:		
1049 cc engine	0.0003 to 0.0006	0.008 to 0.016
1301 cc engine	0.0001 to 0.0003	0.002 to 0.008

	inches	mm
Piston ring side clearance in groove:		
Top compression	0.0018 to 0.0030	0.045 to 0.077
2nd compression	0.0016 to 0.0028	0.040 to 0.072
Oil control	0.0012 to 0.0024	0.030 to 0.062
Piston ring end gap:		
Top compression:		
1049 cc Sport	0.0118 to 0.0197	0.30 to 0.50
All models except 1049 cc Sport	0.0079 to 0.0158	0.20 to 0.40
2nd compression	0.0118 to 0.0197	0.30 to 0.50
Oil control	0.0079 to 0.0138	0.20 to 0.35
Piston ring oversize range	0.0079, 0.0157, 0.0236	0.2, 0.4, 0.6

Crankshaft

	inches	mm
Main bearing journal diameter:		
1049 cc	1.8986 to 1.8994	48.189 to 48.209
1301 cc	1.8999 to 1.8994	48.199 to 48.209
Crankpin diameter:		
1049 cc	1.7331 to 1.7339	43.988 to 44.008
1301 cc	1.7335 to 1.7340	43.998 to 44.009
Main bearing running clearance:		
1049 cc	0.0014 to 0.0032	0.036 to 0.081
1301 cc	0.0014 to 0.0028	0.036 to 0.071
Crankshaft endfloat	0.0022 to 0.0104	0.055 to 0.265

Cylinder head

	inches	mm
Standard valve guide housing bore	0.5886 to 0.5896	14.950 to 14.977
Standard valve outside diameter	0.5921 to 0.5928	15.040 to 15.058
Oversize valve guide	0.002; 0.0039; 0.0098	0.05; 0.10; 0.25
Valve guide interference fit	0.0025 to 0.0043	0.063 to 0.108
Fitted valve guide internal diameter	0.3158 to 0.3165	8.022 to 8.040
Valve stem diameter	0.3139 to 0.3146	7.974 to 7.992
Valve stem fit clearance in guide	0.0012 to 0.0026	0.030 to 0.066
Angle of valve seat	$45^{o} \pm 5'$	
Angle of valve face	$45^{o}\ 30' \pm 5'$	

Valve gear

	inches	mm
Camshaft journal diameter:		
Valve gear end	1.1789 to 1.1795	29.945 to 29.960
Centre	1.0630 to 1.0636	27.000 to 27.015
Flywheel end	0.9843 to 0.9848	25.000 to 25.015
Camshaft bearing bore diameter in head:		
Valve gear end	1.1807 to 1.1817	29.990 to 30.015
Centre	1.0648 to 1.0657	27.045 to 27.070
Flywheel end	0.9860 to 0.9870	25.045 to 25.070
Camshaft endfloat	0.0020 to 0.0110	0.05 to 0.28

Auxiliary shaft

	inches	mm
Bush fit	There must always be an interference fit	
Shaft/bush running clearance (front and rear)	0.0018 to 0.0036	0.046 to 0.091

Lubrication system

Oil filter type:	
903 cc, April 1981 on	Champion C101
1049 cc (except Sport), to March 1981	Champion C117
1049 cc (except Sport), April 1981 on	Champion C106
1049 cc Sport and 1300 GT, 1981 on	Champion C106
Oil pump	Four lobe rotor type
Pump drive	Through auxiliary shaft
Oil pressure relief valve	Incorporated in pump
Pump rotors endfloat	0.0018 to 0.0047 in (0.045 to 0.120 mm)
Rotor-to-pump body clearance	0.0006 to 0.0022 in (0.016 to 0.055 mm)
Inner rotor to outer rotor	0.0010 to 0.0039 in (0.025 to 0.100 mm)
Oil pressure at 100° C (212°F)	50 to 70 lbf/in^2 (3.5 to 5.0 kgf/cm^2)

Torque wrench settings

	lbf ft	Nm
Cylinder head:		
1049 cc:		
1st stage	30	41
2nd stage	45	61
Final stage	61	83
1301 cc (with threads oiled):		
1st stage	15	20
2nd stage	48	65
Final stage	Angle tighten bolts 180° and nuts 80°	
Main bearing cap bolts	59	80

	lbf ft	Nm
Engine mounting securing bolts	43	58
Manifold to head nuts	20	27
Connecting rod big-end nuts	38	52
Flywheel to crankshaft bolts	61	83
Driven gear (plastic)-to-camshaft retaining bolt	87	118
Driven gear (steel)-to-camshaft retaining bolt	87	118
Camshaft cap nuts	14	19
Ignition distributor clamp nut	11	15
Oil pump-to-crankcase bolts	13	18
Cylinder head outlet pipe bolt	16	22
Water pump/alternator drive pulley nut	101	137
Alternator bracket-to-crankcase bolt	20	27
Alternator-to-lower bracket bolt	36	49
Cylinder head upper bracket bolt	20	27
Alternator to upper bracket nut	36	49
Upper bracket securing bolt	13	18
Oil pressure switch	24	33
Coolant temperature switch	36	49
Spark plug	27	37
Sport models only		
Flexible mounting to body (engine side)	65	88
Flexible mounting support (engine side to body)	18	24
Flexible mounting upper support to gearbox	18	24
Engine crossmember to body	18	24
Flexible mounting support nut (gearbox side)	18	24
Flexible mounting support bolt to body (gearbox LH side)	65	88

Cooling system

Radiator fan thermal switch

Cuts in	194° to 201°F (90° to 94°C)
Cuts out	185° to 192°F (85° to 89°C)

Engine coolant thermostat

Starts to open	176° to 183°F (80° to 84°C)
Fully open	205°F (96°C)
Valve travel	0.276 in (7.0 mm)

Impellor vanes-to-pump body fit clearance ... 0.0315 to 0.0512 in (0.8 to 1.3 mm)

Radiator cap relief pressure ... 11.4 lbf/in^2 (0.8 kgf/cm^2)

Fuel system and carburation

Fuel pump

Flow rate capacity	16.5 gal/hr (75.0 litre/hour)

Air cleaner element

1049 cc (except Sport)	Champion W107
1049 cc Sport	Champion W136
1300 GT	Champion type not available

Weber 32 ICEV 16/150

Venturi diameter	0.8465 in (21.5 mm)
Main jet	0.0453 in (1.15 mm)
Air correction jet	0.0728 in (1.85 mm)
Slow running jet	0.0177 in (0.45 mm)
Emulsion tube type	F74
Accelerator pump jet	0.0157 in (0.40 mm)
Needle valve seat	0.0591 in (1.50 mm)
Accelerator pump output (10 strokes)	2 to 3 cm^3
Cold starting device	Automatic choke
Float level	1.402 to 1.421 in (35.6 to 36.1 mm)
CO level	2.5%

Solex C32 TD 1/4

Venturi diameter	0.8465 in (21.5 mm)
Main jet	0.0453 in (1.15 mm)
Air correction jet	0.0768 in (1.95 mm)
Slow running jet	0.0177 in (0.45 mm)
Emulsion tube type	N71
Accelerator pump jet	0.0177 in (0.45 mm)
Needle valve seat	0.0630 in (1.60 mm)
Accelerator pump output (10 strokes)	3 to 5 cm^3
Cold starting device	Automatic choke
Float level	0.158 to 0.197 in (4.0 to 5.0 mm)
CO level	2.5%

Weber 34 DMTR 47/250

	Primary venturi	Secondary venturi
Venturi diameter	0.8661 in (22.0 mm)	0.9449 in (24.0 mm)
Main jet	0.0421 in (1.07 mm)	0.0500 in (1.27 mm)
Air correction jet	0.0728 in (1.85 mm)	0.0866 in (2.20 mm)
Slow running jet	0.0177 in (0.45 mm)	0.0276 in (0.70 mm)
Slow running air bleed	0.0413 in (1.05 mm)	0.0276 in (0.70 mm)
Accelerator pump jet	0.0157 in (0.40 mm)	– –
Needle valve seat	0.0689 in (1.75 mm)	
Accelerator pump output (10 strokes)	8.55 cm^3	
Cold starting device	Manually operated strangler choke	
Float level	0.2657 to 0.2854 in (6.75 to 7.25 mm)	
Auxiliary venturi	0.177 in (4.50 mm)	
Emulsion tube	F30	
Throttle valve fast idle opening	0.032 to 0.034 in (0.80 to 0.85 mm)	
Choke valve mechanical pull-off	0.335 to 0.374 in (8.50 to 9.50 mm)	
Maximum choke valve vacuum pull-off	0.226 to 0.246 in (5.75 to 6.25 mm)	

Weber 30 IBA 22/350

Venturi	0.866 in (22.0 mm)
Main jet	0.047 in (1.2 mm)
Air correction jet	0.071 in (1.8 mm)
Slow running jet	0.018 in (0.45 mm)
Emulsion tube	F50
Accelerator pump output (10 strokes)	3.0 to 3.5 cm^3
Needle valve seat	0.059 in (1.5 mm)
Float level	0.226 to 0.246 in (5.75 to 6.25 mm)

Solex C30 DI 40

Venturi	0.906 in (23.0 mm)
Main jet	0.048 in (1.225 mm)
Air correction jet	0.067 in (1.7 mm)
Slow running jet	0.019 in (0.47 mm)
Emulsion tube	N75
Accelerator pump output (10 strokes)	2.5 to 3.5 cm^3
Needle valve seat	0.051 in (1.3 mm)
Float level	0.315 to 0.354 in (8.0 to 9.0 mm)

Weber 34 DMTR 54/250

	Primary	Secondary
Venturi	0.866 in (22.0 mm)	0.945 in (24.0 mm)
Auxiliary venturi	0.177 in (4.5 mm)	0.177 in (4.5 mm)
Main jet	0.042 in (1.07 mm)	0.045 in (1.15 mm)
Air bleed jet	0.075 in (1.9 mm)	0.087 in (2.2 mm)
Emulsion tube	F30	F30
Idle jet	0.018 in (0.45 mm)	0.028 in (0.70 mm)
Air idle jet	0.041 in (1.05 mm)	0.028 in (0.70 mm)
Pump jet	0.016 in (0.4 mm)	– –
Needle valve	0.069 in (1.75 mm)	
Accelerator pump output (10 strokes)	9.0 to 15.0 cm^3	
Float level	0.276 $\pm$ 0.010 in (7.0 $\pm$ 0.25 mm)	
Throttle valve fast idle opening	0.032 to 0.034 in (0.80 to 0.85 mm)	
Choke valve mechanical pull-off	0.335 to 0.374 in (8.5 to 9.5 mm)	
Choke valve vacuum pull-off	0.226 to 0.246 in (5.75 to 6.25 mm)	

Ignition system

Distributor

Total centrifugal advance:	
1049 cc	25° $\pm$ 2°
1301 cc	27° $\pm$ 2°

Spark plugs

Make and type:	
1049 cc (except Sport)	Champion RN9YCC or RN9YC
1049 cc Sport and 1300 GT	Champion RN7YCC or RN7YC
Electrode gap:	
Champion RN9YCC or RN7YCC	0.032 in (0.8 mm)
Champion RN9YC or RN7YC	0.028 in (0.7 mm)

HT leads

1049 cc	Champion CLS4 (boxed set)
1301 cc	Champion type not available

Clutch 1049 cc and 1301 cc

Lining outer diameter	7.146 in (181.5 mm)
Lining inner diameter	5.0 in (127.0 mm)
Maximum run-out of driven plate linings	0.008 in (0.2 mm)
Travel of release flange, corresponding to a pressure plate displacement of not less than 0.067 in (1.7 mm)	0.335 in (8.5 mm)

Minimum pedal travel

903 cc and 1049 cc	4.724 in (120.0 mm)
1301 cc	4.803 in (122.0 mm)

Transmission

Four-speed gearbox

Final drive ratio (Sport 1049 cc)	4.462 to 1 (13/58)

Five-speed gearbox

Gear ratios:	
First	3.909 to 1
Second	2.055 to 1
Third	1.342 to 1
Fourth	0.964 to 1
Fifth	0.830 to 1
Reverse	3.615 to 1
Final drive	4.071 to 1 (14/57)
Differential bearing preload	0.0047 in (0.12 mm)
Differential gear backlash	0.0039 in (0.10 mm)

Torque wrench settings	**lbf ft**	**Nm**
Starter motor bolts to bellhousing lower support	18	24
Gear selector shaft nut	18	24
Upper gear lever relay lever	22	30
Idler support securing nut	18	24
Differential case flange to gearbox housing	18	24

Braking system (all later models)

Front brakes

Disc nominal thickness	0.421 to 0.429 in (10.7 to 10.9 mm)
Minimum thickness after refacing	0.382 in (9.7 mm)
Minimum thickness (wear)	0.354 in (9.0 mm)
Minimum pad lining thickness	0.059 in (1.5 mm)

Electrical system

Battery

Capacity (1300 GT)	45 amp hr

Alternator	**Marelli**	**Femsa**	**Lucas 24041 E**	**Lucas 18ACR**
Maximum current	47 amp	48 amp	50 amp	50 amp
Cut-in speed	1050 ± 50rpm	1200±50rpm	1200±50rpm	1200±50rpm
Current at 6000 rpm	45 amp	43 rpm	43 amp	43 amp
Field resistance	3.0 to 3.2 ohm	3.8 to 4.2 ohm	3.0 to 3.4 ohm	3.0 to 3.4 ohm
Minimum brush length protrusion	0.3 in (8 mm)	0.3 in (8 mm)	0.3 in (8 mm)	0.3 in (8 mm)
Voltage regulator range	14.3 to 14.0 volt			

Starter motor (Femsa/Marelli)

Nominal power	0.8 kW
Number of poles	4
Armature endfloat	0.004 to 0.020 in (0.1 to 0.5 mm)

Lamps	**Wattage**
Headlamps (Sport and 1300 GT)	55/60
Headlamps (except Sport and 1300 GT)	40/45
Brake and rear light	5/21
Turn indicators	21

	Wattage
Reversing lights	21
Parking lights	5 or 4
Number plate	5
Courtesy light	5
Boot light	5
Turn repeaters	4 or 5
Cigar lighter light	4 or 1.2
Instrument panel	3
Ignition warning	3
Turn indicator warning	1.2
Headlamp warning	1.2
Coolant temperature warning (L models)	1.2
Oil pressure warning	1.2
Fuel warning	1.2
Hazard warning	1.2
Sidelamp (out) warning	3
Brake warning	1.2
Heated rear window warning	1.2
Rear fog-light	21
Rear foglight warning	1.2

Steering

Steering angles

Inner wheel:	
1300 GT	33° 40′ ± 1° 30′
All except 1300 GT	34° 50′
Outer wheel:	
1300 GT	31° 50′
All except 1300 GT	32° 10′

Front wheel alignment	Unladen	Laden
Toe-in:		
903 cc engine models	0.16 to 0 in (4 to 0 mm)	0.06 to 0.10 in (1.5 to 2.5 mm)
1049 cc engine models	0.14 to 0.02 in (3.5 to 0.5 mm)	0.06 to 0.10 in (1.5 to 2.5 mm)
1301 cc engine models	0.04 in toe-in to 0.12 in toe-out (1.0 mm toe in to 3.0 mm toe-out)	0.16 to 0 in (4.0 to 0 mm)
Camber:		
903 cc engine models	1° 35′ ± 30′	50′ ± 30′
1049 cc engine models	1° 25′ ± 30′	45′ ± 30′
1301 cc engine models	0° 55′ to 1° 55′	0° 10′ to 1°10′
Castor:		
903 cc engine models	1° 50′ ± 30′	3° 15′ ± 30′
1049 cc engine models	1° 40′ ± 30′	3° 10′ ± 30′
1301 cc engine models	1° 20′ to 2°20′	2° 40′ to 3° 40′

Torque wrench settings	lbf ft	Nm
Steering wheel	22	30

Suspension

Rear wheel alignment	Unladen	Laden
Toe-in	0 to 0.158 in (0 to 4 mm)	0 to 0.158 in (0 to 4 mm)
Camber	1° 30′ to 0° 30′	3° 50′ to 2° 50′

Roadwheels

Size	4½B x 13	
Tyres	135 SR–13 or 155/70 SR–13	
Tyre pressures (lbf/in²/bar)	**Front**	**Rear**
1300 GT:		
Normal load	29/2.0	29/2.0
Fully laden	29/2.0	32/2.2
Special/Super - normal load	28/1.9	28/1.9

Correct at time of writing – consult your dealer for current recommendations

Torque wrench settings	lbf/ft	Nm
Front wheel bearing ring nut	44	60
Front wheel hub nut	160	217
Roadwheel bolts	64	88
Front suspension track control arm to body	20	27
Front suspension balljoint to hub carrier	40	54
Rear wheel hub nut	160	217

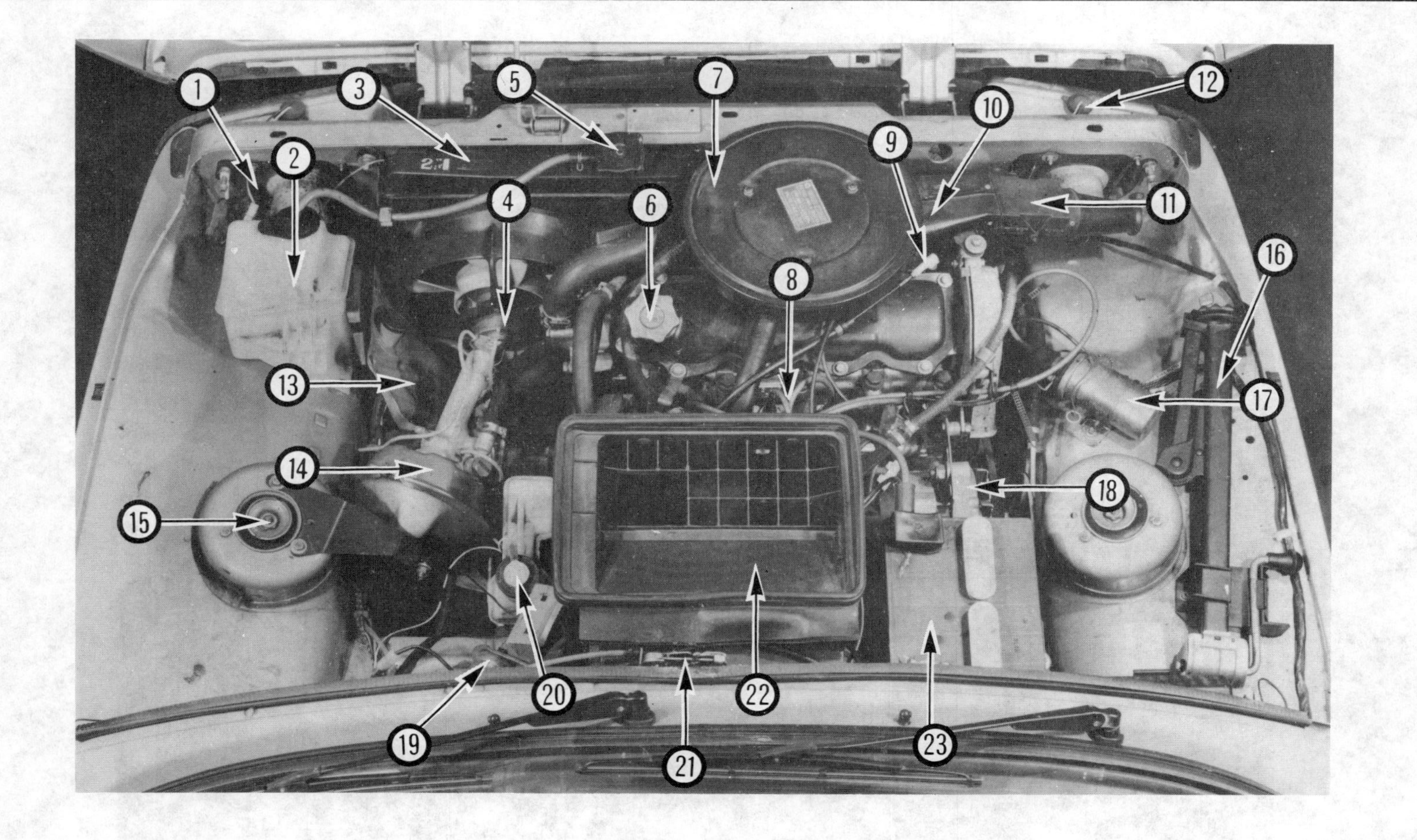

Under-bonnet view (1300 GT)

1 *Headlamp unit*
2 *Cooling system expansion tank*
3 *Radiator*
4 *Cooling fan motor*
5 *Radiator pressure cap*
6 *Engine oil filler cap*
7 *Air cleaner*
8 *HT leads to distributor*
9 *Accelerator cable and end fitting*
10 *Engine oil level dipstick*
11 *Air cleaner temperature control unit*
12 *Sidelamp bulb*
13 *Windscreen washer reservoir*
14 *Brake vacuum servo unit*
15 *Front suspension upper mounting*
16 *Lifting jack*
17 *Ignition coil*
18 *Alternator*
19 *Oil pressure unit*
20 *Brake fluid reservoir filler cap*
21 *Bonnet lock*
22 *Heater unit*
23 *Battery*

View of front underside of car (1300 GT)

1 Gearbox drain plug
2 Suspension strut and coil spring
3 Steering gear
4 Gearbox
5 Windscreen washer pump
6 Radiator bottom hose
7 Engine/gearbox centre mounting
8 Front towing eye
9 Anti-roll bar
10 Engine sump drain plug
11 Brake caliper
12 Driveshaft
13 Suspension control arm
14 Steering knuckle arm
15 Trackrod end
16 Exhaust system
17 Gearchange lever assembly

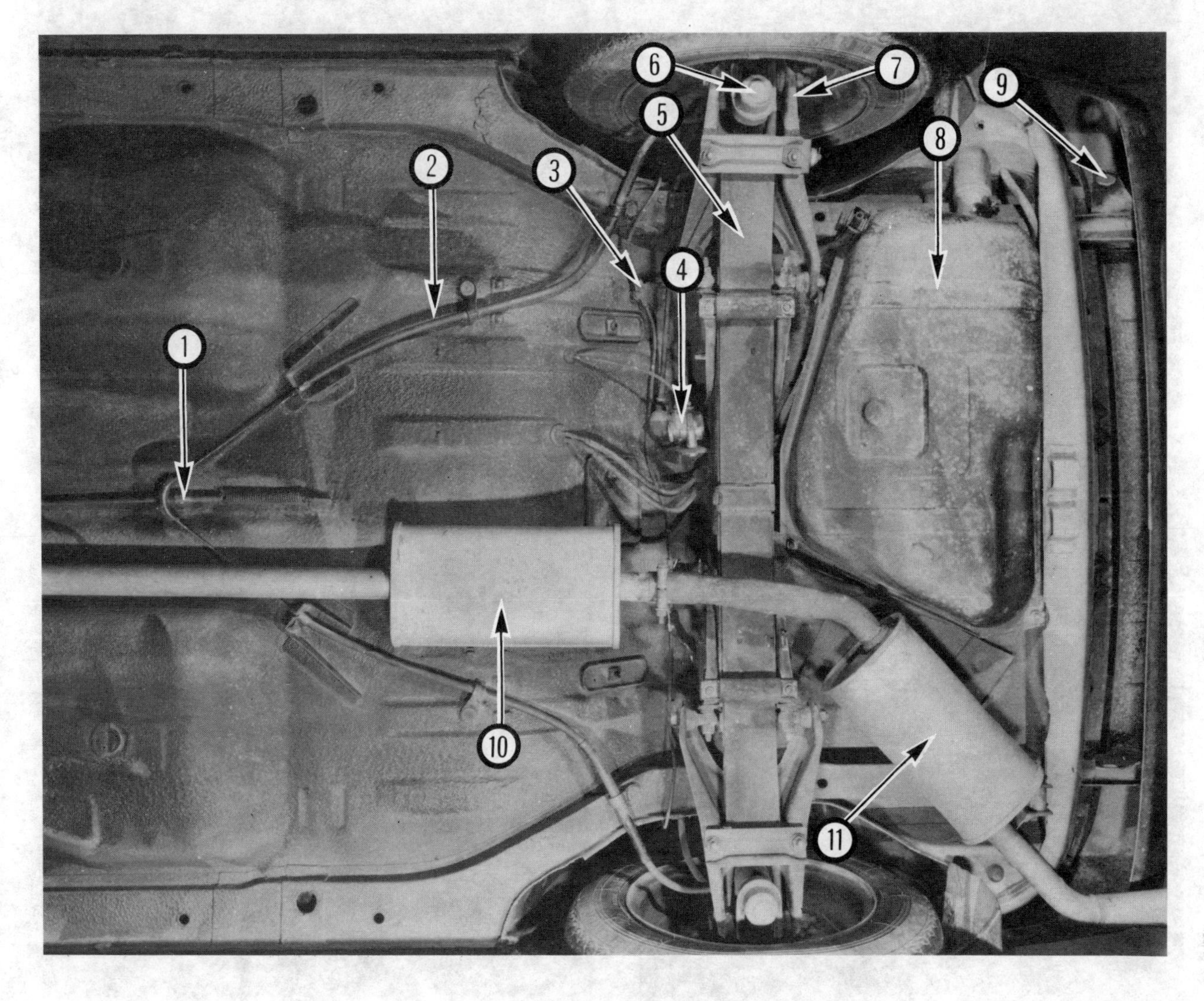

View of rear underside of car (1300 GT)

1 *Handbrake cable compensator and adjustment*
2 *Handbrake cable*
3 *Hydraulic brake pipe 3-way adaptor*
4 *Rear brake pressure regulator*
5 *Rear suspension spring*
6 *Shock absorber*
7 *Suspension control arm*
8 *Fuel tank*
9 *Rear towing eye*
10 *Exhaust intermediate silencer*
11 *Exhaust rear silencer*

3 Engine

1049 cc and 1301 cc engines - general

1 This engine is of overhead camshaft design, using shims for valve clearance adjustment. The crankshaft is supported in five main bearings, the centre one incorporating the thrust washers which control crankshaft endfloat.
2 An auxiliary shaft, driven by the toothed camshaft belt, is used to drive the distributor and the fuel pump.
3 Most major engine components can be removed while the engine is in the car, but operations on the crankshaft, main bearings and flywheel can only be carried out after the engine has been removed.
4 Engine removal and subsequent dismantling follows closely the information given for the overhead valve engine in Chapter 1, but the following sequence for complete engine dismantling is recommended (photos):

(a) Engine ancillaries (alternator, fuel pump, distributor)
(b) Timing belt cover
(c) Water pump
(d) Timing belt tensioner and belt
(e) Manifolds
(f) Cylinder head complete with camshaft
(g) Crankshaft pulley
(h) Auxiliary shaft sprocket
(i) Sump
(j) Oil pump and auxiliary shaft
(k) Connecting rods and pistons
(l) Flywheel and crankshaft oil seal carriers
(m) Crankshaft and main bearings

5 If the cylinder head is to be dismantled, before withdrawing the camshaft have a suitably divided container ready so that the valve clearance adjusting shims can be extracted and kept in strict originally installed order together with their appropriate valves, springs etc.
6 All engine parts must be thoroughly cleaned and examined as explained in Chapter 1. Where required, all defective parts should be renewed before reassembly starts.

3.4A Left-hand side engine mounting on a 1300 GT

3.4B Right-hand side engine mounting on a 1300 GT

3.4C Crossmember mounting on a 1300 GT

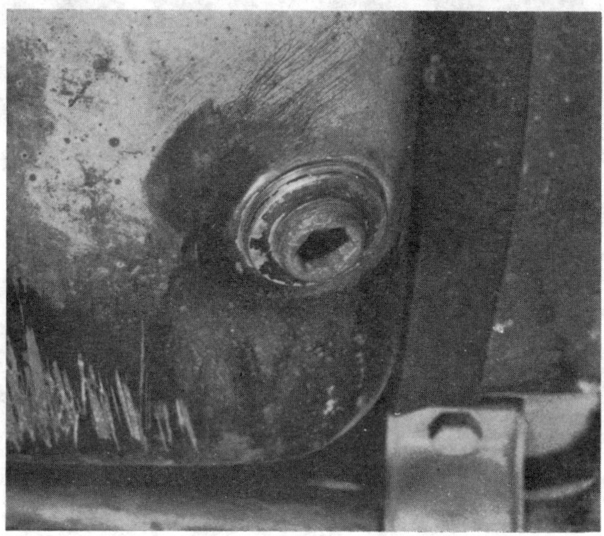

3.4D Engine sump drain plug

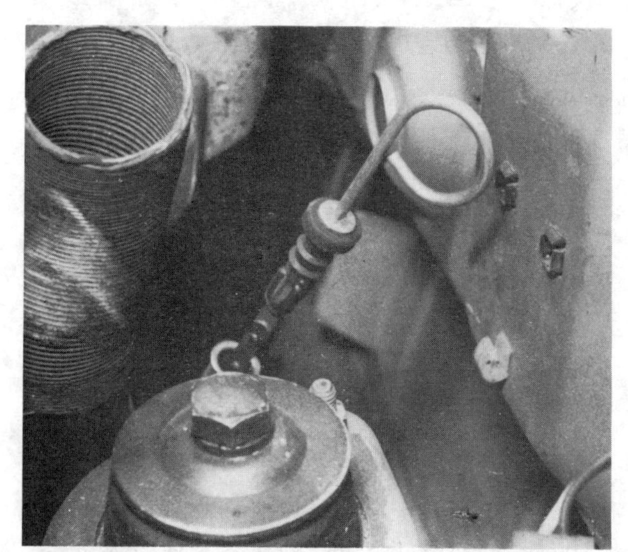

3.4E Engine oil level dipstick

3.4F Engine oil pressure switch unit mounted on the bulkhead

Crankshaft - refitting (OHC engine)

7 Fit the main bearing shells to their seats in the crankcase after making sure that both shells and seats are spotlessly clean and dry (photos).

8 With a light smear of grease, fit the two half thrust washers each side of the centre bearing with the oil grooves in each washer facing away from the bearing shell (photo).

9 Using clean engine oil lubricate the bearing shells and crankshaft main bearing journals (photo).

10 Carefully lower the crankshaft into its bearings in the crankcase after making sure that it is the right way round (photo). Spin the shaft to distribute the oil.

11 Fit the clean and dry bearing shells to the main bearing caps. Oil the bearing face and fit the bearing caps to the crankcase. Make sure that each cap is fitted to its own location by checking the groove marks in the base, and that each cap is the right way round. This is achieved when the axial locating tags in each half bearing shell butt on the same side (photos).

12 Refit the main bearing cap bolts and screw them up finger-tight. Spin the crankshaft and, if all is well, tighten the bolts to the specified torque load and turn the crankshaft again to check freedom of rotation (photo).

13 Measure the crankshaft endfloat with feeler gauges. If this exceeds the specified tolerance, oversize thrust washers will have to be fitted.

3.8 The half thrust washers are fitted to each side of the centre bearing

3.7A Thoroughly clean the bearing shells and seats before assembly

3.9 Oil the bearings and crankshaft journals . . .

3.7B Note that the central bearing shell has no oil groove but all shells are axially located by an offset tag

3.10 . . . and then fit the crankshaft into the crankcase

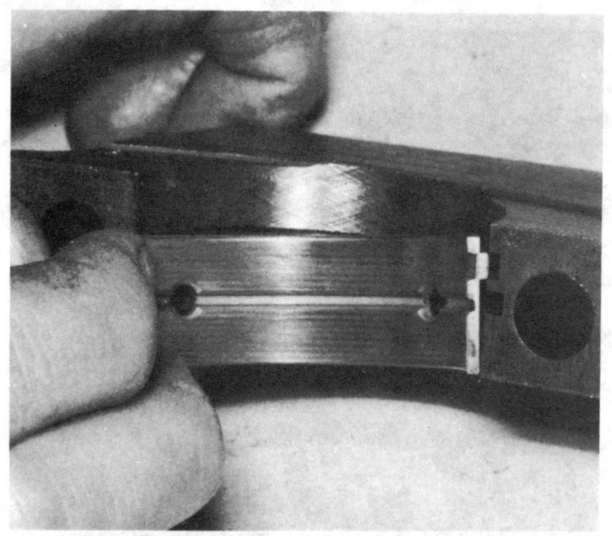
3.11A Fit the bearing shells to the main bearing caps . . .

3.11B . . . and then fit the caps to the crankcase after oiling the shells

3.11C Ensure that each cap is the right way round, in its correct location, and the one with four marks is at the flywheel end of the crankcase

3.12 Tighten the main bearing cap bolts with a torque wrench

Piston rings - refitting (OHC engine)

14 Follow the procedure given in Chapter 1, Section 42.

Pistons and connecting rods - reassembly

15 Follow the procedure given in Chapter 1, Section 43 (photo).

Pistons - refitting (OHC engine)

16 Follow the procedure given in Chapter 1, Section 44.

17 Note that the connecting rods have two oil jet holes leading from the big-end bearing (photo).

18 When the piston is correctly fitted it will have the valve depressions adjacent to the side of the block with the cylinder head studs in (photo).

Connecting rods to crankshaft-reassembly (OHC engine)

19 Follow the procedure given in Chapter 1, Section 45.

20 Torque load the big-end cap nuts to the reading quoted in the Specifications of this Chapter 13 – it is different from the value quoted for the 903 cc engine (photo).

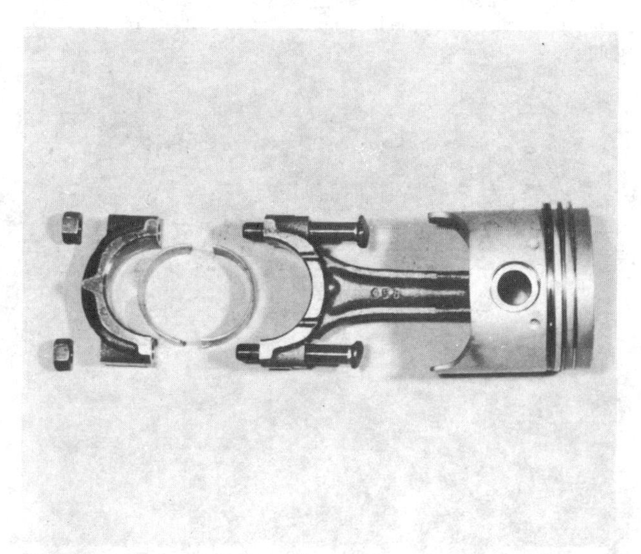
3.15 Piston and connecting rod assembly

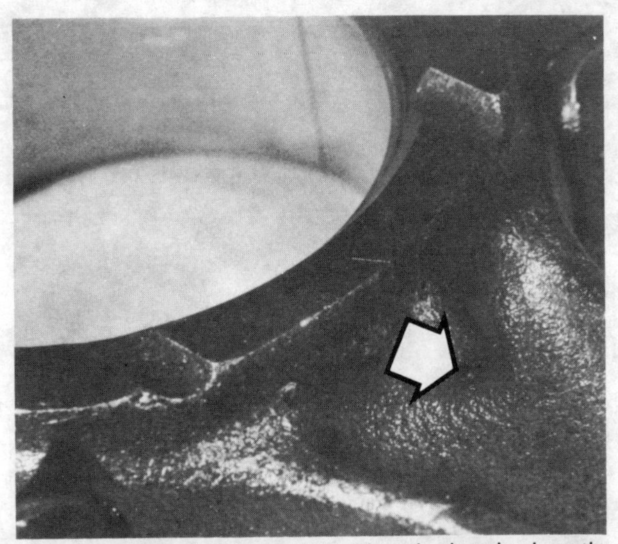

3.17 One of the oil jet holes in the big-end - the other is on the other side of the rod

Oil pump - reassembly (OHC engine)

21 The oil pump is a Hobourn-Eaton rotor type of pump which is quite different from the gear pump fitted to the 903 cc engine. It consists of a four lobed rotor rotating in a five slotted outer rotor which is mounted eccentrically to the inner rotor. As the inner rotor rotates the outer rotor is driven round, and the spaces between the lobes on the inner rotor and the slots in the outer rotor increase and decrease once per revolution. The increasing spaces are connected to the pump inlet and cause oil to be drawn into the pump. The decreasing spaces connect with the pump outlet through which the oil is forced to feed the engine. A spring-loaded relief valve in the outlet of the pump vents excessive oil pressure into the sump. The efficiency of the pump depends on fine clearances and these should be checked during assembly.

22 Fit the outer rotor to the pump body and then fit the inner rotor (photos).

23 Lay an accurate straight edge across the face of the rotors and pump body and measure the clearance with feeler gauges (photo).

24 Insert a feeler gauge between the outer rotor and the pump body to measure the clearance (photo).

25 Similarly insert a feeler gauge between the inner rotor lobe

3.18 The valve depressions in the piston are adjacent to the cylinder head studs in the block

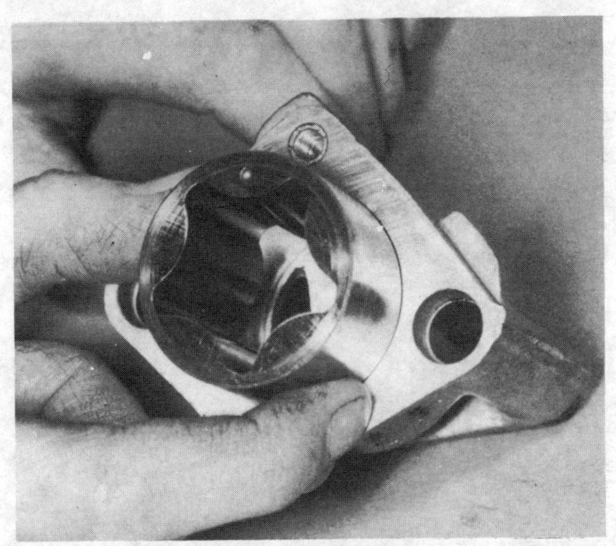

3.22A First fit the outer rotor to the oil pump body . . .

3.20 Torque tighten the big-end cap nuts

3.22B . . . followed by the inner rotor

3.23 Measure the end clearance of the rotors with the pump face

3.25 . . . and the inner-to-outer rotor clearance

3.24 Measure the outer rotor-to-body clearance . . .

3.27A Fit the oil pump pressure relief valve plunger and spring . . .

tip and the outer rotor to measure the clearance between the two (photo).

26 Compare the clearances with the tolerances quoted in the Specifications to this Chapter. Any excessive clearance could result in low oil pressure, and as the inner and outer rotors are matched pairs the only solution is a new or reconditioned pump assembly.

27 Fit the oil pressure relief valve plunger to the pump body followed in turn by the spring, retainer plate and circlip (photos).

28 Lubricate the rotors with clean engine oil and refit the pump lower half body and strainer (photo).

29 With a new gasket, fit the assembled oil pump to the crankcase block and secure with the two long bolts (photo). Tighten to the specified torque load.

30 Fit the oil tube to the crankcase securing it by its bracket to the centre main bearing cap with its retaining bolt (photo).

Crankshaft oil seals and carriers - refitting (OHC engine)

31 Clean the flywheel mounting spigot on the end of the crankshaft and lubricate the crankshaft seal with clean engine oil. Fit a new gasket.

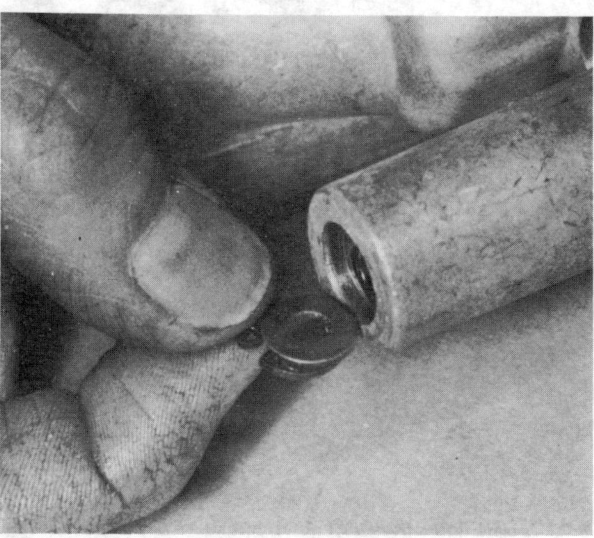
3.27B . . . followed by the retainer plate and circlip

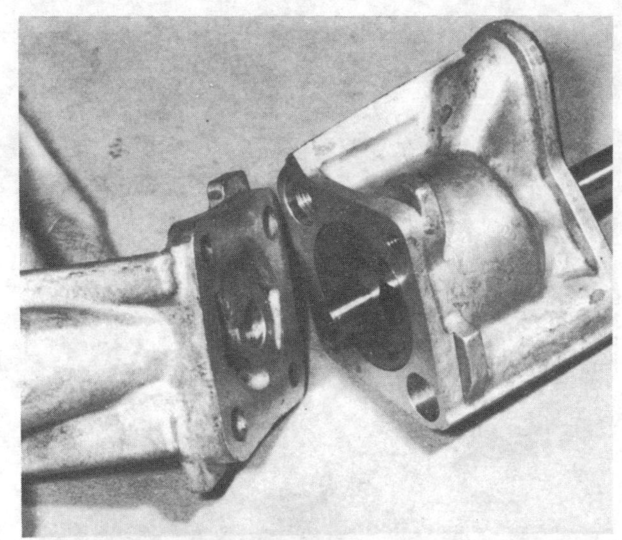
3.28 Refit the pump lower half body

3.32 Fitting the crankshaft seal and carrier at the flywheel end . . .

3.29 Fit the oil pump to the crankcase

32 Carefully ease the lip of the seal onto the spigot and secure the carrier with the bolts and washers (photo).
33 Similarly clean the crankshaft at the timing belt end and fit the seal and carrier (photo). Retain by the two bottom bolts.
34 Put the timing indicator bracket over the two top bolt holes in the seal carrier and fit the two top bolts (photo).

Sump - refitting (OHC engine)

35 Make sure that there are no remnants of the old gasket on the sump flange and fit a new gasket using a little grease to hold it in position. Check that it is bedded down evenly all round the flange.
36 Fit the sump to the crankcase (photo). Put the load spreading washers on each bolt and screw into the crankcase.
37 Tighten the bolts evenly to avoid warping the flange.

Auxiliary shaft and seal - refitting (OHC engine)

38 Clean the auxiliary shaft bearings and lubricate with clean engine oil.
39 Insert the shaft into the crankcase bushes and rotate the shaft to spread the oil (photo).

3.30 The oil tube is secured by a bracket to the centre main bearing cap with a bolt and washer

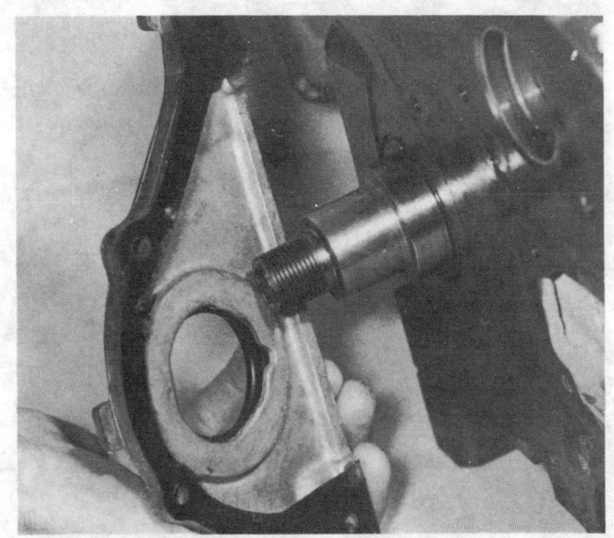
3.33 . . . and at the timing cover end

3.34 The timing indicator bracket (arrowed) is located behind the two top bolts of the seal carrier

3.40 . . . and its seal and carrier

3.36 Fit the sump to the crankcase

40 Lubricate the auxiliary shaft seal in its carrier and carefully ease the seal over the shaft spigot (photo).

41 Fit the seal carrier retaining bolts and washers and tighten.

Belt pulleys and tensioner - refitting (OHC engine)

42 Fit the toothed pulley to the auxiliary shaft. The recess in the pulley fits on the auxiliary shaft with the dowel on the shaft in the hole in the pulley (photo). Fit the retaining bolt and washer and partially tighten, as it will be necessary to wait until the drivebelt has been fitted before finally tightening this bolt. Alternatively, it is possible to hold the auxiliary shaft carefully in a vice, fit the seal and carrier to the shaft, followed by the toothed pulley and its retaining bolt and washer, and then tighten the bolt fully before fitting the complete assembly to the block.

43 The belt tensioner bracket can now be fitted. Clean off all traces of old gasket from the bracket and block and use a new gasket on assembly. Fit the retaining bolts and washers and tighten (photo).

44 Insert the spring-loaded plunger assembly into the tensioner bracket (photo).

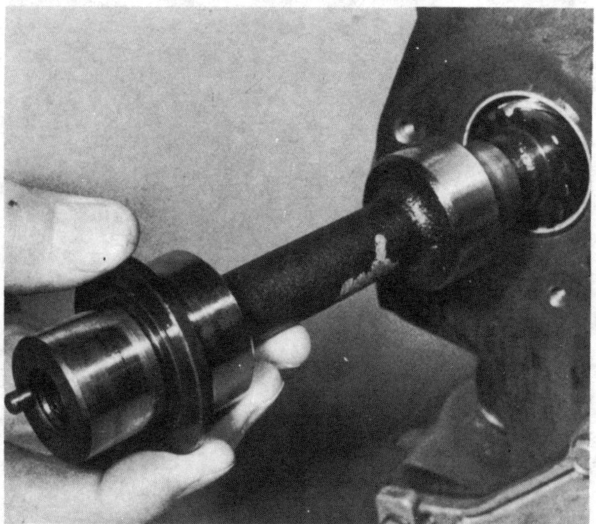
3.39 Fitting the auxiliary shaft . . .

3.42 This face of the pulley fits towards the shaft

3.43 Fit the belt tensioner bracket with a new gasket to the block . . .

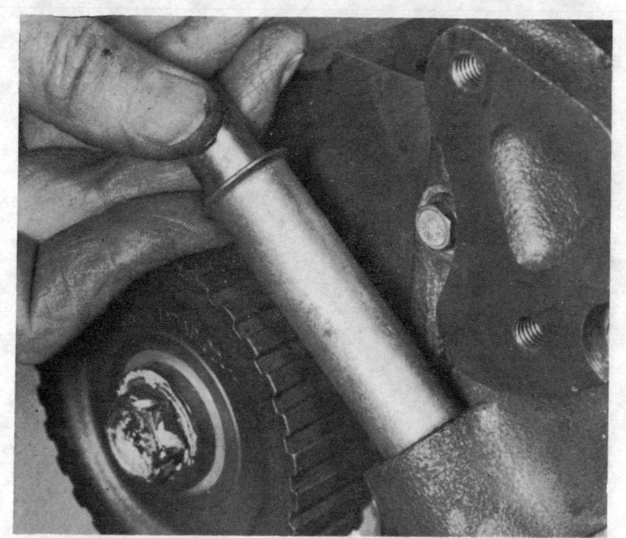

3.44 . . . followed by the spring-loaded plunger . . .

3.45 . . . and then the tensioning wheel

45 The tensioning wheel in its carrier can now be fitted (photo). Put the top bolt and washer in first and then bear down to compress the spring and fit the bottom bolt through the kidney-shaped slot. Note that this latter bolt has an additional large washer against the wheel carrier. Temporarily tighten the two bolts – they will have to be retightened after the belt has been fitted.

46 If the key had previously been removed from the pulley end of the crankshaft, clean the key slot and refit the key. Then slide on the bevelled washer, making sure that the bevel is on the side away from the crankcase (photo).

47 Refit the crankshaft toothed pulley (photo). Then the V-belt pulley can be refitted together with its retaining nut and washer. Final tightening of this nut can wait until the flywheel has been fitted, when a 'gag' can be fitted to the flywheel to hold the crankshaft whilst tightening the nut – see Chapter 1, photo 23.2A.

3.46 The bevelled washer is fitted with the bevel outwards and is followed by . . .

3.47 . . . the crankshaft toothed pulley

Cylinder head and valve gear - reassembly (OHC engine)

48 It is assumed that the valves will have already been examined and renovated as described in Chapter 1, Section 31. Follow the procedure given in Chapter 1, Section 51 to reassemble the valves, but note that new oil seals should be fitted to this engine

when the valves have been inserted in the guides and before the springs are fitted (photos).

49 Each tappet bucket contains a shim in the head which is used to control the valve clearance. Before assembling the buckets to their valves, prise out each shim and take a note of the thickness. This is etched on the lower face of the shim and indicates the thickness in millimetres to two decimal places. If the number has worn off, use a micrometer to check the shim thickness. Make a table showing each valve by number and the thickness of shim on assembly. Reassemble each shim to its bucket and after lubricating with clean engine oil fit the buckets to their respective valves (photos).

50 Lubricate the two camshaft bearings in the cylinder head and carefully thread the camshaft through the driving end hole and lower it onto its bearings. The cams will rest on the tappet buckets and the camshaft should now be turned so that the two cams over No. 1 cylinder (driving end) are pointing upwards (compression/firing stroke). This is to reduce the bending load on the camshaft as the two bearing halves are being tightened down (photo).

3.48C . . . the lower spring seat and springs . . .

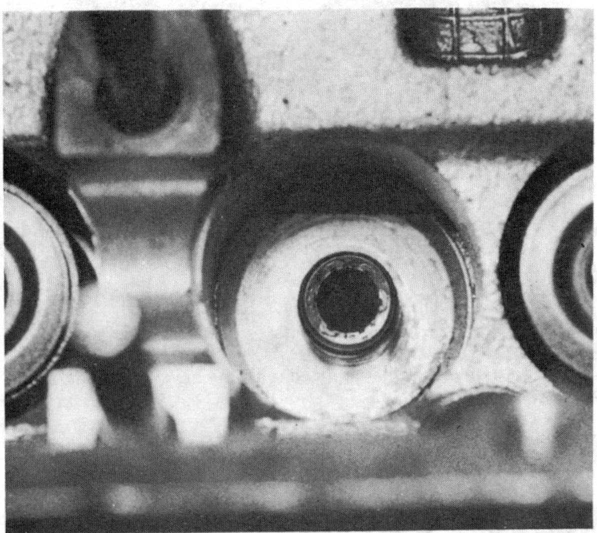

3.48A Valve guide oil seal

3.48D . . . and the upper spring seat

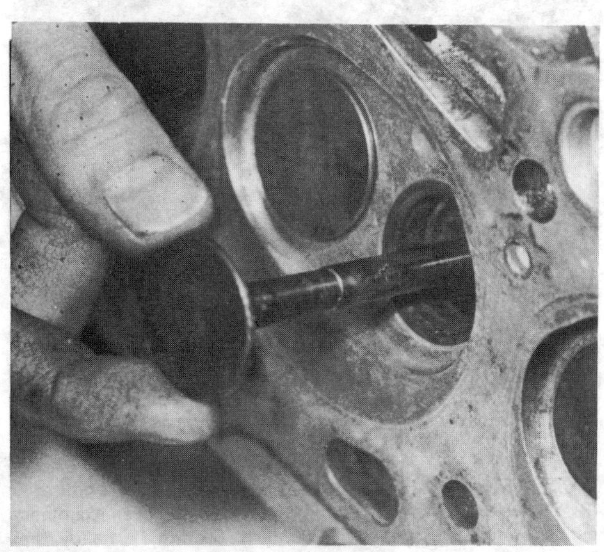

3.48B Inserting a valve followed by . . .

3.48E A magnet is useful for refitting the split collets

3.48F After assembly a sharp tap on the valve stem will help to bed the parts in

3.49A Use a small pointed tool to . . .

3.49B . . . prise out the valve clearance adjusting shim

3.49C This shim is 4.15 mm thick

3.49D Refit each tappet bucket with its shim to its valve

3.50 Position the camshaft in the cylinder head . . .

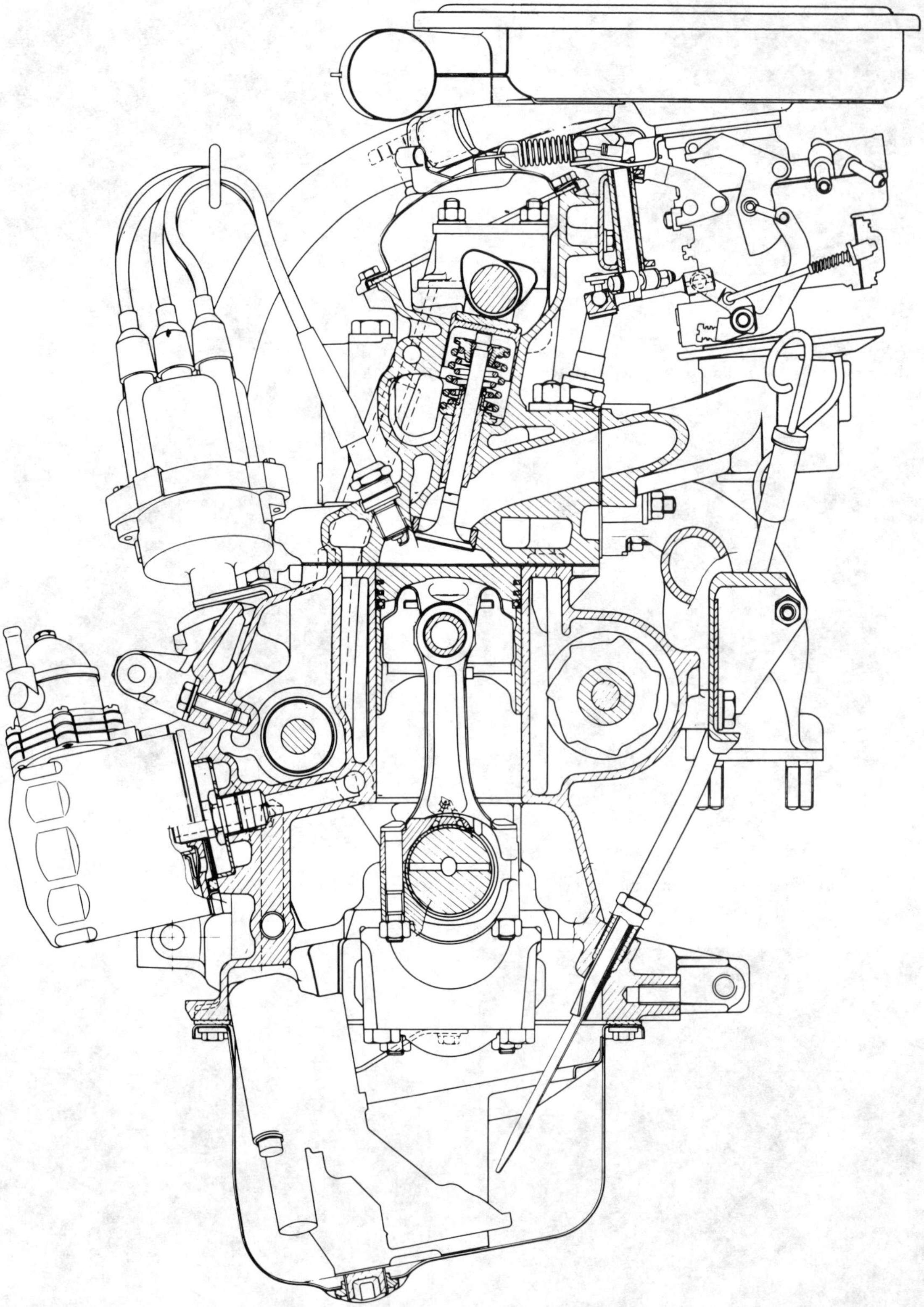

Fig. 13.1 1049 cc and 1301 cc engine, cross-section (Sec 3)

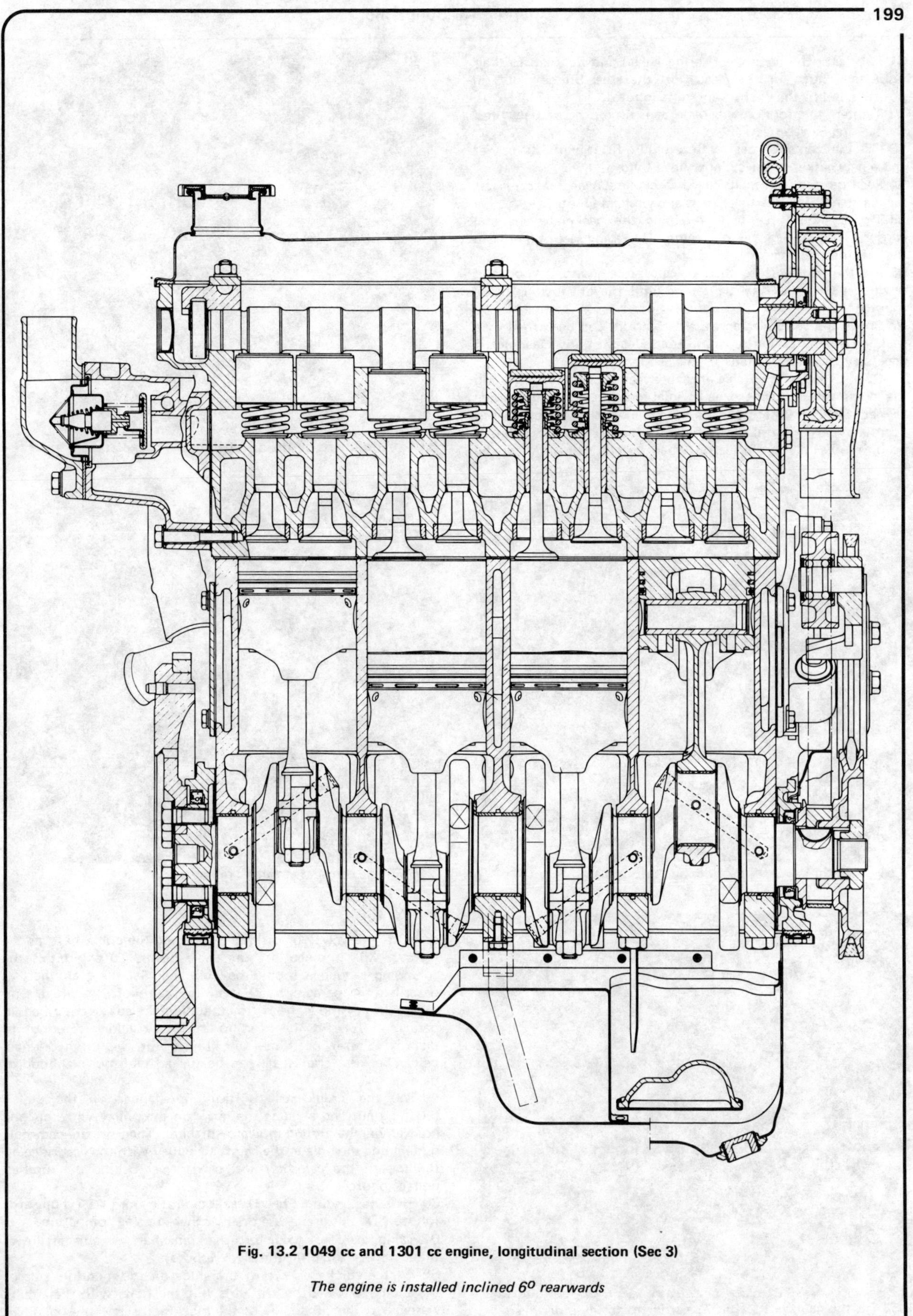

Fig. 13.2 1049 cc and 1301 cc engine, longitudinal section (Sec 3)

The engine is installed inclined 6° rearwards

51 Lubricate the camshaft bearing halves and fit them to their respective studs in the head. Put the steel bridge plates in position and fit the washers and nuts (photo).
52 Tighten the four nuts a little at a time progressively until the bearing halves meet.
53 Oil the camshaft seal and carefully fit it with its carrier and a new gasket to the cylinder head (photo).
54 Position the drivebelt guard backplate over the camshaft seal carrier and fit the three bolts and washers (photo).
55 Tighten the three bolts retaining the backplate and seal carrier. Torque load the four camshaft bearing securing nuts to the specified setting.
56 Fit the camshaft toothed driving pulley with its bolt and washer. Leave the final tightening until the drivebelt is fitted (photo).
57 Check the valve clearances; the camshaft can be turned by a spanner on the pulley retaining bolt. Readjust any clearance if necessary, as described later in this Section.
58 Owing to the small clearance between a piston at TDC and the open valves during exhaust/inlet overlap, it is imperative to get the crankshaft and camshaft in their correct related positions before turning the engine after fitting the cylinder head. If this relationship is out then serious damage could be done to the valves or pistons by turning the crankshaft, as the pistons will impinge on the valve heads. To avoid this the following sequence should be observed. Set the crankshaft as described in paragraph 59, and the camshaft (before assembling the cylinder head to the block) also as described in paragraph 59. Fit the head to the block taking care not to disturb the set positions and, finally, fit the toothed drivebelt. Then the crankshaft can be turned with no likelihood of damage.
59 Set the crankshaft by using a spanner on the pulley retaining nut and aligning the mark on the pulley with the long pointer on the timing indicator bracket. Then set the camshaft by aligning the hole in the camshaft pulley with the cast ridge on the top of the camshaft seal carrier just behind the toothed pulley (photo).
60 Fit a new cylinder head gasket with the word ALTO upwards (photo). Do not use grease or any other jointing compound.
61 Taking care not to disturb the crankshaft or camshaft lower the cylinder head onto the block (photo).
62 Fit the thick washers to the studs on the manifold side of the block followed by the nuts. Fit the bolts with their thick washers to the other side of the block and screw the bolts and nuts down to lightly compress the gasket. The bolts and nuts

3.54 . . . followed by the drivebelt guard backplate

3.51 . . . and fit the half bearings and bridge plates

3.56 Fit the camshaft toothed pulley

3.53 Fit the camshaft seal in its carrier to the cylinder head . . .

3.59 Hole in pulley aligned with cast ridge, viewed from behind the wheel

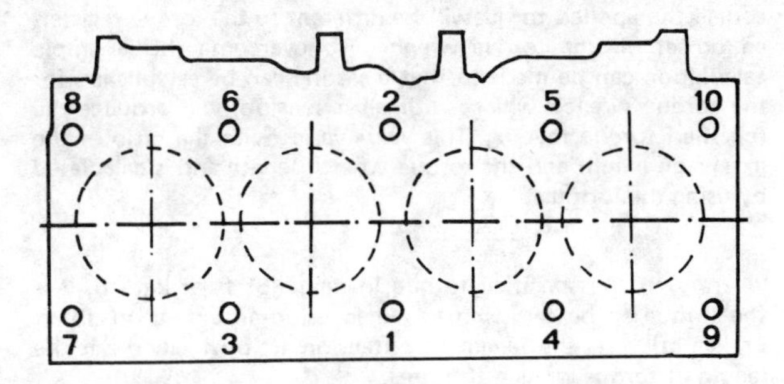

Fig. 13.3 1049 cc and 1301 cc engine cylinder head nut and bolt tightening sequence (Sec 3)

must now be torque loaded by stages (see Specifications) in sequence, to the specified load and in the order shown in Fig. 13.3 (photo). This may present some difficulty as, owing to the shape of the cylinder head, it is not possible to get a socket spanner fitted to a torque wrench over the cylinder head nuts (photo). These can be tightened using a ring or open-ended

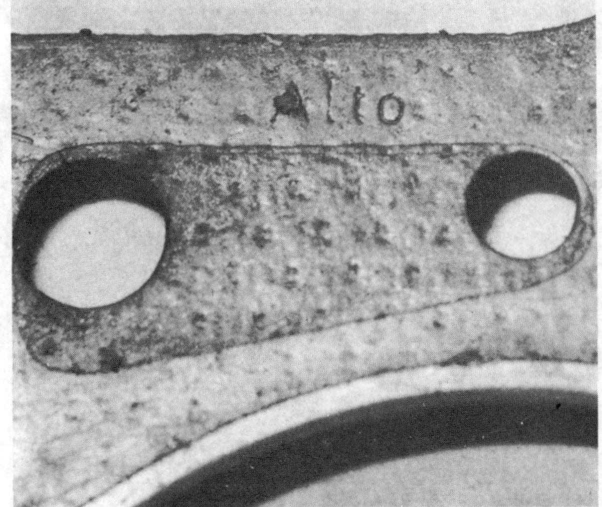

3.60 The cylinder head gasket is fitted with 'Alto' upwards

3.62A Tighten the cylinder head bolts and nuts to the specified torque

3.61 Lower the cylinder head onto the block

3.62B These cylinder head nuts might be difficult to torque tighten

spanner as an extension to the torque wrench, but then, of course, the applied torque will be different to the torque registered or set on the torque wrench. To overcome this a simple calculation can be made so that a setting can be established for the torque wrench which, with an extension, will produce the specified torque loading. This value varies with the ratio of the extension length and the torque wrench length and is calculated by using the formula:

$$B = \frac{Az}{y + z}$$

where A is the specified torque loading (lbf ft or kgf m), B is the torque to be set, or read, when using an extension (lbf ft or kgf m), y is the length of extension (ft or m), and z is the length of torque wrench (ft or m).

Example: Specified torque is 60 lbf ft, length of extension is 6 in, and length of torque wrench is 2 ft, then B = (60 x 2) ÷ (6/12 + 2) = 120 ÷ 2½ = 48 lbf ft; torque wrench setting or indication.

When measuring, use centres of bolts/nuts, and centre of torque wrench drive square. When calculating, keep values constant, that is, don't mix inches and feet for example.

63 Fit the drivebelt as described later in this Section. When fitted, the auxiliary shaft pulley bolt and the camshaft pulley bolt can be torque loaded to the specified settings if these were not done on assembly.

Distributor - refitting (OHC engine)

64 The distributor is mounted nearly vertical on the oil filter side of the engine and is driven by skew gears from the auxiliary shaft (photo). In turn the distributor shaft also drives the oil pump through a splined coupling.

65 It is more convenient to set the contact breaker gap before fitting the distributor to the engine. Adjust to the specified clearance (photo).

66 Turn the crankshaft in the normal direction of rotation until the line on the crankshaft pulley is adjacent to the first of the three pointers on the timing bracket. This is 10° BTDC, the second pointer is 5° BTDC and the large, third pointer is TDC. Note which cylinder, either 1 or 4, is on the compression stroke. This is indicated by both of the inlet and exhaust cams pointing upwards causing their relative valves to be shut.

67 Rotate the distributor shaft until the rotor is opposite the terminal in the distributor cap serving the same cylinder number as the one which was on compression stroke and with the contact breaker points just breaking.

68 Lubricate the distributor skew gear with clean engine oil and insert the distributor into the engine block (photo). Watch the rotor carefully to see how much it turns as the skew gears mesh. Then withdraw the distributor, reset the rotor and then preset it the same amount that it turned when initially inserted; then reinsert it into the engine. Ideally the contact points should be just breaking with the rotor opposite the correct numbered terminal in the cap. If this is not the case, repeat on a trial and error basis until this is achieved.

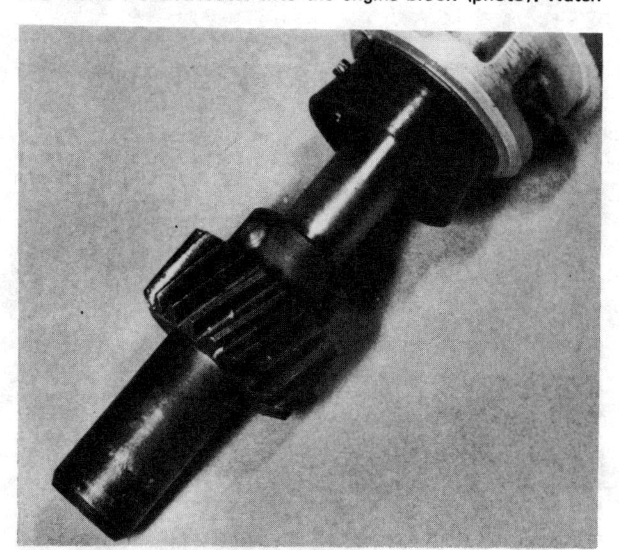

3.64 Distributor driveshaft skew gear

3.65 Checking the contact breaker gap

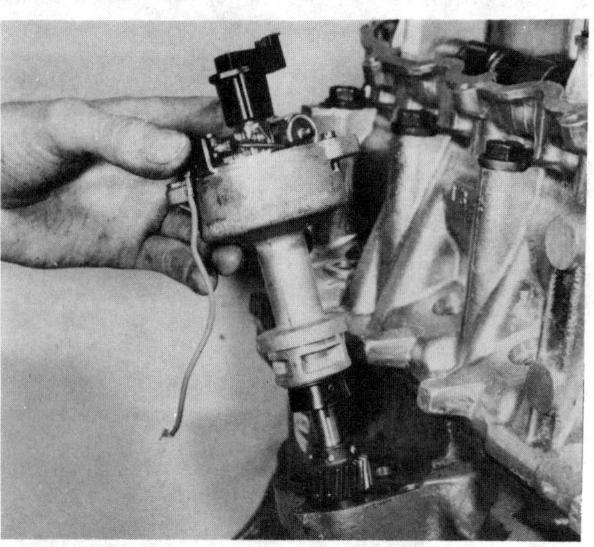

3.68 Fitting the distributor

69 Fit the clamp washer and nut and lightly tighten temporarily.

70 With a lamp and battery across the contact breaker points and the clamp nut just loosened, move the distributor body round its axis until the light just goes out. Tighten the clamp nut (photo).

Toothed drivebelt - removal, refitting and adjustment (OHC engine)

Note: *If adjustment is necessary on an old belt or if the belt is removed for any reason, always change the belt for a new one, never adjust using the old belt.*

71 The toothed drivebelt should be renewed at 36,000 miles (60,000 km). This can be done with the engine in the car.

72 Using a spanner on the crankshaft pulley nut turn the engine over until the timing mark on the crankshaft pulley is aligned with the TDC pointer (long one).

3.70 Distributor clamp nut

3.78 Setting the belt tensioner

73 Remove the drivebelt cover and the alternator/water pump drivebelt.
74 Check that the camshaft pulley timing hole is aligned with the cast ridge on the seal housing; refer to paragraph 59 of this Section. If it isn't, turn the engine over one revolution to get it lined up.
75 Before removing the drivebelt it must be remembered that neither the camshaft nor the crankshaft must be moved with the belt off. If this precaution is not observed the pistons and valves could impact, causing serious damage.
76 Release the tension on the drivebelt by slackening the bolt in the kidney-shaped slot on the tensioner bracket, loosening the other (pivot) bolt, pushing the tensioner wheel against the spring unit and tightening both bolts. Slide the drivebelt off the pulleys.
77 Fit the new belt. Start at the crankshaft drive pulley and, taking care not to kink or strain the belt, ease it into place over the auxiliary shaft pulley and the camshaft pulley. It might be necessary to slightly turn the camshaft to get the belt to mesh. This should always be done in the direction of least movement to achieve a mesh. Fit the belt on the tensioner pulley last. If this is difficult do not lever or force the belt on but recheck it and try again (photo).
78 Slacken the tensioner bolts to tension the belt (photo).

3.77 Fitting the toothed drivebelt - crankshaft V-belt pulley removed for clarity

79 Turn the engine anti-clockwise for two complete revolutions to even out belt tension then tighten the tensioner bolts. Never rock the camshaft when tensioning the belt, as slack could develop in the belt and it might jump a tooth.
80 Refit and tension the alternator/water pump V-belt. Refit the drivebelt cover.

Valve clearances - adjustment (OHC engine)

81 Checking the valve clearances should be done at the 6,000 mile servicing, or whenever the cylinder head has been removed and refitted for any reason. It is important that the clearances are set correctly, otherwise the timing will be wrong and the engine performance will be poor. If there is no clearance at all, the valves and seats will soon get burnt. Set the clearances with the engine cold.
82 Remove the camshaft cover. The engine can be turned over by either using a spanner on the crankshaft pulley nut, or by jacking up a front wheel, engaging top gear and using the wheel to turn the engine.
83 Each tappet must be checked when its operating cam is pointing upwards, 180° away from the tappet. Check the clearances in the firing order, No. 1 cylinder first and then 3, 4 and 2. Do the exhaust of one cylinder and the inlet of the one after; do this at the same time to minimise the amount of engine turning. Counting from the timing belt end, exhaust valves are 1–4–5–8, inlet valves 2–3–6–7.
84 Insert the feeler gauge for the appropriate valve. See the Specifications for correct settings. The feeler should slide in readily between cam and shim, but with slight frictional drag (photo). Try one a size thicker and one a size thinner. The thick one should not go in and the thinner one should be too loose.
85 If the clearance is wrong, measure the clearance and write it down with the number of the valve. When all the clearances have been checked, it will be necessary to remove those shims which are fitted where the clearances are wrong, and renew them with different thickness shims. If a clearance is too big, use a thicker shim. If a clearance is too small use a thinner shim. Calculate by simple subtraction.
86 To change a shim, turn the engine until the relevant cam is pointing upwards, then turn the tappet in its housing so that the slot in the rim is accessible.
87 The manufacturer provides special tools for depressing tappets (Nos. A60480 and A60443). These make the job easier if they can be borrowed, but it is possible to do the job without them. The best way is to make up a tool from a piece of steel plate shaped as shown in the photograph. Alternatively use a screwdriver to lever the tappet down and another one, on edge,

3.84 Checking a valve clearance

3.88A With the tappet depressed . . .

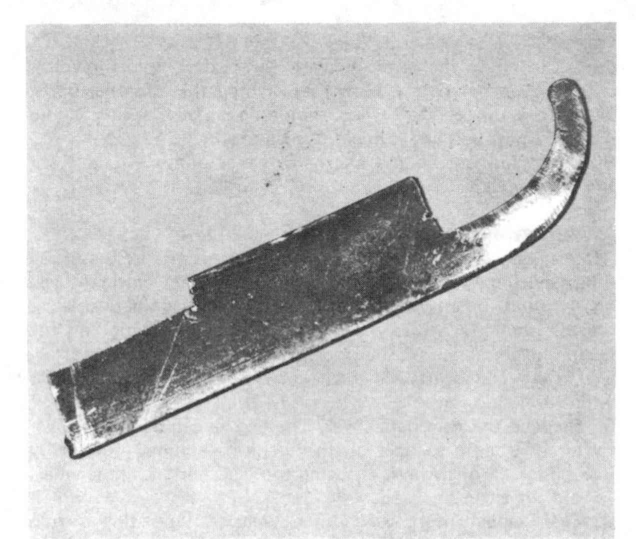

3.87 Made-up tool for depressing tappet buckets

3.88B . . . the shim can be removed

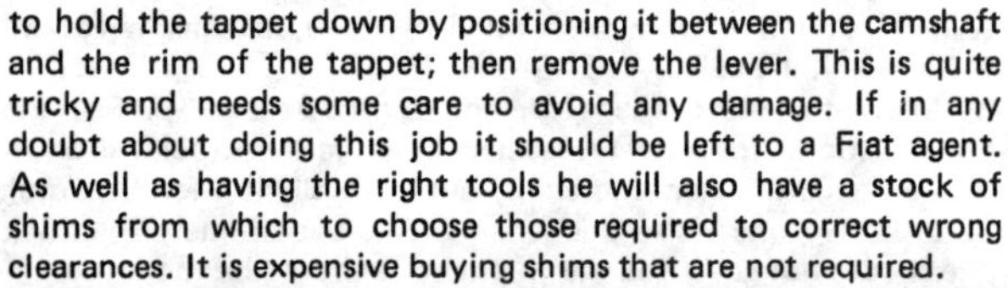

to hold the tappet down by positioning it between the camshaft and the rim of the tappet; then remove the lever. This is quite tricky and needs some care to avoid any damage. If in any doubt about doing this job it should be left to a Fiat agent. As well as having the right tools he will also have a stock of shims from which to choose those required to correct wrong clearances. It is expensive buying shims that are not required.

88 With the tappet held down, prise the shim out with a thin screwdriver. The Fiat way is to lift the shim using compressed air, which is effective if an air line is available. The shims are held in quite strongly by the oil film and they must be lifted up square or they will jam. Remove the shim with long-nosed pliers (photos).

89 When new, shims have their thickness marked in millimetres on their undersides. This marking may wear off and then it will be necessary to measure their thickness with a micrometer. From the thickness of the shim and the error in the clearance, calculate the size of shim required to produce the correct clearance.

90 Insert the new shim, numbered side down towards the tappet. Remove the tools used for depressing the tappet and repeat the operation until all clearances are correct.

91 Using clean engine oil lubricate all moving and sliding parts

3.91 Fit the camshaft cover with a new gasket

in the camshaft/tappet assembly. Fit a new gasket and refit the camshaft cover, together with the nine bolts and washers. Tighten the bolts progressively (photo).

Engine - final assembly (OHC engine)

92 Completing assembly of the 1049 cc and 1301 cc engines follows broadly the procedure for that of the smaller engine which is dealt with in Chapter 1, Section 56.

93 The manifold gasket is supplied in two parts for the overhead cam engine (photo). These serve both the inlet and exhaust manifolds which are on the same side of the engine (photo).

Engine removal - 127 Special and 127 from chassis number 1104290

94 The removal of the engine assembly from these models is slightly different from that described in Chapter 1, owing to redesigned engine mountings.

95 For the later type of mounting, disconnection of the engine is achieved by removal of the engine suspension mounting securing bolts after taking the weight of the assembly on a hoist.

3.93A The two-piece gasket for the manifold joint . . .

3.93B . . . and the manifolds fitted

96 On refitting the assembly tighten the securing bolts to the specified torque load.

Valve clearances - adjustment (all engines)

97 When adjusting the valve clearances note that the valves are arranged in the following order:

EX IN IN EX EX IN IN EX
EX Exhaust
IN Inlet
The order is identical from either end of the engine.

Timing chain (903 cc engine) - modification

98 As from engine number 6523590 (late 1982) the timing chain has been modified, together with the gears having teeth with a new profile. New and old components should not be mixed.

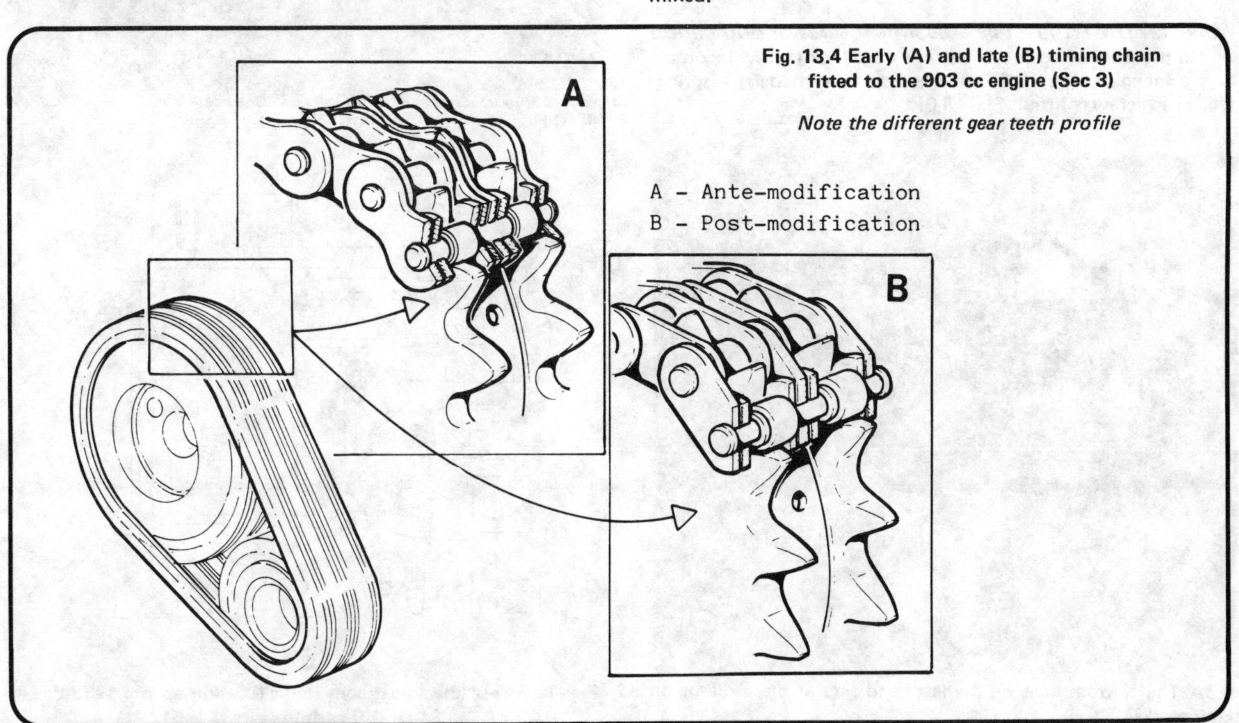

Fig. 13.4 Early (A) and late (B) timing chain fitted to the 903 cc engine (Sec 3)

Note the different gear teeth profile

Cylinder head (1301 cc engines) - refitting

99 The cylinder head nuts and bolts should be lightly lubricated with engine oil before being fitted and tightened.

100 Tighten the nuts and bolts to the initial torque wrench settings as given in the Specifications using the sequence shown in Fig. 13.3, then angle tighten them by the specified amounts again using the correct sequence.

101 It is not necessary to retighten the nuts and bolts after a settling down period.

4 Cooling system

Cooling system - draining

1 On some later models there is no drain plug at the bottom of the radiator, and in this case the system must be drained by disconnecting the bottom hose.

Water pump - overhaul

2 On later models the water pump shaft and bearings are manufactured as one unit and it is not possible to renew the components separately. Furthermore it is not possible to fit the later shaft and bearings to the earlier pump body.

Radiator - modifications

3 On later models the bottom outlet is located on the left-hand side of the radiator and the expansion tank is also on the left-hand side of the engine compartment. The upper mountings have nuts facing inwards (photos).

5 Fuel system and carburation

Fuel pump - general

1 On later models it is not possible to dismantle the fuel pump for overhaul as it is a sealed unit. If the unit is proved faulty it must be renewed complete.

Carburettors - general

2 Several additional carburettors have been fitted as listed in the Specifications, and the applicable procedures are given in the following paragraphs.

Weber 34 DMTR-fast idle adjustment screw modification

3 On pre 1983 models the fast idle adjustment screw incorporates a locknut (Fig. 13.5) however, on later models, a spring-tensioned screw is fitted (Fig. 13.6).

4.3B Expansion tank location on later models

4.3C Radiator upper mounting on later models

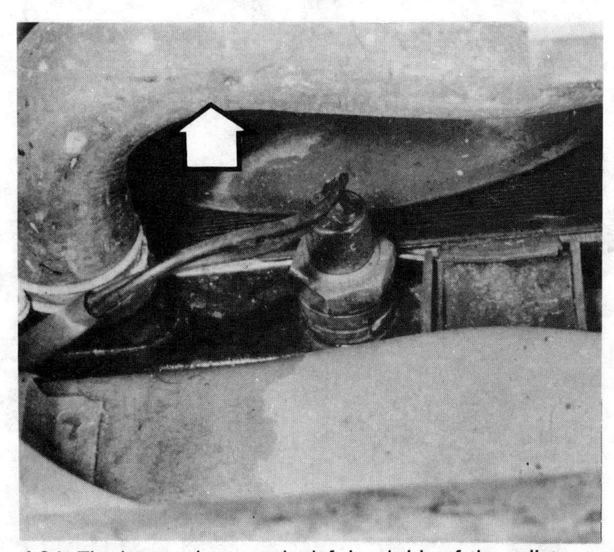

4.3A The bottom hose on the left-hand side of the radiator on later models

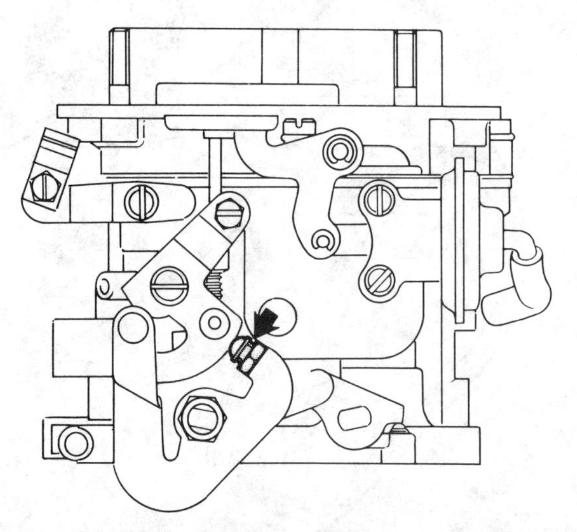

Fig. 13.5 Fast idle adjustment screw location on pre-1983 Weber 34 DMTR carburettors (Sec 5)

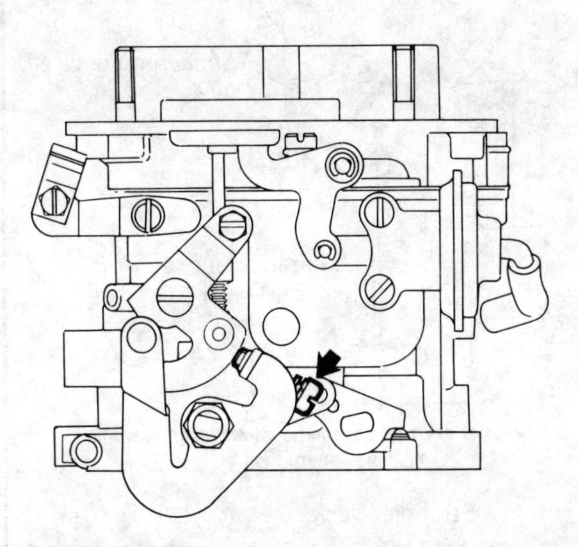

Fig. 13.6 Fast idle adjustment screw location on 1983-on Weber 34 DMTR carburettors (Sec 5)

Solex C30 DI 40 - adjustments

4 To adjust the fast idle throttle valve setting refer to Fig. 13.7 and set the fast idle lever on the first step of the cam using an elastic band to simulate the action of the bi-metal spring (cover removed). Using a twist drill, check that dimension A is between 0.032 and 0.034 in (0.80 and 0.85 mm) and, if necessary, adjust the screw on the fast idle stop rod. Note that the dimension must be taken from the progression hole side of the throttle valve.

5 To adjust the fast idle choke valve setting refer to Fig. 13.8 and set the fast idle lever on the second step of the cam. Using a twist drill, check that dimension B is between 0.118 and 0.138 in (3.0 and 3.5 mm) and, if necessary, turn the adjusting nut as required. Note that the dimension must be taken from the float side of the choke valve.

6 To adjust the automatic choke lever clearance refer to Fig. 13.9 then fully close the choke valve and check that the dimension C is between 0.008 and 0.016 in (0.2 and 0.4 mm). If adjustment is necessary bend the tab on the weakening device lever.

7 To adjust the choke valve minimum pull-off setting refer to Fig. 13.10 then fully close the choke valve and move the pull-off rod to fully depress the diaphragm. If necessary a screw can be inserted in the special hole provided in order to move the pull-off rod. Movement of the pull-off rod will open the choke valve to provide dimension D which should be 0.148 to 0.168 in (3.75 to 4.25 mm). If adjustment is necessary remove the plug and adjust the screw on the end of the diaphragm unit. Note that the dimension must be taken from the float side of the choke valve.

8 To check the choke valve maximum pull-off setting refer to Fig. 13.11 and open the valve against the pull-off rod. The dimension E should be 0.275 to 0.315 in (7.0 to 8.0 mm), but if not the automatic choke components are worn excessively and the unit should be renewed.

Solex C32 TD 1/4 - adjustment

9 To adjust the choke valve pull-off opening refer to Fig. 13.12 and fully close the choke valve then fully depress the adjustment screw against the diaphragm spring and use a twist drill to check that the choke valve opening is 0.110 to 0.126 in (2.8 to 3.2 mm). If necessary turn the adjustment screw as required.

10 To adjust the throttle valve fast idle setting refer to Fig. 13.13, close the choke valve and position the adjustment screw on the highest step of the cam. Check that dimension B is 0.032 to 0.036 in (0.80 to 0.90 mm) using a twist drill and, if necessary, turn the adjustment screw as required.

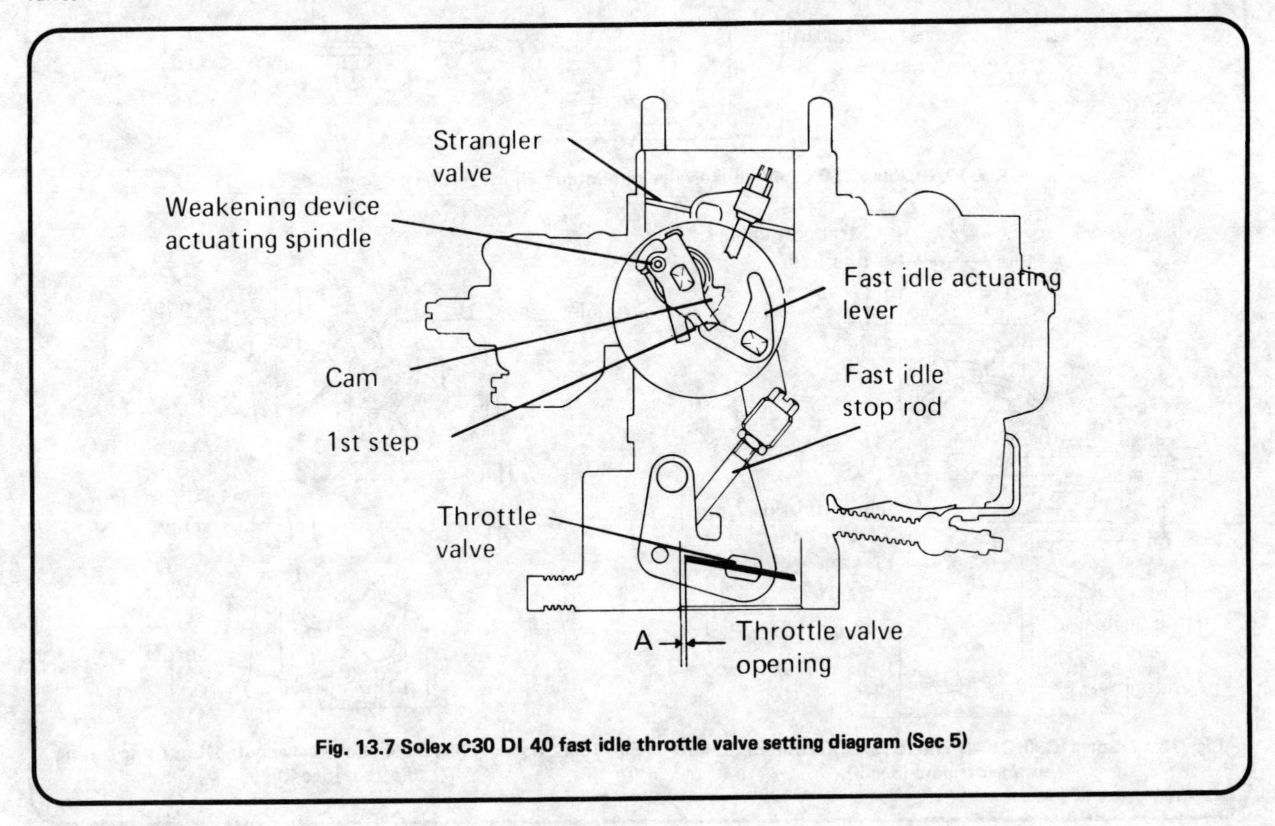

Fig. 13.7 Solex C30 DI 40 fast idle throttle valve setting diagram (Sec 5)

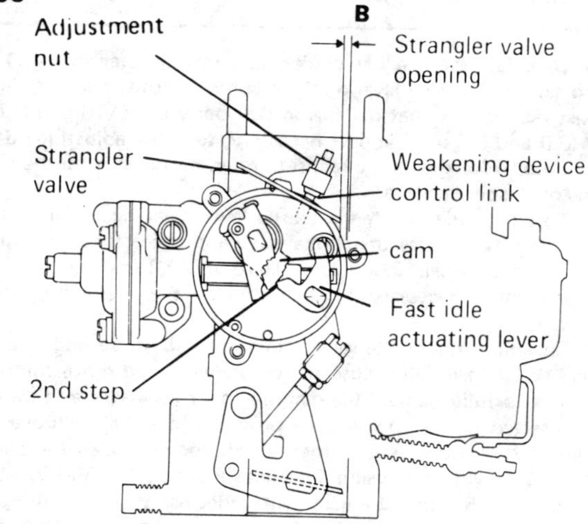

Fig. 13.8 Solex C30 DI 40 fast idle choke valve setting diagram (Sec 5)

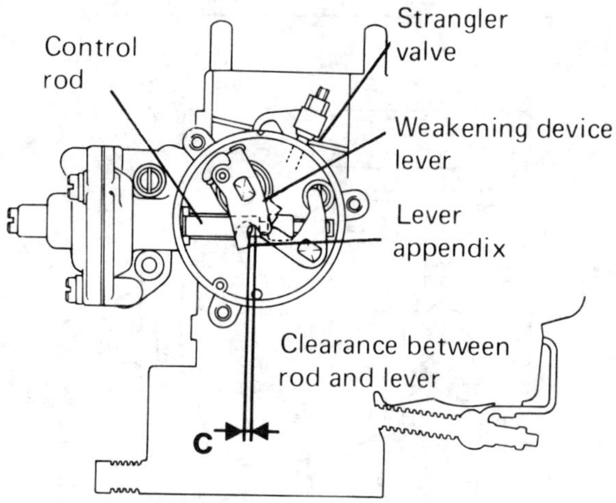

Fig. 13.9 Solex C30 DI 40 automatic choke lever clearance setting diagram (Sec 5)

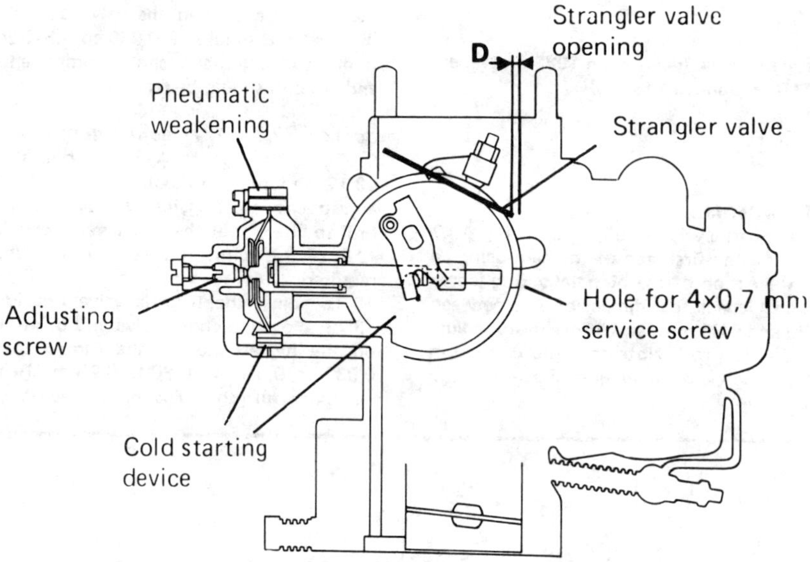

Fig. 13.10 Solex C30 DI 40 choke valve minimum pull-off setting diagram (Sec 5)

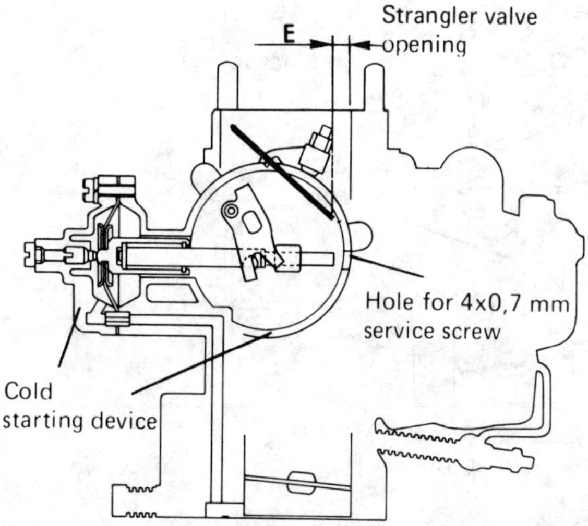

Fig. 13.11 Solex C30 DI 40 choke valve maximum pull-off setting diagram (Sec 5)

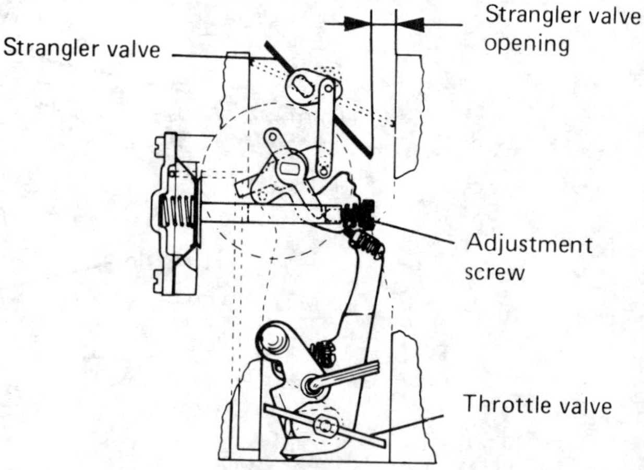

Fig. 13.12 Solex C32 TD1/4 choke valve pull-off opening setting diagram (Sec 5)

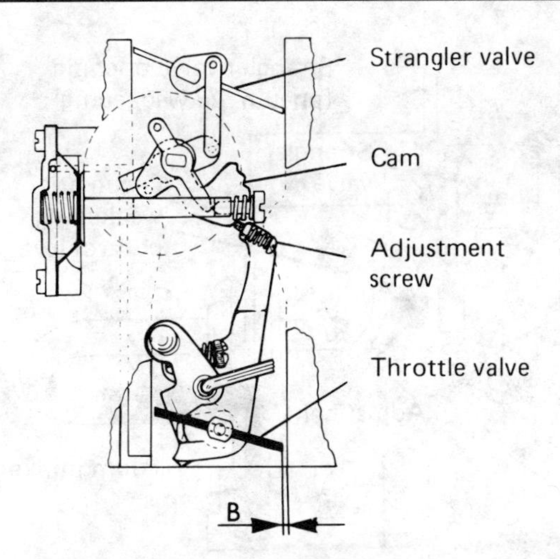

Fig. 13.13 Solex C32 TD 1/4 throttle valve fast idle setting diagram (Sec 5)

Weber 30 IBA 22/450 - adjustments

11 To adjust the choke valve opening setting first fully close the choke valve using the operating lever then open the valve against the spring tension. Using a twist drill check that the dimension between the edge of the valve and the float side of the barrel is 0.158 to 0.177 in (4.0 to 4.5 mm). If not, bend the operating lever as required.

12 To adjust the throttle valve fast idle opening fully close the choke valve using the operating lever then use a twist drill to check that the dimension between the edge of the throttle valve and the progression hole side of the barrel is 0.035 to 0.037 in (0.90 to 0.95 mm). If not, adjust the linkage.

Weber 34 DMTR 54/250 - adjustments

13 To adjust the float level remove the carburettor cover, together with the gasket, and hold it vertical with the float hanging down so that the needle valve is closed but the spring-tensioned ball not depressed. Refer to Fig. 13.14 and check that the distance from the 'top' of the float to the fitted gasket is 0.266 to 0.286 in (6.75 to 7.25 mm). If not, bend the float arm tab A as necessary.

14 To check the float travel hold the cover horizontal and check that the distance from the bottom of the float to the fitted gasket is 1.673 to 1.713 in (42.5 to 43.5 mm). If not, bend the tab on the end of the float arm as necessary.

15 To check the accelerator pump delivery the carburettor must be removed and the float chamber full of petrol (photo). Fully operate the throttle lever several times to ensure that the pump channels are full, then locate the carburettor **vertically** over a calibrated container and operate the throttle fully ten times, pausing briefly at the full open and closed positions. The delivery should be between 9.0 cc and 15.0 cc.

16 To adjust the throttle fast idle setting refer to Fig. 13.15 and fully close the choke lever (photo). Using a twist drill check that dimension A is between 0.032 and 0.034 in (0.80 and 0.85 mm) and if necessary turn the adjustment screw. Note that the dimension is taken at the progression hole side of the barrel.

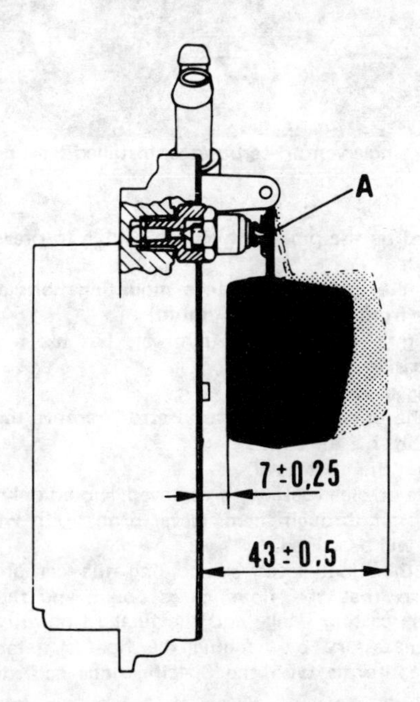

Fig. 13.14 Weber 34 DMTR 54/250 float level and travel setting diagram (Sec 5)

A Float arm tab *Measurements in mm*

5.15 View of the Weber DMTR carburettor showing the accelerator pump (arrowed)

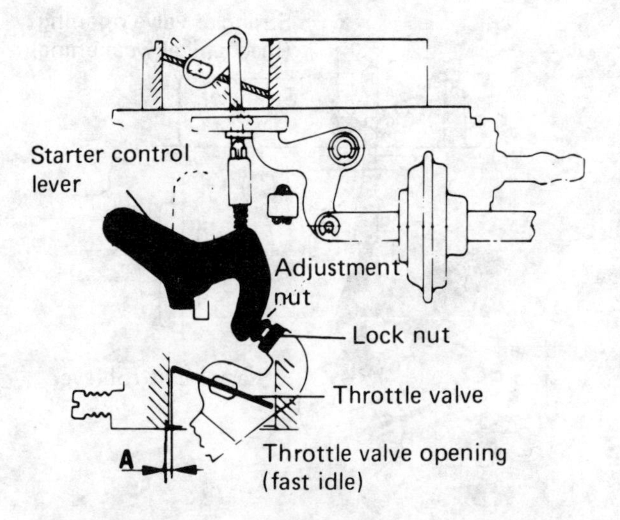

Fig. 13.15 Weber 34 DMTR 54/250 throttle fast idle setting diagram (Sec 5)

5.16 View of the Weber DMTR carburettor showing the fast idle adjustment screw and vacuum pull-off unit (arrows)

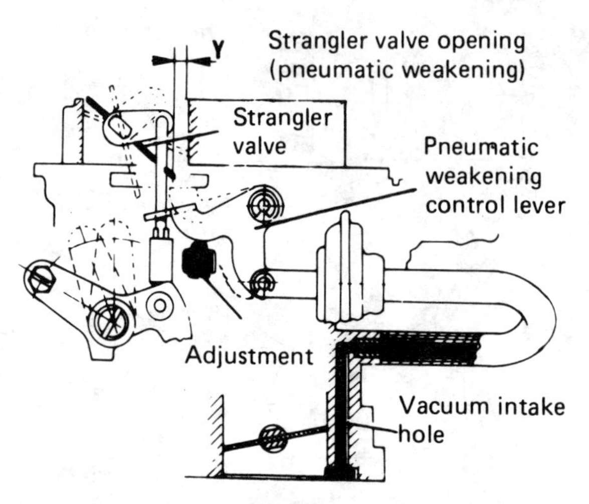

Fig. 13.17 Weber 34 DMTR 54/250 choke valve vacuum pull-off dimension diagram (Sec 5)

17 To check the choke valve mechanical pull-off dimension refer to Fig. 13.16 and fully close the choke valve using the lever, then open the valve itself against the spring tension and use a twist drill to check that dimension X is between 0.335 and 0.374 in (8.5 and 9.5 mm). If not, bend the stop tab on the choke lever as necessary.

18 To check the choke valve vacuum pull-off dimension refer to Fig. 13.17 and fully close the choke valve using the lever. Turn the vacuum unit control lever and check that dimension Y is between 0.246 and 0.295 in (6.25 and 7.5 mm) using a twist drill. If necessary turn the adjustment screw as required.

Weber or Solex single venturi carburettors

Removal and refitting

19 Remove the air cleaner.

20 Disconnect the throttle cable.

21 Disconnect the distributor vacuum pipe (photo).

22 Disconnect the fuel flow and return hoses.

23 Disconnect the coolant hoses from the automatic choke

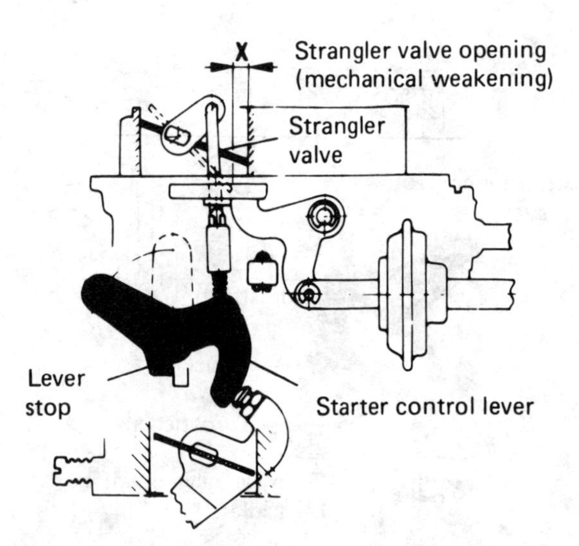

Fig. 13.16 Weber 34 DMTR 54/250 choke valve mechanical pull-off dimension diagram (Sec 5)

5.21 Weber single venturi carburettor installed (vacuum pipe arrowed)

housing and tie the pipes as high as possible to prevent coolant loss (photo).

24 Unbolt the carburettor flange mounting bolts and lift the carburettor from the manifold (photo).

25 Refitting is a reversal of removal, but use a new flange mounting gasket.

Dismantling and adjustment

26 Clean the outside of the carburettor, remove the top cover screws and lift off the cover.

27 Clean out the float bowl.

28 Any jets or bleed screws, if removed, should only be cleaned by blowing air through them. Never probe with wire or their calibration will be ruined.

29 Check the float level by holding the carburettor cover vertically so that the float hangs down, and then measure between the casting flange and the furthest point of the float. Adjust if necessary by bending the float arm tab until the clearance conforms with the Specifications according to carburettor.

30 The automatic choke will not normally require dismantling, but if the cover is removed, make sure that the centre index marks on the cover and housing are in alignment before tightening the choke cover screws.

5.23 Weber single venturi carburettor showing coolant hoses (arrowed) to automatic choke housing. Note fuel hoses connected to top right of carburettor

5.24 Weber 32 ICEV carburettor

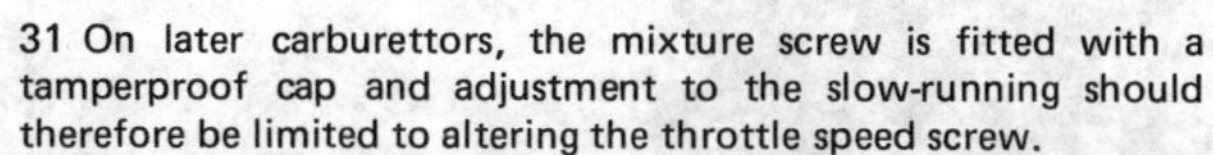

31 On later carburettors, the mixture screw is fitted with a tamperproof cap and adjustment to the slow-running should therefore be limited to altering the throttle speed screw.

32 On earlier carburettors, or if new components have been fitted to carburettors having a tamperproof screw, then the slow-running and mixture can be altered, but only if an exhaust gas analyser is available.

33 On earlier units, use a device such as a mixture adjustment aid or a vacuum gauge fitted in accordance with the manufacturers' instructions to obtain the correct slow-running adjustment.

34 On later units, an exhaust gas analyser, again used in accordance with the maker's instructions, will ensure that the CO level is within the limits specified. Fit a new tamperproof cap on completion.

Weber dual venturi carburettor

Removal and refitting

35 These operations are similar to the operations described in paragraphs 19 to 25 of this Section, but the unit is of manually operated choke design and will therefore not have coolant hose attachments. A choke cable is fitted instead (photo).

Fig. 13.18 Solex carburettor adjustment screws (Sec 5)

A Throttle speed screw *B Sealed mixture screw*

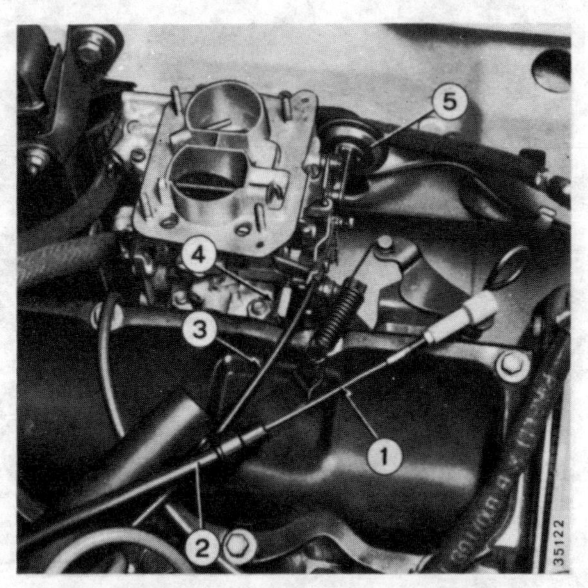

Fig. 13.19 Weber dual venturi carburettor (Sec 5)

1 Throttle cable
2 Cable conduit
3 Choke cable
4 Throttle speed screw
5 Throttle control diaphragm

5.35 Throttle cable and choke cable connections to a Weber dual venturi carburettor

Dismantling and adjustment

36 The operations are similar to those described for the single venturi type carburettor, once the upper body has been removed (six screws).

37 If jets are extracted, identify them as to which side of the carburettor (primary or secondary) they came from.

38 To check the float setting, hold the carburettor cover vertically with the float hanging down. The distance between the float and the cover (with gasket fitted) should be .0266 to 0.286 in (6.75 to 7.25 mm). Where necessary, bend the float tongue.

39 This type of carburettor is fitted with a tamperproof cap on the mixture screw, and unless an exhaust gas analyser is available, restrict any adjustment to the slow-running to turning the throttle speed screw (photos).

5.39A Weber DMTR carburettor idle speed adjustment screw location (arrowed)

5.39B Weber DMTR carburettor mixture adjustment screw location (arrowed)

Carburettor overhaul - general

40 Whenever it is decided to overhaul a carburettor, always obtain the appropriate repair kit in advance. If throttle or choke valve plates or their spindles are worn, do not dismantle them but obtain a new carburettor. A new carburettor is an economical proposition if the original one has seen service to a high mileage. Very often a generous allowance is made for the old unit.

Air cleaner (1300 GT) - description

41 On the 1300 GT model the air cleaner incorporates an air temperature control to stabilise the temperature of the incoming air. In the 'Winter' position the air temperature is maintained at 22 to 24° C (71 to 75° F) and in the 'Summer' position at 27 to 29° C (80 to 84° F).

Fig. 13.20 Location of mixture control screw (1) (Weber dual venturi carburettor) (Sec 5)

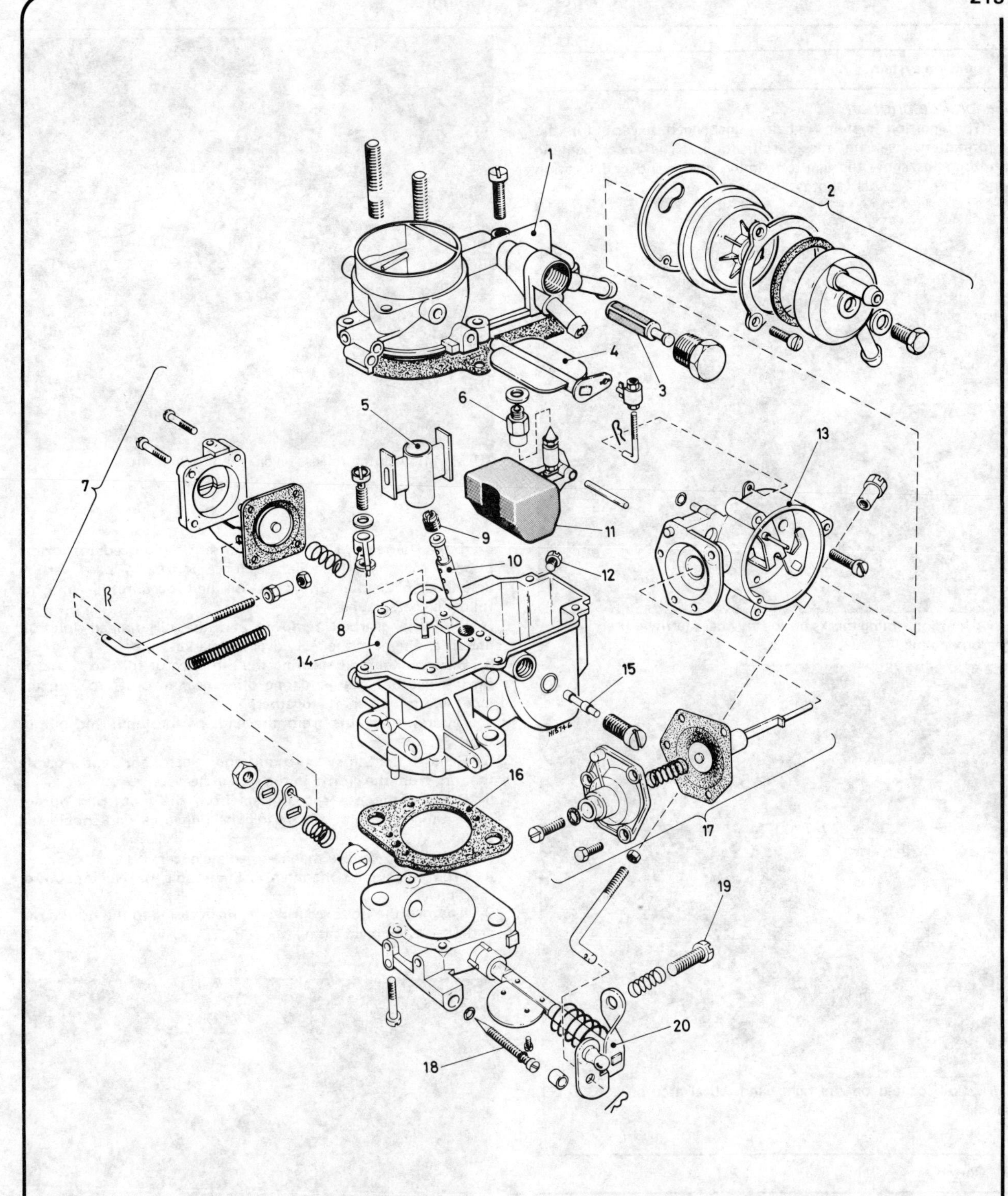

Fig. 13.21 Exploded view of the Weber 32 ICEV 16/150 carburettor (Sec 5)

1 Cover
2 Automatic choke cover components
3 Filter
4 Choke valve
5 Auxiliary venturi
6 Needle valve
7 Accelerator pump components
8 Accelerator pump jet
9 Air correction jet
10 Emulsion tube
11 Float
12 Main jet
13 Automatic choke body
14 Carburettor body
15 Idle jet
16 Gasket
17 Vacuum pull-off unit components
18 Mixture adjustment screw
19 Idle speed adjustment screw
20 Throttle valve components

6 Ignition system

General description

1 The ignition system remains unchanged except for the information given in the Specifications. The accompanying photographs show the ignition timing marks and coil location on a 1300 GT model (photos).

6.1A Ignition timing marks on the flywheel (arrowed) and gearbox casing

6.1B Coil located on the right-hand wheel arch on a 1300 GT model

7 Clutch

Release lever - modification

1 The end of the release lever has been modified on later models to accommodate a half round seat for the cable end fitting (photo). However, the adjustment procedure is identical to that given in Chapter 5.

7.1 Modified clutch release lever fitted to later models

8 Transmission

Five-speed gearbox - general description, dismantling and reassembly

1 As from 1982 certain models are fitted with a five-speed gearbox. The removal, overhaul and refitting procedures for the 5-speed gearbox are similar to those for the 4-speed version described in Chapter 6, except for the procedures given in the following paragraphs.

2 With the gearbox removed and the oil drained (photo), unbolt and remove the rear cover and gasket.

3 Unscrew the bolt securing the 5th selector fork to the shaft then select 5th gear and one other gear in order to lock the input and output shafts together.

4 Unscrew the nuts from the ends of the input and output shafts.

5 Remove the 5th synchro unit and selector fork and separate the fork from the groove in the sliding sleeve.

6 Remove the 5th speed gears from the input and output shafts, together with the needle bearing cage and shouldered bush.

7 Unbolt and remove the intermediate cover.

8 From the side of the main casing unbolt the selector control shaft cover.

9 Unscrew the bolts securing the main casing to the front cover and lift off the main casing.

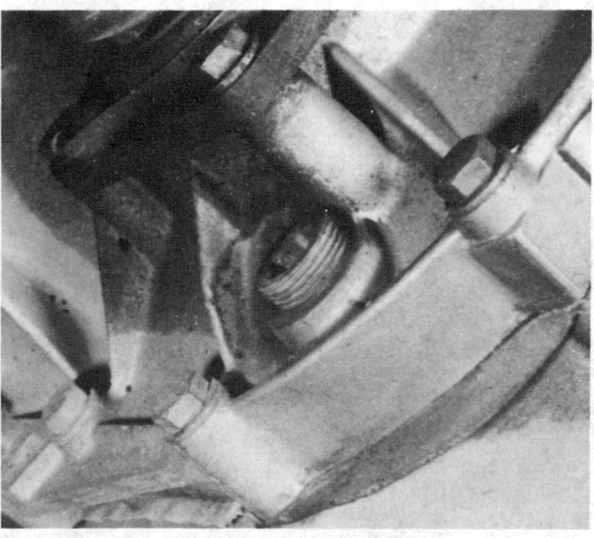
8.2 Gearbox drain plug location

10 The remaining procedure is similar to that for the 4-speed version described in Chapter 6.

Gear linkage (5-speed gearbox) - removal and refitting

11 Jack up the front of the car and support on axle stands. Apply the handbrake.
12 Unscrew the nuts and withdraw the cover from the bottom of the gear lever assembly, then slide the rubber bolt along the linkage rods.
13 Extract the clip and disconnect the lower linkage rod.
14 Unbolt the mounting bracket and lower it, then remove the pivot pin and disconnect the upper linkage rod.
15 Working inside the car, release the gaiter and withdraw the gear lever.
16 From under the car, unscrew the nuts and withdraw the linkage relay bracket from the gearbox and disconnect the linkage as necessary (photo).
17 Check the linkage relay bracket for excessive wear and renew it if necessary.
18 Refitting is a reversal of removal. There is no provision for adjustment.

8.16 Front ends of the gearchange linkage rods

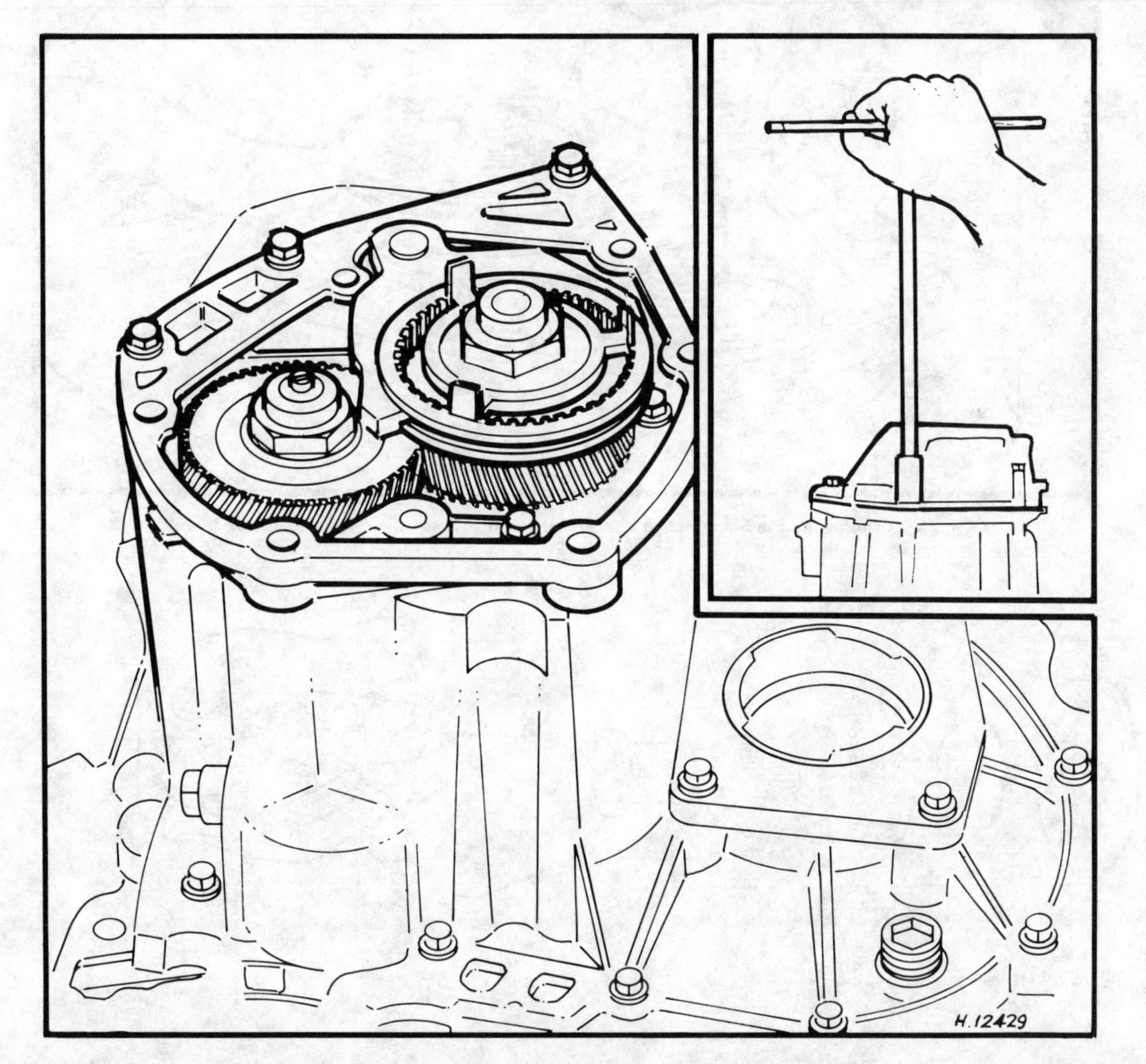

Fig. 13.22 Removing the rear cover on the 5-speed gearbox (Sec 8)

Fig. 13.23 Removing the 5th synchro unit and selector fork on the 5-speed gearbox (Sec 8)

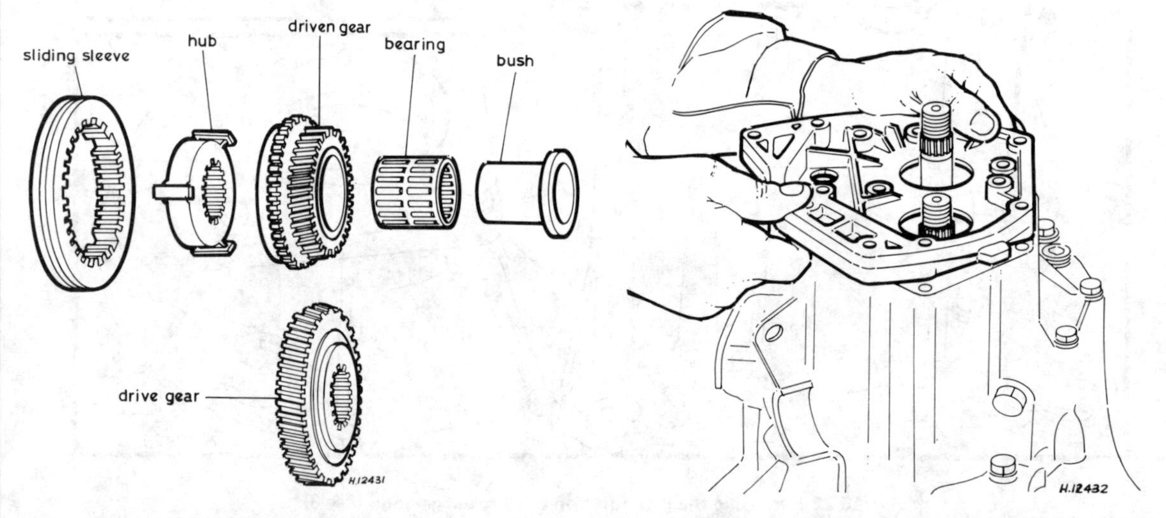

Fig. 13.24 5th speed gear components on the 5-speed gearbox (Sec 8)

Fig. 13.25 Removing the intermediate cover on the 5-speed gearbox (Sec 8)

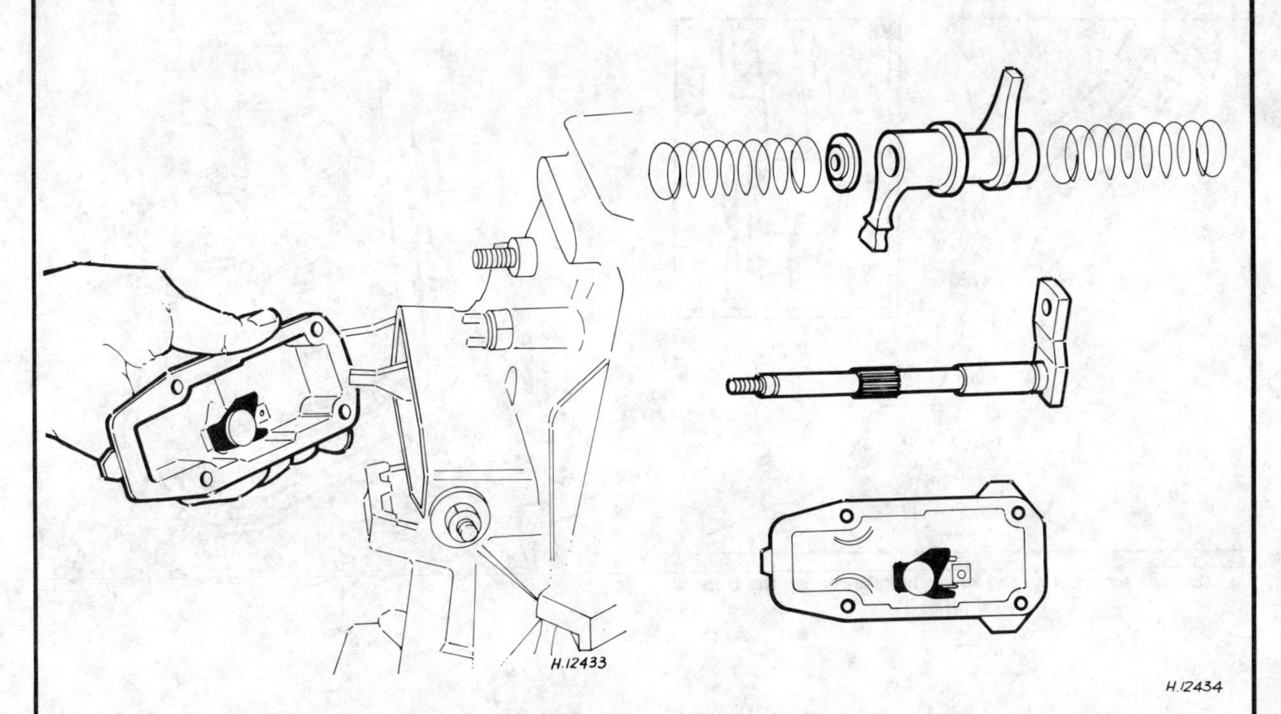

Fig. 13.26 Removing the selector control shaft cover on the 5-speed gearbox (Sec 8)

Fig. 13.27 Gear selector control shaft components (Sec 8)

Fig. 13.28 Removing the main casing on the 5-speed gearbox (Sec 8)

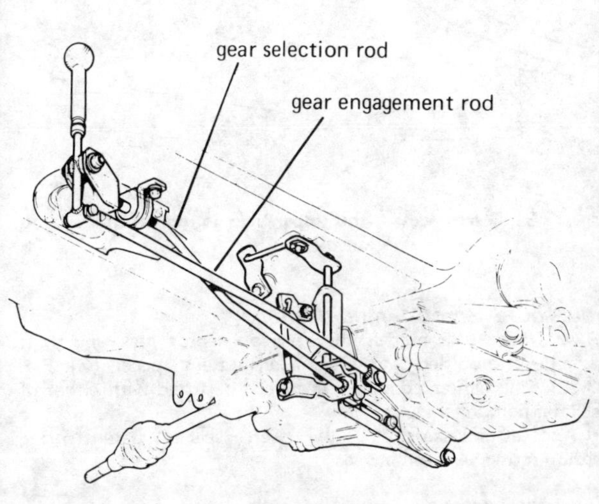

Fig. 13.29 Gear linkage components on the 5-speed gearbox (Sec 8)

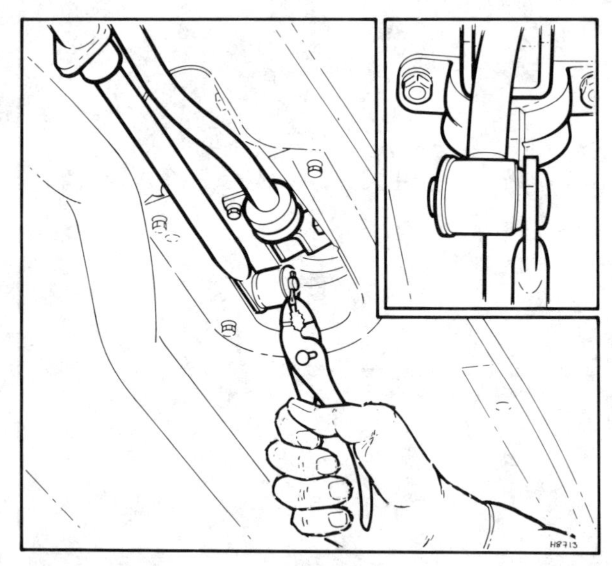

Fig. 13.30 Disconnecting the lower linkage rod on the 5-speed gearbox (Sec 8)

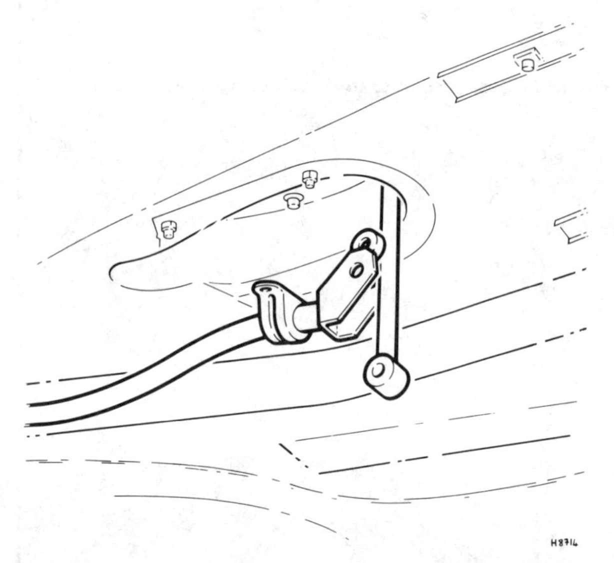

Fig. 13.31 Disconnecting the upper linkage rod on the 5-speed gearbox (Sec 8)

Four-speed gearbox - modifications

19 As from 1982 models the reverse selector fork and shaft have been modified and now incorporate a spacer (see Fig. 13.32). The earlier components cannot be fitted with either of the later components.

20 Also, at the same time, the reverse selector detent spring tension rating has been increased.

Five-speed gearbox - modifications

21 As from 1982 models the reverse selector detent spring tension rating has been increased, but the 5th/reverse selector control spring tension rating has been decreased.

9 Driveshafts

Driveshafts - modifications

1 As from chassis numbers 3647051 and 3649373 the drive-

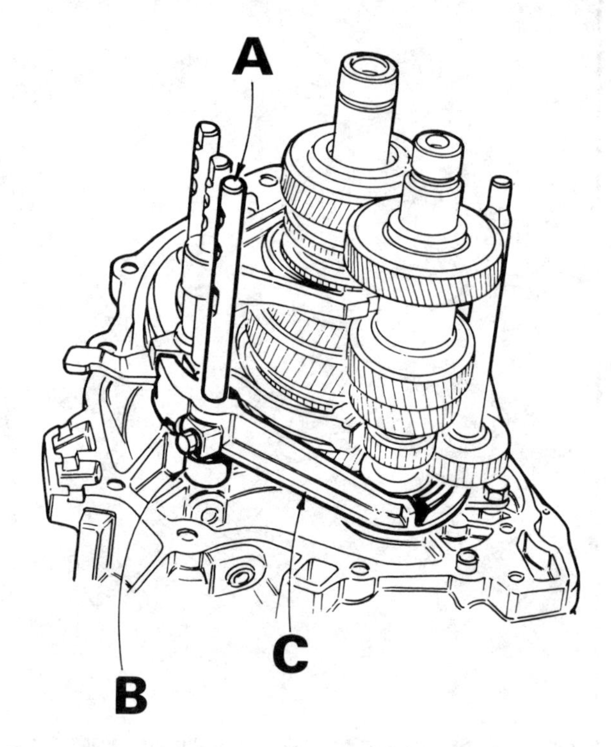

Fig. 13.32 The modified reverse selector fork and shaft on the 4-speed gearbox (Sec 8)

A Shaft
B Spacer
C Selector fork

shafts incorporate larger constant velocity joints and, in conjunction with this, new sun gears are fitted to the differential unit.

10 Braking system

Brake fluid reservoir - modifications

1 On Sport models fitted with the 1049 cc engine the brake fluid reservoir is divided into two sections. Each of the two reservoir caps incorporates a level warning device which actuates a warning lamp on the instrument panel if the fluid level should fall below the minimum mark.

2 Once a week, turn the ignition key to MAR and depress the reservoir caps one at a time to test that the warning lamp is operating.

Brake pads (front) end calipers - modifications

3 On later models the front brake pads and calipers incorporate small modifications as shown in the photographs. The renewal procedures remain unchanged as given in Chapter 8 (photos).

Brake fluid reservoir cap - modifications

4 On all later models the brake fluid reservoir cap incorporates a level warning device and test facility, as described in paragraphs 1 and 2 of this Section (photo).

Hydraulic pipes and hoses - general

5 The accompanying photographs show the locations of various hydraulic pipes and hoses (photos).

Remote vacuum servo unit - general

6 The remote vacuum servo unit is located on the left-hand side of the engine compartment behind the radiator and electric cooling fan (photo).

7 When bleeding the hydraulic circuit the servo unit cylinder

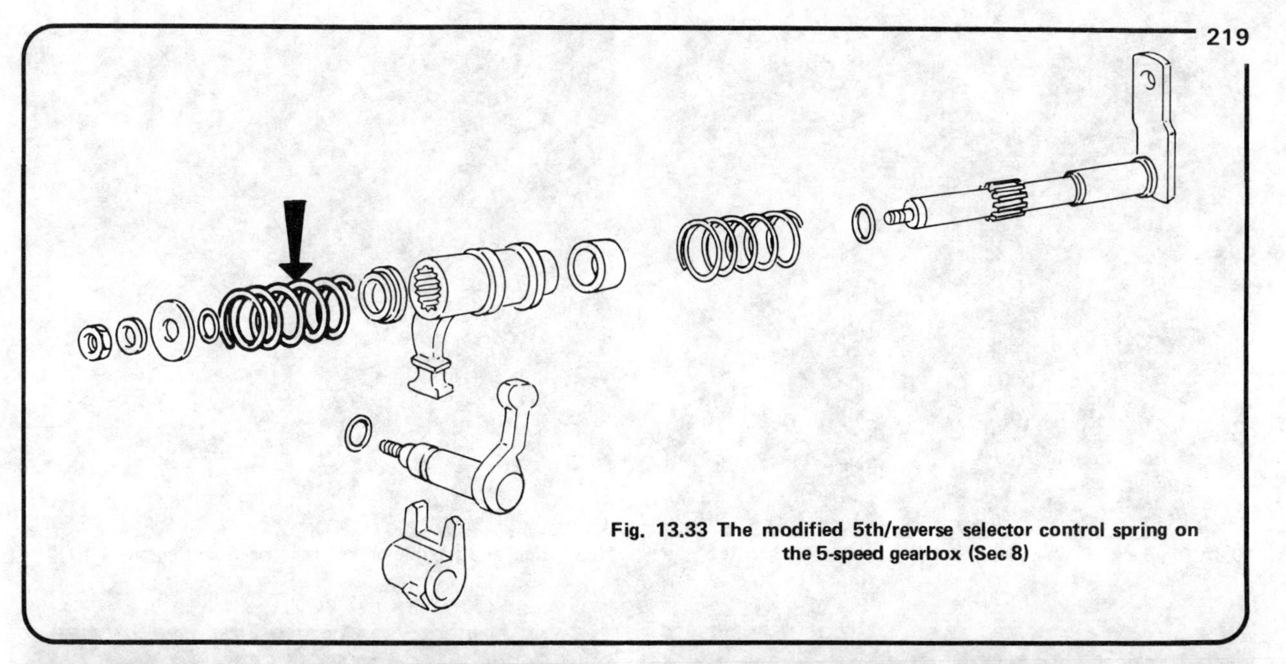

Fig. 13.33 The modified 5th/reverse selector control spring on the 5-speed gearbox (Sec 8)

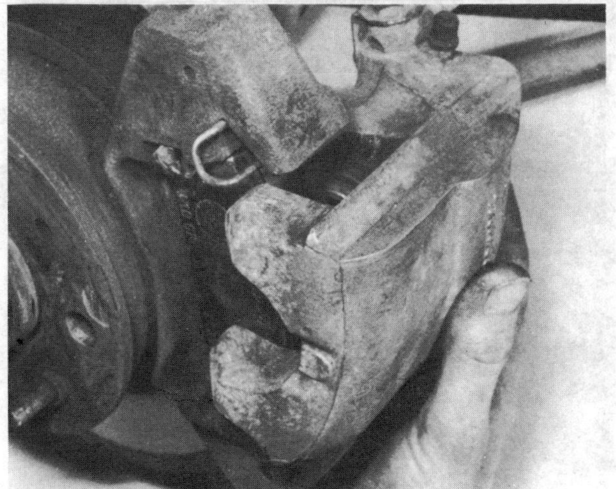

10.3A Removing the later type front brake caliper

10.3B Removing the later type front brake pads

Fig. 13.34 The brake fluid reservoir caps on Sport models fitted with the 1049 cc engine (Sec 10)

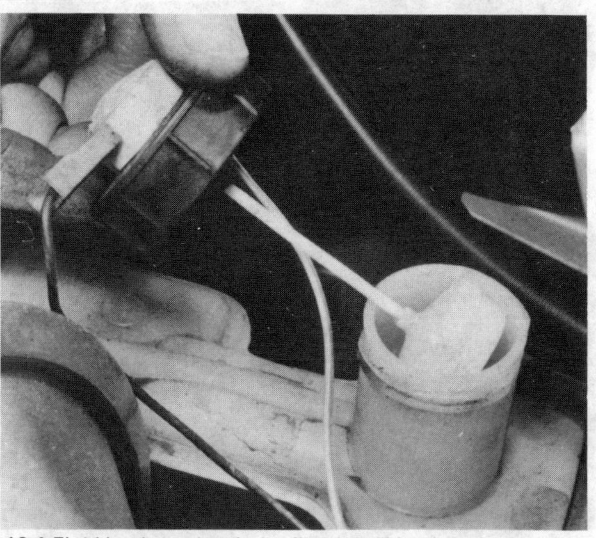

10.4 Fluid level warning device fitted to all later models

10.5A Rigid and flexible brake hose front mounting bracket

10.5B Rigid and flexible brake hose rear mounting bracket

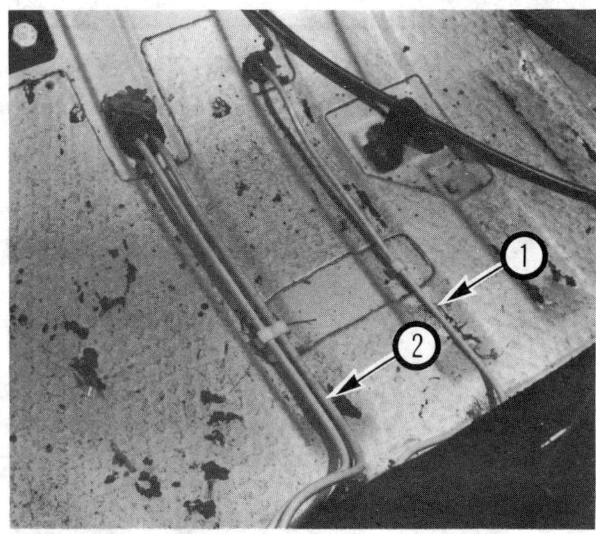

10.5C Rigid rear brake pipe (1) and fuel pipes (2) located beneath the rear seat cushion

10.6 Remote vacuum servo unit

10.7 The bleed nipple on the servo unit cylinder (arrowed)

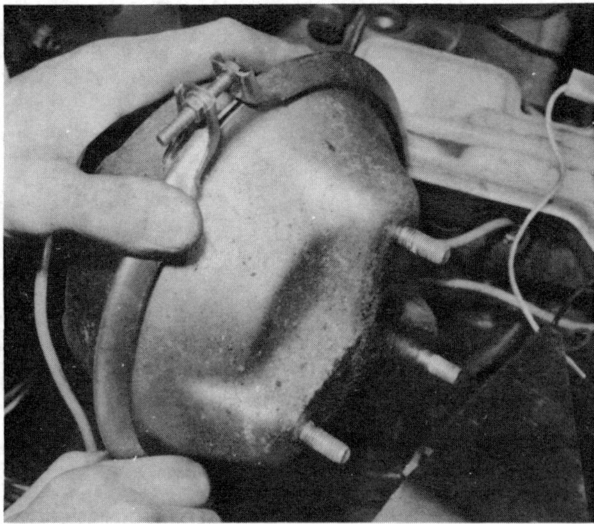

10.8 Removing the vacuum servo unit and cylinder assembly

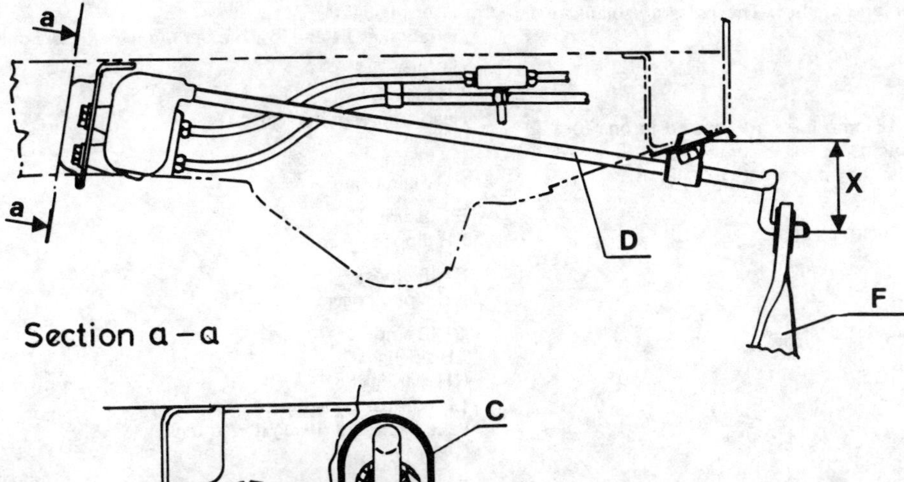

Fig. 13.35 Diagram and cross-section of the rear brake regulator (Sec 10)

A Mounting bolt
B Mounting bolt
C Rubber boot
D Torsion bar
D1 Torsion bar end
E Piston
F Link
G Pin
X = Setting dimension

must be bled before the remaining circuits by using the bleed nipple provided (photo).

8 To remove the vacuum servo unit disconnect the hydraulic pipes and vacuum hose, then unscrew the mounting nuts and withdraw the unit from the bracket (photo). Refitting is a reversal of removal, but finally bleed the hydraulic circuit with reference to Chapter 8.

Rear brake regulator (later models) - adjustment

9 On 1982 on 1300 GT models the adjustment procedure for the rear brake regulator is different to that described in Chapter 8. First jack up the rear of the car and support on axle stands. Chock the front wheels.

10 Refer to Fig. 13.35 and disconnect the link (F) from the torsion bar (D).

11 Prise the rubber boot (C) from the regulator.

12 With the link end of the torsion bar (D) positioned 2.162 to 2.562 in (55 to 65 mm) from the seat of the mounting rubber (dimension X in Fig. 13.35) check that the torsion bar end (D1) is just touching the piston (F).

13 If adjustment is necessary, loosen the regulator mounting bolts (A and B), pivot the regulator as necessary then tighten the bolts.

14 After making the adjustment, lubricate the contact surfaces of the piston (E), torsion bar end (D1), and the pin (G) with multi-purpose grease.

15 Reconnect the link (F) to the torsion bar (D).

16 Lower the car to the ground.

11 Electrical system

Part A - general equipment

Alternator - general

1 Different alternators have been fitted to later models, as listed in the Specifications (photo). The Lucas 18 ACR alternator incorporates an integral electronic voltage regulator fitted on the brush holder as one unit. To remove this unit first remove the alternator, as described in Chapter 9.

11.1 Alternator and drivebelt

2 Remove the two screws and withdraw the end cover.

3 Disconnect the connector from the rectifier pack, and remove the brush moulding and regulator case retaining screws. Remove the regulator and brush holder.

4 Check that the brushes protrude from the moulding by no less than the specified minimum amount. Renew the brushes if necessary.

5 Refitting of the regulator and brush holder is a reversal of removal, but make sure that the brushes move freely, and before fitting the unit clean the alternator rotor slip rings using fine glasspaper and a fuel-soaked cloth. Refer to Chapter 9 for the alternator refitting procedure.

Tailgate wiper and washer - fitting

6 As from 1979 holes are provided on all new cars for the fitting of a tailgate wiper and washer. The holes are blanked with plastic plugs.

Fuses (127 Special)

7 Eight 8 amp and two 16 amp fuses are located in one fuse box in the engine compartment.
8 The fuses and the circuits they protect are as follows:

1 (8 amp)
Direction indicators and warning light
Windscreen wiper motor
Windscreen washer pump

2 (8 amp)
Oil pressure warning light
Water temperature gauge
Fuel gauge and warning light
Brake lights
Heater fan motor

3 (8 amp)
LH main beam
Headlamp warning light

4 (8 amp)
RH main beam

5 (8 amp)
LH dipped beam

6 (8 amp)
RH dipped beam

7 (8 amp)
LH front sidelight
RH rear sidelight
LH number plate light
Instrument panel light
Reversing light

8 (8 amp)
RH front sidelight
LH rear sidelight
RH number plate light
Cigar lighter light

9 (16 amp)
Interior light, incorporated in rear view mirror
Horn
Radiator fan motor

10 (16 amp)
Cigar lighter
Heated rear window and indicator light (when fitted)

9 All the 8 amp fused circuits are energised when the ignition switch is on.
10 The following circuits have no fuse protection: generator, ignition, starter, charge indicator light, and the radiator fan relay coil.

Fuses (early 1049 cc models)

11 Eight 8 amp fuses and two 16 amp fuses are located in one fuse box in the engine compartment.
12 The fuses and the circuits they protect are as follows:

A (8 amps)
Heated rear window relay coil
Windscreen wiper motor
Direction indicators and warning light
Windscreen washer pump
Rear screen washer and wiper (Sports only)

B (8 amps)
Heater fan motor
Brake lights
Water temperature gauge
Fuel gauge and warning light
Oil pressure warning light
Reversing lights
Handbrake ON and low brake fluid level (Sport only)

C (8 amps)
LH main beam
Headlamp warning light

D (8 amps)
RH main beam

E (8 amps)
LH dipped beam

F (8 amps)
RH dipped beam

G (8 amps)
LH sidelight
RH rear light
LH number plate light
Panel light and sidelight warning

H (8 amps)
RH sidelight
LH rear light
RH number plate light
Cigar lighter light
Boot light (not Sport)

I (16 amps)
Horn
Courtesy light
Radiator fan motor

L (16 amps)
Cigar lighter
Heated rear window
Hazard warning light
Clock (Sport)

Fuses (Mk 3 Special and Super)

Fuses (Mk 3 Special and Super)

13 Eight 8 amp fuses and two 16 amp fuses are located in one fuse box in the engine compartment.
14 The fuses and the circuits they protect are as follows:

A (8 amps)
Heated rear window relay winding
Direction indicators and warning light
Heater fan
Reversing lights
Oil pressure warning light
Water temperature gauge (Super)
Fuel gauge and warning light
Handbrake/brake fluid level warning light
Water temperature warning light
Hazard warning lights

B (8 amps)
Windscreen wiper motor
Stop lights
Windscreen washer pump
Rear screen wiper and washer (if fitted)

C (8 amps)
LH main beam
Main beam warning light

D (8 amps)
RH main beam

E (8 amps)
LH dipped beam
Rear foglights and warning light

F (8 amps)
RH dipped beam

G (8 amps)
Instrument panel lights
Sidelight warning light
LH side and RH rear lights
Number plate lights

H (8 amps)
RH side and LH rear lights
Cigar lighter illumination light

I (16 amps)
Horns
Radiator cooling fan

L (16 amps)
Cigar lighter
Courtesy light
Clock (Super)
Radio (if fitted)
Heated rear window (if fitted)

Fuses (1300 GT)

15 Eight 8 amp fuses and two 16 amp fuses are located in one fuse box in the engine compartment (photo).

11.15 Fuse box on a 1300 GT

16 The fuses and the circuits they protect are as follows:

A (8 amps)
Heated rear window relay winding
Direction indicators and warning light
Heater fan
Reversing lights
Oil pressure warning light
Water temperature gauge
Fuel gauge and warning light
Handbrake/brake fluid level warning light
Tachometer
Digital clock illumination
Oil pressure gauge

B (8 amps)
Windscreen wiper and washer
Rear screen wiper and washer
Stop iights

C (8 amps)
LH main beam
Main beam warning light

D (8 amps)
RH main beam

E (8 amps)
LH dipped beam
Rear foglights and warning light

F (8 amps)
RH dipped beam

G (8 amps)
Instrument panel lights
LH side and RH rear lights
Number plate lights

H (8 amps)
RH side and LH rear lights
Cigar lighter illumination light
Oil pressure gauge light
Clock light dimmer

I (16 amps)
Horns
Radiator cooling fan

L (16 amps)
Cigar lighter
Radio (if fitted)
Heated rear window (if fitted)
Hazard warning lights

Lamp bulbs (later models) - renewal

17 To renew a headlamp bulb pull off the connector and remove the rubber cover, then remove the retainer by depressing the lugs and turning them anti-clockwise. Extract the bulb (photo). When fitting the new bulb make sure that the raised shoulder engages the slot provided.
18 To renew a front sidelamp bulb pull up the rubber retainer, extract the bulbholder and remove the bulb (photo).
19 To renew an interior light bulb prise the lamp from the digital clock/rearview mirror console, then extract the festoon type bulb from the spring contacts (photo).
20 To renew a front direction indicator bulb remove the cross-head screws, release the lamp from the upper tab, and withdraw the lamp. Release the spring clip and remove the bulb holder, then depress and twist the bulb to remove it (photos).
21 To renew a side repeater lamp bulb first remove the plastic wheel arch guard, if fitted, then remove the bulb holder from under the wing and extract the bulb (where applicable).
22 To renew a tail lamp bulb unscrew the lens retaining knob through the access hole in the rear compartment using an Allen key, if necessary. Pull out the outer end of the lens to release the inner tabs, then depress and twist the relevant bulb to remove it (photos).

11.17A Pull off the connector . . .

11.17B . . . remove the cover . . .

11.17C . . . release the retainer . . .

11.17D . . . and extract the headlamp bulb

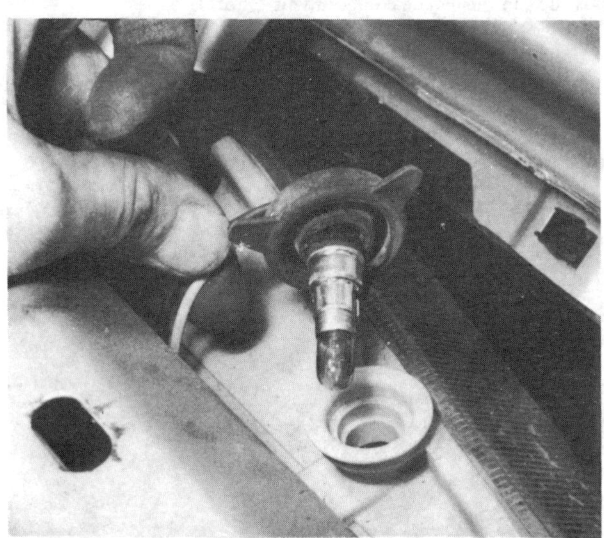

11.18 Removing a front sidelamp bulb

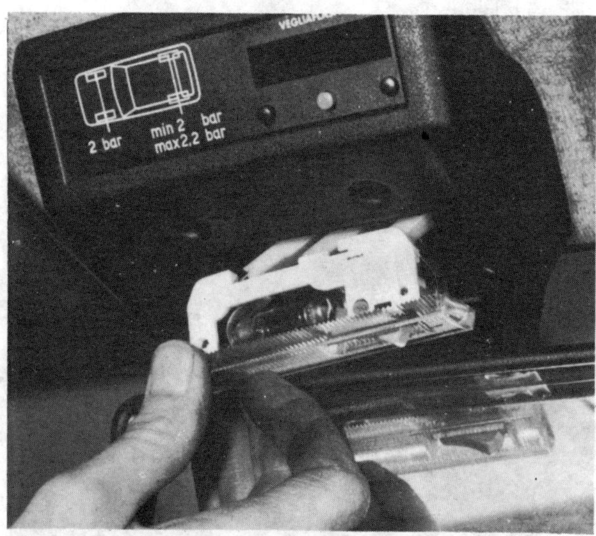

11.19 Removing the front interior light bulb

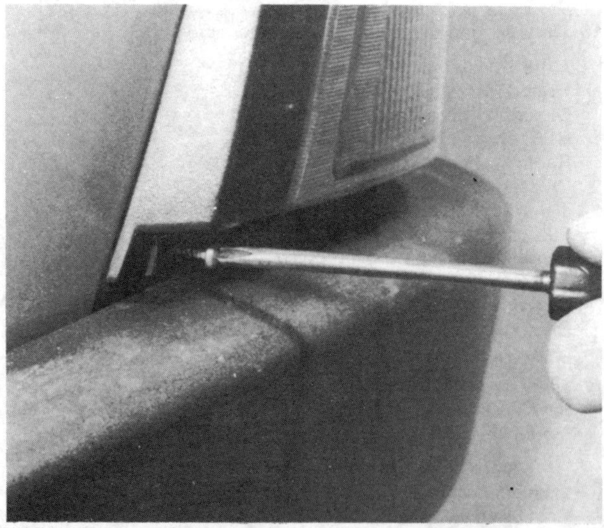

11.20A Remove the screws . . .

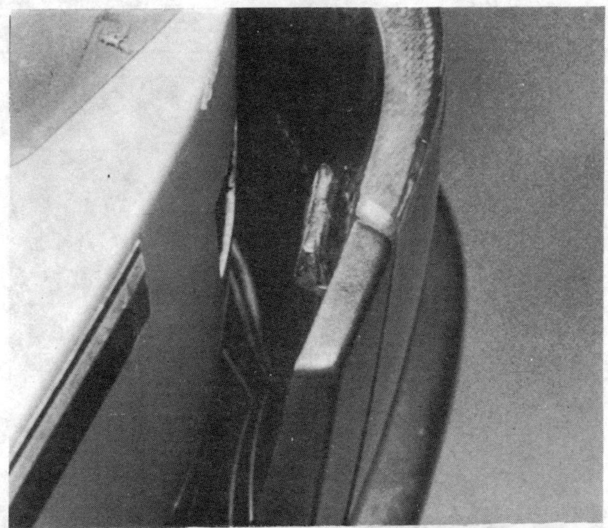

11.20B . . . release the tront direction indicator lamp from the upper tab . . .

11.20C . . . release the clip . . .

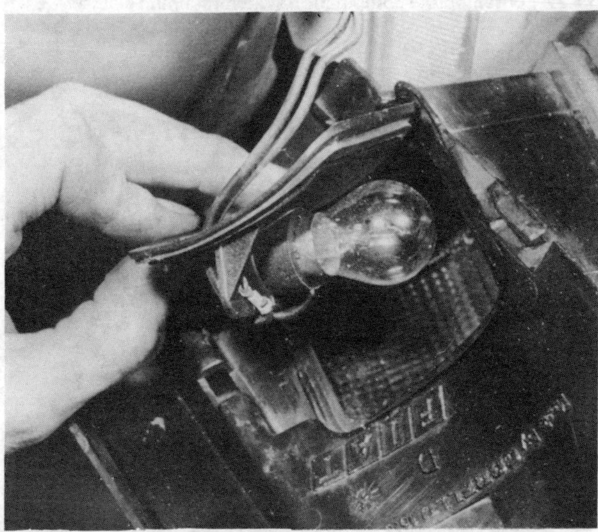

11.20D . . . and withdraw the bulb-holder and bulb

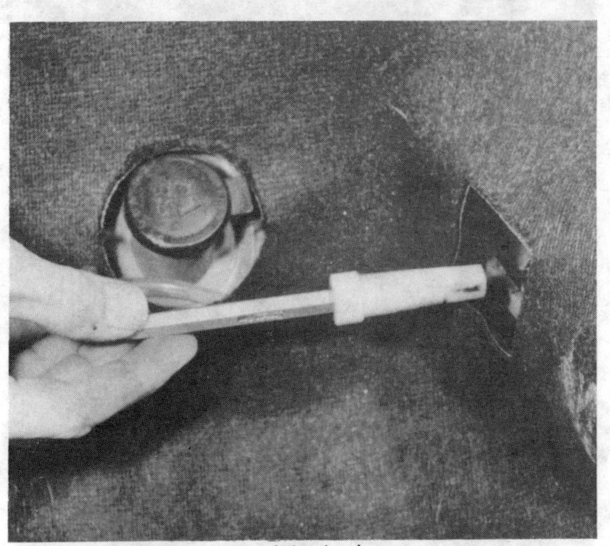

11.22A Unscrew the lens retaining knob . . .

11.22B . . . and remove the lens . . .

11.22C . . . for access to the tail lamp bulbs

23 To renew a number plate bulb, raise the tailgate and remove the cross-head screws in order to withdraw the lamp and lens. Extract the festoon type bulb from the spring contacts (photo).
24 To renew an instrument panel warning light bulb first remove the instrument panel. Twist the relevant bulb holder through 90° and remove it, then pull out the wedge type bulb (photo).

11.23 Removing the number plate bulb

11.24 Removing an instrument panel warning light bulb

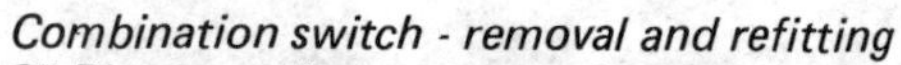

Combination switch - removal and refitting

25 Disconnect the battery negative lead.
26 Remove the steering wheel.
27 Using a socket through the hole in the bottom of the steering column shroud loosen the switch retaining clamp (photo).
28 Remove the shroud mounting screws and disconnect the wiring connectors to the switch.
29 Withdraw the switch and shroud from the steering column (photo).
30 Refitting is a reversal of removal.

Instrument panel - removal and refitting

31 For better access it is recommended that the steering wheel is first removed.
32 Disconnect the battery negative lead.
33 Remove the cross-head screws from the sides of the instrument panel (photo).

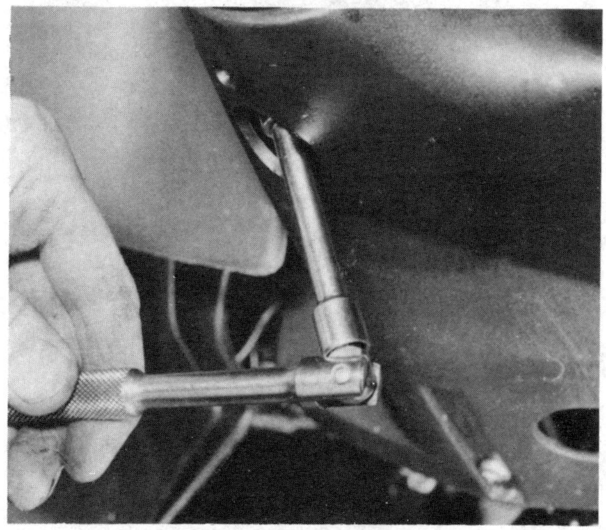
11.27 Loosening the combination switch retaining clamp

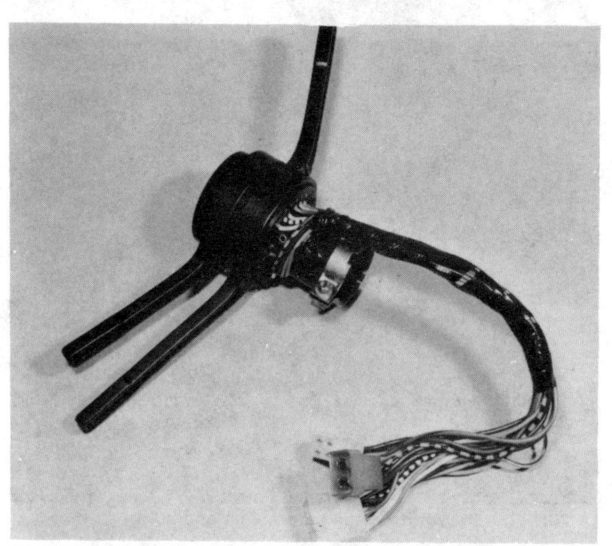
11.29 Combination switch and wiring

11.33 Removing the instrument panel retaining screws

11.34A Disconnecting the speedometer cable (arrowed) from the instrument panel

11.34B Front and rear views of the instrument panel

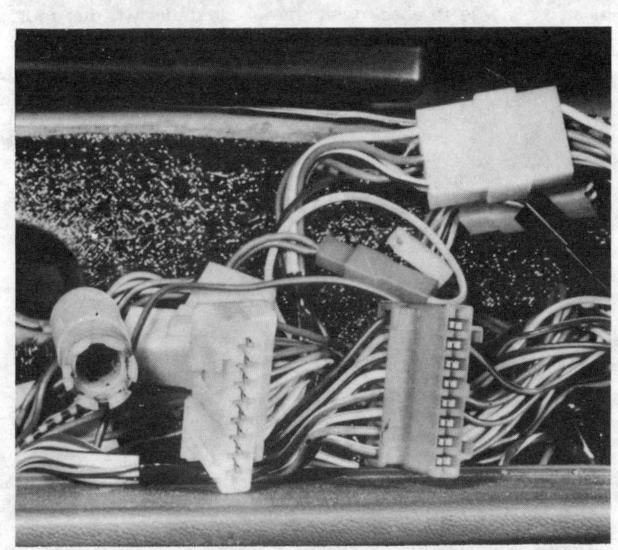

11.34C Wiring connectors and speedometer cable in the instrument panel recess

34 Ease the instrument panel from the facia sufficiently to disconnect the speedometer cable and wiring. Remove the panel (photos).
35 Refitting is a reversal of removal.

Windscreen washer pump - general

36 The windscreen washer pump is attached to the bottom of the reservoir bag located on the left-hand side of the engine compartment (photo).

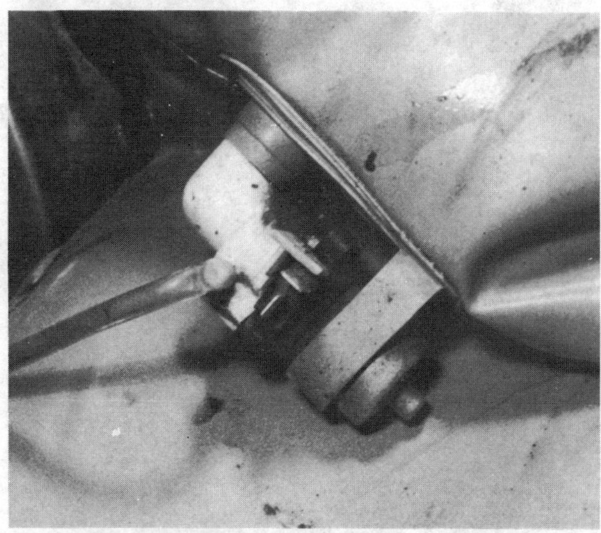

11.36 Windscreen washer pump

Digital clock console - removal and refitting

37 Disconnect the battery negative lead.
38 Prise the interior lamp from the console and disconnect the wiring.
39 Using a socket through the lamp recess, unscrew the mounting bolts, then lower the console and disconnect the wiring (photos).
40 If necessary, remove the two cross-head screws and separate the digital clock from the console (photo).
41 Refitting is a reversal of removal.

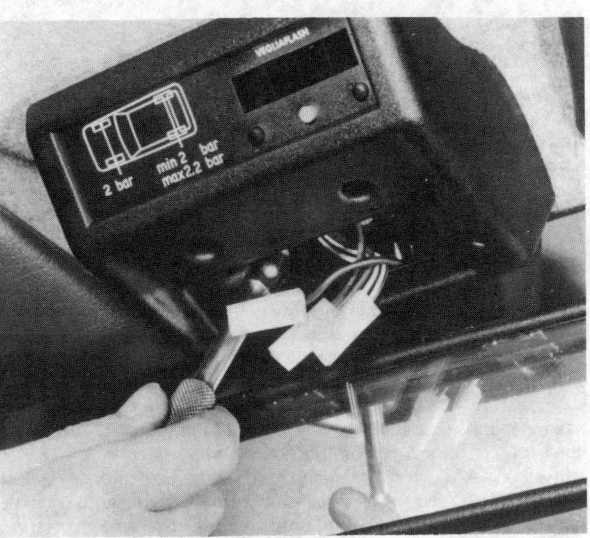

11.39A Remove the mounting bolts . . .

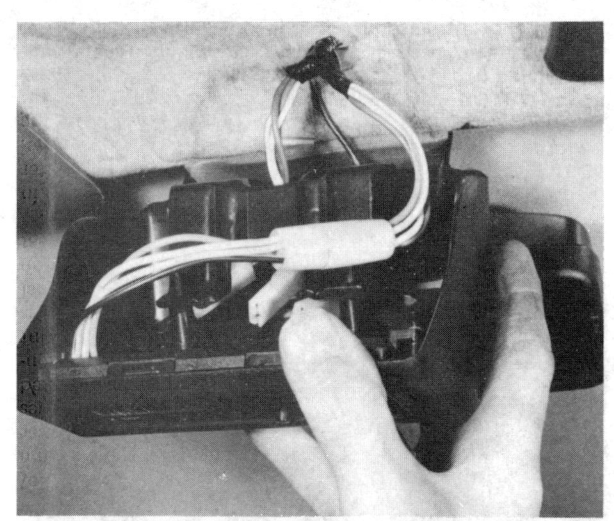

11.39B . . . and lower the digital clock console

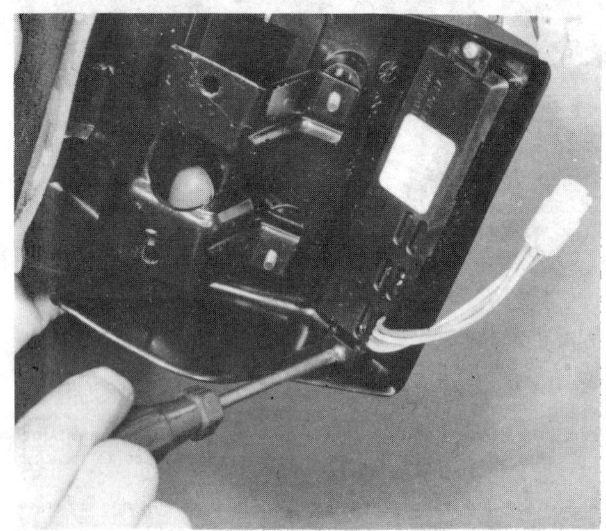

11.40 Removing the digital clock from the console

Part B - mobile radio equipment

Aerials – selection and fitting

The choice of aerials is now very wide. It should be realised that the quality has a profound effect on radio performance, and a poor, inefficient aerial can make suppression difficult.

A wing-mounted aerial is regarded as probably the most efficient for signal collection, but a roof aerial is usually better for suppression purposes because it is away from most interference fields. Stick-on wire aerials are available for attachment to the inside of the windscreen, but are not always free from the interference field of the engine and some accessories.

Motorised automatic aerials rise when the equipment is switched on and retract at switch-off. They require more fitting space and supply leads, and can be a source of trouble.

There is no merit in choosing a very long aerial as, for example, the type about three metres in length which hooks or clips on to the rear of the car, since part of the aerial will inevitably be located in an interference field. For VHF/FM radios the best length of aerial is about one metre. Active aerials have a transistor amplifiermounted at the base and this serves to boost the received signal. The aerial rod is sometimes rather shorter than normal passive types.

A large loss of signal can occur in the aerial feeder cable, especially over the Very High Frequency (VHF) bands. The design of feeder cable is invariably in the co-axial form, ie a centre conductor surrounded by a flexible copper braid forming the outer (earth) conductor. Between the inner and outer conductors is an insulator material which can be in solid or stranded form. Apart from insulation, its purpose is to maintain the correct spacing and concentricity. Loss of signal occurs in this insulator, the loss usually being greater in a poor quality cable. The quality of cable used is reflected in the price of the aerial with the attached feeder cable.

The capacitance of the feeder should be within the range 65 to 75 picofarads (pF) approximately (95 to 100 pF for Japanese and American equipment), otherwise the adjustment of the car radio aerial trimmer may not be possible. An extension cable is necessary for a long run between aerial and receiver. If this adds capacitance in excess of the above limits, a connector containing a series capacitor will be required, or an extension which is labelled as 'capacity-compensated'.

Fitting the aerial will normally involve making a 7/8 in (22 mm) diameter hole in the bodywork, but read the instructions that come with the aerial kit. Once the hole position has been selected, use a centre punch to guide the drill. Use sticky masking tape around the area for this helps with marking out and drill location, and gives protection to the paintwork should the drill slip. Three methods of making the hole are in use.:

(a) Use a hole saw in the electric drill. This is, in effect, a circular hacksaw blade wrapped round a former with a centre pilot drill.
(b) Use a tank cutter which also has cutting teeth, but is made to shear the metal by tightening with an Allen key.
(c) The hard way of drilling out the circle is using a small drill, say 1/8 in (3 mm), so that the holes overlap. The centre metal drops out and the hole is finished with round and half-round files.

Whichever method is used, the burr is removed from the body metal and paint removed from the underside. The aerial is fitted tightly ensuring that the earth fixing, usually a serrated washer, ring or clamp, is making a solid connection. *This earth connection is important in reducing interference.* Cover any bare metal with primer paint and topcoat, and follow by underseal if desired.

Aerial feeder cable routing should avoid the engine compartment and areas where stress might occur, eg under the carpet where feet will be located. Roof aerials require that the headlining be pulled back and that a path is available down the door pillar. It is wise to check with the vehicle dealer whether roof aerial fitting is recommended.

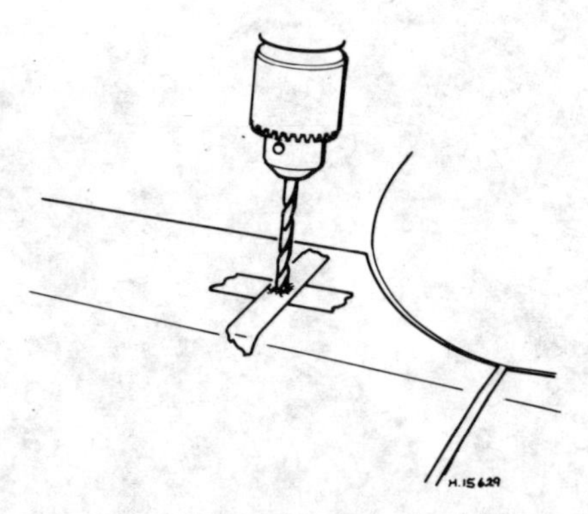

Fig. 13.36 Drilling the bodywork for aerial mounting (Sec 11)

Loudspeakers

Speakers should be matched to the output stage of the equipment, particularly as regards the recommended impedance. Power transistors used for driving speakers are sensitive to the loading placed on them.

Before choosing a mounting position for speakers, check whether the vehicle manufacturer has provided a location for them. Generally door-mounted speakers give good stereophonic reproduction, but not all doors are able to accept them. The next best position is the rear parcel shelf, and in this case speaker apertures can be cut into the shelf, or pod units may be mounted.

For door mounting, first remove the trim, which is often held on by 'poppers' or press studs, and then select a suitable gap in the inside door assembly. Check that the speaker would not obstruct glass or winder mechanism by winding the window up and down. A template is often provided for marking out the trim panel hole, and then the four fixing holes must be drilled through. Mark out with chalk and cut cleanly with a sharp knife or keyhole saw. Speaker leads are then threaded through the door and door pillar, if necessary drilling 10 mm diameter holes. Fit grommets in the holes and connect to the radio or tape unit correctly. Do not omit a waterproofing cover, usually supplied with door speakers. If the speaker has to be fixed into the metal of the door itself, use self-tapping screws, and if the fixing is to the door trim use self-tapping screws and flat spire nuts.

Rear shelf mounting is somewhat simpler but it is necessary to find gaps in the metalwork underneath the parcel shelf. However, remember that the speakers should be as far apart as possible to give a good stereo effect. Pod-mounted speakers can be screwed into position through the parcel shelf material, but it is worth testing for the best position. Sometimes good results are found by reflecting sound off the rear window.

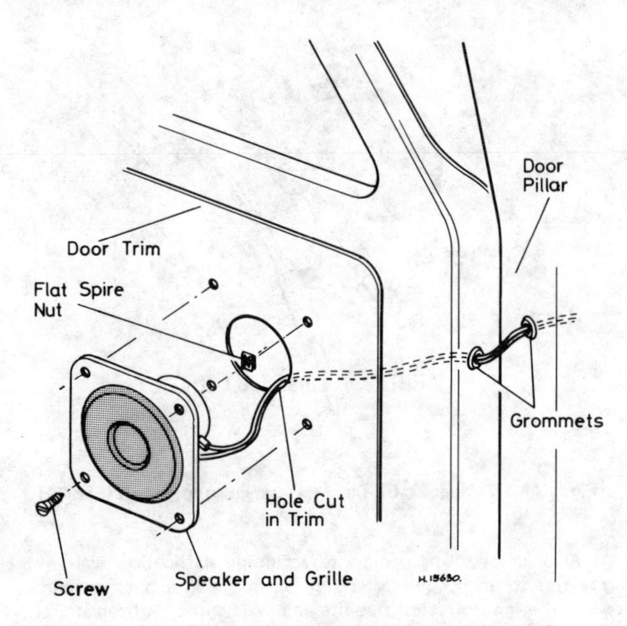

Fig. 13.37 Door-mounted speaker installation (Sec 11)

Unit installation

Many vehicles have a dash panel aperture to take a radio/audio unit, a recognised international standard being 189.5 mm x 60 mm. Alternatively a console may be a feature of the car interior design and this, mounted below the dashboard, gives more room. If neither facility is available a unit may be mounted on the underside of the parcel shelf; these are frequently non-metallic and an earth wire from the case to a good earth point is necessary. A three-sided cover in the form of a cradle is obtainable from car radio dealers and this gives a professional appearance to the installation; in this case choose a position where the controls can be reached by a driver with his seat belt on.

Installation of the radio/audio unit is basically the same in all cases, and consists of offering it into the aperture after removal of all knobs (*not* push buttons) and the trim plate. In some cases a special mounting plate is required to which the unit is attached. It is worthwhile supporting the rear end in cases where sag or strain may occur, and it is usually possible to use a length of perforated metal strip attached between the unit and

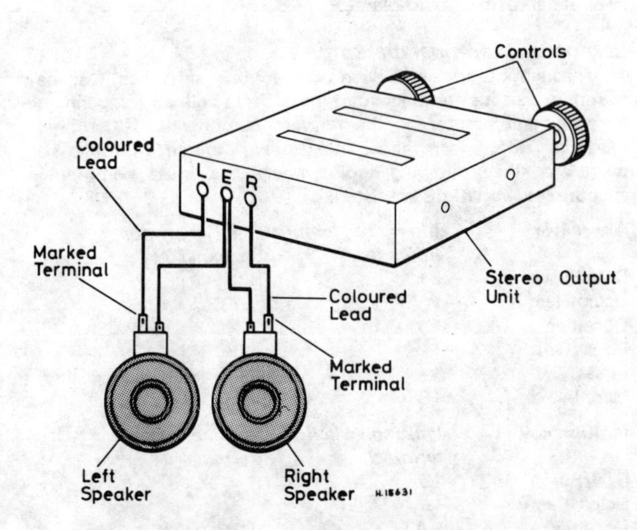

Fig. 13.38 Speaker connections must be correctly made as shown (Sec 11)

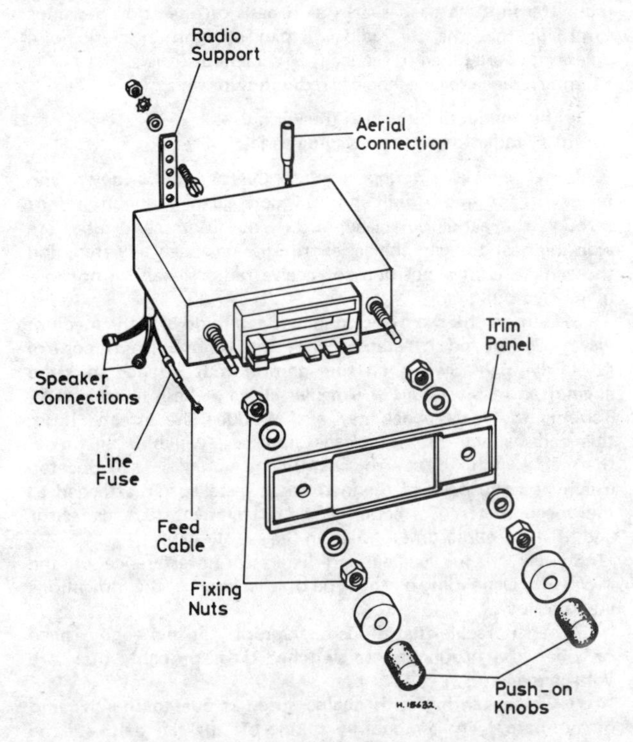

Fig. 13.39 Mounting component details for radio/cassette unit (Sec 11)

a good support point nearby. In general it is recommended that tape equipment should be installed at or nearly horizontal.

Connections to the aerial socket are simply by the standard plug terminating the aerial downlead or its extension cable. Speakers for a stereo system must be matched and correctly connected, as outlined previously.

Note: *While all work is carried out on the power side, it is wise to disconnect the battery earth lead.* Before connection is made to the vehicle electrical system, check that the polarity of the unit is correct. Most vehicles use a negative earth system, but radio/audio units often have a reversible plug to convert the set to either + or – earth. *Incorrect connection may cause serious damage.*

The power lead is often permanently connected inside the unit and terminates with one half of an in-line fuse carrier. The other half is fitted with a suitable fuse (3 or 5 amperes) and a wire which should go to a power point in the electrical system. This may be the accessory terminal on the ignition switch, giving the advantage of power feed with ignition or with the ignition key at the 'accessory' position. Power to the unit stops when the ignition key is removed. Alternatively, the lead may be taken to a live point at the fusebox with the consequence of having to remember to switch off the unit before leaving the vehicle.

Before switching on for initial test, be sure that the speaker connections have been made, for running without load can damage the output transistors. Switch on next and tune through the bands to ensure that all sections are working, and check the tape unit if applicable. The aerial trimmer should be adjusted to give the strongest reception on a weak signal in the medium wave band, at say 200 metres.

Interference

In general, when electric current changes abruptly, unwanted electrical noise is produced. The motor vehicle is filled with electrical devices which change electric current rapidly, the most obvious being the contact breaker.

When the spark plugs operate, the sudden pulse of spark current causes the associated wiring to radiate. Since early radio transmitters used sparks as a basis of operation, it is not surprising that the car radio will pick up ignition spark noise unless steps are taken to reduce it to acceptable levels.

Interference reaches the car radio in two ways:

(a) by conduction through the wiring.
(b) by radiation to the receiving aerial.

Initial checks presuppose that the bonnet is down and fastened, the radio unit has a good earth connection *(not* through the aerial downlead outer), no fluorescent tubes are working near the car, the aerial trimmer has been adjusted, and the vehicle is in a position to receive radio signals, ie not in a metal-clad building.

Switch on the radio and tune it to the middle of the medium wave (MW) band off-station with the volume (gain) control set fairly high. Switch on the ignition (but do not start the engine) and wait to see if irregular clicks or hash noise occurs. Tapping the facia panel may also produce the effects. If so, this will be due to the voltage stabiliser, which is an on-off thermal switch to control instrument voltage. It is located usually on the back of the instrument panel, often attached to the speedometer. Correction is by attachment of a capacitor and, if still troublesome, chokes in the supply wires.

Switch on the engine and listen for interference on the MW band. Depending on the type of interference, the indications are as follows.

A harsh crackle that drops out abruptly at low engine speed or when the headlights are switched on is probably due to a voltage regulator.

A whine varying with engine speed is due to the dynamo or alternator. Try temporarily taking off the fan belt – if the noise goes this is confirmation.

Regular ticking or crackle that varies in rate with the engine speed is due to the ignition system. With this trouble in particular and others in general, check to see if the noise is entering the receiver from the wiring or by radiation. To do this, pull out the aerial plug, (preferably shorting out the input socket or connecting a 62 pF capacitor across it). If the noise disappears it is coming in through the aerial and is *radiation noise.* If the noise persists it is reaching the receiver through the wiring and is said to be *line-borne.*

Interference from wipers, washers, heater blowers, turn-indicators, stop lamps, etc is usually taken to the receiver by wiring, and simple treatment using capacitors and possibly chokes will solve the problem. Switch on each one in turn (wet the screen first for running wipers!) and listen for possible interference with the aerial plug in place and again when removed.

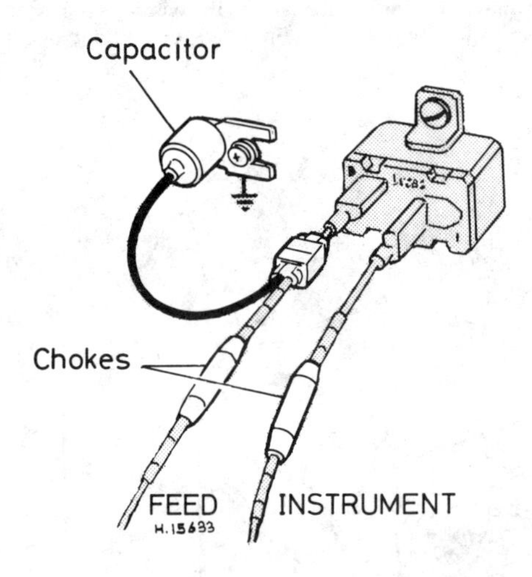

Fig. 13.40 Voltage stabiliser interference suppression (Sec 11)

Electric petrol pumps are now finding application again and give rise to an irregular clicking, often giving a burst of clicks when the ignition is on but the engine has not yet been started. It is also possible to receive whining or crackling from the pump.

Note that if most of the vehicle accessories are found to be creating interference all together, the probability is that poor aerial earthing is to blame.

Component terminal markings

Throughout the following sub-sections reference will be found to various terminal markings. These will vary depending on the manufacturer of the relevant component. If terminal markings differ from those mentioned, reference should be made to the following table, where the most commonly encountered variations are listed.

Alternator	*Alternator terminal (thick lead)*	*Exciting winding terminal*
DIN/Bosch	B+	DF
Delco Remy	+	EXC
Ducellier	+	EXC
Ford (US)	+	DF
Lucas	+	F
Marelli	+B	F
Ignition coil	*Ignition switch terminal*	*Contact breaker terminal*
DIN/Bosch	15	1
Delco Remy	+	–
Ducellier	BAT	RUP
Ford (US)	B/+	CB/–
Lucas	SW/+	–
Marelli	BAT/+B	D

Voltage regulator	*Voltage input terminal*	*Exciting winding terminal*
DIN/Bosch	B+/D+	DF
Delco Remy	BAT/+	EXC
Ducellier	BOB/BAT	EXC
Ford (US)	BAT	DF
Lucas	+/A	F
Marelli		F

Suppression methods – ignition

Suppressed HT cables are supplied as original equipment by manufacturers and will meet regulations as far as interference to neighbouring equipment is concerned. It is illegal to remove such suppression unless an alternative is provided, and this may take the form of resistive spark plug caps in conjunction with plain copper HT cable. For VHF purposes, these and 'in-line' resistors may not be effective, and resistive HT cable is preferred. Check that suppressed cables are actually fitted by observing cable identity lettering, or measuring with an ohmmeter – the value of each plug lead should be 5,000 to 10,000 ohms.

A 1 microfarad capacitor connected from the LT supply side of the ignition coil to a good nearby earth point will complete basic ignition interference treatment. *NEVER fit a capacitor to the coil terminal which connects to the contact breaker – the result would be burnt out points in a short time.*

If ignition noise persists despite the treatment above, the following sequence should be followed:

(a) Check the earthing of the ignition coil; remove paint from fixing clamp.
(b) If this does not work, lift the bonnet. Should there be no change in interference level, this may indicate that the bonnet is not electrically connected to the car body. Use a proprietary braided strap across a bonnet hinge ensuring a first class electrical connection. If, however, lifting the bonnet increases the interference, then fit resistive HT cables of a higher ohms-per-metre value.
(c) If all these measures fail, it is probable that re-radiation from metallic components is taking place. Using a braided strap between metallic points, go round the vehicle systematically – try the following: engine to body, exhaust system to body, front suspension to engine and to body, steering column to body (especially French and Italian cars), gear lever to engine and to body (again especially French and Italian cars), Bowden cable to body, metal parcel shelf to body. When an offending component is located it should be bonded with the strap permanently.
(d) As a next step, the fitting of distributor suppressors to each lead at the distributor end may help.
(e) Beyond this point is involved the possible screening of the distributor and fitting resistive spark plugs, but such advanced treatment is not usually required for vehicles with entertainment equipment.

Electronic ignition systems have built-in suppression components, but this does not relieve the need for using suppressed HT leads. In some cases it is permitted to connect a capacitor on the low tension supply side of the ignition coil, but not in every case. Makers' instructions should be followed carefully, otherwise damage to the ignition semiconductors may result.

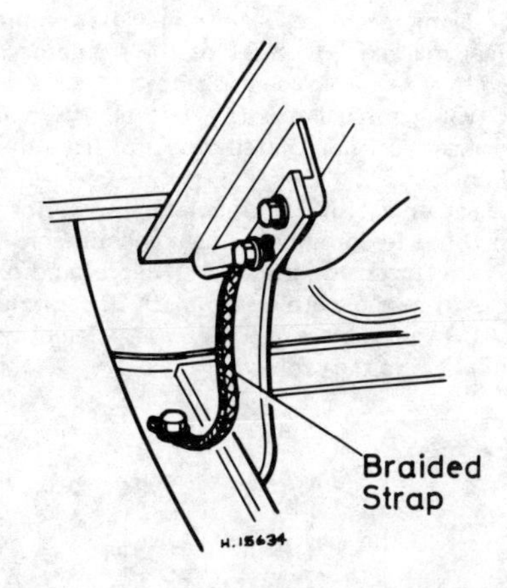

Fig. 13.41 Braided earth strap between bonnet and body (Sec 11)

Suppression methods – generators

For older vehicles with dynamos a 1 microfarad capacitor from the D (larger) terminal to earth will usually cure dynamo whine. Alternators should be fitted with a 3 microfarad capacitor from the B+ main output terminal (thick cable) to earth. Additional suppression may be obtained by the use of a filter in the supply line to the radio receiver.

It is most important that:

(a) Capacitors are never connected to the field terminals of either a dynamo or alternator.
(b) Alternators must not be run without connection to the battery.

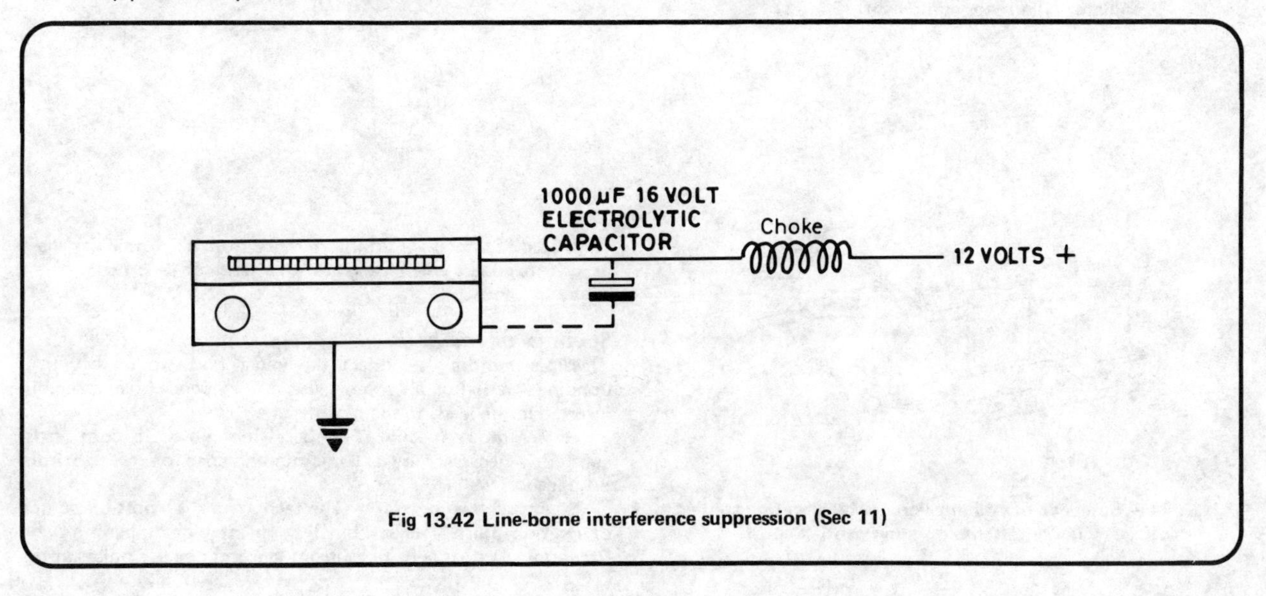

Fig 13.42 Line-borne interference suppression (Sec 11)

Suppression methods – voltage regulators

Voltage regulators used with DC dynamos should be suppressed by connecting a 1 microfarad capacitor from the control box D terminal to earth.

Alternator regulators come in three types:

(a) Vibrating contact regulators separate from the alternator. Used extensively on continental vehicles.
(b) Electronic regulators separate from the alternator.
(c) Electronic regulators built-in to the alternator.

In case (a) interference may be generated on the AM and FM (VHF) bands. For some cars a replacement suppressed regulator is available. Filter boxes may be used with non-suppressed regulators. But if not available, then for AM equipment a 2 microfarad or 3 microfarad capacitor may be mounted at the voltage terminal marked D+ or B+ of the regulator. FM bands may be treated by a feed-through capacitor of 2 or 3 microfarad.

Electronic voltage regulators are not always troublesome, but where necessary, a 1 microfarad capacitor from the regulator + terminal will help.

Integral electronic voltage regulators do not normally generate much interference, but when encountered this is in combination with alternator noise. A 1 microfarad or 2 microfarad capacitor from the warning lamp (IND) terminal to earth for Lucas ACR alternators and Femsa, Delco and Bosch equivalents should cure the problem.

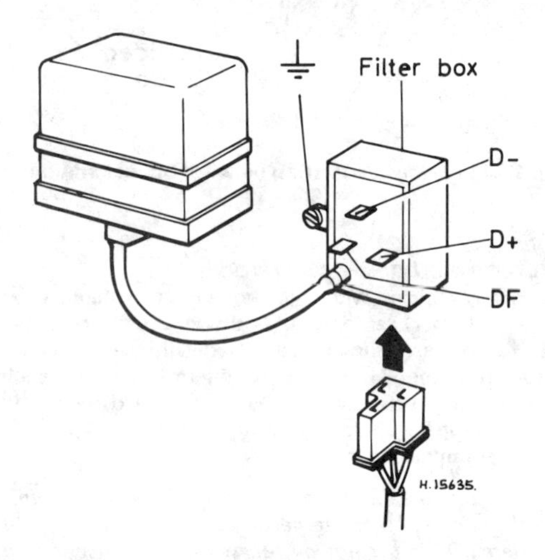

Fig. 13.43 Typical filter box for vibrating contact voltage regulator (alternator equipment) (Sec 11)

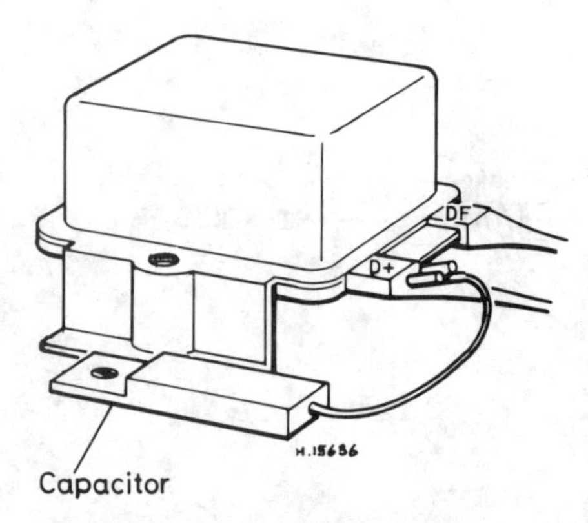

Fig. 13.44 Suppression of AM interference by vibrating contact voltage regulator (alternator equipment) (Sec 11)

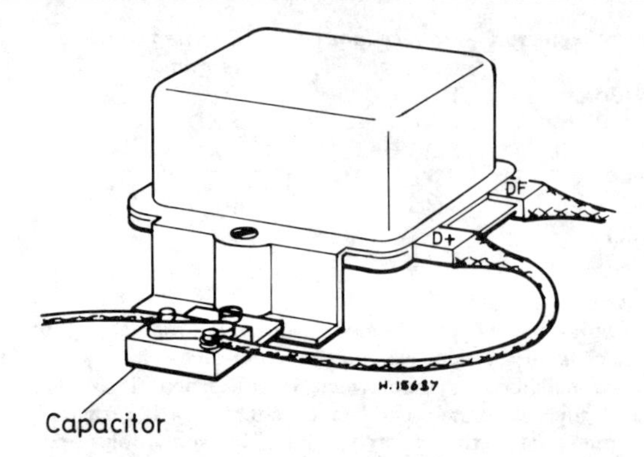

Fig. 13.45 Suppression of FM interference by vibrating contact voltage regulator (alternator equipment) (Sec 11)

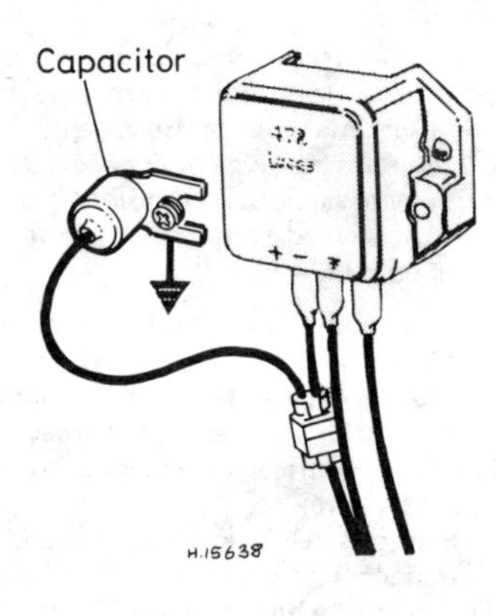

Fig. 13.46 Electronic voltage regulator suppression (Sec 11)

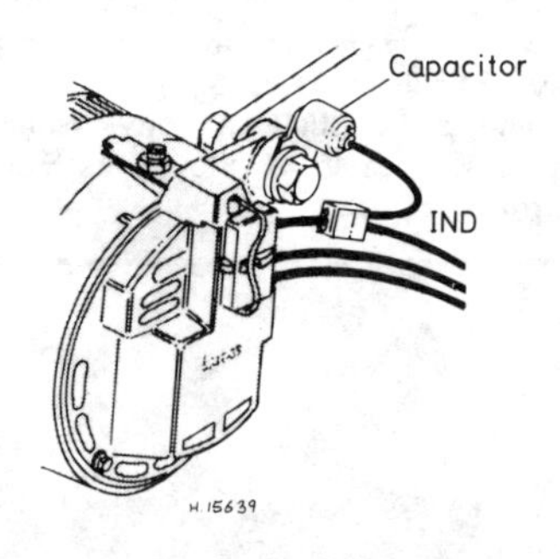

Fig. 13.47 Suppression of interference from electronic voltage regulator when integral with alternator (Sec 11)

Suppression methods – other equipment

Wiper motors – connect the wiper body to earth with a bonding strap. For all motors use a 7 ampere choke assembly inserted in the leads to the motor.

Heater motors – Fit 7 ampere line chokes in both leads, assisted if necessary by a 1 microfarad capacitor to earth from both leads.

Electronic tachometer – The tachometer is a possible source of ignition noise – check by disconnecting at the ignition coil CB terminal. It usually feeds from ignition coil LT pulses at the

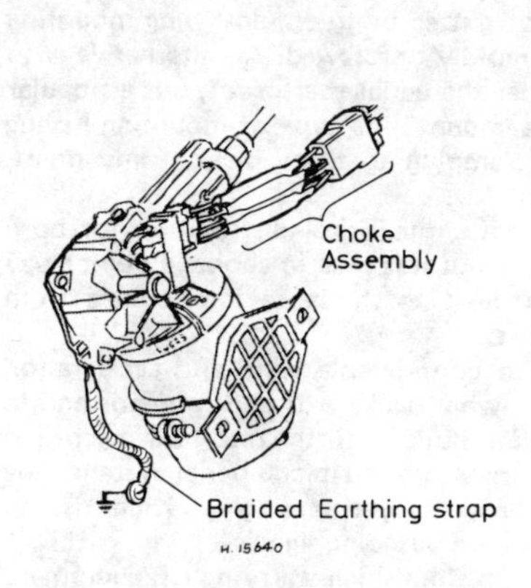

Fig. 13.48 Wiper motor suppression (Sec 11)

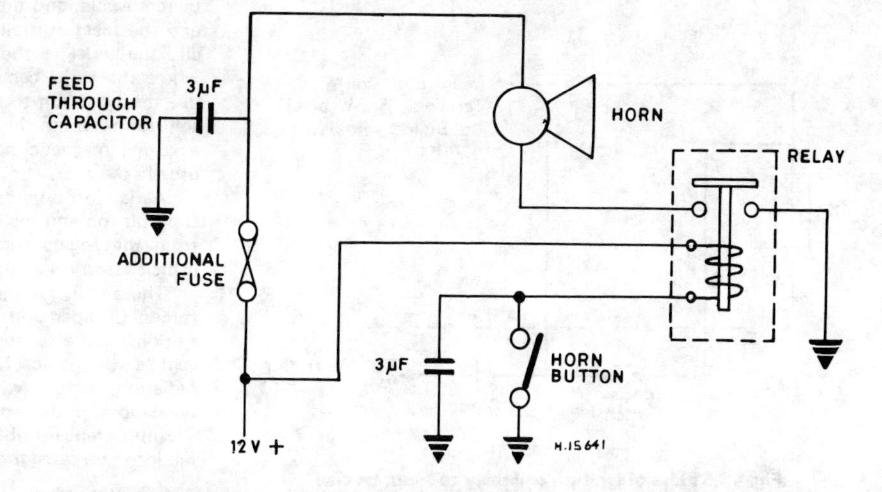

Fig. 13.49 Use of relay to reduce horn interference (Sec 11)

contact breaker terminal. A 3 ampere line choke should be fitted in the tachometer lead at the coil CB terminal.

Horn – A capacitor and choke combination is effective if the horn is directly connected to the 12 volt supply. The use of a relay is an alternative remedy, as this will reduce the length of the interference-carrying leads.

Electrostatic noise – Characteristics are erratic crackling at the receiver, with disappearance of symptoms in wet weather. Often shocks may be given when touching bodywork. Part of the problem is the build-up of static electricity in non-driven wheels and the acquisition of charge on the body shell. It is possible to fit spring-loaded contacts at the wheels to give good conduction between the rotary wheel parts and the vehicle frame. Changing a tyre sometimes helps – because of tyres' varying resistances. In difficult cases a trailing flex which touches the ground will cure the problem. If this is not acceptable it is worth trying conductive paint on the tyre walls.

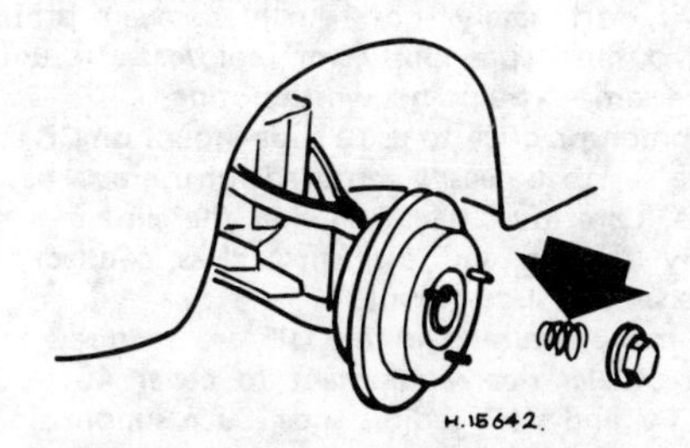

Fig. 13.50 Use of spring contacts at wheels (Sec 11)

Fuel pump – Suppression requires a 1 microfarad capacitor between the supply wire to the pump and a nearby earth point. If this is insufficient a 7 ampere line choke connected in the supply wire near the pump is required.

Fluorescent tubes – Vehicles used for camping/caravanning frequently have fluorescent tube lighting. These tubes require a relatively high voltage for operation and this is provided by an inverter (a form of oscillator) which steps up the vehicle supply voltage. This can give rise to serious interference to radio reception, and the tubes themselves can contribute to this interference by the pulsating nature of the lamp discharge. In such situations it is important to mount the aerial as far away from a fluorescent tube as possible. The interference problem may be alleviated by screening the tube with fine wire turns spaced an inch (25 mm) apart and earthed to the chassis. Suitable chokes should be fitted in both supply wires close to the inverter.

Radio/cassette case breakthrough

Magnetic radiation from dashboard wiring may be sufficiently intense to break through the metal case of the radio/ cassette player. Often this is due to a particular cable routed too close and shows up as ignition interference on AM and cassette play and/or alternator whine on cassette play.

The first point to check is that the clips and/or screws are fixing all parts of the radio/cassette case together properly. Assuming good earthing of the case, see if it is possible to re-route the offending cable – the chances of this are not good, however, in most cars.

Next release the radio/cassette player and locate it in different positions with temporary leads. If a point of low interference is found, then if possible fix the equipment in that area. This also confirms that local radiation is causing the trouble. If re-location is not feasible, fit the radio/cassette player back in the original position.

Alternator interference on cassette play is now caused by radiation from the main charging cable which goes from the battery to the output terminal of the alternator, usually via the + terminal of the starter motor relay. In some vehicles this cable is routed under the dashboard, so the solution is to provide a direct cable route. Detach the original cable from the alternator output terminal and make up a new cable of at least 6 mm^2 cross-sectional area to go from alternator to battery with the shortest possible route. *Remember – do not run the engine with the alternator disconnected from the battery.*

Ignition breakthrough on AM and/or cassette play can be a difficult problem. It is worth wrapping earthed foil round the offending cable run near the equipment, or making up a deflector plate well screwed down to a good earth. Another possibility is the use of a suitable relay to switch on the ignition coil. The relay should be mounted close to the ignition coil; with this arrangement the ignition coil primary current is not taken into the dashboard area and does not flow through the ignition switch. A suitable diode should be used since it is possible that at ignition switch-off the output from the warning lamp alternator terminal could hold the relay on.

Capacitors are usually supplied with tags on the end of the lead, while the capacitor body has a flange with a slot or hole to fit under a nut or screw with washer.

Connections to feed wires are best achieved by self-stripping connectors. These connectors employ a blade which, when squeezed down by pliers, cuts through cable insulation and makes connection to the copper conductors beneath.

Chokes sometimes come with bullet snap-in connectors fitted to the wires, and also with just bare copper wire. With connectors, suitable female cable connectors may be purchased

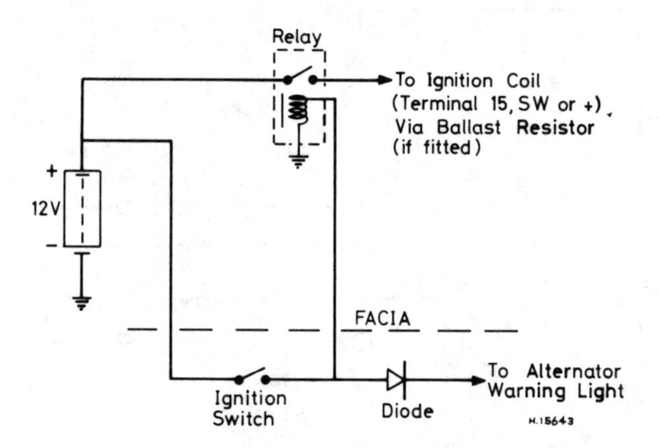

Fig. 13.51 Use of ignition coil relay to suppress case breakthrough (Sec 11)

from an auto-accessory shop together with any extra connectors required for the cable ends after being cut for the choke insertion. For chokes with bare wires, similar connectors may be employed together with insulation sleeving as required.

VHF/FM broadcasts

Reception of VHF/FM in an automobile is more prone to problems than the medium and long wavebands. Medium/long wave transmitters are capable of covering considerable distances, but VHF transmitters are restricted to line of sight, meaning ranges of 10 to 50 miles, depending upon the terrain, the effects of buildings and the transmitter power.

Because of the limited range it is necessary to retune on a long journey, and it may be better for those habitually travelling long distances or living in areas of poor provision of transmitters to use an AM radio working on medium/long wavebands.

When conditions are poor, interference can arise, and some of the suppression devices described previously fall off in performance at very high frequencies unless specifically designed for the VHF band. Available suppression devices include reactive HT cable, resistive distributor caps, screened plug caps, screened leads and resistive spark plugs.

For VHF/FM receiver installation the following points should be particularly noted:

(a) Earthing of the receiver chassis and the aerial mounting is important. Use a separate earthing wire at the radio, and scrape paint away at the aerial mounting.
(b) If possible, use a good quality roof aerial to obtain maximum height and distance from interference generating devices on the vehicle.
(c) Use of a high quality aerial downlead is important, since losses in cheap cable can be significant.
(d) The polarisation of FM transmissions may be horizontal, vertical, circular or slanted. Because of this the optimum mounting angle is at 45° to the vehicle roof.

Citizens' Band radio (CB)

In the UK, CB transmitter/receivers work within the 27 MHz and 934 MHz bands, using the FM mode. At present interest is concentrated on 27 MHz where the design and manufacture of equipment is less difficult. Maximum transmitted power is 4 watts, and 40 channels spaced 10 kHz apart within the range 27.60125 to 27.99125 MHz are available.

Aerials are the key to effective transmission and reception. Regulations limit the aerial length to 1.65 metres including the loading coil and any associated circuitry, so tuning the aerial is necessary to obtain optimum results. The choice of a CB aerial is dependent on whether it is to be permanently installed or removable, and the performance will hinge on correct tuning and the location point on the vehicle. Common practice is to clip the aerial to the roof gutter or to employ wing mounting where the aerial can be rapidly unscrewed. An alternative is to use the boot rim to render the aerial theftproof, but a popular solution is to use the 'magmount' – a type of mounting having a strong magnetic base clamping to the vehicle at any point, usually the roof.

Aerial location determines the signal distribution for both transmission and reception, but it is wise to choose a point away from the engine compartment to minimise interference from vehicle electrical equipment.

The aerial is subject to considerable wind and acceleration forces. Cheaper units will whip backwards and forwards and in so doing will alter the relationship with the metal surface of the vehicle with which it forms a ground plane aerial system. The radiation pattern will change correspondingly, giving rise to break-up of both incoming and outgoing signals.

Interference problems on the vehicle carrying CB equipment fall into two categories:

(a) Interference to nearby TV and radio receivers when transmitting.
(b) Interference to CB set reception due to electrical equipment on the vehicle.

Problems of breakthrough to TV and radio are not frequent, but can be difficult to solve. Mostly trouble is not detected or reported because the vehicle is moving and the symptoms rapidly disappear at the TV/radio receiver, but when the CB set is used as a base station any trouble with nearby receivers will soon result in a complaint.

It must not be assumed by the CB operator that his equipment is faultless, for much depends upon the design. Harmonics (that is, multiples) of 27 MHz may be transmitted unknowingly and these can fall into other users' bands. Where trouble of this nature occurs, low pass filters in the aerial or supply leads can help, and should be fitted in base station aerials as a matter of course. In stubborn cases it may be necessary to call for assistance from the licensing authority, or, if possible, to have the equipment checked by the manufacturers.

Interference received on the CB set from the vehicle equipment is, fortunately, not usually a severe problem. The precautions outlined previously for radio/cassette units apply, but there are some extra points worth noting.

It is common practice to use a slide-mount on CB equipment enabling the set to be easily removed for use as a base station, for example. Care must be taken that the slide mount fittings are properly earthed and that first class connection occurs between the set and slide-mount.

Vehicle manufacturers in the UK are required to provide suppression of electrical equipment to cover 40 to 250 MHz to protect TV and VHF radio bands. Such suppression appears to be adequately effective at 27 MHz, but suppression of individual items such as alternators/dynamos, clocks, stabilisers, flashers, wiper motors, etc, may still be necessary. The suppression capacitors and chokes available from auto-electrical suppliers for entertainment receivers will usually give the required results with CB equipment.

Other vehicle radio transmitters

Besides CB radio already mentioned, a considerable increase in the use of transceivers (ie combined transmitter and receiver units) has taken place in the last decade. Previously this type of equipment was fitted mainly to military, fire, ambulance and police vehicles, but a large business radio and radio telephone usage has developed.

Generally the suppression techniques described previously will suffice, with only a few difficult cases arising. Suppression is carried out to satisfy the 'receive mode', but care must be taken to use heavy duty chokes in the equipment supply cables since the loading on 'transmit' is relatively high.

Glass-fibre bodied vehicles

Such vehicles do not have the advantage of a metal box surrounding the engine as is the case, in effect, of conventional vehicles. It is usually necessary to line the bonnet, bulkhead and wing valances with metal foil, which could well be the aluminium foil available from builders merchants. Bonding of sheets one to another and the whole down to the chassis is essential.

Wiring harness may have to be wrapped in metal foil which again should be earthed to the vehicle chassis. The aerial base and radio chassis must be taken to the vehicle chassis by heavy metal braid. VHF radio suppression in glass-fibre cars may not be a feasible operation.

In addition to all the above, normal suppression components should be employed, but special attention paid to earth bonding. A screen enclosing the entire ignition system usually gives good improvement, and fabrication from fine mesh perforated metal is convenient. Good bonding of the screening boxes to several chassis points is essential.

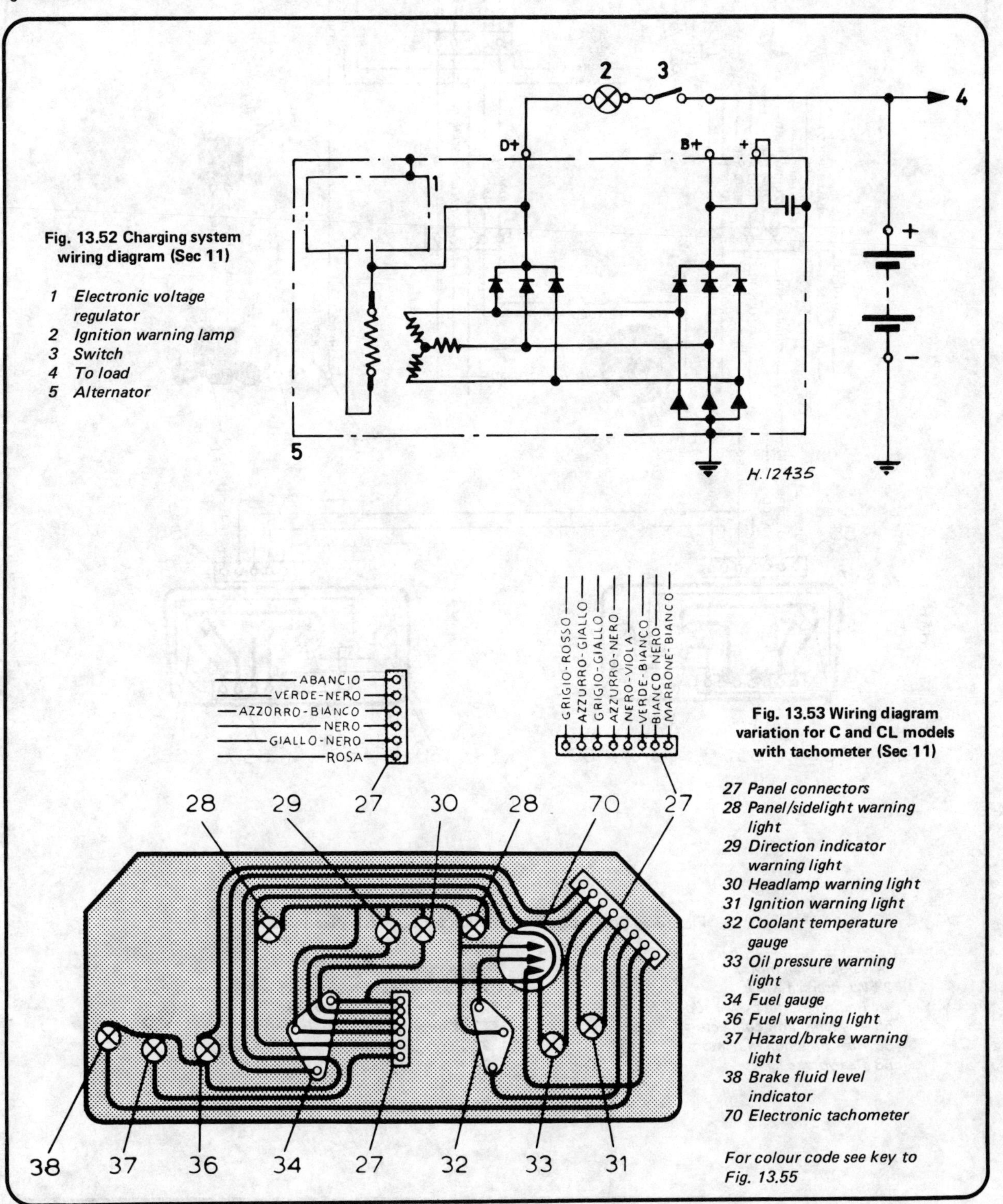

Fig. 13.52 Charging system wiring diagram (Sec 11)

1 Electronic voltage regulator
2 Ignition warning lamp
3 Switch
4 To load
5 Alternator

Fig. 13.53 Wiring diagram variation for C and CL models with tachometer (Sec 11)

27 Panel connectors
28 Panel/sidelight warning light
29 Direction indicator warning light
30 Headlamp warning light
31 Ignition warning light
32 Coolant temperature gauge
33 Oil pressure warning light
34 Fuel gauge
36 Fuel warning light
37 Hazard/brake warning light
38 Brake fluid level indicator
70 Electronic tachometer

For colour code see key to Fig. 13.55

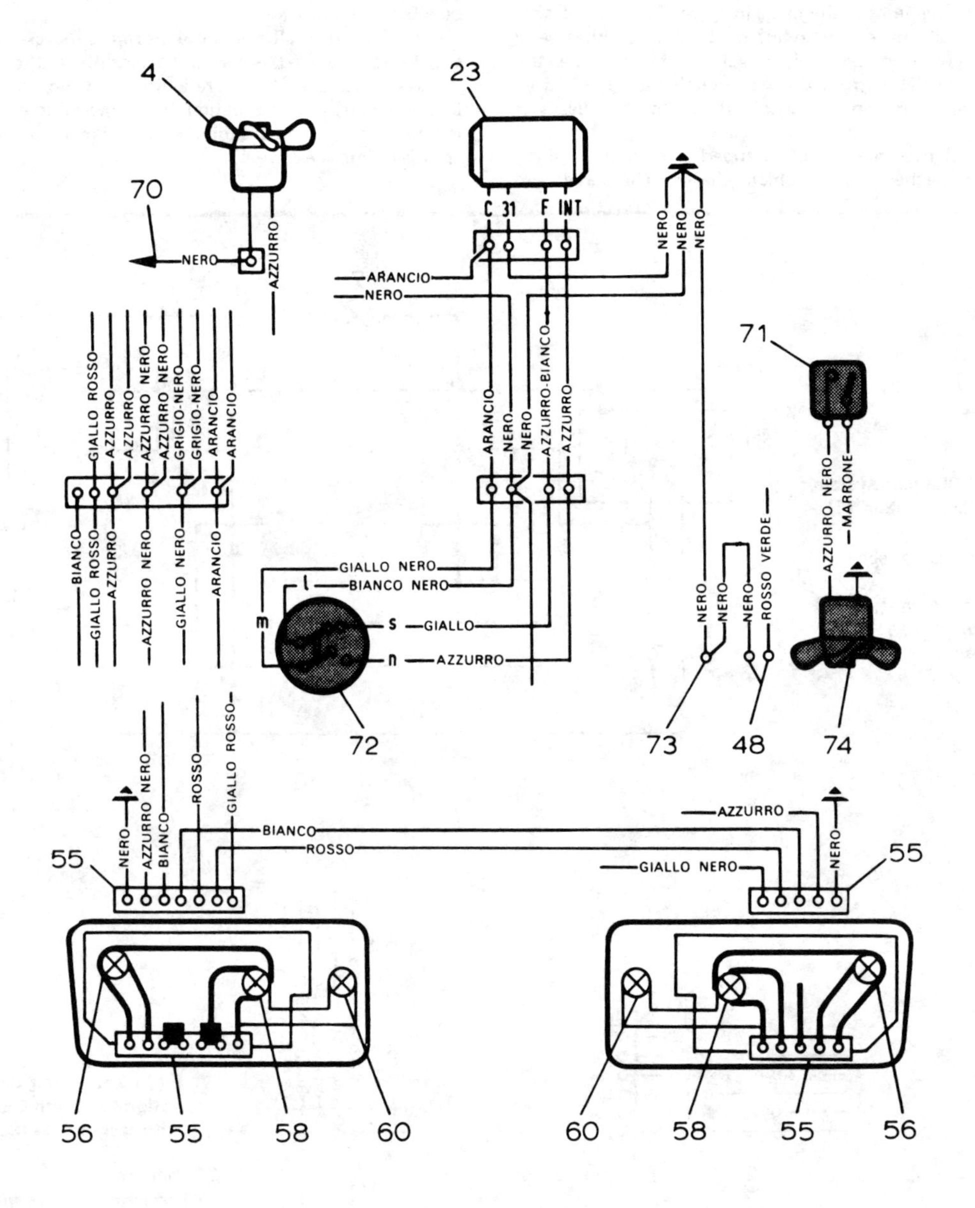

Fig. 13.54 Wiring diagram variation for L models (Sec 11)

4 Coolant fan motor
23 Windscreen wiper
48 Radio/cassette power supply leads
55 Tail lamp unit connectors
56 Rear direction indicators
58 Rear/brake lights
60 Number plate lights
70 To side/front direction indicator LH terminal
71 Heater fan motor switch
72 Windscreen wiper switch
73 Earth terminal
74 Heater fan motor

For colour code see key to Fig. 13.55

Key to wiring diagram for C and CL models with 1049 cc engine (Sec 11)

Explanatory note: Each wire section end has an identification number. Complete wiring run can be traced by following the wire number in the adjacent shaded strip where the run is resumed.

1 Front direction indicator
2 Sidelights
3 Headlamp - main and dipped beams
4 Radiator fan motor
5 Radiator fan control switch
6 Horns
7 Repeater lights
8 Windscreen washer pump (C and CL versions)
9 Brake fluid level sender
10 Starter
11 Alternator
12 Ignition coil
13 Tachometer lead (optional) (C and CL versions)
14 Coolant temperature gauge sender (thermal switch on L version)
15 Battery
16 Ignition distributor
17 Reversing lights switch (C and CL versions)
18 Spark plugs
19 Oil pressure sender
20 Hazard warning flasher
21 Direction indicator flasher
22 Heated rear window relay
23 Windscreen wiper motor
24 Windscreen wiper intermitter (C and CL versions)
25 Radio in-line fuse
26 Fuse block
27 Panel connectors
28 Panel lights and sidelights warning light
29 Direction indicator warning light
30 Headlamp warning light
31 Ignition warning light
32 Coolant temperature gauge (C and CL versions) and warning light (L version)
33 Oil pressure warning light
34 Fuel gauge
35 Ignition switch
36 Fuel warning light
37 Hazard warning light
38 Brake fluid level indicator
39 Hazard warning switch
40 Heated rear window switch and warning light
41 Lighting/panel light switch
42 Windscreen wiper/washer switch
43 Headlamp switch
44 Direction indicator switch
45 Horn switch
46 Heater fan switch
47 Cigar lighter and light (C and CL versions)
48 Radio/cassette player power supplies
49 Door pillar switch
50 Heater fan motor
51 Courtesy light and switch
52 Brake fluid level indicator/handbrake check switch
53 Brake light switch
54 Heated rear window (optional)
55 Tail lamp unit connectors
56 Rear direction indicators
57 Boot light (C and CL two-door versions)
58 Rear and brake lights
59 Reversing lights
60 Number plate lights
61 Fuel gauge sender

Wiring colour code

Arancio	= Amber	Blu	= Dark blue	Marrone	= Brown	Rosso	= Red
Azzurro	= Light blue	Giallo	= Yellow	Nero	= Black	Verde	= Green
Bianco	= White	Grigio	= Grey	Rosa	= Pink	Viola	= Violet

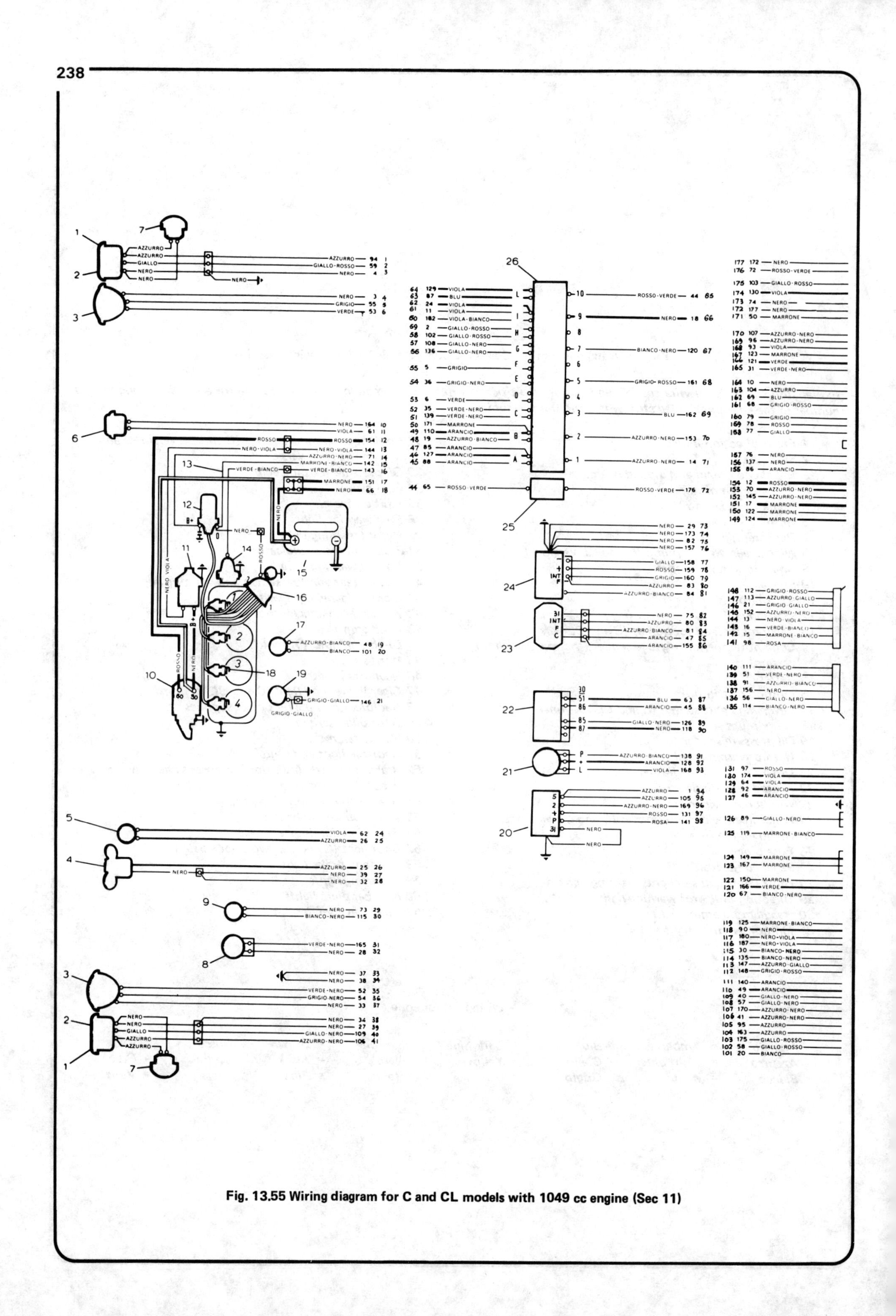

Fig. 13.55 Wiring diagram for C and CL models with 1049 cc engine (Sec 11)

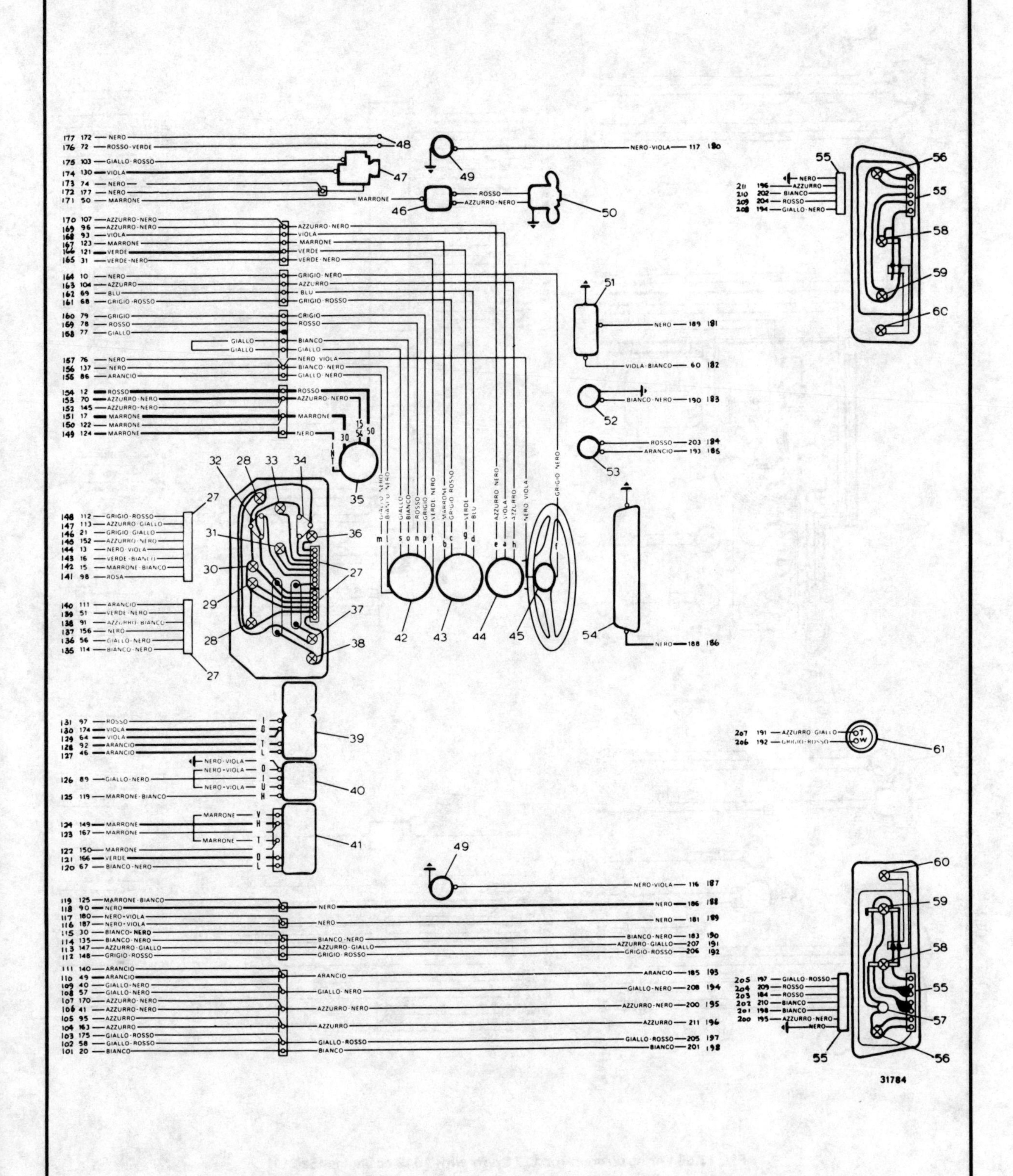

Fig. 13.55 Wiring diagram for C and CL models with 1049 cc engine (Sec 11) (continued)

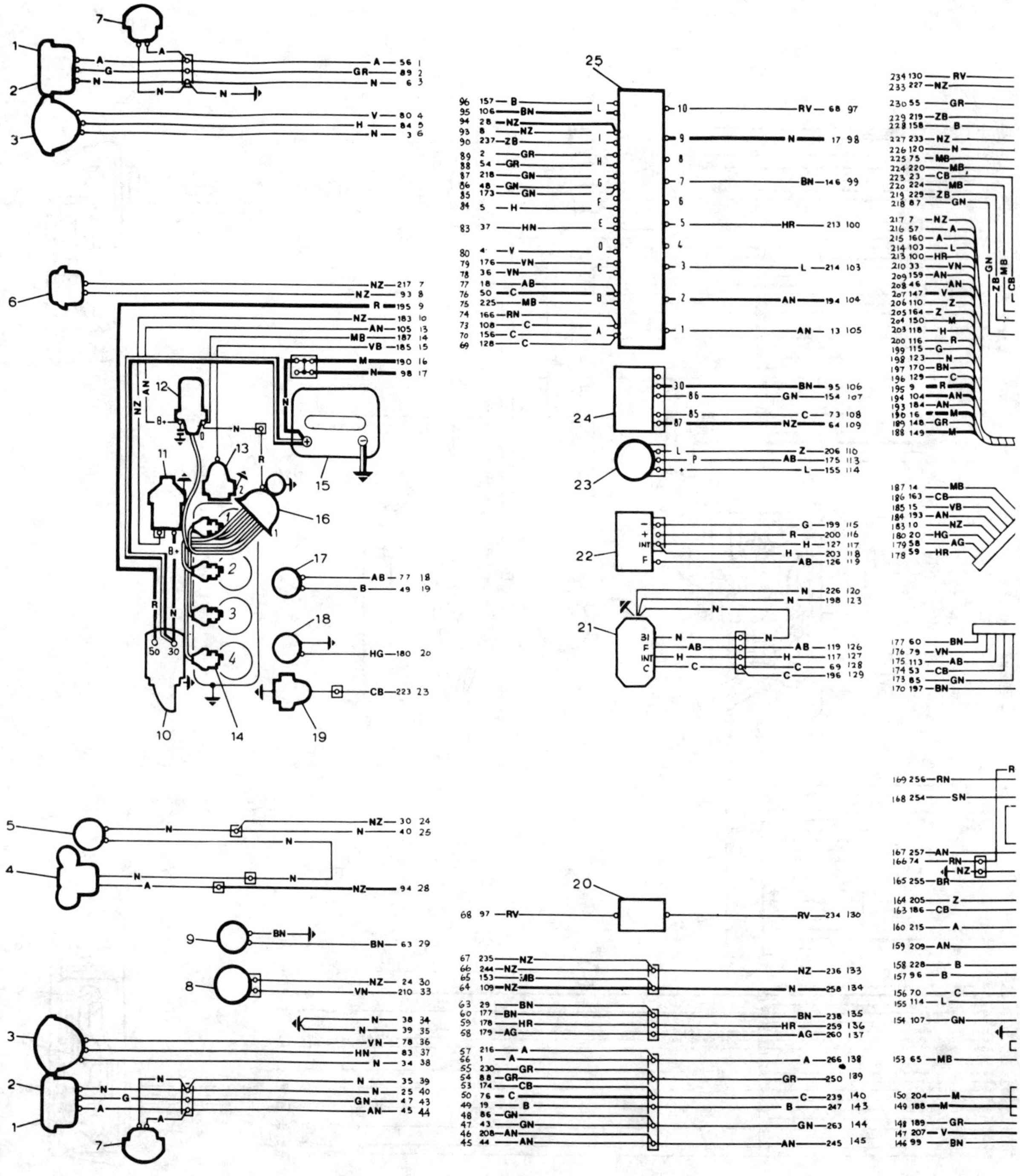

Fig. 13.56 Wiring diagram for 127 Sport with 1049 cc engine (Sec 11)

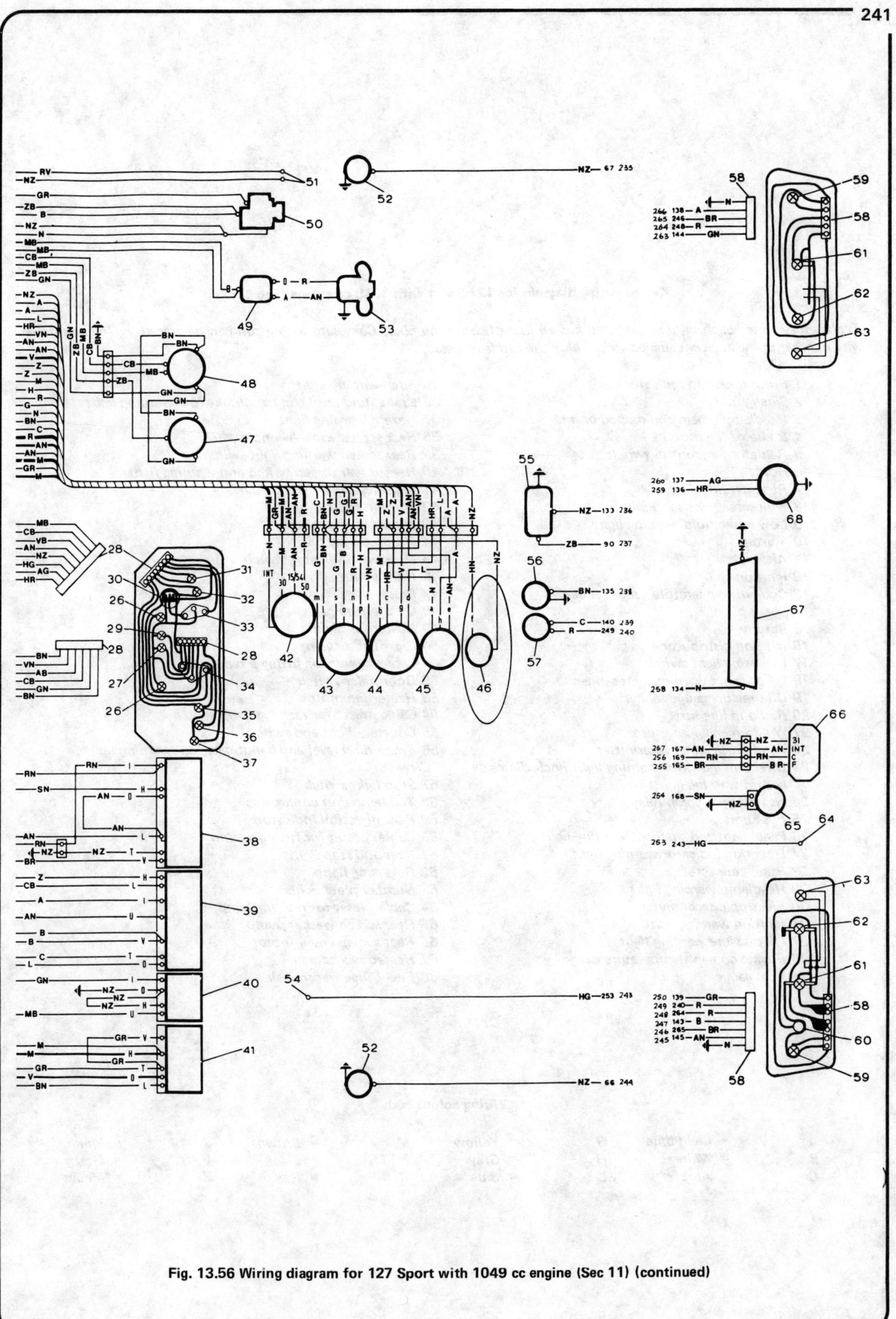

Fig. 13.56 Wiring diagram for 127 Sport with 1049 cc engine (Sec 11) (continued)

Key to wiring diagram for 127 Sport with 1049 cc engine (Sec 11)

Explanatory note: Each wire section end has an identification number. Complete wiring run can be traced by following the wire number in the adjacent shaded strip where the run is resumed.

1 Front direction indicator
2 Sidelights
3 Headlamp - main and dipped beams
4 Radiator fan motor
5 Radiator fan control switch
6 Horns
7 Repeater lights
8 Windscreen washer pump
9 Low brake fluid warning light
10 Starter
11 Alternator
12 Ignition coil
13 Coolant temperature gauge sender
14 Spark plugs
15 Battery
16 Ignition distributor
17 Reversing light switch
18 Oil pressure warning light sender
19 Oil pressure gauge sender
20 Radio in-line fuse
21 Windscreen wiper motor
22 Windscreen wiper intermitter
23 Direction indicator warning light (including hazard warning light)
24 Heated rear screen relay
25 Fuse unit
26 Panel light and sidelight warning light
27 Direction indicator warning light
28 Panel connectors
29 Headlamp warning light
30 Electronic tachometer
31 Ignition warning light
32 Oil pressure warning light
33 Engine coolant temperature gauge
34 Fuel gauge
35 Fuel warning light
36 Brake fluid level and handbrake 'on' warning light
37 Hazard warning light
38 Rear screen wiper/washer switch
39 Rear screen warning light switch
40 Heated rear screen switch and warning light
41 Lighting/panel light switch
42 Ignition switch
43 Wiper/washer switch
44 Headlamp switch
45 Direction indicator switch
46 Horn control
47 Clock
48 Oil pressure gauge
49 Heater fan switch
50 Cigar lighter/light
51 Cables preset for fitting a radio set
52 Door pillar switch
53 Heater fan motor
54 Cable preset for rear fog lamp switch
55 Courtesy light and switch
56 Brake fluid level and handbrake 'on' warning light switch
57 Stop light switch
58 Tail lamp unit connectors
59 Rear direction indicators
60 Cables preset for fitting a lamp
61 Rear and stop lights
62 Reversing lights
63 Number plate lights
64 Cable preset for rear fog lamp
65 Rear screen washer pump
66 Rear screen wiper motor
67 Heated rear screen
68 Fuel gauge transmitter

Wiring colour code

A	= Light blue	G	= Yellow	M	= Brown	S	= Pink
B	= White	H	= Grey	N	= Black	V	= Green
C	= Amber	L	= Blue	R	= Red	Z	= Mauve

Key to wiring diagram for later C and CL models (Sec 11)

1 Front direction indicators (21W, Spherical)
2 Sidelights (5W, Spherical)
3 Headlamp main and dipped beams (40/45W, Spherical)
4 Radiator fan motor
5 Radiator fan control switch
6 Horns
7 Repeater light (4W, Tubular)
8 Windscreen washer pump
9 Starter motor
10 Alternator
11 Ignition coil
12 Cable preset for fitting tachometer
13 Engine coolant temperature transmitter (control switch on L version)
14 Spark plugs
15 Oil pressure warning transmitter
16 Reversing light switch
17 Ignition distributor
18 Battery
19 Direction indicator flasher
20 Heated backlight relay
21 Windscreen wiper motor
22 Windscreen wiper intermitter
23 Fuse holder preset for fitting a radio set
24 Fuses
25 Panel connectors
26 Panel/side light w/l bulbs (3W, w/b)
27 Direction indicator w/l (1.2W, w/b)
28 Headlamp w/l (1.2W, w/b)
29 Ignition w/l (3W, w/b)
30 Coolant temperature gauge (w/l for L version - 1.2W - w/b)
31 Engine oil pressure w/l (1.2W, w/b)
32 Fuel gauge
33 Ignition switch
34 Fuel w/l (1.2W, w/b)
35 Connections preset for fitting hazard warning light
36 Connections preset for fitting low brake fluid level w/l
37 Heated backlight switch and light
38 Lighting/panel light switch
39 Screen wiper/washer switch
40 Headlamp control lever
41 Direction indicator switch
42 Horn control
43 Door switch
44 Heater fan motor switch
45 Cigar lighter/light (4W, Tubular)
46 Cables preset for fitting a radio/cassette player
47 Heater fan motor
48 Courtesy light/switch (5W, Tubular)
49 Stop light switch
50 Heated backlight (if fitted)
51 Tail lamp unit connectors
52 Rear direction indicators (21W)
53 Boot light (5W bulb) (Two-door versions)
54 Rear/stop lights 5/21W, Spherical)
55 Reversing lights (21W, Spherical)
56 Fuel gauge transmitter

Explanatory note

Each wire section ends in an identification number. Actual wiring can be traced by looking for the wire number in the adjacent shaded strip, where wire path is resumed.

Cable colour code

Arancio	*= Amber*	*Giallo*	*= Yellow*	*Rosa*	*= Pink*
Azzurro	*= Light blue*	*Grigio*	*= Grey*	*Rosso*	*= Red*
Bianco	*= White*	*Marrone*	*= Brown*	*Verde*	*= Green*
Blu	*= Dark blue*	*Nero*	*= Black*	*Viola*	*= Mauve*

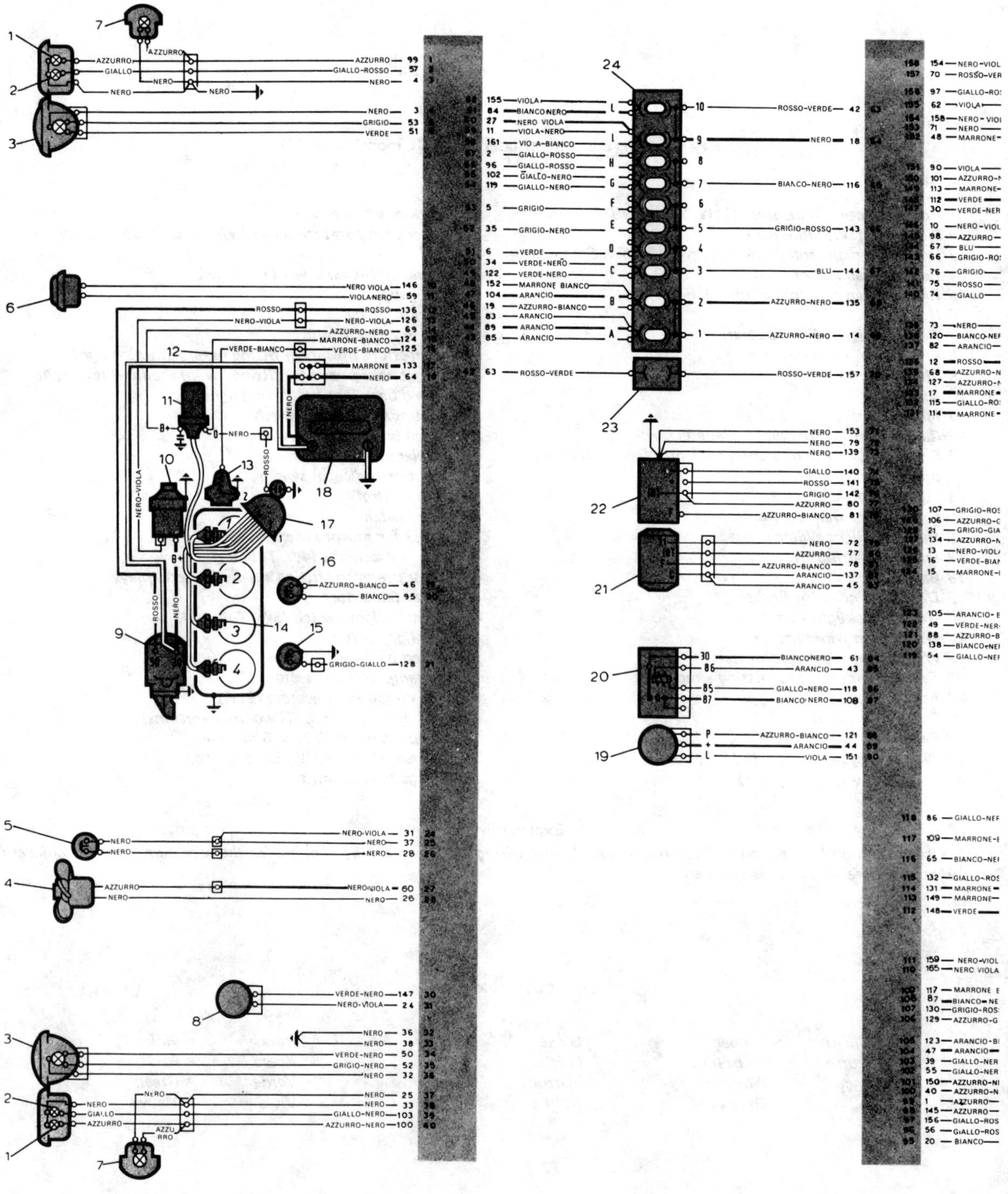

Fig. 13.57 Wiring diagram for later C and CL models (Sec 11)

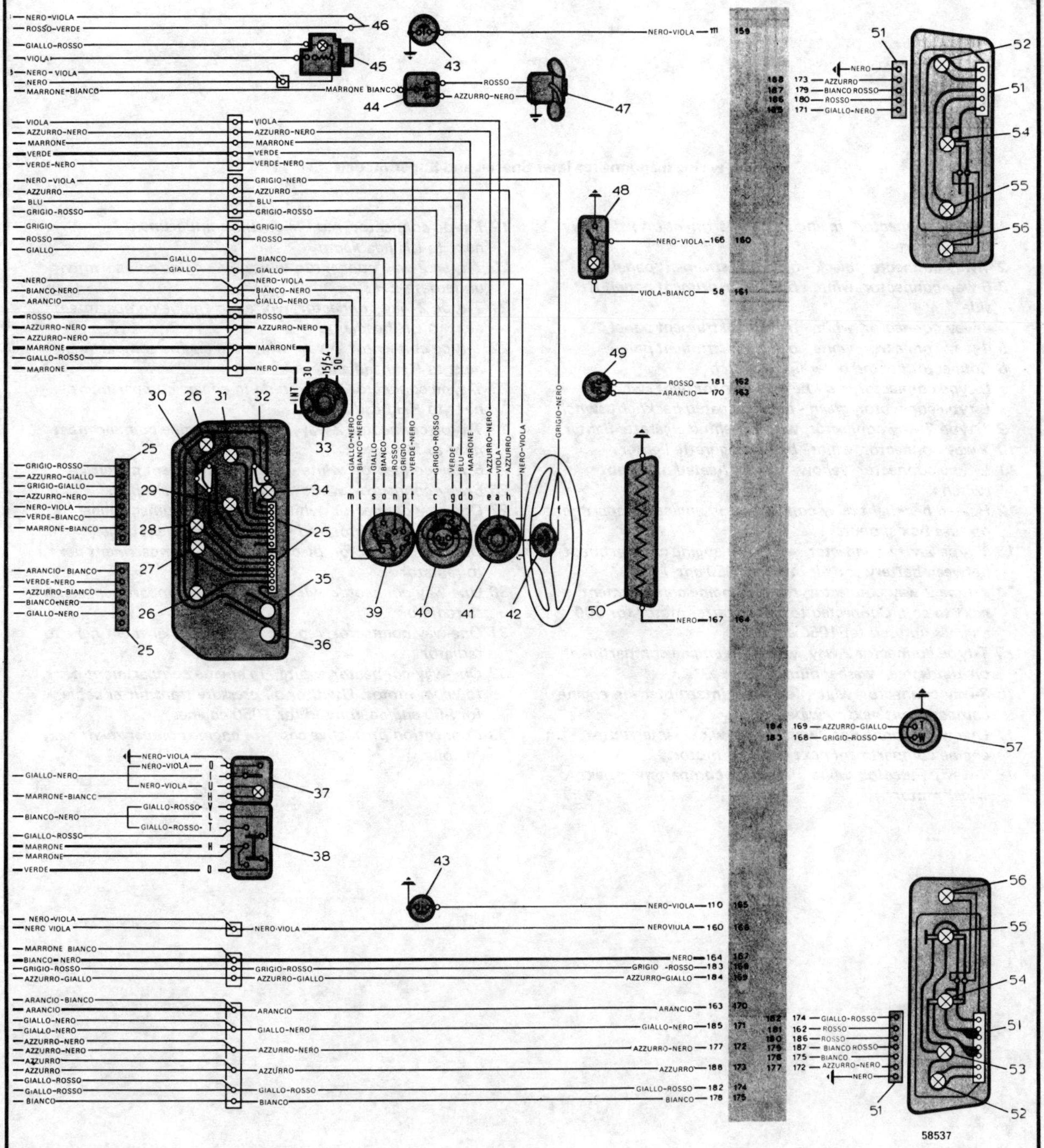

58537

Fig. 13.57 Wiring diagram for later C and CL models (Sec 11) (continued)

Key to wiring diagrams for later Special and Super models (Sec 11)

1 *H-type connector - in engine compartment on fuse/relay box support*
2 *4-way connector, black - behind instrument panel*
3 *6-way connector, white - behind instrument panel, RH side*
4 *4-way connector, white - behind instrument panel*
5 *4-way connector, white - behind instrument panel*
6 *Connector behind outer lights switch*
7 *L-type connector, red - behind rear for light switch*
8 *L-type connector, green - behind heated backlight switch*
9 *T-type 2-way connector, white - behind cigarette lighter*
10 *1-way connector, white - behind cigarette lighter*
11 *L-type connector, yellow - behind heated backlight switch*
12 *Heated backlight relay connector - in engine compartment on fuse box support*
13 *T-type 2-way connector, white - in engine compartment between battery and air intake (1050 cm^3)*
14 *T-type 2-way connector, red - in engine compartment next to coil. Connected to 50 of starter motor for 900 engines, unused for 1050 engines*
15 *T-type connector 2-way, white - in engine compartment on windscreen washer pump*
16 *3-way connector, white, for wiper interrupter - in engine compartment next to wiper motor*
17 *One-way connector, white, for wiper interrupter - in engine compartment next to wiper motor*
18 *4-way connector, white - in engine compartment next to wiper motor*
19 *T-type connector, red - in engine compartment next to LH headlamp*
20 *T-type 2-way connector, white - in engine compartment underdash LH side*
21 *T-type 2-way connector, white - in engine compartment next to LH headlamp*
22 *T-type connector 2-way, white - in engine compartment next to RH headlamp*
23 *T-type connector 2-way, red - in engine compartment next to RH headlamp*
24 *T-type connector 2-way, white - in engine compartment next to fusebox*
25 *6-way connector, white - behind instrument panel*
26 *6-way connector, white - behind instrument panel*
27 *One-way connector, white - behind instrument panel*
28 *One-way connector, red - behind instrument panel*
29 *One-way connector, black - in engine compartment next to radiator*
30 *One-way connector, white - in engine compartment next to radiator*
31 *One-way connector, red - in engine compartment next to radiator*
32 *One-way connector, white - in engine compartment next to wiper motor. Used for oil pressure transmitter cable for 900 engine, unused for 1050 engines*
33 *Connection protective case - in engine compartment next to coil*

Key to wiring diagrams for later Special and Super models (Sec 11) (continued)

35 Connection protective case - in engine compartment next to coil
36 Connection protective case - in engine compartment next to coil
37 4-way connector, white - under rear LH pillar trim panel
38 One-way connector, white - under rear LH pillar trim panel
39 6-way connector, white - inside LH tail lamp unit
40 T-type 2-way connector, white - under trim panel next to licence plate light
41 T-type 2-way connector, white - in luggage compartment under trim panel, RH side
42 6-way connector, white - inside RH tail lamp unit
43 L-type connector, white - at the back on instrument panel
44 L-type connector, red - at the back of instrument panel
45 One-way connector, white - at the back of instrument panel
46 T-type 3-way connector - inside tailgate next to rear screen wiper motor
47 6-way connector - at the back on warning light switch
52 Connection protective case - in engine compartment next to engine oil pressure transmitter
53 Connection protective case - underdash next to radio receptacle

Location of users

70 Engine water temperature gauge sending unit - on cylinder head
71 Engine oil pressure sending unit - on crankcase
72 Engine cooling fan thermostatic switch - on radiator
73 Reversing light switch - on gearbox
74 Rear screen washer pump - in luggage compartment under RH side trim panel
75 Windscreen washer pump - on windscreen washer reservoir
76 Windscreen wiper interrupter - in engine compartment, underdash, LH side
77 Heated backlight relay - on fuse box support
78 Stop light switch - on pedal assembly
79 Low brake fluid level sending unit - on brake fluid reservoir
80 Direction indicator/hazard warning flasher - underdash LH side
81 Handbrake ON indicator switch - under handbrake lever

Earth points

A In engine compartment under battery container
B On engine and connected to point A
C In engine compartment, underdash next to windscreen wiper interrupter
D In engine compartment, on coil mounting screw
E In passenger compartment, on rear view mirror
F On door pillars, through courtesy light mounting screw
G On switch under handbrake lever
H On fuel tank through float mounting screw
I In luggage compartment under RH trim panel next to rear screen washer pump
L On tailgate through rear screen wiper mounting screw
N On tailgate next to glazing
O In engine compartment on air intake
P On horns mounting point on body work

Ignition - Charging - Starting - Engine oil pressure - Engine coolant temperature

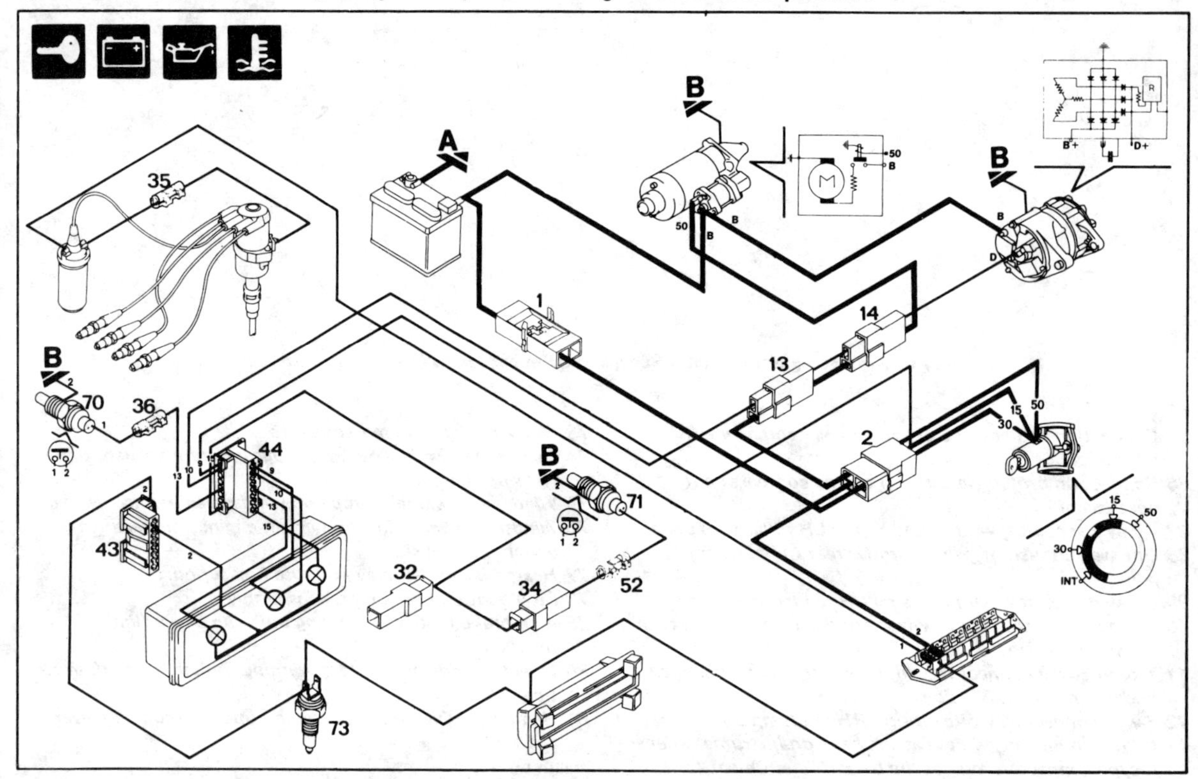

Side lights - License plate lights - Instrument panel lights

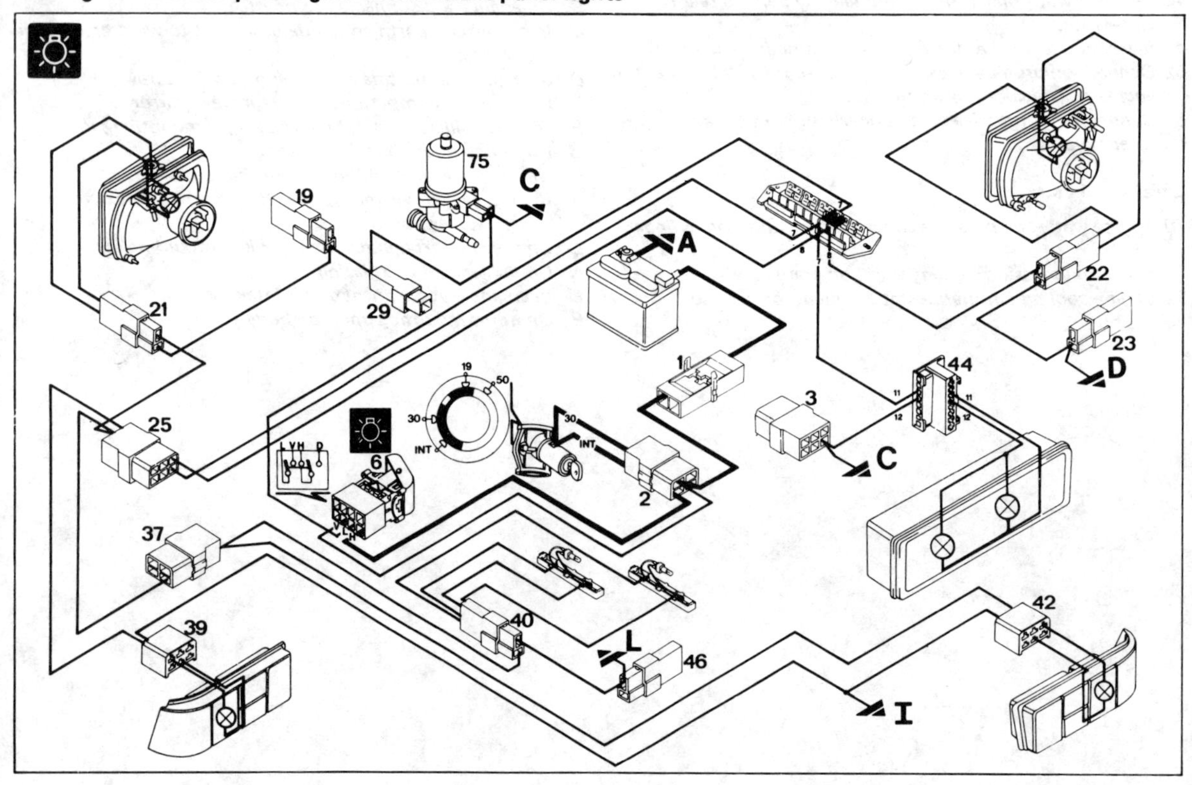

Fig. 13.58 Wiring diagram for later Special and Super models (Sec 11)

Heated backlight - Courtesy lights - Fuel level indication - Horns

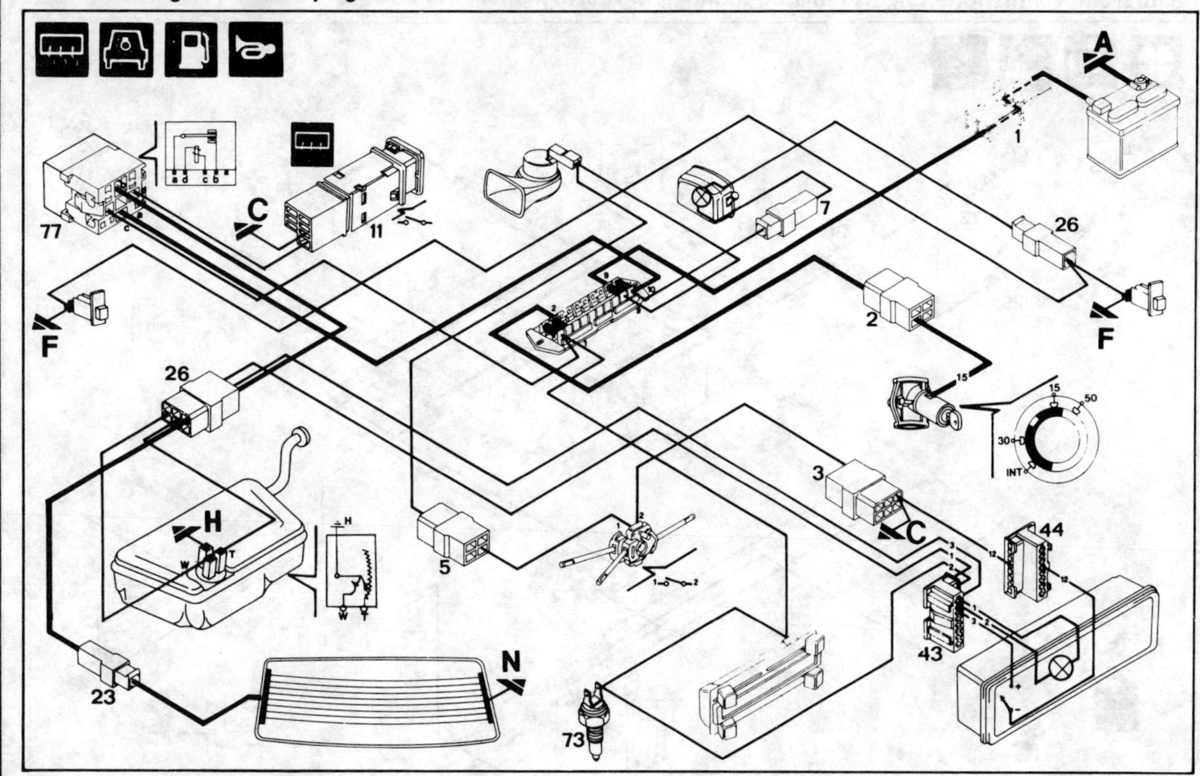

Stop lights - Reversing light - Clock - Low brake fluid level/handbrake ON indication

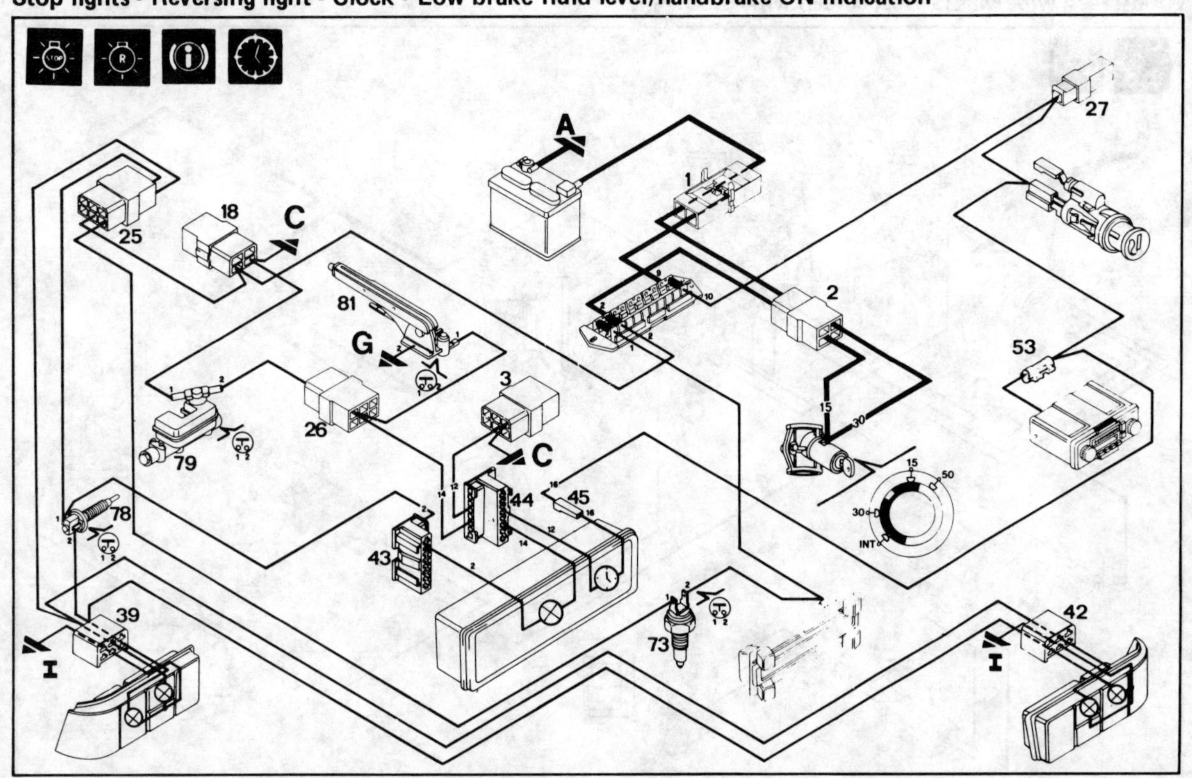

Fig. 13.58 Wiring diagram for later Special and Super models (continued) (Sec 11)

Heating and ventilation - Engine radiator cooling - Cigarette lighter - Radio

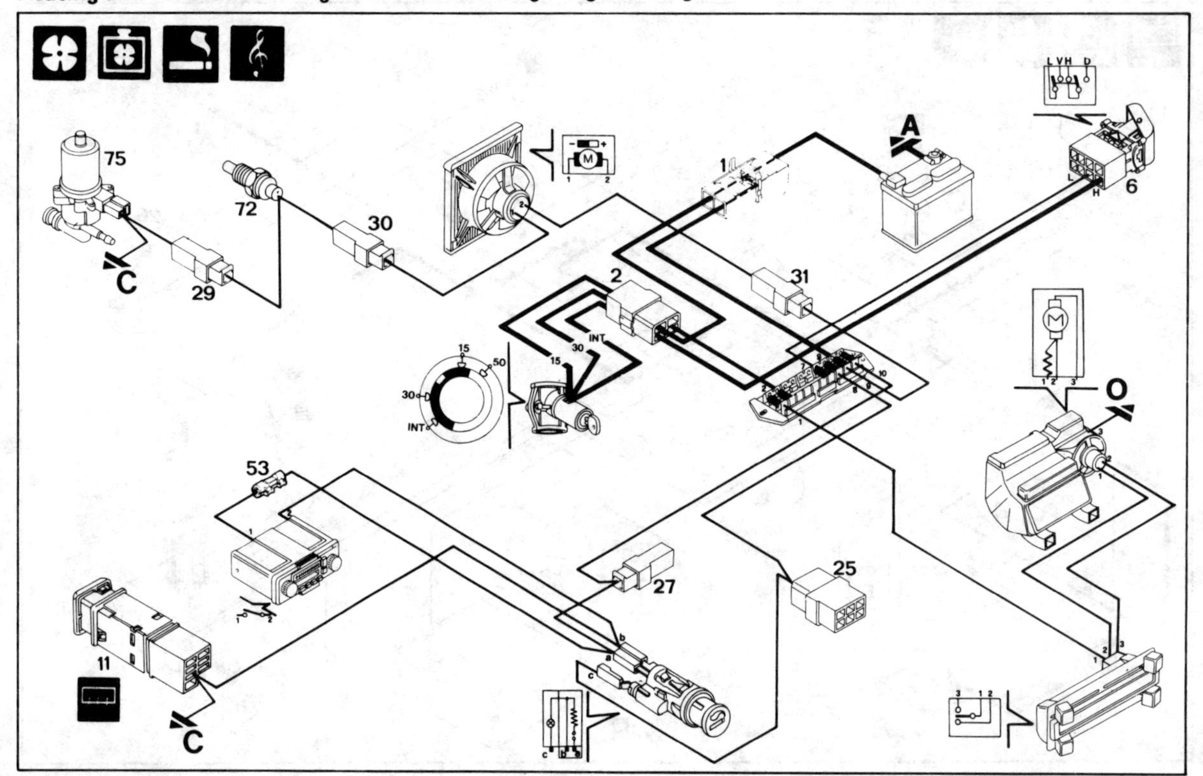

Windscreen wipe/wash - Rear screen wipe/wash

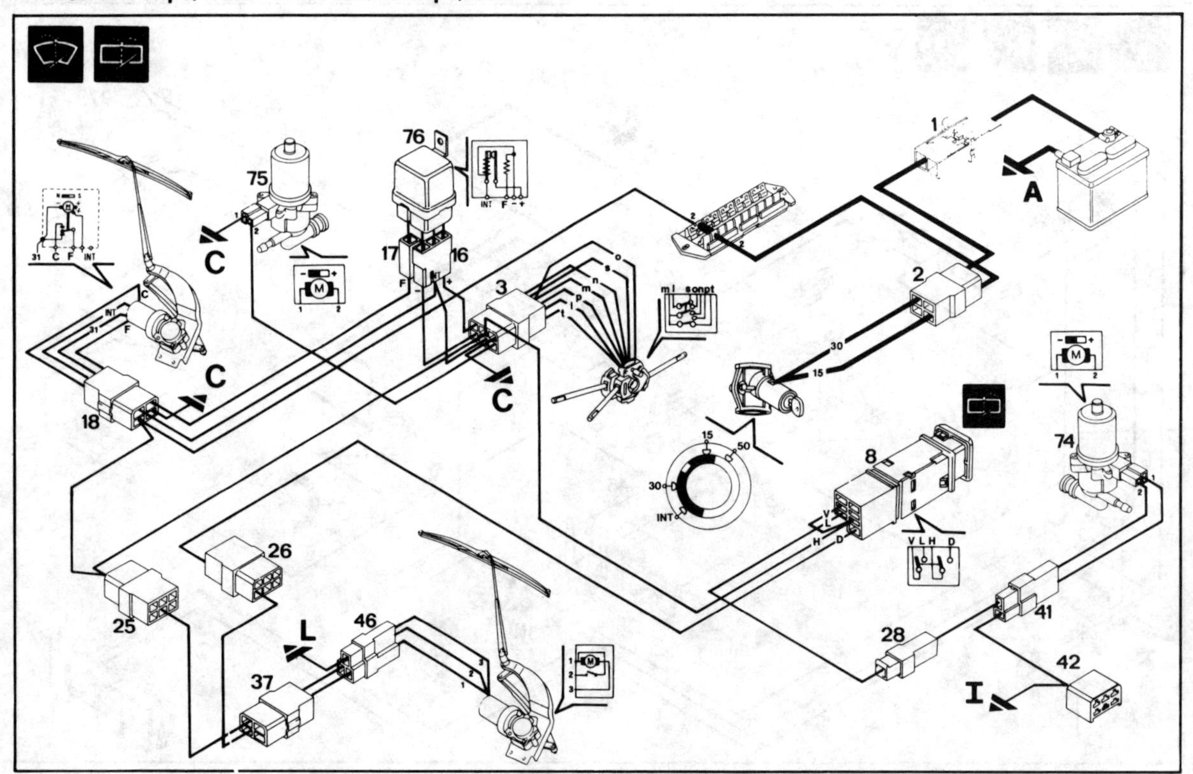

Fig. 13.58 Wiring diagrams for later Special and Super models (continued) (Sec 11)

Direction indicators - Hazard warning lights

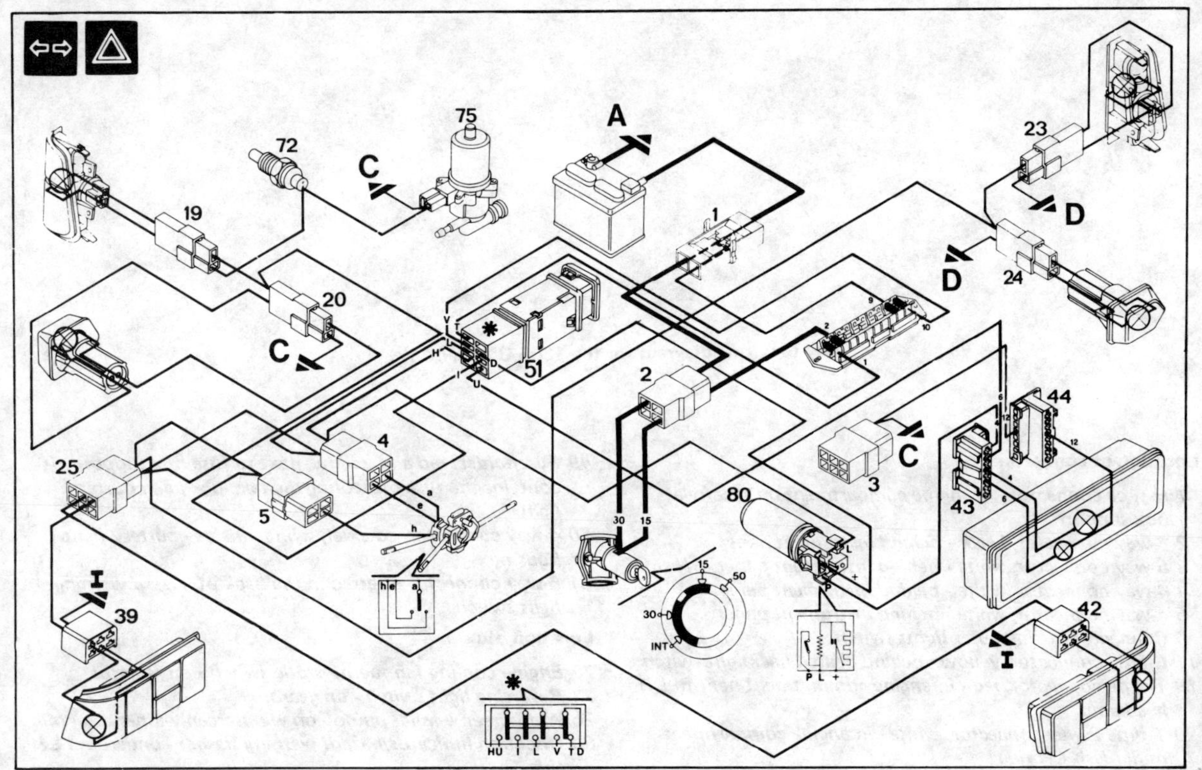

Headlamps - Rear fog light

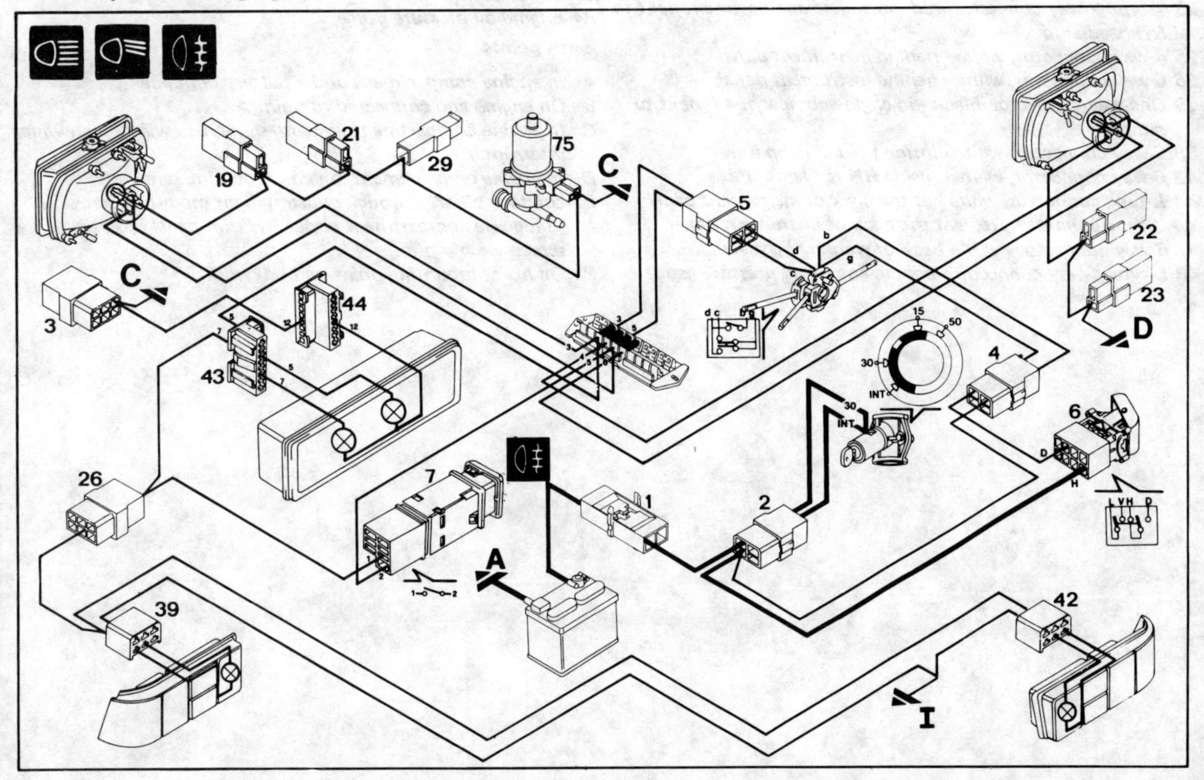

Fig. 13.58 Wiring diagrams for later Special and Super models (continued) (Sec 11)

Key to wiring diagram for the 1300 GT (Sec 11)

Location of connectors

1 H-type connector - in engine compartment on fuse/relay box support
2 4-way connector, black - behind instrument panel
3 6-way connector, white - behind instrument panel, RH side
4 4-way connector, white - behind instrument panel
5 4-way connector, white - behind instrument panel
6 Connector behind outer lights switch
11 L-type connector, yellow - behind heated backlight switch
19 T-type connector, red - in engine compartment next to LH headlamp
20 T-type 2-way connector, white - in engine compartment underdash LH side
21 T-type 2-way connector, white - in engine compartment next to LH headlamp
23 T-type 2-way connector, red - in engine compartment next to RH headlamp
25 6-way connector, white - behind instrument panel
26 6-way connector, white - behind instrument panel
29 One-way connector, black - in engine compartment next to radiator
39 6-way connector, white - inside LH tail lamp unit
42 6-way connector, white - inside RH tail lamp unit
43 L-type connector, white - at the back of instrument panel
44 L-type connector, red - at the back of instrument panel
47 6-way connector - at the back of warning light switch
48 L-type 4-way connector - behind engine oil pressure gauge
49 Fuseholder, red and yellow, next to fuse box support - It contains the fuse protecting digital clock and courtesy lights
50 4-way connector, co-axial, white - inside courtesy light housing
51 6-way connector inserted in the back of hazard warning light switch

Location of users

72 Engine cooling fan thermostatic switch - on radiator
73 Reversing light switch - on gearbox
75 Windscreen washer pump - on windscreen washer reservoir
80 Direction indicator/hazard warning flasher - underdash LH side
82 Horns relay - on fuse/relay box support
83 Digital lock
84 Engine oil pressure gauge

Earth points

A In engine compartment under battery container
B On engine and connected to point A
C In engine compartment, underdash next to windscreen wiper interrupter
D In engine compartment, on coil mounting screw
F On door pillars, through courtesy light mounting screw
I In luggage compartment under RH trim panel next to rear screen washer pump
P On horns mounting point on bodywork

Direction indicators - Radio - Hazard warning light

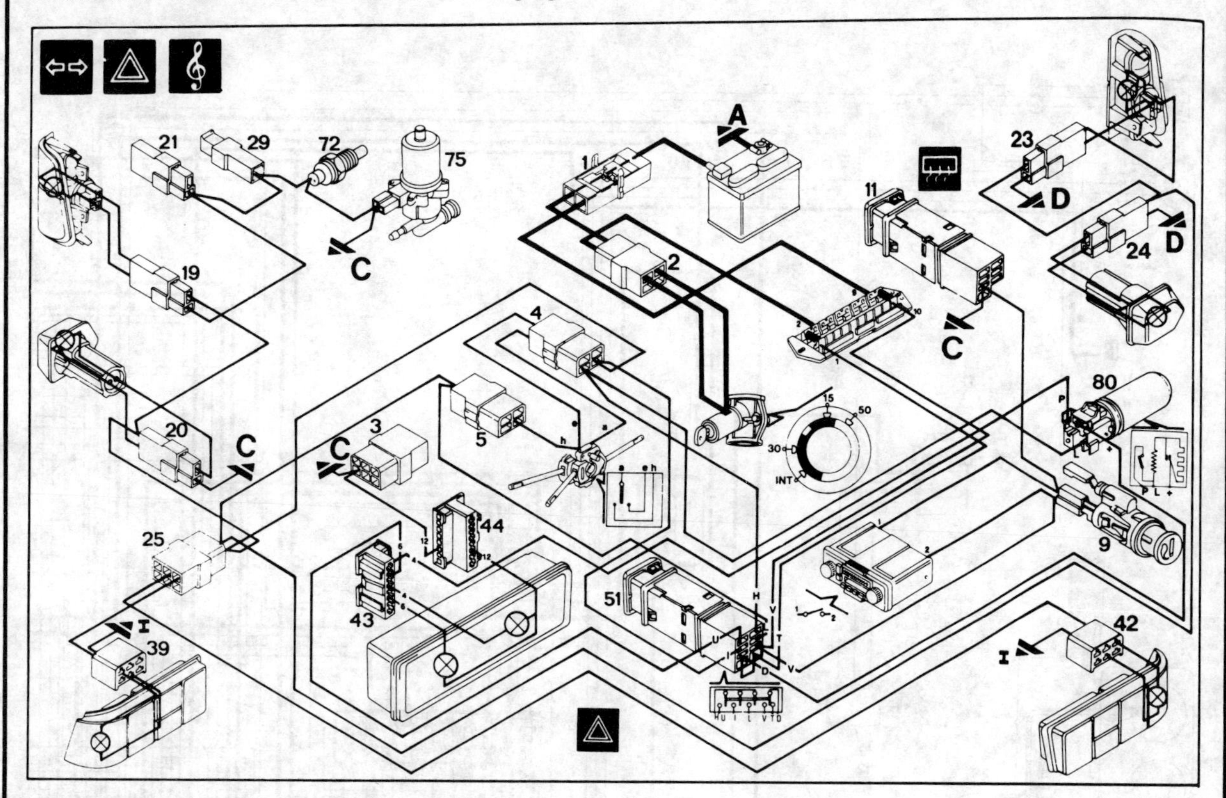

Courtesy lights - Horns - Tachometer - Digital clock

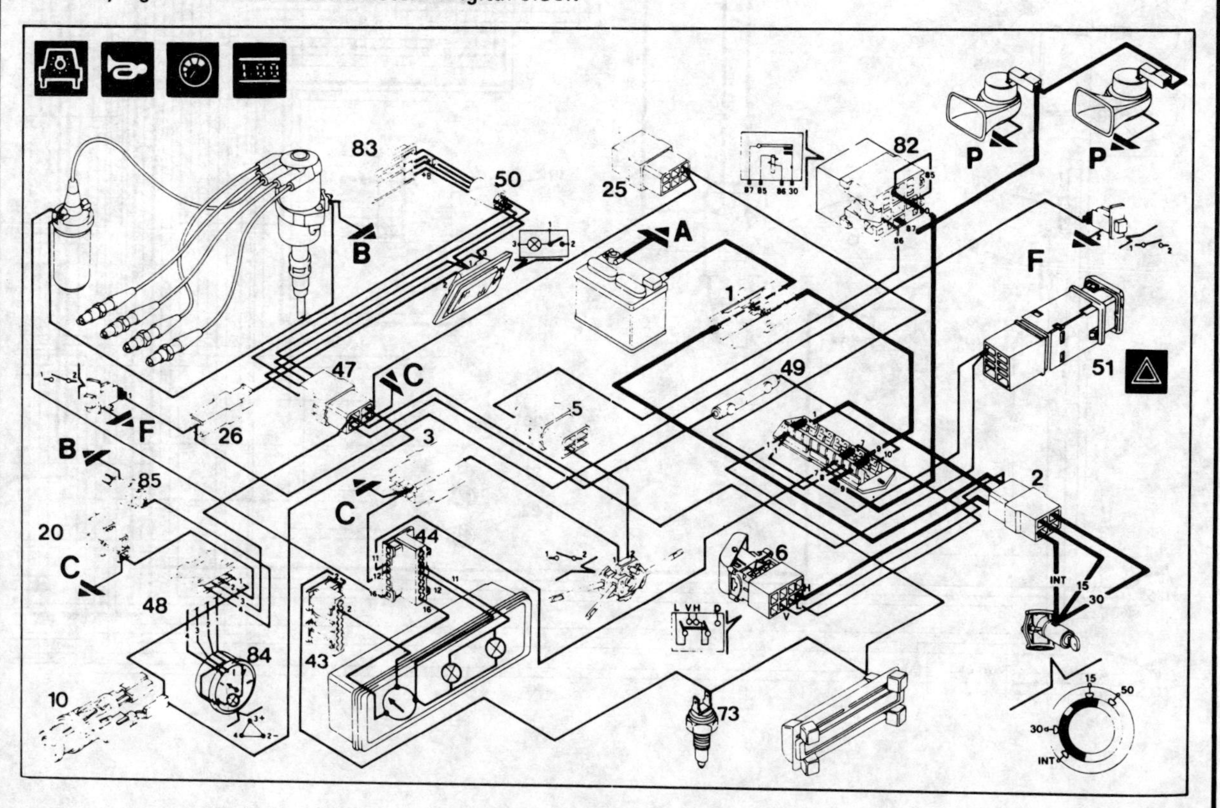

Fig. 13.59 Wiring diagram for the 1300 GT (Sec 11)

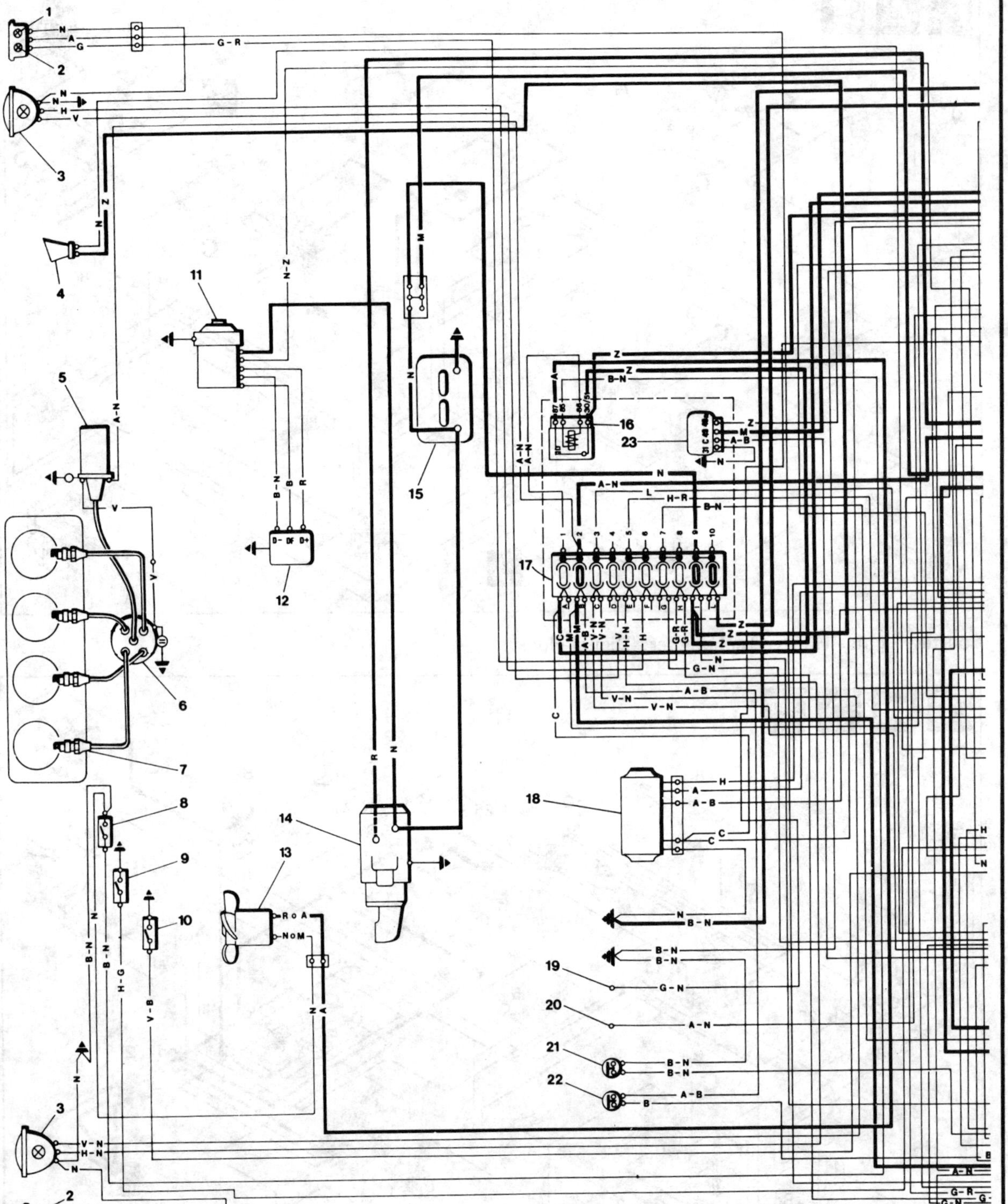

Fig. 13.60 Wiring diagram for typical late Fiorino models (Sec 11)

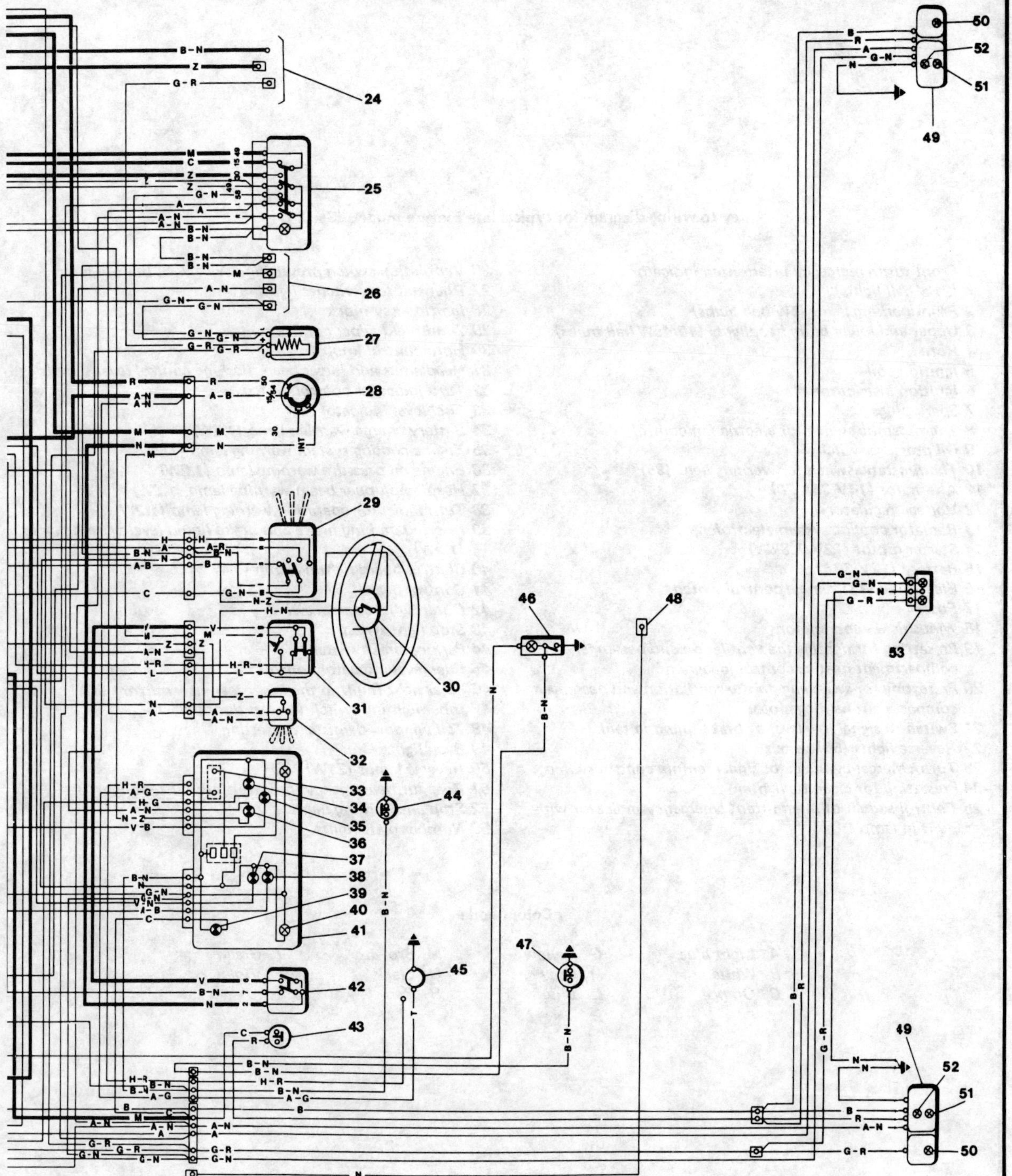

Fig. 13.60 Wiring diagram for typical late Fiorino models (Sec 11) (continued)

Key to wiring diagram for typical late Fiorino models (Sec 11)

1 *Front turn/emergency intermitten indicator (21W ball lights)*
2 *Front parking lights (5W ball bulbs)*
3 *Upper and lower beam headlights (40/45W ball bulbs)*
4 *Horn*
5 *Ignition coil*
6 *Ignition distributor*
7 *Spark plugs*
8 *Thermostatic switch for electric fan control*
9 *Oil pressure switch*
10 *Thermostatic switch for warning light (35)*
11 *Alternator (14V 35A 20)*
12 *Voltage regulator*
13 *Radiator cooling system electrofan*
14 *Starter motor (12V 0, 8kW)*
15 *Battery (12V 36Ah)*
16 *Electrofan (13) remote control switch*
17 *Fuses*
18 *Windshield wiper motor*
19 *Presetting for lighting the ventilation and passenger compartment heating control ideogram*
20 *Presetting for switching on the ventilation and passenger compartment heating motor*
21 *Switch to signal low level of brake liquid in tank*
22 *Reverse lights (50) switch*
23 *Turn/emergency indicator signal remote control switch*
24 *Presetting for cigarette lighter*
25 *Control switch of intermittent emergency indicator with built-in lamp*
26 *Ventilation switch presetting device with built-in lamp*
27 *Rheostat for switches lighting*
28 *Ignition key block*
29 *Windshield wiper control switch*
30 *Horn control knob*
31 *Headlamps and lower beam flashing control lever switch*
32 *Turn indicator control switch*
33 *Fuel level indicator*
34 *Battery charge warning lamp (1.2W)*
35 *Engine cooling system warning lamp (1.2W)*
36 *Engine oil pressure warning lamp (1.2W)*
37 *Headlamp upper beam warning lamp (1.2W)*
38 *Turn indicator operation warning lamp (1.2W)*
39 *Inserted parking brake and brake liquid level warning lamp (1.2W)*
40 *Control board lighting lamp (3W)*
41 *Control board*
42 *Outer lights control switch*
43 *Stop light switch*
44 *Parking brake switch*
45 *Fuel level indicator control*
46 *Inner light (built in the inner rear view mirror - 5W)*
47 *Inner lighting switch (on left door pillar)*
48 *Tail window demister presetting*
49 *Back lights*
50 *Reverse lamps (21W)*
51 *Rear turn/emergency indicator intermittent lights*
52 *Stop and parking lights*
53 *Number plate lights*

Colour code

A Light blue	*G Yellow*	*M Brown*	*V Green*
B White	*H Grey*	*N Black*	*Z Violet*
C Orange	*L Blue*	*R Red*	

12 Steering

Trackrod balljoint - removal

1 When removing the trackrod balljoint a purpose-made separator tool should be used to prevent possible damage to the steering arm (photo).

Steering wheel (1300 GT) - removal

2 When removing the steering wheel on the 1300 GT, prise the horn push out then unscrew the centre nut and follow the procedure given in Chapter 10 (photo).

Ignition switch/steering column lock - removal

3 The accompanying photograph shows the ignition switch and steering column lock, and additional information is given in Chapter 10 (photo).

Steering gear - lubrication

4 As from late 1979, chassis number 2803650, Fiat grease type K854 (Duckhams Adgear 00) is used to lubricate the steering gear instead of oil.

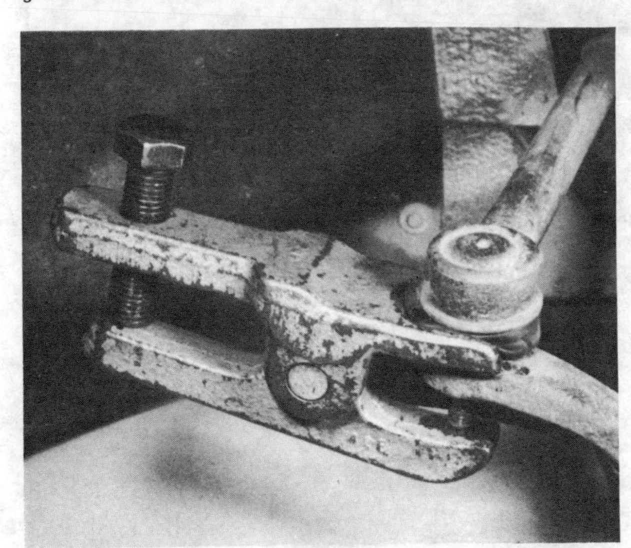

12.1 Using a balljoint separator tool

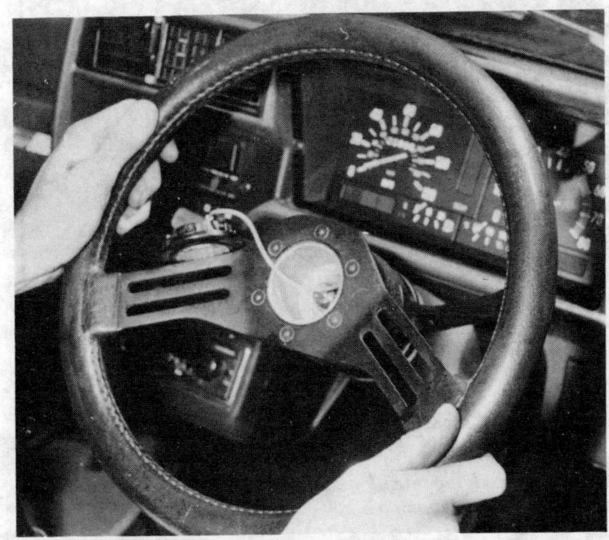

12.2 Removing the steering wheel on a 1300 GT

12.3 Ignition switch and steering column lock

13 Bodywork and fittings

Wing mouldings and front spoiler (Sport) - removal and refitting

1 Refer to Fig. 13.61. Extract the retaining screw (1).
2 Remove the front spoiler screw (2).
3 Remove the retaining bracket (3) and lift away the wing moulding.
4 To remove the front spoiler refer to Fig. 13.62; first detach both wing mouldings then extract the spoiler screws (2) and (3). Remove the spoiler.
5 Refitting is a reversal of removal.

Rear spoiler (Sport) - removal and refitting

6 Refer to Fig. 13.63. The rear spoiler can be simply removed from the rear of the roof by extracting the screws (1) which hold the spoiler brackets to the drip moulding. Remove the brackets followed by the spoiler.
7 Refitting is a reversal of removal.

Door trim pad - removal and refitting

8 Remove the window winder handle by pushing in the escutcheon and prising out the ends of the spring clip, then removing the handle from the splines (photo).
9 Prise the plugs from the armrest/pocket then remove the screws and withdraw the armrest (photos).
10 Release the surround from the interior door handle (photo).
11 Using a wide-bladed screwdriver, prise the trim pad from the door, taking care not to break the retaining clips or their locations in the trim pad.
12 Refitting is a reversal of removal, but before fitting the window winder handle locate it on the outer splines and fully close the window, then reposition the handle as required and press it fully onto the splines until the spring clip engages the groove.

Window regulator - removal and refitting

13 Remove the trim pad, as described in paragraphs 8 to 12.
14 With the window bottom channel positioned in the access hole, unscrew the cross-head screws securing the channel clamp to the control cable (photo).
15 Unscrew the nuts and withdraw the regulator through the access hole, at the same time releasing the control cable from the pulleys (photos).
16 Refitting is a reversal of removal, but adjust the cable tension by moving the bottom pulley within the slot and adjust the cable position in the channel clamp as required.

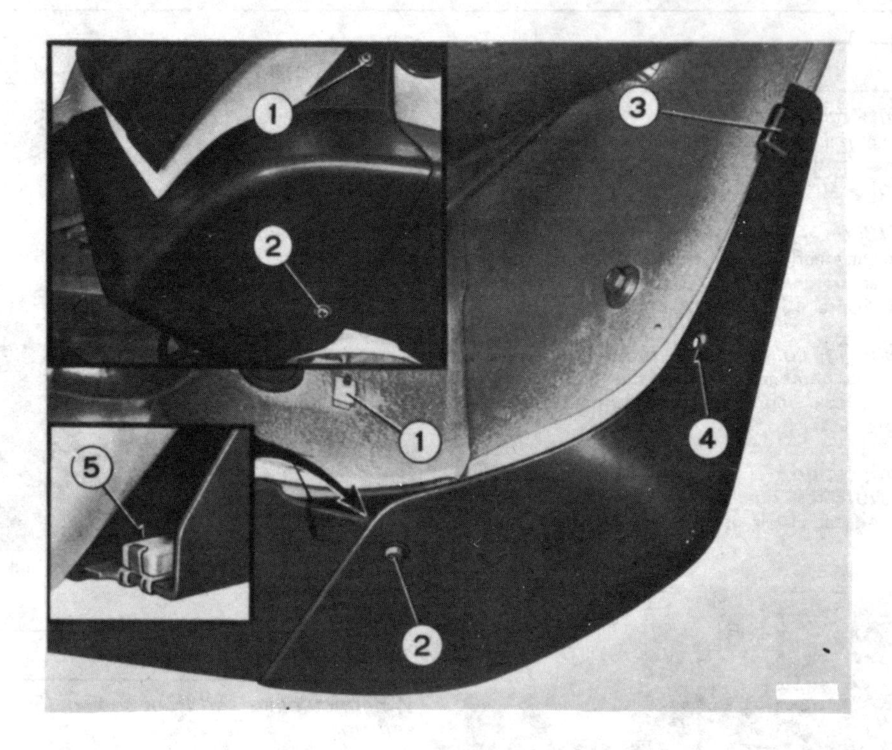

Fig. 13.61 Wing mouldings on Sport and GT models (Sec 13)

1 *Retaining screw*
2 *Retaining screw*
3 *Clip*
4 *Clip positioning hole*
5 *Threaded block*

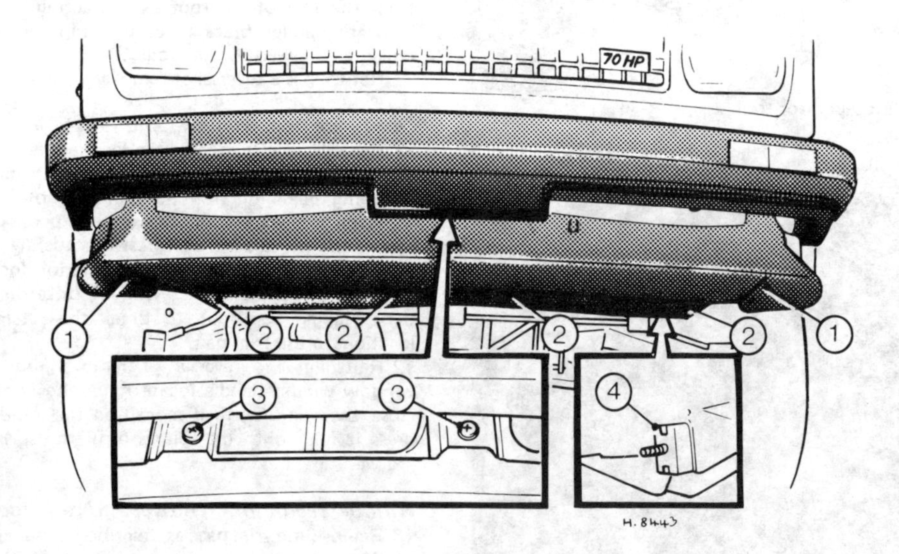

Fig. 13.62 Front spoiler attachment on Sport and GT models (Sec 13)

1 *Front wing mouldings*
2 *Spoiler-to-flange retaining screws*
3 *Spoiler-to-lining retaining screws*
4 *Flange screw retaining block*

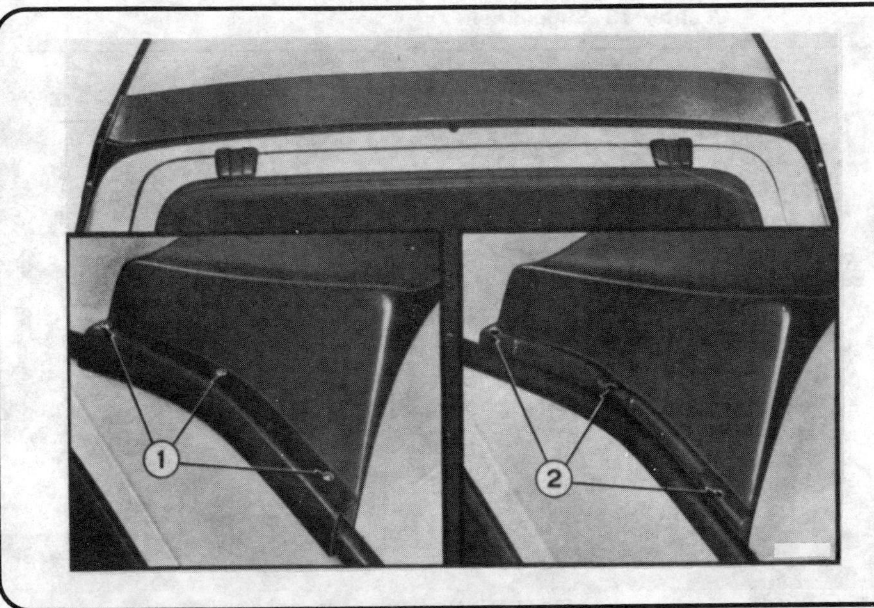

Fig. 13.63 Rear spoiler attachment on Sport and GT models (Sec 13)

1 Drip moulding bracket screws
2 Spoiler screw holes

13.8 Removing the window winder handle

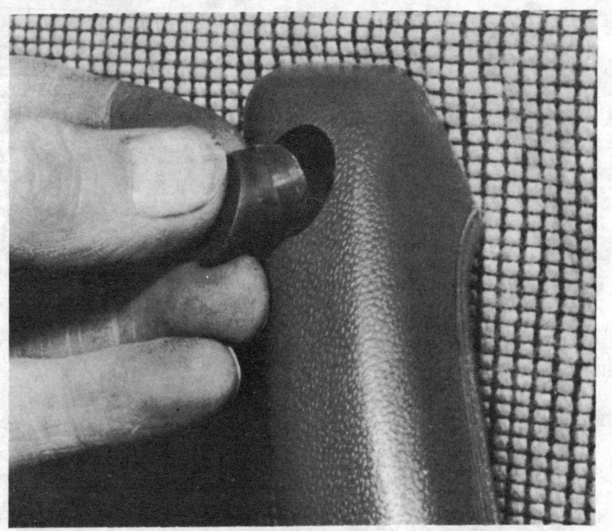

13.9A Removing the armrest/pocket plug . . .

13.9B . . . front screw . . .

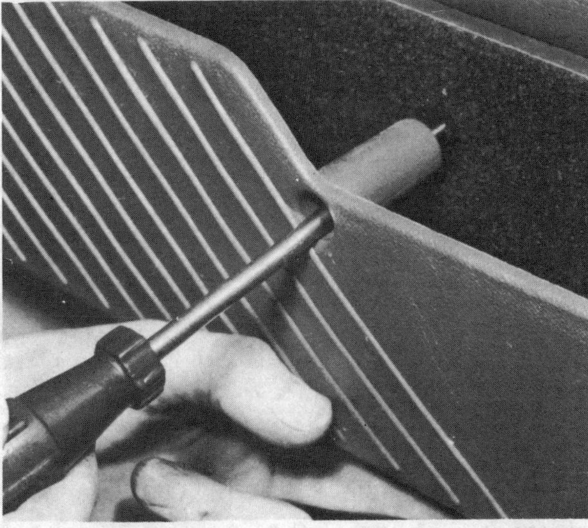

13.9C . . . and rear screw

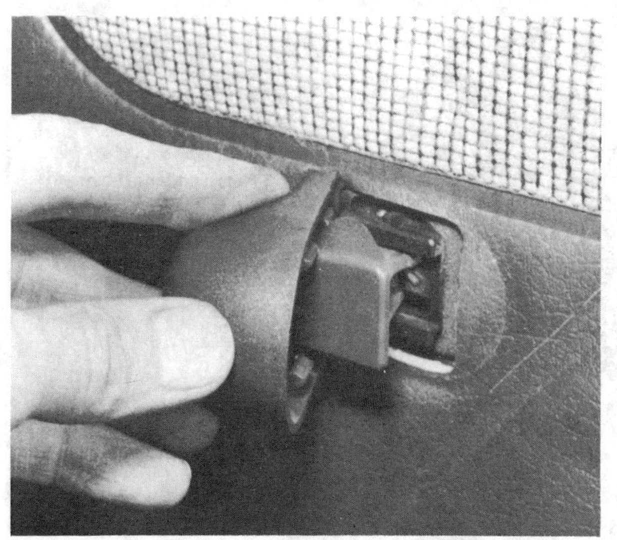
13.10 Removing the interior door handle surround

13.15B Window regulator cable and pulleys

13.14 Window channel clamp and cable

Door locks - removal and refitting

17 Remove the trim pad, as described in paragraphs 8 to 12.
18 To remove the door lock, disconnect the control rods and unscrew the cross-head mounting screws, then withdraw the lock through the access hole (photo).
19 To remove the exterior handle and private lock, disconnect the control rod and unscrew the mounting nuts, then withdraw the assembly from the outside (photo).
20 Refitting is a reversal of removal.

Dashboard (1300 GT) - general

21 The dashboard fitted to 1300 GT models differs from the one described in Chapter 12 (photo). Removal of the instrument panel is described in Section 11.

Bonnet lock - removal and refitting

22 Mark the position of the lock on the bulkhead then unscrew the mounting bolts (photo).
23 Disconnect the control cable.
24 Refitting is a reversal of removal, but if necessary adjust the lock position so that the striker is correctly aligned with it.

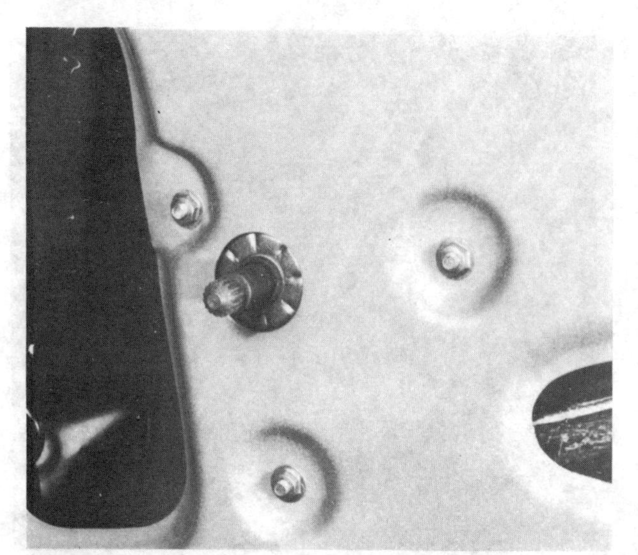
13.15A Window regulator control location

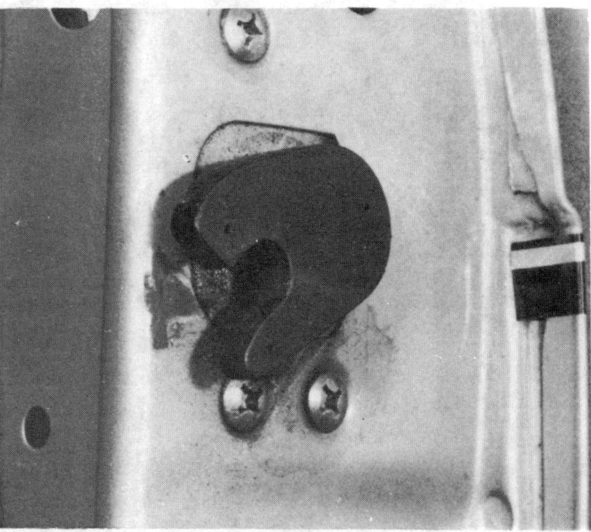
13.18 Door lock and mounting screws

13.19 Inner view of the door handle and private lock

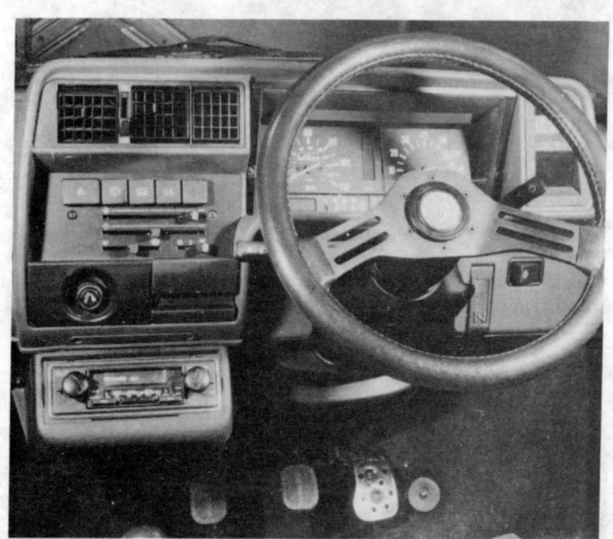

13.21 Dashboard on a 1300 GT

13.22 Bonnet lock located on the bulkhead

Tailgate - removal and refitting

25 Open the tailgate and mark the position of the hinges on the tailgate.

26 Support the tailgate and disconnect the struts either from the tailgate or body (photos).

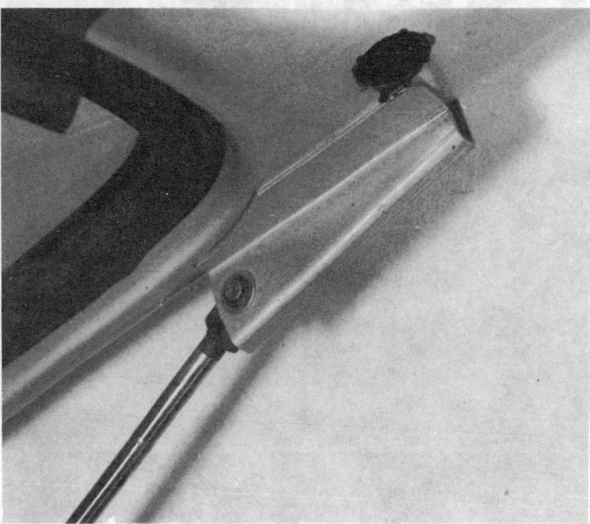

13.26A Tailgate strut mounting on the tailgate

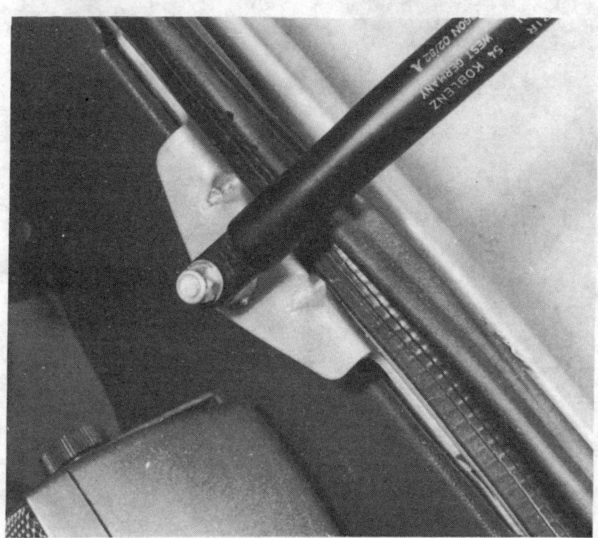

13.26B Tailgate strut mounting on the body

27 Disconnect the wiring then unscrew the hinge nuts and withdraw the tailgate from the car (photo).

28 Refitting is a reversal of removal, but make sure that the tailgate is positioned centrally within the body aperture and, if necessary, adjust the striker position so that the lock engages with it correctly (photo).

Tailgate wiper motor - removal and refitting

29 Disconnect the battery negative lead.

30 Remove the rear screen wiper arm, with reference to Chapter 9.

31 Unscrew the spindle nut and remove the spacer.

32 Open the tailgate and disconnect the wiper motor wiring.

33 Unscrew the mounting bolts and withdraw the wiper motor, noting that the earth lead is secured by one of the bolts (photo).

34 Refitting is a reversal of removal.

13.27 A tailgate hinge

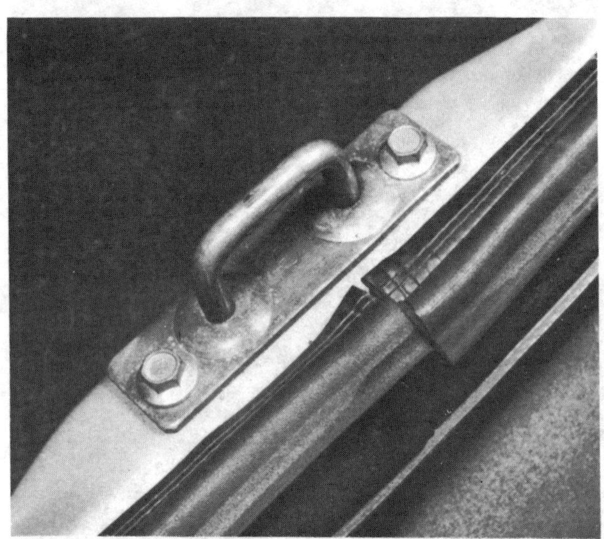

13.28 Tailgate lock striker

13.33 Tailgate wiper motor

Hinged side window - general

35 Some models are fitted with a hinged side window instead of the fixed type. Removal is straightforward once the opening latch has been disconnected. Refitting is a reversal of removal (photo).

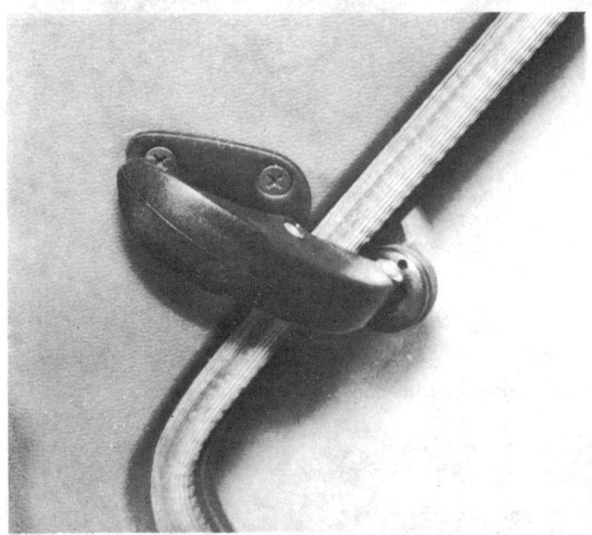

13.35 Hinged side window latch

Front seats - removal and refitting

36 To remove a front seat, adjust it fully rearward and remove the front cross-head mounting screws, then adjust the seat forward and remove the rear screws (photo).

37 Refitting is a reversal of removal.

13.36 A front seat front mounting screw

Seat belts - general

38 The seat belts should be examined regularly for wear and security, and if fraying is evident they must be renewed (photos).

13.38A Seat belt centre anchorage

13.38B Seat belt side anchorage

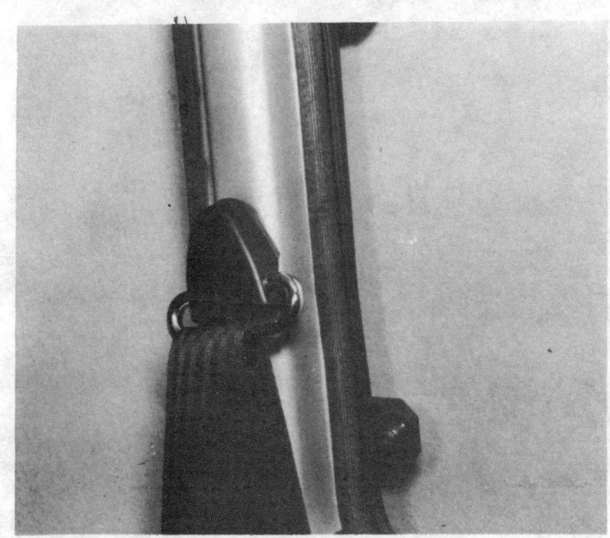
13.38C Seat belt pillar anchorage

Folding sunroof - removal and refitting

39 Open the sunroof a short distance.
40 Remove the plugs from the rear of the sunroof aperture, then remove the mounting screws and lift out the sunroof.
41 If it is required to remove the aperture sections, first drill out the corner rivets then remove the intermediate screws.
42 Refitting is a reversal of removal, using new rivets for the aperture sections where necessary.

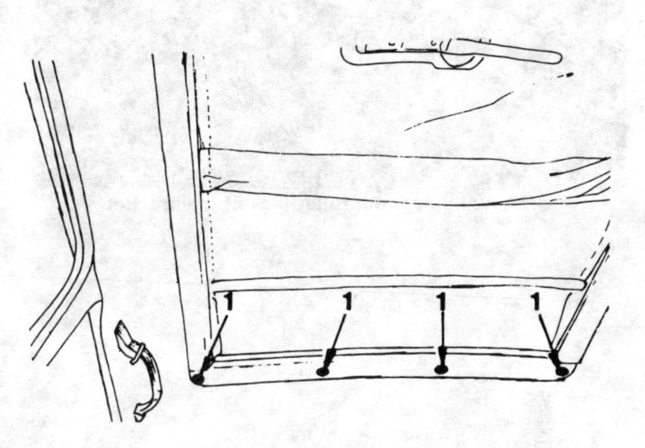

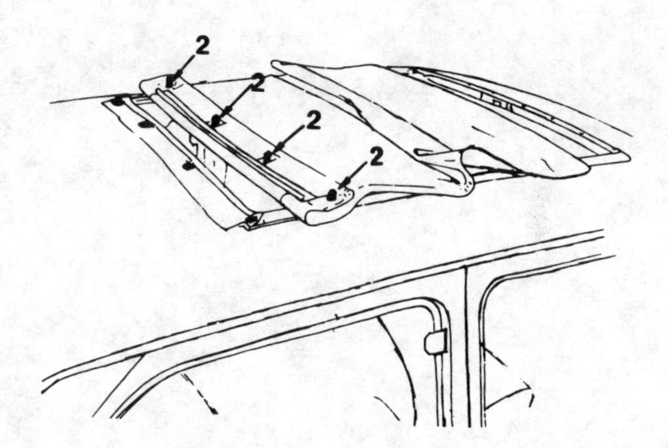

Fig. 13.64 Folding sunroof removal (Sec 13)

1 Plugs *2 Mounting points*

Jacking and towing

The jack supplied with the car should only be used for changing roadwheels, and the jack arm must always be inserted fully into the sockets provided beneath each side sill. When using a trolley jack, position the jack head beneath the central brackets provided at the front and rear of the car, and additionally support the body with axle stands.

Towing eyes are provided at the centre front and rear sides of the car (photos). When being towed the ignition key must always be inserted and, where applicable, turned to the 'garage' position to ensure that the steering is unlocked.

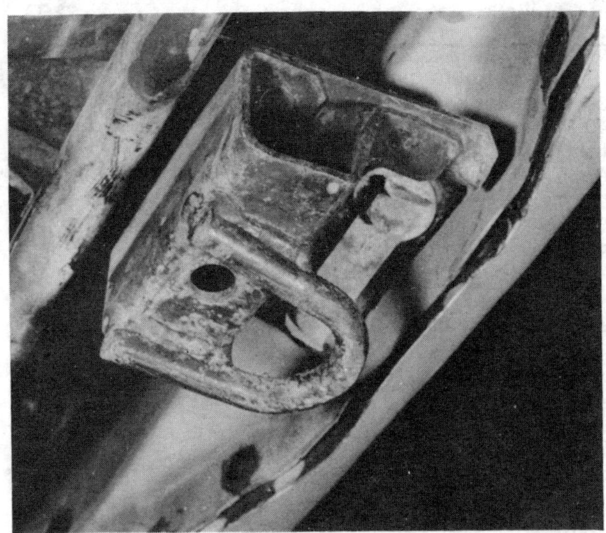
Front towing eye and trolley jack lifting bracket

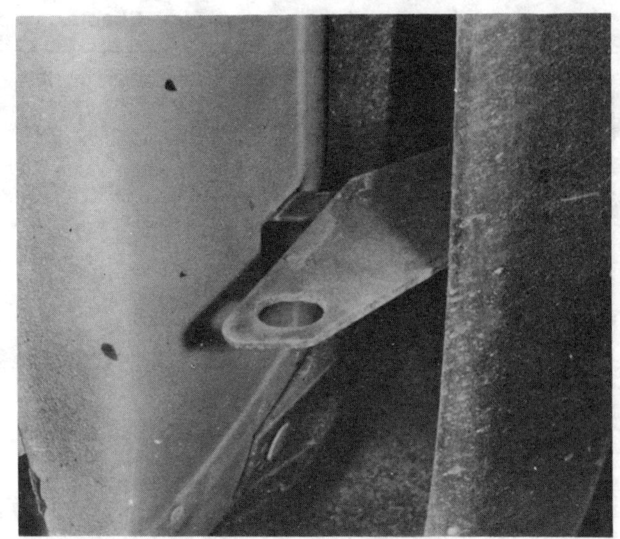
A rear towing eye

Fault diagnosis

Introduction

The vehicle owner who does his or her own maintenance according to the recommended schedules should not have to use this section of the manual very often. Modern component reliability is such that, provided those items subject to wear or deterioration are inspected or renewed at the specified intervals, sudden failure is comparatively rare. Faults do not usually just happen as a result of sudden failure, but develop over a period of time. Major mechanical failures in particular are usually preceded by characteristic symptoms over hundreds or even thousands of miles. Those components which do occasionally fail without warning are often small and easily carried in the vehicle.

With any fault finding, the first step is to decide where to begin investigations. Sometimes this is obvious, but on other occasions a little detective work will be necessary. The owner who makes half a dozen haphazard adjustments or replacements may be successful in curing a fault (or its symptoms), but he will be none the wiser if the fault recurs and he may well have spent more time and money than was necessary. A calm and logical approach will be found to be more satisfactory in the long run. Always take into account any warning signs or abnormalities that may have been noticed in the period preceding the fault – power loss, high or low gauge readings, unusual noises or smells, etc – and remember that failure of components such as fuses or spark plugs may only be pointers to some underlying fault.

The pages which follow here are intended to help in cases of failure to start or breakdown on the road. There is also a Fault Diagnosis Section at the end of each Chapter which should be consulted if the preliminary checks prove unfruitful. Whatever the fault, certain basic principles apply. These are as follows:

Verify the fault. This is simply a matter of being sure that you know what the symptoms are before starting work. This is particularly important if you are investigating a fault for someone else who may not have described it very accurately.

Don't overlook the obvious. For example, if the vehicle won't start, is there petrol in the tank? (Don't take anyone else's word on this particular point, and don't trust the fuel gauge either!) If an electrical fault is indicated, look for loose or broken wires before digging out the test gear.

Cure the disease, not the symptom. Substituting a flat battery with a fully charged one will get you off the hard shoulder, but if the underlying cause is not attended to, the new battery will go the same way. Similarly, changing oil-fouled spark plugs for a new set will get you moving again, but remember that the reason for the fouling (if it wasn't simply an incorrect grade of plug) will have to be established and corrected.

Don't take anything for granted. Particularly, don't forget that a 'new' component may itself be defective (especially if it's been rattling round in the boot for months), and don't leave components out of a fault diagnosis sequence just because they are new or recently fitted. When you do finally diagnose a difficult fault, you'll probably realise that all the evidence was there from the start.

Electrical faults

Electrical faults can be more puzzling than straightforward mechanical failures, but they are no less susceptible to logical analysis if the basic principles of operation are understood. Vehicle electrical wiring exists in extremely unfavourable conditions – heat, vibration and chemical attack – and the first things to look for are loose or corroded connections and broken or chafed wires, especially where the wires pass through holes in the bodywork or are subject to vibration.

All metal-bodied vehicles in current production have one pole of the battery 'earthed', ie connected to the vehicle bodywork, and in nearly all modern vehicles it is the negative (–) terminal. The various electrical components – motors, bulb holders etc – are also connected to earth, either by means of a lead or directly by their mountings. Electric current flows through the component and then back to the battery via the bodywork. If the component mounting is loose or corroded, or if a good path back to the battery is not available, the circuit will be incomplete and malfunction will result. The engine and/or

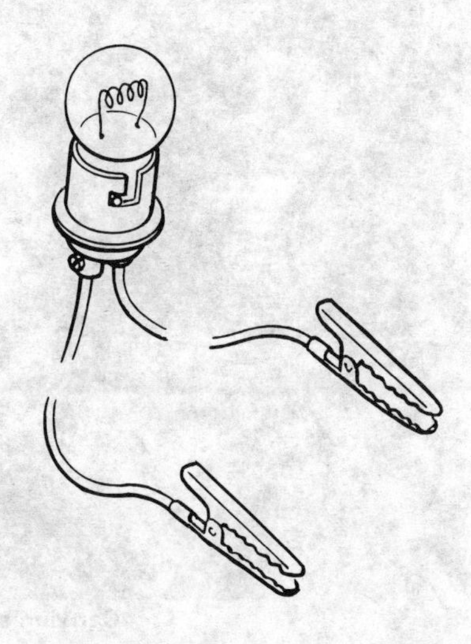

Simple test lamp is useful for tracing electrical faults

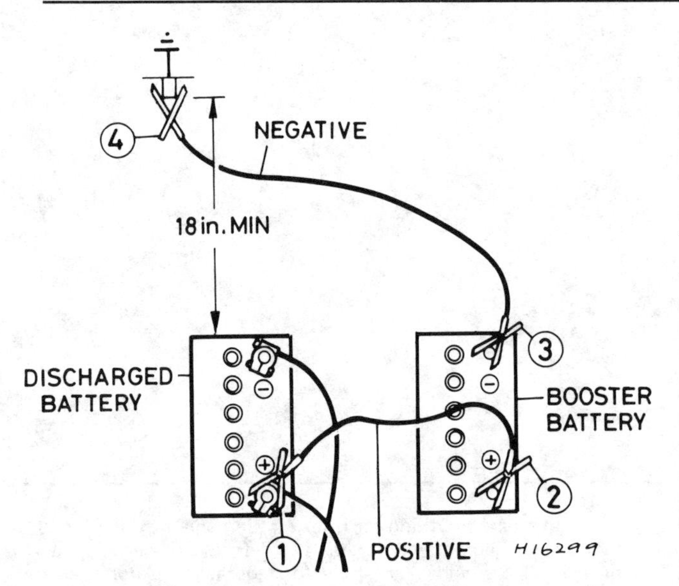

Jump start lead connections for negative earth vehicles – connect leads in order shown

gearbox are also earthed by means of flexible metal straps to the body or subframe; if these straps are loose or missing, starter motor, generator and ignition trouble may result.

Assuming the earth return to be satisfactory, electrical faults will be due either to component malfunction or to defects in the current supply. Individual components are dealt with in Chapter 9. If supply wires are broken or cracked internally this results in an open-circuit, and the easiest way to check for this is to bypass the suspect wire temporarily with a length of wire having a crocodile clip or suitable connector at each end. Alternatively, a 12V test lamp can be used to verify the presence of supply voltage at various points along the wire and the break can be thus isolated.

If a bare portion of a live wire touches the bodywork or other earthed metal part, the electricity will take the low-resistance path thus formed back to the battery: this is known as a short-circuit. Hopefully a short-circuit will blow a fuse, but otherwise it may cause burning of the insulation (and possibly further short-circuits) or even a fire. This is why it is inadvisable to bypass persistently blowing fuses with silver foil or wire.

Spares and tool kit

Most vehicles are supplied only with sufficient tools for wheel changing; the *Maintenance and minor repair* tool kit detailed in *Tools and working facilities,* with the addition of a hammer, is probably sufficient for those repairs that most motorists would consider attempting at the roadside. In addition a few items which can be fitted without too much trouble in the event of a breakdown should be carried. Experience and available space will modify the list below, but the following may save having to call on professional assistance:

Spark plugs, clean and correctly gapped
HT lead and plug cap – long enough to reach the plug furthest from the distributor
Distributor rotor, condenser and contact breaker points
Drivebelt(s) – emergency type may suffice
Spare fuses
Set of principal light bulbs
Tin of radiator sealer and hose bandage
Exhaust bandage
Roll of insulating tape
Length of soft iron wire
Length of electrical flex
Torch or inspection lamp (can double as test lamp)
Battery jump leads
Tow-rope
Ignition water dispersant aerosol
Litre of engine oil
Sealed can of hydraulic fluid
Emergency windscreen
Worm drive clips

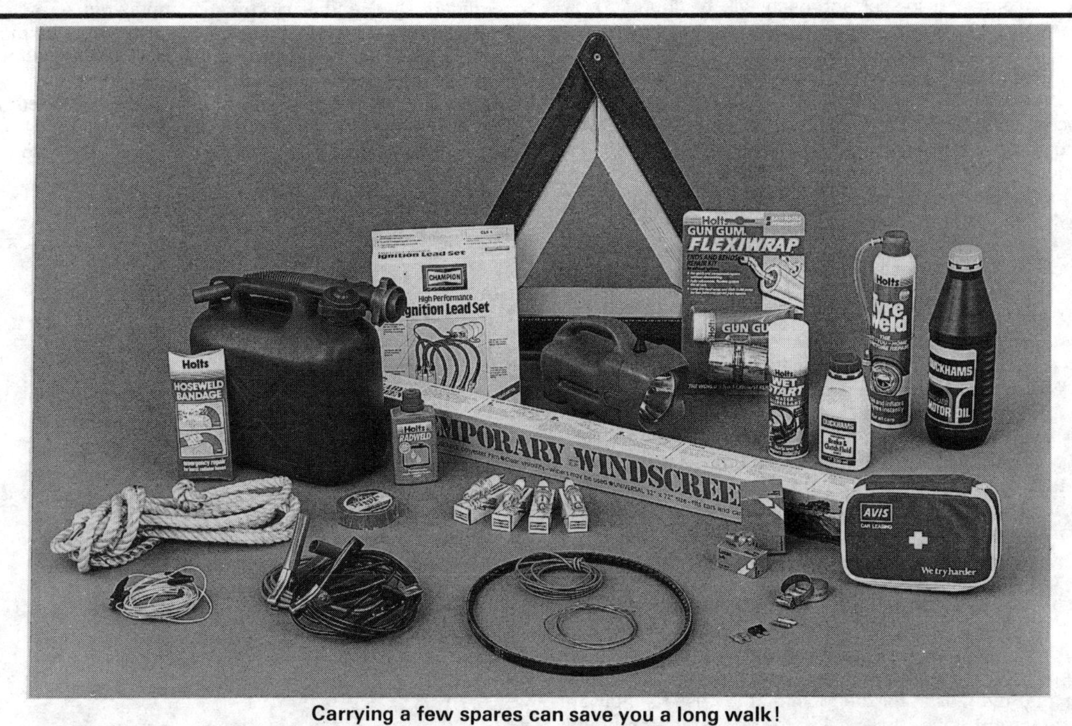

Carrying a few spares can save you a long walk!

If spare fuel is carried, a can designed for the purpose should be used to minimise risks of leakage and collision damage. A first aid kit and a warning triangle, whilst not at present compulsory in the UK, are obviously sensible items to carry in addition to the above.

When touring abroad it may be advisable to carry additional spares which, even if you cannot fit them yourself, could save having to wait while parts are obtained. The items below may be worth considering:

Clutch and throttle cables
Cylinder head gasket
Dynamo or alternator brushes
Fuel pump repair kit
Tyre valve core

One of the motoring organisations will be able to advise on availability of fuel etc in foreign countries.

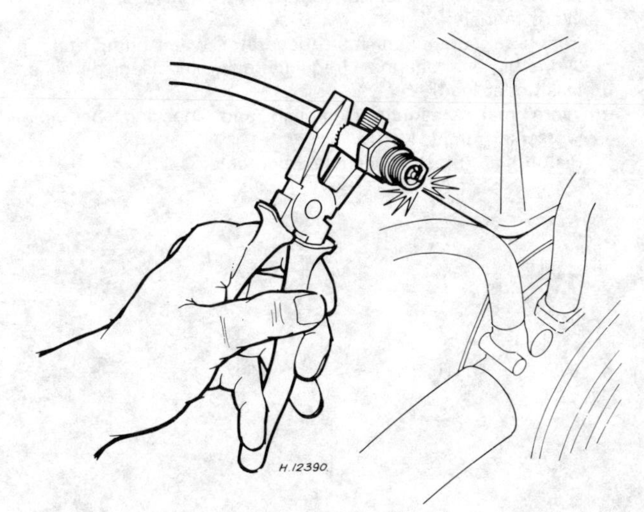

Crank engine and check for spark. Note use of insulated tool to hold plug lead

Engine will not start

Engine fails to turn when starter operated
Flat battery (recharge, use jump leads, or push start)
Battery terminals loose or corroded
Battery earth to body defective
Engine earth strap loose or broken
Starter motor (or solenoid) wiring loose or broken
Automatic transmission selector in wrong position, or inhibitor switch faulty
Ignition/starter switcn faulty
Major mechanical failure (seizure)
Starter or solenoid internal fault (see Chapter 9)

Starter motor turns engine slowly
Partially discharged battery (recharge, use jump leads, or push start)
Battery terminals loose or corroded
Battery earth to body defective
Engine earth strap loose
Starter motor (or solenoid) wiring loose
Starter motor internal fault (see Chapter 9)

Starter motor spins without turning engine
Flat battery
Starter motor pinion sticking on sleeve
Flywheel gear teeth damaged or worn
Starter motor mounting bolts loose

Engine turns normally but fails to start
Damp or dirty HT leads and distributor cap – crank engine and check for spark (photo), or try a moisture dispersant such as Holts Wet Start
Dirty or incorrectly gapped distributor points (if applicable)
No fuel in tank (check for delivery at carburettor)
Excessive choke (hot engine) or insufficient choke (cold engine)
Fouled or incorrectly gapped spark plugs – remove and regap, or renew as a set
Other ignition system fault (see Chapter 4)
Other fuel system fault (see Chapter 3)
Poor compression (see Chapter 1)
Major mechanical failure (eg camshaft drive)

Engine fires but will not run
Insufficient choke (cold engine)
Air leaks at carburettor or inlet manifold
Fuel starvation (see Chapter 3)
Ballast resistor defective, or other ignition fault (see Chapter 4)

Engine cuts out and will not restart

Engine cuts out suddenly – ignition fault
Loose or disconnected LT wires
Wet HT leads or distributor cap (after traversing water splash)
Coil or condenser failure (check for spark)
Other ignition fault (see Chapter 4)

Engine misfires before cutting out – fuel fault
Fuel tank empty
Fuel pump defective or filter blocked (check for delivery)
Fuel tank filler vent blocked (suction will be evident on releasing cap)
Carburettor needle valve sticking
Carburettor jets blocked (fuel contaminated)
Other fuel system fault (see Chapter 3)

Engine cuts out – other causes
Serious overheating
Major mechanical failure (eg camshaft drive)

Engine overheats

Ignition (no-charge) warning light illuminated
Slack or broken drivebelt – retension or renew (Chapter 9)

Ignition warning light not illuminated
Coolant loss due to internal or external leakage (see Chapter 2)
Thermostat defective
Low oil level
Brakes binding
Radiator clogged externally or internally
Electric cooling fan not operating correctly
Engine waterways clogged
Ignition timing incorrect or automatic advance malfunctioning
Mixture too weak

Note: *Do not add cold water to an overheated engine or damage may result*

Low engine oil pressure

Gauge reads low or warning light illuminated with engine running

Oil level low or incorrect grade
Defective gauge or sender unit
Wire to sender unit earthed
Engine overheating
Oil filter clogged or bypass valve defective
Oil pressure relief valve defective
Oil pick-up strainer clogged
Oil pump worn or mountings loose
Worn main or big-end bearings

Note: *Low oil pressure in a high-mileage engine at tickover is not necessarily a cause for concern. Sudden pressure loss at speed is far more significant. In any event, check the gauge or warning light sender before condemning the engine.*

Engine noises

Pre-ignition (pinking) on acceleration

Incorrect grade of fuel
Ignition timing incorrect
Distributor faulty or worn
Worn or maladjusted carburettor
Excessive carbon build-up in engine

Whistling or wheezing noises

Leaking vacuum hose
Leaking carburettor or manifold gasket
Blowing head gasket

Tapping or rattling

Incorrect valve clearances
Worn valve gear
Worn timing chain or belt
Broken piston ring (ticking noise)

Knocking or thumping

Unintentional mechanical contact (eg fan blades)
Worn fanbelt
Peripheral component fault (generator, water pump etc)
Worn big-end bearings (regular heavy knocking, perhaps less under load)
Worn main bearings (rumbling and knocking, perhaps worsening under load)
Piston slap (most noticeable when cold)

General repair procedures

Whenever servicing, repair or overhaul work is carried out on the car or its components, it is necessary to observe the following procedures and instructions. This will assist in carrying out the operation efficiently and to a professional standard of workmanship.

Joint mating faces and gaskets

Where a gasket is used between the mating faces of two components, ensure that it is renewed on reassembly, and fit it dry unless otherwise stated in the repair procedure. Make sure that the mating faces are clean and dry with all traces of old gasket removed. When cleaning a joint face, use a tool which is not likely to score or damage the face, and remove any burrs or nicks with an oilstone or fine file.

Make sure that tapped holes are cleaned with a pipe cleaner, and keep them free of jointing compound if this is being used unless specifically instructed otherwise.

Ensure that all orifices, channels or pipes are clear and blow through them, preferably using compressed air.

Oil seals

Whenever an oil seal is removed from its working location, either individually or as part of an assembly, it should be renewed.

The very fine sealing lip of the seal is easily damaged and will not seal if the surface it contacts is not completely clean and free from scratches, nicks or grooves. If the original sealing surface of the component cannot be restored, the component should be renewed.

Protect the lips of the seal from any surface which may damage them in the course of fitting. Use tape or a conical sleeve where possible. Lubricate the seal lips with oil before fitting and, on dual lipped seals, fill the space between the lips with grease.

Unless otherwise stated, oil seals must be fitted with their sealing lips toward the lubricant to be sealed.

Use a tubular drift or block of wood of the appropriate size to install the seal and, if the seal housing is shouldered, drive the seal down to the shoulder. If the seal housing is unshouldered, the seal should be fitted with its face flush with the housing top face.

Screw threads and fastenings

Always ensure that a blind tapped hole is completely free from oil, grease, water or other fluid before installing the bolt or stud. Failure to do this could cause the housing to crack due to the hydraulic action of the bolt or stud as it is screwed in.

When tightening a castellated nut to accept a split pin, tighten the nut to the specified torque, where applicable, and then tighten further to the next split pin hole. Never slacken the nut to align a split pin hole unless stated in the repair procedure.

When checking or retightening a nut or bolt to a specified torque setting, slacken the nut or bolt by a quarter of a turn, and then retighten to the specified setting.

Locknuts, locktabs and washers

Any fastening which will rotate against a component or housing in the course of tightening should always have a washer between it and the relevant component or housing.

Spring or split washers should always be renewed when they are used to lock a critical component such as a big-end bearing retaining nut or bolt.

Locktabs which are folded over to retain a nut or bolt should always be renewed.

Self-locking nuts can be reused in non-critical areas, providing resistance can be felt when the locking portion passes over the bolt or stud thread.

Split pins must always be replaced with new ones of the correct size for the hole.

Special tools

Some repair procedures in this manual entail the use of special tools such as a press, two or three-legged pullers, spring compressors etc. Wherever possible, suitable readily available alternatives to the manufacturer's special tools are described, and are shown in use. In some instances, where no alternative is possible, it has been necessary to resort to the use of a manufacturer's tool and this has been done for reasons of safety as well as the efficient completion of the repair operation. Unless you are highly skilled and have a thorough understanding of the procedure described, never attempt to bypass the use of any special tool when the procedure described specifies its use. Not only is there a very great risk of personal injury, but expensive damage could be caused to the components involved.

Safety first!

Professional motor mechanics are trained in safe working procedures. However enthusiastic you may be about getting on with the job in hand, do take the time to ensure that your safety is not put at risk. A moment's lack of attention can result in an accident, as can failure to observe certain elementary precautions.

There will always be new ways of having accidents, and the following points do not pretend to be a comprehensive list of all dangers; they are intended rather to make you aware of the risks and to encourage a safety-conscious approach to all work you carry out on your vehicle.

Essential DOs and DON'Ts

DON'T rely on a single jack when working underneath the vehicle. Always use reliable additional means of support, such as axle stands, securely placed under a part of the vehicle that you know will not give way.
DON'T attempt to loosen or tighten high-torque nuts (e.g. wheel hub nuts) while the vehicle is on a jack; it may be pulled off.
DON'T start the engine without first ascertaining that the transmission is in neutral (or 'Park' where applicable) and the parking brake applied.
DON'T suddenly remove the filler cap from a hot cooling system – cover it with a cloth and release the pressure gradually first, or you may get scalded by escaping coolant.
DON'T attempt to drain oil until you are sure it has cooled sufficiently to avoid scalding you.
DON'T grasp any part of the engine, exhaust or catalytic converter without first ascertaining that it is sufficiently cool to avoid burning you.
DON'T allow brake fluid or antifreeze to contact vehicle paintwork.
DON'T syphon toxic liquids such as fuel, brake fluid or antifreeze by mouth, or allow them to remain on your skin.
DON'T inhale dust – it may be injurious to health (see *Asbestos* below).
DON'T allow any spilt oil or grease to remain on the floor – wipe it up straight away, before someone slips on it.
DON'T use ill-fitting spanners or other tools which may slip and cause injury.
DON'T attempt to lift a heavy component which may be beyond your capability – get assistance.
DON'T rush to finish a job, or take unverified short cuts.
DON'T allow children or animals in or around an unattended vehicle.
DO wear eye protection when using power tools such as drill, sander, bench grinder etc, and when working under the vehicle.
DO use a barrier cream on your hands prior to undertaking dirty jobs – it will protect your skin from infection as well as making the dirt easier to remove afterwards; but make sure your hands aren't left slippery. Note that long-term contact with used engine oil can be a health hazard.
DO keep loose clothing (cuffs, tie etc) and long hair well out of the way of moving mechanical parts.
DO remove rings, wristwatch etc, before working on the vehicle – especially the electrical system.
DO ensure that any lifting tackle used has a safe working load rating adequate for the job.
DO keep your work area tidy – it is only too easy to fall over articles left lying around.
DO get someone to check periodically that all is well, when working alone on the vehicle.
DO carry out work in a logical sequence and check that everything is correctly assembled and tightened afterwards.
DO remember that your vehicle's safety affects that of yourself and others. If in doubt on any point, get specialist advice.
IF, in spite of following these precautions, you are unfortunate enough to injure yourself, seek medical attention as soon as possible.

Asbestos

Certain friction, insulating, sealing, and other products – such as brake linings, brake bands, clutch linings, torque converters, gaskets, etc – contain asbestos. *Extreme care must be taken to avoid inhalation of dust from such products since it is hazardous to health.* If in doubt, assume that they *do* contain asbestos.

Fire

Remember at all times that petrol (gasoline) is highly flammable. Never smoke, or have any kind of naked flame around, when working on the vehicle. But the risk does not end there – a spark caused by an electrical short-circuit, by two metal surfaces contacting each other, by careless use of tools, or even by static electricity built up in your body under certain conditions, can ignite petrol vapour, which in a confined space is highly explosive.

Always disconnect the battery earth (ground) terminal before working on any part of the fuel or electrical system, and never risk spilling fuel on to a hot engine or exhaust.

It is recommended that a fire extinguisher of a type suitable for fuel and electrical fires is kept handy in the garage or workplace at all times. Never try to extinguish a fuel or electrical fire with water.

Note: *Any reference to a 'torch' appearing in this manual should always be taken to mean a hand-held battery-operated electric lamp or flashlight. It does NOT mean a welding/gas torch or blowlamp.*

Fumes

Certain fumes are highly toxic and can quickly cause unconsciousness and even death if inhaled to any extent. Petrol (gasoline) vapour comes into this category, as do the vapours from certain solvents such as trichloroethylene. Any draining or pouring of such volatile fluids should be done in a well ventilated area.

When using cleaning fluids and solvents, read the instructions carefully. Never use materials from unmarked containers – they may give off poisonous vapours.

Never run the engine of a motor vehicle in an enclosed space such as a garage. Exhaust fumes contain carbon monoxide which is extremely poisonous; if you need to run the engine, always do so in the open air or at least have the rear of the vehicle outside the workplace.

If you are fortunate enough to have the use of an inspection pit, never drain or pour petrol, and never run the engine, while the vehicle is standing over it; the fumes, being heavier than air, will concentrate in the pit with possibly lethal results.

The battery

Never cause a spark, or allow a naked light, near the vehicle's battery. It will normally be giving off a certain amount of hydrogen gas, which is highly explosive.

Always disconnect the battery earth (ground) terminal before working on the fuel or electrical systems.

If possible, loosen the filler plugs or cover when charging the battery from an external source. Do not charge at an excessive rate or the battery may burst.

Take care when topping up and when carrying the battery. The acid electrolyte, even when diluted, is very corrosive and should not be allowed to contact the eyes or skin.

If you ever need to prepare electrolyte yourself, always add the acid slowly to the water, and never the other way round. Protect against splashes by wearing rubber gloves and goggles.

When jump starting a car using a booster battery, for negative earth (ground) vehicles, connect the jump leads in the following sequence: First connect one jump lead between the positive (+) terminals of the two batteries. Then connect the other jump lead first to the negative (–) terminal of the booster battery, and then to a good earthing (ground) point on the vehicle to be started, at least 18 in (45 cm) from the battery if possible. Ensure that hands and jump leads are clear of any moving parts, and that the two vehicles do not touch. Disconnect the leads in the reverse order.

Mains electricity and electrical equipment

When using an electric power tool, inspection light etc, always ensure that the appliance is correctly connected to its plug and that, where necessary, it is properly earthed (grounded). Do not use such appliances in damp conditions and, again, beware of creating a spark or applying excessive heat in the vicinity of fuel or fuel vapour. Also ensure that the appliances meet the relevant national safety standards.

Ignition HT voltage

A severe electric shock can result from touching certain parts of the ignition system, such as the HT leads, when the engine is running or being cranked, particularly if components are damp or the insulation is defective. Where an electronic ignition system is fitted, the HT voltage is much higher and could prove fatal.

Conversion factors

Length (distance)

Inches (in)	X	25.4	=	Millimetres (mm)	X	0.0394	=	Inches (in)
Feet (ft)	X	0.305	=	Metres (m)	X	3.281	=	Feet (ft)
Miles	X	1.609	=	Kilometres (km)	X	0.621	=	Miles

Volume (capacity)

Cubic inches (cu in; in^3)	X	16.387	=	Cubic centimetres (cc; cm^3)	X	0.061	=	Cubic inches (cu in; in^3)
Imperial pints (Imp pt)	X	0.568	=	Litres (l)	X	1.76	=	Imperial pints (Imp pt)
Imperial quarts (Imp qt)	X	1.137	=	Litres (l)	X	0.88	=	Imperial quarts (Imp qt)
Imperial quarts (Imp qt)	X	1.201	=	US quarts (US qt)	X	0.833	=	Imperial quarts (Imp qt)
US quarts (US qt)	X	0.946	=	Litres (l)	X	1.057	=	US quarts (US qt)
Imperial gallons (Imp gal)	X	4.546	=	Litres (l)	X	0.22	=	Imperial gallons (Imp gal)
Imperial gallons (Imp gal)	X	1.201	=	US gallons (US gal)	X	0.833	=	Imperial gallons (Imp gal)
US gallons (US gal)	X	3.785	=	Litres (l)	X	0.264	=	US gallons (US gal)

Mass (weight)

Ounces (oz)	X	28.35	=	Grams (g)	X	0.035	=	Ounces (oz)
Pounds (lb)	X	0.454	=	Kilograms (kg)	X	2.205	=	Pounds (lb)

Force

Ounces-force (ozf; oz)	X	0.278	=	Newtons (N)	X	3.6	=	Ounces-force (ozf; oz)
Pounds-force (lbf; lb)	X	4.448	=	Newtons (N)	X	0.225	=	Pounds-force (lbf; lb)
Newtons (N)	X	0.1	=	Kilograms-force (kgf; kg)	X	9.81	=	Newtons (N)

Pressure

Pounds-force per square inch (psi; lbf/in^2; lb/in^2)	X	0.070	=	Kilograms-force per square centimetre (kgf/cm^2; kg/cm^2)	X	14.223	=	Pounds-force per square inch (psi; lbf/in^2; lb/in^2)
Pounds-force per square inch (psi; lbf/in^2; lb/in^2)	X	0.068	=	Atmospheres (atm)	X	14.696	=	Pounds-force per square inch (psi; lbf/in^2; lb/in^2)
Pounds-force per square inch (psi; lbf/in^2; lb/in^2)	X	0.069	=	Bars	X	14.5	=	Pounds-force per square inch (psi; lbf/in^2; lb/in^2)
Pounds-force per square inch (psi; lbf/in^2; lb/in^2)	X	6.895	=	Kilopascals (kPa)	X	0.145	=	Pounds-force per square inch (psi; lbf/in^2; lb/in^2)
Kilopascals (kPa)	X	0.01	=	Kilograms-force per square centimetre (kgf/cm^2; kg/cm^2)	X	98.1	=	Kilopascals (kPa)
Millibar (mbar)	X	100	=	Pascals (Pa)	X	0.01	=	Millibar (mbar)
Millibar (mbar)	X	0.0145	=	Pounds-force per square inch (psi; lbf/in^2; lb/in^2)	X	68.947	=	Millibar (mbar)
Millibar (mbar)	X	0.75	=	Millimetres of mercury (mmHg)	X	1.333	=	Millibar (mbar)
Millibar (mbar)	X	0.401	=	Inches of water (inH$_2$O)	X	2.491	=	Millibar (mbar)
Millimetres of mercury (mmHg)	X	0.535	=	Inches of water (inH$_2$O)	X	1.868	=	Millimetres of mercury (mmHg)
Inches of water (inH$_2$O)	X	0.036	=	Pounds-force per square inch (psi; lbf/in^2; lb/in^2)	X	27.68	=	Inches of water (inH$_2$O)

Torque (moment of force)

Pounds-force inches (lbf in; lb in)	X	1.152	=	Kilograms-force centimetre (kgf cm; kg cm)	X	0.868	=	Pounds-force inches (lbf in; lb in)
Pounds-force inches (lbf in; lb in)	X	0.113	=	Newton metres (Nm)	X	8.85	=	Pounds-force inches (lbf in; lb in)
Pounds-force inches (lbf in; lb in)	X	0.083	=	Pounds-force feet (lbf ft; lb ft)	X	12	=	Pounds-force inches (lbf in; lb in)
Pounds-force feet (lbf ft; lb ft)	X	0.138	=	Kilograms-force metres (kgf m; kg m)	X	7.233	=	Pounds-force feet (lbf ft; lb ft)
Pounds-force feet (lbf ft; lb ft)	X	1.356	=	Newton metres (Nm)	X	0.738	=	Pounds-force feet (lbf ft; lb ft)
Newton metres (Nm)	X	0.102	=	Kilograms-force metres (kgf m; kg m)	X	9.804	=	Newton metres (Nm)

Power

Horsepower (hp)	X	745.7	=	Watts (W)	X	0.0013	=	Horsepower (hp)

Velocity (speed)

Miles per hour (miles/hr; mph)	X	1.609	=	Kilometres per hour (km/hr; kph)	X	0.621	=	Miles per hour (miles/hr; mph)

*Fuel consumption**

Miles per gallon, Imperial (mpg)	X	0.354	=	Kilometres per litre (km/l)	X	2.825	=	Miles per gallon, Imperial (mpg)
Miles per gallon, US (mpg)	X	0.425	=	Kilometres per litre (km/l)	X	2.352	=	Miles per gallon, US (mpg)

Temperature

Degrees Fahrenheit = (°C x 1.8) + 32

Degrees Celsius (Degrees Centigrade; °C) = (°F - 32) x 0.56

**It is common practice to convert from miles per gallon (mpg) to litres/100 kilometres (l/100km), where mpg (Imperial) x l/100 km = 282 and mpg (US) x l/100 km = 235*

Index

C

D

E

F

G

H

I

J

L

M

O

P

R

S

T

V

W